Microbiomes and the Global Climate Change

Showkat Ahmad Lone • Abdul Malik
Editors

Microbiomes and the Global Climate Change

Springer

Editors
Showkat Ahmad Lone
Post-Graduate Programme
in Microbiology, Centre of Research
for Development (CORD)
University of Kashmir
Srinagar, Jammu and Kashmir, India

Abdul Malik
Department of Agricultural Microbiology,
Faculty of Agricultural Sciences
Aligarh Muslim University
Aligarh, Uttar Pradesh, India

ISBN 978-981-33-4507-2 ISBN 978-981-33-4508-9 (eBook)
https://doi.org/10.1007/978-981-33-4508-9

This Springer imprint is published by the registered company Springer Nature Singapore Pte Ltd.
The registered company address is: 152 Beach Road, #21-01/04 Gateway East, Singapore 189721, Singapore

Preface

Microorganisms are indispensable to the sustenance of life on the earth because they drive the flow of matter in the biogeochemical cycling of carbon, nitrogen, and other important nutrients. Being the most primordial inhabitants, they have dominated our planet and have been shaping the subsequent inhabitants for over 3.5 billion years. The calamitous shift from non-oxidizing atmosphere dominated by greenhouse gases like methane (CH_4) and carbon dioxide (CO_2) to the oxygen (O_2) rich environment was mediated by the microorganisms capable of carrying out oxygenic photosynthesis, the latter evolved to be the present day Cyanobacteria. Microorganisms have been enjoying harmonious co-existence with all other life forms, but of late this equilibrium is disturbed by natural and more often by anthropogenic interventions. The geometric progression of the human population which is currently estimated to be around 7.2 billion is putting enormous pressure on the biosphere and the microbiome associated with it. The challenges associated with the increased population that need to be addressed are food, energy, and the chemicals necessary for life. Meeting these challenges would certainly put pressure on the natural resources like hydrosphere, lithosphere, and ultimately on the entire biosphere which may subsequently lead to climate change. The climate change has far reaching implications on the health of the environment and the life forms associated with it including the human beings. Therefore, the challenges are to be met without any compromise on the sustainability aspects of the ecosystem.

The proposed book aims at providing the holistic understanding of the risks associated with climate change, their implications on the environment, life forms including the human health, and the role of planetary microbiomes in addressing the issue of global climate change. This book shall serve as a handy resource to the students, teachers, researchers, and policymakers in the area of climate change.

Rise in average surface temperature, ocean warming, glacial retreat, decreased snow cover, declining arctic sea ice, extreme weather events, and ocean acidification are the scientific evidences of global climate change. Climate change is a global concern which has the potential to directly affect the agriculture, environment, and human health and indirectly the tourism and employment. Microbiomes associated with the biosphere have the central role to play in the cycling of important nutrients like carbon and nitrogen and thereby have direct effect on the earth's climate. Moreover, the health implications of the climate change also affect the disease epidemiology, as the epidemic in one particular country may quickly be spread to other countries due to displacement caused by, say, drought or floods. Therefore, a coordinated effort by global partners must be in place to minimize the risk of health emergencies for the common good of all inhabitants of the planet.

The proposed book shall cover the contemporary environmental issues facing the life on the planet and the role of planetary microbiome in addressing or aggravating such issues. Moreover, it is unclear whether the microbial processes have net favourable or adverse effect on the climate change. We intend to cover both the spontaneous and anthropocentric events having impact on climate change, their effect on life on the planet, and the role of microorganism in alleviating or escalating the global climate change. The book shall cover the ecological significance of microbiomes associated with the kingdoms Plantae and Animalia with respect to climate change, natural and anthropogenic causes of climate change, microbial interactions in nature, planetary microbiomes and food security, climate change in relation to disease epidemiology and human health, and engineering microorganisms to mitigate the consequences of climate change. The individual chapters in the book shall provide to the readers both theoretical and practical exposure to the current issues and future challenges of climate change in relation to the microbiomes. The chapters shall serve as ready reference to the researchers working in the area to reshape their future research in addressing the challenges of global climate change.

Srinagar, Jammu and Kashmir, India — Showkat Ahmad Lone
Aligarh, Uttar Pradesh, India — Abdul Malik

Contents

Editors and Contributors

About the Editors

Showkat Ahmad Lone is an Assistant Professor in Microbiology at the Centre of Research for Development (CORD), University of Kashmir, Srinagar since July 2017. He has been teaching to Post Graduate students of Microbiology for more than four years and has research experience of five years. Dr. Lone has published papers in a number of reputed peer reviewed journals published by leading publication houses like Nature Publishing Group, Elsevier, Springer, and Wiley. He has been awarded various fellowships during Ph.D. Dr. Lone has presented papers in many international conferences and has received a number of best paper awards including one "Young Scientist Award" by Bacillus ACT Steering committee, United States for his significant contribution in *Bacillus thuringiensis* research. His major area of research is molecular microbiology and microbial ecology.

Abdul Malik is Professor and Head, Department of Agricultural Microbiology at Aligarh Muslim University, Aligarh, India and has teaching experience of more than 25 years to Post Graduate students and research experience of more than 30 years. Dr. Malik has been awarded several international fellowships, awards, and honours including DBT CREST Awards of Govt. of India, DAAD fellowship (Germany), TUBITAK Fellowship for visiting scientists, Turkey, BOYSCAST overseas fellowship, Jawaharlal Nehru Memorial Fund award, etc., and has worked/visited several universities including Ohio State University, Columbus, USA, University of Strathclyde, Glasgow, UK, Albert Ludwigs University of Freiburg, Germany, Technical University Berlin, Germany, University of Minho, Braga, Portugal, Cukurova University, Adana, Turkey. He has received research grants from several funding agencies of the Govt. of India including DST, CSIR, UGC as Principal Investigator. Dr. Malik has edited nine books published by the leading international Publishers. His major areas of research are environmental microbiology, ecotoxicology, and agricultural microbiology. Dr. Malik has published a number of research papers in leading scientific journals of the world with high impact factor.

Contributors

Saghir Ahmad Department of Post Harvest Engineering and Technology, Faculty of Agricultural Sciences, Aligarh Muslim University, Aligarh, India

Syed Nasir Ahmad Centre of Research for Development (CORD), University of Kashmir, Srinagar, India

Insha Amin Centre of Research for Development (CORD), University of Kashmir, Srinagar, India

Mohammad Anas Department of Agricultural Microbiology, Faculty of Agricultural Sciences, Aligarh Muslim University, Aligarh, Uttar Pradesh, India

Waquar Akhter Ansari National Bureau of Agriculturally Important Microorganisms, Mau, UP, India

Koushik Biswas Department of Biotechnology, Visva-Bharati University (A Central University), Santiniketan, Bolpur, WB, India

Devashish R. Chobe Legumes Pathology, International Crop Research Institute for the Semi-Arid Tropics, Patancheru, Telangana, India

Ram Kishor Fagodiya ICAR-Central Soil Salinity Research Institute, Karnal, India

BashirAhmad Ganai Centre of Research for Development (CORD), University of Kashmir, Srinagar, India

Jeelani Gousia Centre of Research for Development (CORD), University of Kashmir, Srinagar, India

Dipak Kumar Gupta ICAR-Indian Agricultural Research Institute, Gauria Karma Barhi, Jharkhand, India

Burhan Hamid Centre of Research for Development (CORD), University of Kashmir, Srinagar, Jammu and Kashmir, India

Shah Ishfaq Department of Environmental Science, University of Kashmir, Srinagar, India

Fahad Khan Department of Biotechnology, Noida Institute of Engineering and Technology, Greater Noida, India

Irfan Khan Department of Post Harvest Engineering and Technology, Aligarh Muslim University, Aligarh, UP, India

Sadia Khan Department of Environmental Engineering, NED University of Engineering and Technology, Karachi, Pakistan

Sana Khan Department of Agricultural Microbiology, Faculty of Agricultural Sciences, Aligarh Muslim University, Aligarh, Uttar Pradesh, India

Shakeel A. Khan Centre for Environment Science and Climate Resilient Agriculture, ICAR-Indian Agricultural Research Institute, New Delhi, India

Khalida Khatoon Department of Agricultural Microbiology, Aligarh Muslim University, Aligarh, UP, India

Charles Knapp Department of Civil and Environmental Engineering, University of Strathclyde, Glasgow, UK

Abhay Kumar ICAR-National Research Centre on Litchi, Muzaffarpur, Bihar, India

Amit Kumar Central Muga Eri Research and Training Institute, Central Silk Board, Lahdoigarh, Jorhat, Assam, India

Om Kumar Centre for Environment Science and Climate Resilient Agriculture, ICAR-Indian Agricultural Research Institute, New Delhi, India

Smita S. Kumar J.C. Bose University of Science and Technology, YMCA, Faridabad, Haryana, India

Showkat Ahmad Lone Centre of Research for Development (CORD), University of Kashmir, Srinagar, India

Nilanjana Majumder Department of Biotechnology, Visva-Bharati University (A Central University), Santiniketan, Bolpur, WB, India

Abdul Malik Department of Agricultural Microbiology, Faculty of Agricultural Sciences, Aligarh Muslim University, Aligarh, Uttar Pradesh, India

Sandeep K. Malyan Research Management and Outreach Division, National Institute of Hydrology, Roorkee, Uttarakhand, India

Farhana Masood Department of Post Harvest Engineering and Technology, Faculty of Agricultural Sciences, Aligarh Muslim University, Aligarh, India

Tanveer Ahmad Mir Centre of Research for Development (CORD), University of Kashmir, Srinagar, India

Murugan Kumar National Bureau of Agriculturally Important Microorganisms, Kushmaur Mau, Uttar Pradesh, India

Ruqeya Nazir Centre of Research for Development (CORD), University of Kashmir, Srinagar, India
Department of Environmental Science, University of Kashmir, Srinagar, India

Sasmita Pattnaik Department of Microbiology, Government Medical College Srinagar, Srinagar, India

Mariya Amin Qurieshi Department of Community Medicine, Government Medical College, Srinagar, Jammu and Kashmir, India

T. Radhakrishnan ICAR-Directorate of Groundnut Research, Junagadh, Gujarat, India

Mahendra VikramSingh Rajawat National Bureau of Agriculturally Important Microorganisms, Mau, UP, India

Mehbooba Rasool Department of Community Medicine, Government Medical College, Srinagar, Jammu and Kashmir, India

Beenish Saleem Amity Institute of Biotechnology, Amity University, Noida, Uttar Pradesh, India

Emmanuel Salifu Department of Civil and Environmental Engineering, University of Strathclyde, Glasgow, UK

Mohammad Asim Sami Department of Agricultural Microbiology, Aligarh Muslim University, Aligarh, UP, India

Anil Kumar Saxena National Bureau of Agriculturally Important Microorganisms, Mau, UP, India

Talib Shareef Centre of Research for Development (CORD), University of Kashmir, Srinagar, India

Mamta Sharma Legumes Pathology, International Crop Research Institute for the Semi-Arid Tropics, Patancheru, Telangana, India

Zarreena Siddiqui Department of Agricultural Microbiology, Aligarh Muslim University, Aligarh, UP, India

Ajeet Singh Division of Biochemistry, ICAR-Indian Agricultural Research Institute, New Delhi, India

Bansh Narayan Singh National Bureau of Agriculturally Important Microorganisms, Mau, UP, India

Devendra Singh National Bureau of Agriculturally Important Microorganisms, Mau, UP, India

Mark Stillings Department of Civil and Environmental Engineering, University of Strathclyde, Glasgow, UK

Avijit Tarafdar Legumes Pathology, International Crop Research Institute for the Semi-Arid Tropics, Patancheru, Telangana, India

Rebecca Tonner Department of Civil and Environmental Engineering, University of Strathclyde, Glasgow, UK

Ronald Turner Department of Civil and Environmental Engineering, University of Strathclyde, Glasgow, UK

Baba Uqab Department of Environmental Science, University of Kashmir, Srinagar, India

Ajar Nath Yadav Akal College of Agriculture, Eternal University, Baru Sahib, Sirmour, Himachal Pradesh, India

Reena Yadav ICAR-Directorate of Groundnut Research, Junagadh, Gujarat, India

Riasa Zaffar Department of Environmental Science, University of Kashmir, Srinagar, India

Mohammad Tarique Zeyad National Bureau of Agriculturally Important Microorganisms, Mau, UP, India
Department of Agricultural Microbiology, Aligarh Muslim University, Aligarh, UP, India

Part I

Climate Change and Microbial Ecology

Microbes and Climate: A Tangled Relation

1

Irfan Khan, Fahad Khan, Saghir Ahmad, Pratibha Pandey, and Mohammad Mustufa Khan

Abstract

In the present scenario, the undesirable changes in the climatic conditions impact different components of ecosystem such as microbes, plants, animals, soil, etc. In this chapter, an effort has been made to summarize the relation between microbes and climate and their impact on each other. There are very few relationships that could be as complex as that of microorganisms and climate. More recent evidence has attempted to explain the role of microorganisms in regulating the fluxes of greenhouse gases such as carbon dioxide, methane, and nitrous oxide. Microbes play an integral role in modulating the cycles of various elements such as carbon, nitrogen, and oxygen. They directly as well as indirectly influence the release of greenhouse gases into the environment as well as the consumption of these gases from the atmosphere. As such, the application of utilizing these microbes for mitigating climate change has started receiving greater attention. This chapter enumerates the various beneficial microbes and their individual role in amelioration of the damage to the climate, laying emphasis on the potential to exploit these microbes as a technique for remedying the adverse climatic conditions.

I. Khan (✉) · S. Ahmad
Department of Post Harvest Engineering and Technology, Aligarh Muslim University, Aligarh, Uttar Pradesh, India

F. Khan · P. Pandey
Department of Biotechnology, Noida Institute of Engineering and Technology, Greater Noida, India

M. M. Khan
School of Biological Engineering & Sciences, Shobhit University, Saharanpur, India

S. A. Lone, A. Malik (eds.), *Microbiomes and the Global Climate Change*,
https://doi.org/10.1007/978-981-33-4508-9_1

Keywords

Microbes · Climate · Ecosystem · Agriculture · Agrochemicals · Anthropogenic · Global warming

1.1 Brief History and Introduction of Microbial Evolution and Climate Change

The most important issues of the current environmental situation are global warming and climate change. The problems arise due to higher greenhouse gas concentrations in the atmosphere that exert the warming effect. While both the anthropogenic causes and impacts of these gases have gained great attention, the role and effects of microorganisms have remained overlooked. This perhaps creates a gap to establish the correlation between microbiomes and climate change. The chapter highlights this neglected feature of terrestrial and aquatic ecosystems specifically the function and responses of microbes in this regard. That is why, an effort has been made to summarize the tangled relation between microorganisms and climate change in this chapter. To discuss the impacts of microbiomes on climate change effectively, it is mandatory to light on the evolutionary aspects of microorganisms during the path of time. The reader must have to know the basic information of microbial evolution to learn the correlation between climate change and microbiomes. Microorganisms are the living organisms with dimensions of micron level (10^{-6}) unicellular and multicellular living organisms. The microorganisms are classified into bacteria, fungi, protozoa, viruses, and algae. Each and every class of microorganisms directly or indirectly influences the ecosystem and thereby leads to climatic change. The existence of microorganisms is claimed 4000 million years back in history. While first time their presence was experimentally approved by Antonie van Leeuwenhoek with his pioneering work of microscopy in 1673. Later in 1822–1895, Louis Pasteur confirmed the existence of microorganisms with scientific arguments. Several theories have been given in reference to evolution and their relevant causes. Among most of the evolution theories, Darwinism was the most accepted speculation to define the evolution of the living organisms in response to environmental factors, viz. pressure, temperature, stress, nutrition, and many more. Darwinism theory supports the concept of survival of the fittest, firstly given by Dr. Herbert Spencer. Evolution is the continuous process that goes on at its natural pace. The reason behind the evolution could also be understood with Le Chatelier's principle given in reference to chemical reactions which stated that the equilibrium will shift in such a direction as to reduce the stress. In a similar way, microorganisms try to exist in that state where they feel minimum stress. Evolution is the process where the microorganisms change themselves as to accommodate according to climatic conditions. As discussed earlier, a number of climatic factors are there which force the microorganisms to change relatively. The microorganisms change their metabolic processes in response to the climatic change to survive. Perhaps, the most important reason why microorganism gets resistance against a particular type of

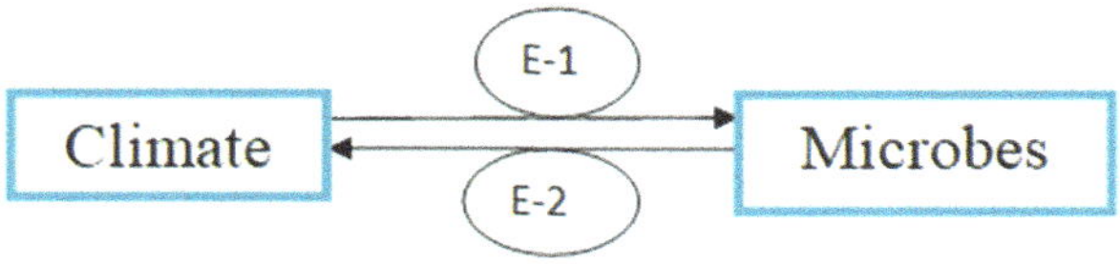

Fig. 1.1 Mutual effect of climate and microbes on each other

antibiotic within the stipulated time period. In the 1970s, the chloramphenicol antibiotic drug was discovered to treat typhoid against the Salmonella typhi bacteria. In the mid of 1980–1990, the *Salmonella typhi* got resistant against chloramphenicol which ultimately influenced the scientific community to discover another way of drug that might treat the diseases effectively. This is the big question for the scientist community to design such drugs against which the microorganisms would never acquire resistance. Today, antibiotic resistance problems are so common that increased the mortality rate of subjects. The typical example of antibiotic resistance is seen in the case of tuberculosis where the threatening conditions of MDR (multiple drug resistance) and TDR (total drug resistance) are responsible for millions of deaths throughout the globe (statistical data). All such problems need sustainable solutions in the coming future. In the very first stage the climatic conditions affect the microorganisms, while the affected microbes influenced the climate at the second stage. The microorganisms play a crucial role in sustainable climatic conditions and thereby affect human life in various means. The tangled relation between microbes and climate influences soil health, agriculture, food that ultimately affect the health of human beings and other living organisms (Fig. 1.1).

Where E-1 and E-2 represent the effect of climate on microorganisms and effect of microorganism on climate. The microbes and the climate affect mutually since the origin of the earth. The microorganisms adapt themselves against the change in the environmental conditions. Having a shorter generation time, the microorganisms are the ideal entities to discover the effect of climatic conditions on the biological system. The generation time of microorganisms varies from a few minutes to a few hours. Perhaps, the microorganisms would be suitable models to find sustainable solutions to various problems. The association of world scientists alerts human beings about the impact of human activities on climate through changes in microorganisms.

1.2 Climatic Change and Microorganisms

The current environmental situation is marked by global warming and global climate change. The major problem of the twenty-first century is climate change that is affecting human life by several means. Climate change is putting plants and humans both at more risk for microbial diseases. Although the management and the use of microbial processes can contribute to climate change mitigation and adaptation. The major climatic issue is the emission of greenhouse gases (GHGs) released in the environment by natural and anthropogenic activities. Although there are several factors that influence the climate, in this chapter the main focus is the microorganisms and their relation with climate. The microorganisms have been accepted in lowering the greenhouse gases, viz. N_2O, CH_4, and CO_2 (Singh et al. 2010).

The microorganisms play a crucial role in determining greenhouse gases as users and producers. The emission of greenhouse gases increased as a result of the industrial revolution in the last few decades. A continuous effort is being made to reduce the GHGs from fuel burning by innovating alternate energy technologies (Edenhofer et al. 2012). The microbial communities such as methylotrophic bacteria present in ocean, wetlands, rumen, and termites play a crucial role in releasing GHGs into the environment. This undesirable change in the climate disturbed biotic and abiotic components of ecosystem, the microbial community is one of them. Climate change is the main function of anthropogenic activities. Adverse human activities are the main causative factors that play a crucial role in the changes in climatic condition and microorganisms. Microbes are the backbone of every ecological system on earth. Microbes control the biogeochemical cycle by limiting carbon and nitrogen elements. The microorganisms produce and consume a variety of greenhouse gases, viz. methane (CH_4), nitrous oxide (N_2O), and carbon dioxide (CO_2). Microbes play important role in nutrient cycling, agriculture, plant, and animal health, also in global food web. To understand the tangled relation between microorganisms and climate, an example of terrestrial and marine ecosystem has been taken. The anthropogenic effects on microbes are characterized well to understand their respective effects. The uncontrolled and non-prescribed use of antibiotics causes the development of resistance in the microorganisms that ultimately disturbed the natural cycle and negatively affecting agriculture, animals, and other biotic and abiotic factors. These changes in microbial biodiversity and behaviors will certainly affect the resilience of other organisms to climate change (Maloy et al. 2017). To understand the real time effects of the above discussed factors on microbes and their sustainable solution, it is mandatory to enlist all the factors responsible for change in microbial diversity. First, the problems should be identified, then the solutions could be formulated.

1.3 Impact of Climate Change on Microbial Community

No single factor can be considered to enumerate the impact on microbial biodiversity. Both biotic and abiotic factors would be there to influence the biogeochemical cycles and ultimately affecting the microbiomes. On a wide note, the responsible factors are categorized into two major categories, viz. anthropogenic factors and natural factors. Fig. 1.2 clearly represents the correlation among different components of climate and microbes. Temperature plays a significant role in the command of microorganism development and growth (Rose 1967).

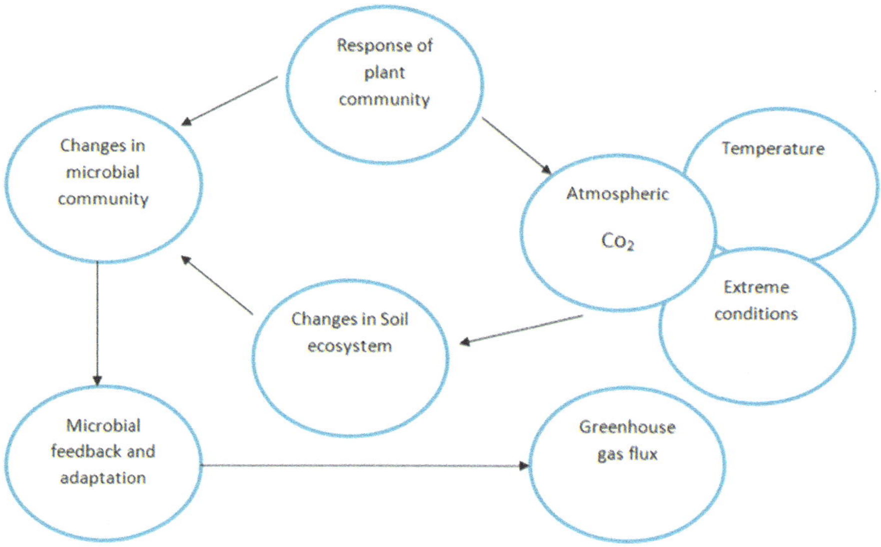

Fig. 1.2 Effect of climate change on microbes (Dutta and Dutta 2016)

1.3.1 Anthropogenic Factors

The anthropogenic factors define the harmful human activities that directly or indirectly influence the resilience of microorganisms in negative sense. Global warming takes place at unprecedented levels and much of the warming is responsible for increase in the anthropogenic greenhouse gas concentrations (IPCC 2007). The atmospheric CO_2 levels increase at a rate of 0.4% each year, with anthropogenic activities including fossil fuel combustion and land-use changes predicting to double by 2100 (Lal 2005; IPCC 2007).

1.3.1.1 Use of Antibiotics and Antifungal Agents

The most potent side of these anthropogenic factors is the unauthorized consumption of antibiotics, fungicides, other types of chemical agents being used in daily human lives. Since the antibiotics came into existence till today, there is a long list of antibacterial and antifungal drugs against which the bacteria and fungi, respectively, got resistance. To get the desired results the physician sometimes increase the dose or replace the drug with another advanced salt that ultimately increases the load on human organelles. The side effects of the same could be seen in the form of lethal/ incurable diseases such as cancer, kidney failure, liver failure, etc. Today, antibiotic resistance is a big issue and life threatening conditions like MDR (multiple drug resistance) and TDR (total drug resistance cases) continuously observed particularly during the treatment of tuberculosis. Once the microbe got resistant to a particular drug, the mutated gene continuously transfers through horizontal gene transfer among the microbial community. This way the anthropogenic activities affected the microbial biodiversity and influenced every ecosystem.

1.3.1.2 Use of Agrochemicals

The switching from organic farming to chemical farming always results in harmful complications. Just to increase the productivity of crop, scientists made hybrid and genetically modified crop plants which were less immune than conventional crop plants. So to sustain their health, a lot of precautions are taken, viz. application of agrochemicals to treat plant diseases and increase crop yield. The sprayed agrochemicals pollute the soil, so as the soil microflora. Similar to human beings, the drug resistance could also be seen in crop plants also which were sprayed with chemicals. The use of agrochemicals affects the microflora in greater fraction than the use of antibiotics and antifungal in human beings. The impact of harmful human activities and agriculture has been shown in Fig. 1.3.

1.3.1.3 Disposing Untreated Waste by Scientific Laboratories and Industries

A number of scientific laboratories lack proper disposing facilities and the treated samples disposed directly into the environment that pose threats for the biotic and abiotic components of the ecosystem. Although many sophisticated scientific laboratories are there at major scale that have proper disposing facility. Particularly in low income countries, there is lack of proper laboratory infrastructure along with trained scientific laboratory assistants. This is the burning issue in low income countries. The concerned statutory organizations of every country should focus on the safety aspects of scientific laboratories and a periodic auditing must be there to ensure safe and ethical research activities. Untreated samples in the chemical laboratories, biological laboratories directly disseminated into the environment affecting drinking water, croplands, environment, and livestock in several means.

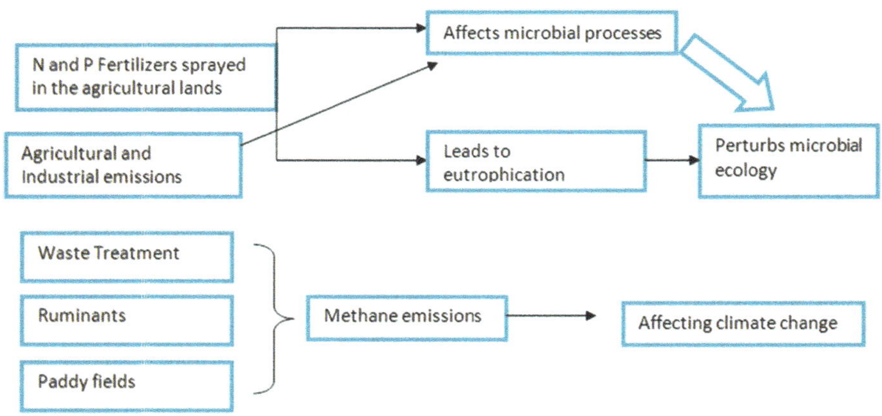

Fig. 1.3 Impact of human activities and agriculture on microorganisms (Cavicchioli et al. 2019)

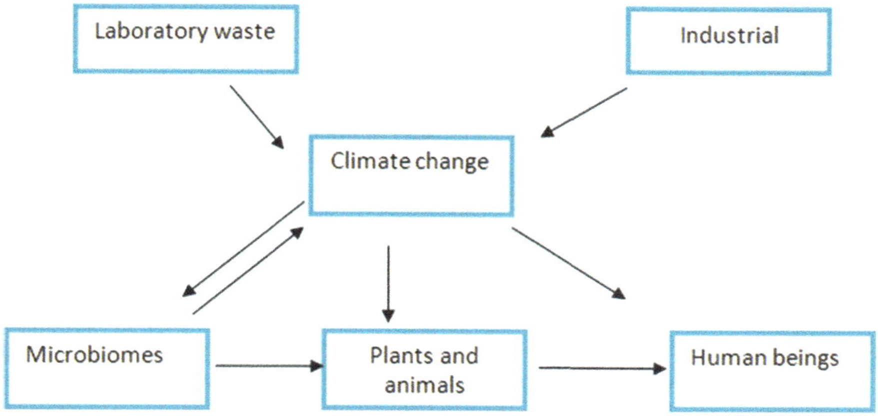

Fig. 1.4 Laboratory and industrial waste versus climate change

The industries are also disposing industrial effluents and disturbing the ecosystem by polluting nature in different ways. The toxic compounds present in the waste cause mutation in the microbial genome and develop more resistant microbial strains. This is the common scientific understanding that microbial strains develop resistance against environmental stress to survive. These resistant strains transfer the mutated gene among the microbial biomes and ultimately affect climate thereby plants, animals, and human beings. Fig. 1.4 shows the mutual effect of climate change and microbiomes on each other.

The climate changes influence the microbiomes accordingly and microbiomes affect climate reversibly. The undesirable changes in the environment and microbial community affect the plants, animals, and human beings adversely. The responses of microbes to biotic and abiotic stimuli are clearly demonstrated and therefore the effects of climate change are reliable for these organisms.

1.4 Microorganisms, Agriculture, and Global Warming

The soil microorganisms are vital entity for several ecological processes such as decaying of plant matter, biogeochemical cycles, metal transformation (Panikov 1999). Soil microorganisms make a major contribution to greenhouse gas production and use, which includes methane, carbon dioxide (CO_2), nitric oxide, and nitrous oxide. Understanding the role that soil microbes play in climate change, as well as its reaction, will enable us to decide if they can be used to minimize or drive us to climate catastrophe any faster. The soil microorganisms may have its role in declining global warming. Its role in global warming is yet to be found experimentally.

A greater fraction of atmospheric carbon dioxide is mitigated by the terrestrial forest (Schlesinger and Andrews 2000). Soil has twice as high as atmospheric carbon, which makes them one of the biggest sinks for atmospheric CO_2 and organic carbon (Jenkinson et al. 1991; Melillo et al. 2002; Willey et al. 2009). In agriculture, microorganisms play a major role as soil microbes including bacteria and fungi are important for the degradation of organic matter and recycling of old plant materials. Agricultural microorganisms may act as symbionts, including nitrogen fixing microorganisms and mycorrhizae, fungi that establish relationships with plant roots. They can also enhance plant defense against drought. The soil microbes are important to impact the production of sufficient food and agricultural products to feed and fulfill the need globally. Perhaps this is the reason that continuous efforts are being made to understand the physiology, mechanisms of agricultural microorganisms. So that new ways could be devised to tackle out the problems of food and nutritional security throughout the world. The increasing use of chemical agents in the crop distorted the structure of soil and also made unwanted changes in the soil microflora that somehow affect the climate in negative sense (Gelfand et al. 2011; Levine et al. 2011). To overcome these negative effects, a practice of using probiotics agricultural microbes is in trend to mitigate the climate changes and promote plant health. The farmers use synthetic nitrogen to fulfill the nitrogen requirement of plants such as urea that provide 46% of nitrogen to the plants. These practices somehow increased the burden of greenhouse gases which is a threat to future of agriculture. Simultaneously, the graph of organic farming and zero budget natural farming practices is increasing in pursue of sustainable agriculture. Many chemical fertilizer industries have moved their production culture towards biofertilizer and some other nanotechnology based chemical fertilizers to improve soil health and reduce further soil deterioration. In India, IFFCO (Indian Farmers Fertiliser Cooperative Limited 2019) recently launched three nano-based products, viz. Nano Nitrogen, Nano Zinc, and Nano Copper considering adverse climatic effects (www.iffco.in). According to IFFCO, these are environment friendly products that will minimize the usage of chemical fertilizers by 50% besides raising crop yield by 15–30% (www.iffco.in). Some soil fungi and bacteria establish relationships with plant roots that provide important nutrients such as nitrogen, phosphorus, etc. In response to increasing world population, efforts should be made to double food production by 2050 to feed the world population. The target of increased crop yield could be achieved by using beneficial soil microorganisms (Singh et al. 2017). During the cultivation of crops, microorganisms also play an important role. By affecting geochemical processes like the nitrogen cycle, they aid plant growth directly or indirectly. The fabaceae family plants such as pulses have nitrogen fixing bacteria in their root nodules. These plants fix atmospheric nitrogen with the help of nitrogen fixing bacteria such as *Rhizobium leguminosarum*, Mycorrhiza, *Azospirillum species*, *Bacillus species*, *Pseudomonas species* (Singh et al. 2017). Perhaps, the most significant reason that the legumes grow in specific geographical region is the presence of nitrogen fixing bacteria in soil. In the context of India, Madhya Pradesh and Rajasthan are the dominating states in the production of leguminous crops due to richness in nitrogen fixing bacteria. This is very certain, if the microbiomes would be affected by climate change, it will ultimately affect the crops and their productivity. A number of other bacterial species such as

Table 1.1 Methylotrophic bacteria involved in greenhouse gas mitigation

S. No.	Methylotrophs	GHG mitigation	References
1	Rhizobium sp.	Methane	Baesman et al. (2015)
2	Methylobacterium Organophilum	Methane	Patt et al. (1974)
3	Methylobacterium populi	Methane	Aken et al. (2004)
4	Bacillus subtilis Bacillus megaterium Bacillus aerius Rhizobium sp. Paenibacillus illinoisensis Methylobacterium extorquens	Methane	Jhala et al. (2014, 2015)
5	Methylobacterium organophilum cz-2	Methane	Zuniga et al. (2011)
6	Methylosinus trichosporium OB3b	Methane	Rodriguez et al. (2009), Zuniga et al. (2011)
7	Methylobacter sp.	Methane	Oshkin et al. (2014)
8	Type I or II methanotrophs	Methane	Strong et al. (2017)
9	Type I (Gammaproteobacteria) Type II (Alphaproteobacteria)	Methane	Cai et al. (2016)
8	Starkeya novella	Carbon sequestration	Kappler and Nouwens (2013)
9	Acidomonas methanolica MB58	Carbon dioxide fixation	Mitsui et al. (2015)
10	SAR 11 Alphaproteobacteria	Methanol mitigation	Murrell and McDonald (2000)

GHG = Greenhouse gases (Source: Kumar et al. 2018)

methylotrophic strains are exploited as the plant growth promoter by several researchers (Long et al. 1996; Baldani et al. 1997; Sy et al. 2001; Lidstrom and Chistoserdova 2002; Omer et al. 2004; Raja et al. 2006; Glick et al. 2007; Anitha 2010; Jayashree et al. 2011; Yim et al. 2012; Agafonova et al. 2013; Jeyajothi et al. 2014). It is clear from the above discussion that microbes play important role in food production and if any changes would be there in microbial genome that will surely be reflected in terms of crop productivity. The mutation frequency is greater in microbes than large animals due to lower generation time. The generation time in case of bacteria lies in the range of 15–20 min. The effects of mutation are observed quickly in microbes due to lower generation time. Perhaps the most significant reason why bacteria are employed as models for various researches. The methylotrophic bacteria play important role in greenhouse mitigation (Table 1.1).

As shown in Table 1.1, a number of methylotrophic bacteria mitigate methane, methanol, carbon dioxide, etc. Several efforts or methods are in practice towards

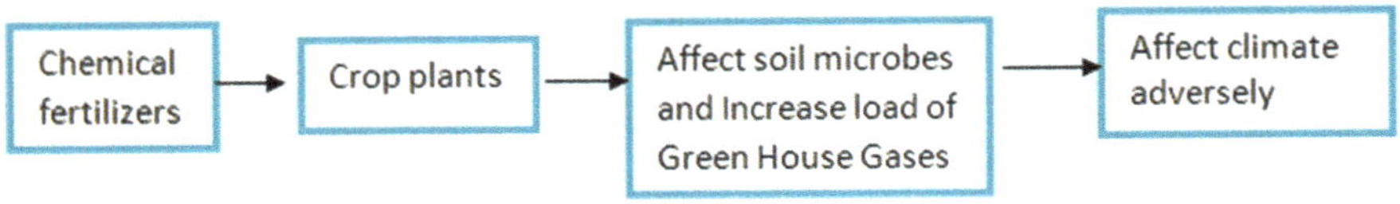

Fig. 1.5 How the use of chemical fertilizers affects the climate

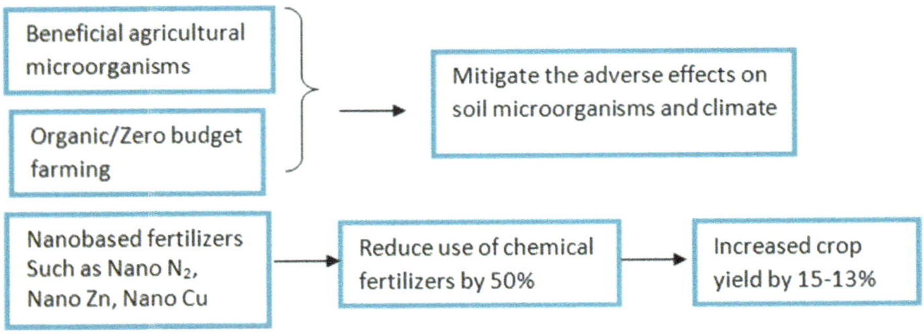

Fig. 1.6 Efforts in the pursuance of sustainable agriculture

sustainable agriculture, viz. organic farming, zero budget natural farming, etc. Using natural methods without any chemical fertilizer, soil health could be improved through the path of period. The application of natural methods rejuvenates the distorted soil structure and declined soil microflora which ultimately helps in higher crop productivity, comparatively safe and healthy food production along with mitigation of greenhouse gases. This way, natural agricultural practices help to establish a more reliant and stable relationship between microorganisms and the climate.

As shown in Fig. 1.5, the chemical fertilizers when applied to the crop plants affect the quality of soil in terms of soil texture, porosity, microflora, add toxic compounds in food chain that ultimately affect human health. In recent times, cases of liver failure and kidney failure become common perhaps due to consumption of food carrying toxic ingredients. Also, the application of chemical fertilizers increases the load of greenhouse gases globally that ultimately put adverse effects on climatic conditions. These unwanted changes in the climate disturb the natural biogeochemical cycle and the cost is paid by flora and fauna.

Figure 1.6 shows the agricultural practices in pursuit of sustainable agriculture, i.e. higher productivity, food production without toxic ingredients and maintaining the soil health as well. The chemical fertilizer companies have started to find new ways as to relatively less impact to soil as well as climate. The typical example of this is Indian based fertilizer company, i.e. IFFCO (Indian Farmer's Fertilizer Cooperation Limited) has moved towards the minimum fertilizer concept and they recently launched nanotechnology based fertilizers as mentioned in Fig. 1.5 that helps to reduce the fertilizer use by 50% and will increase the crop yield by 15–30%.

This may have negligible effects on the soil microorganisms and thereby reduce further load of greenhouse gases.

1.5 Conclusion

The nature of the microbial communities living beneath is difficult to discern the various responses that soil microbes may have to global warming because of their various ways of dealing with their environments. Microorganisms may play a remarkable role in assessing greenhouse gas atmospheric concentration. It is clear that microbes can have a major impact on future climatic scene and ecosystematic responses to climate change. Soil respiration has a key role to play in this because of the high levels of methane and carbon dioxide emitted during process, the dependence of carbon stocks on the breathing rates in soils, and the soil's original sensitivity to elevated air temperature. Further analysis of long-term effects on climate change of soil respiration will help us to better understand the global climate change effects, including the ability of terrestrial forests to remove excess CO_2 from the atmosphere. The change in climate would certainly increase temperatures and also modify the precipitation patterns in drylands throughout the world, influencing their structure and function. In the coming future, new methods should be devised to decrease any change in the climate particularly through anthropogenic activities. So that health and sustainable future could be weaved for mankind.

References

Agafonova NV, Kaparullina EN, Doronina NV, Trotsenko YA (2013) Phosphate-solubilizing activity of aerobic methylobacteria. Microbiology 82:864–867

Aken VB, Peres C, Doty S, Yoon J, Schnoor J (2004) Methylobacterium populi sp. nov., a novel aerobic, pink-pigmented, facultatively methylotrophic, methane-utilising bacterium isolated from poplar trees (Populus deltoids × nigra DN34). Int J Syst Evol Microbiol 54:1191–1196

Anitha KG (2010) Enhancing seed germination of mono and dicotyledons through IAA production of PPFM trends. Soil Sci Plant Nutr 1:14–18

Baesman SM, Miller LG, Wei JH, Cho Y, Matys ED, Summons RE, Welander PV, Oremland RS (2015) Methane oxidation and molecular characterization of methanotrophs from a former mercury mine impoundment. Microorganisms 3:290–309

Baldani JI, Caruso L, Baldani VLD, Goi SR, Dobereiner J (1997) Recent advances in BNF with non legume plants. Soil Biol Biochem 29:911–922

Cai Y, Zheng Y, Bodelier PLE, Conrad R, Jia Z (2016) Conventional methanotrophs are responsible for atmospheric methane oxidation in paddy soils. Nat Commun 7:11728. https://doi.org/10.1038/ncomms11728

Cavicchioli R et al (2019) Scientists' warning to humanity: microorganisms and climate change. Nat Rev Microbiol 17:569

Dutta H, Dutta A (2016) The microbial aspect of climate change. Energ Ecol Environ 1(4):209–232. https://doi.org/10.1007/s40974-016-0034-7

Edenhofer O, Pichs-Madruga R, Sokona Y (2012) Renewable energy sources and climate change mitigation: special report of the intergovernmental panel on climate change. Cambridge University Press, Cambridge, UK

Gelfand I, Zenone T, Jasrotia P, Chen J, Hamilton SK, Robertson GP (2011) Carbon debt of conservation reserve program (CRP) grasslands converted to bioenergy production. Proc Natl Acad Sci U S A 108:13864–13869. https://doi.org/10.1073/pnas.1017277108
Glick BR, Cheng Z, Czarny J, Duan J (2007) Promotion of plant growth by ACC deaminase-producing soil bacteria. Eur J Plant Pathol 119:329–339
Indian Farmer's Fertilizer Cooperation Limited (2019) Production of nanotechnology based chemical fertilizers. www.iffco.in
IPCC (2007) Climate change 2007: synthesis report. http://www.ipcc.ch/publications_and_data/publications_ipcc_fourth_assessment_report_synthesis_report.Htm
Jayashree S, Vadivukkarasi P, Anand K, Kato Y, Seshadri S (2011) Evaluation of pink pigmented facultative methylotrophic bacteria for phosphate solubilization. Arch Microbiol 193:543–552
Jenkinson DS, Adams DE, Wild A (1991) Model estimates of CO2 emissions from soil in response to global warming. Nature 351:304–306
Jeyajothi R, Subbalakshmi L, Nalliah D (2014) Effect of PPFM application on microbial population in transplanted rice. Trends Biosci 7:3573–3574
Jhala YK, Rajababu VV, Shelat Harsha N, Patel HK, Patel HK, Patel KT (2014) Isolation and characterization of methane utilizing bacteria from wetland paddy ecosystem. World J Microbiol Biotechnol 30(6):1845–1860
Jhala YK, Rajababu VV, Panpatte Deepak G, Shelat Harsha N (2015) Rapid methods for isolation and screening of methane degrading bacteria. J Bioremed Biodegr 7:322
Kappler U, Nouwens AS (2013) Metabolic adaptation and trophic strategies of soil bacteria C1-metabolism and sulfur chemolithotrophy in Starkeya novella. Front Microbiol 4:1–12
Kumar M, Saxena R, Tomar RS, Rai PK, Paul D (2018) Role of methylotrophic bacteria in climate change mitigation. In: Kashyap PL, Srivastava AK, Tiwari SP, Kumar S (eds) Microbes for climate resilient agriculture, 1st edn. Wiley, Hoboken, NJ
Lal R (2005) Forest soils and carbon sequestration. For Ecol Manag 220:242–258
Levine UY, Teal TK, Robertson GP, Schmidt TM (2011) Agriculture's impact on microbial diversity and associated fluxes of carbon dioxide and methane. ISMEJ 5:1683–1691. https://doi.org/10.1038/ismej.2011.40
Lidstrom ME, Chistoserdova L (2002) Plants in the pink: cytokinin production by Methylobacterium. J Bacteriol 184:1818
Long RLG, Holland MA, Stebbins N, Morris RO, Polacco JC (1996) Evidence for cytokinin production by plant-associated methylotrophs. Plant Physiol 111:316
Maloy S, Moran MA, Mulholland MR, Sosik HM, Spear JR (2017) Microbes and climate change: report on an American Academy of Microbiology and American Geophysical Union Colloquium held in Washington, DC
Melillo JM, Steudler PA, Aber JD, Newkirk K, Lux H, Bowles FP, Catricala C, Magill A, Ahrens T, Morrisseau S (2002) Soil warming and carbon-cycle feedbacks to the climate system. Science 298:2173–2176
Mitsui R, Katayama H, Tanaka M (2015) Requirement of carbon dioxide for initial growth of facultative methylotroph, Acidomonas methanolica MB58. J Biosci Bioeng 120(1):31–35
Murrell C, McDonald I (2000) Methylotrophy. In: Lederberg J III (ed) Encyclopedia of microbiology. Academic Press, New York, pp 245–255
Omer ZS, Tomboloni R, Broberg A, Gerhardson B (2004) Indole-3-acetic acid production by pink-pigmented facultative methylotrophic bacteria. Plant Growth Regul 43:93–96
Oshkin IY, Beck DAC, Lamb AE, Tchesnokova V, Benuska G, Benuska G, McTaggart TL, Kalyuzhnaya MG, Dedysh SN, Lidstrom ME, Chistoserdova L (2014) Methane-fed microbial microcosms show differential community dynamics and pinpoint taxa involved in communal response. ISME J 9(5):1119–1129
Panikov NS (1999) Understanding and prediction of soil microbial community dynamics under global change. Appl Soil Ecol 11:161–176

Patt TE, Cole GC, Bland J, Hanson RS (1974) Isolation and characterization of bacteria that grow on methane and organic compounds as sole sources of carbon and energy. J Bacteriol 120:955–964

Raja P, Uma S, Sundaram S (2006) Non-nodulating pink-pigmented facultative Methylobacterium sp. with a functional nifH gene. World J Microbiol Biotechnol 22:1381–1384

Rodríguez AS, Valdman B, Madereiros SA (2009) Analysis of methane biodegradation by Methylosinus trichosporium OB3b. Braz J Microbiol 40:301–307

Rose AH (1967) Thermobiology. Academic Press, London

Schlesinger WH, Andrews JA (2000) Soil respiration and the global carbon cycle. Biogeochemistry 48:7–20

Singh BK, Bardgett RD, Smith P, Reay DS (2010) Microorganisms and climate change: terrestrial feedbacks and mitigation options. Nat Rev Microbiol 8:779–790

Singh S, Singh V, Pal K (2017) Importance of micro organisms in agriculture. Climate and environmental changes: Impact, challenges and solutions. pp 93–117. https://www.researchgate.net/publication/322926367

Strong PJ, Karthikeyan OP, Zhu J, Clarke W, Wu W (2017) Methanotrophs: methane mitigation, denitrification and bioremediation. In: Agro-environmental sustainability. Springer, New York, pp 19–40

Sy A, Giraud E, Jourand P, Garcia N, Willems A, de Lajudie P, Prin Y, Neyra M, Gillis M, Boivin-Masson C, Dreyfus B (2001) Methylotrophic methylobacterium bacteria nodulate and fix nitrogen in symbiosis with legumes. J Bacteriol 183:214–220

Willey JM, Sherwood LM, Woolverton CJ (2009) Prescott's principles of microbiology. McGraw-Hill, New York

Yim W, Woo S, Kim K, Sa T (2012) Regulation of ethylene emission in tomato (Lycopersicon esculentum mill.) and red pepper (Capsicum annuum L.) inoculated with ACC deaminase producing Methylobacterium spp. Korean J Soil Sci Fert 45:37–42

Zúniga C, Morales M, Borgne LS, Revah S (2011) Production of poly-hydroxybutyrate (PHB) by Methylobacterium organophilum isolated from a methanotrophic consortium in a two-phase partition bioreactor. J Hazard Mater 190:876–882

Carbon Sequestration in Aquatic System Using Microbial Pump

2

Syed Nasir Ahmad, Tanveer Ahmad Mir, Talib Shareef, Sasmita Pattnaik, and Showkat Ahmad Lone

Abstract

The amount of carbon stored in the oceans is far greater than what is present in the atmosphere. Understanding the mechanism of absorption of carbon dioxide by oceans is crucial in deciphering the role of CO_2 in the context of climate change on micro- and macroclimatic level. Transformation and sequestration of dissolved organic carbon (DOC) involves the production of refractory dissolved organic carbon (RDOC) from labile dissolved organic carbon (LDOC) and this process is mediated primarily by microorganisms. Microbial carbon pump mechanism excludes the carbon from atmosphere by producing a pool of carbon that is recalcitrant to remineralization and thus cannot be reverted to CO_2. The microbial carbon pump (MCP) is a biological phenomenon driven by microbes that involves transformation and sequestration of carbon in the ocean. The major carbon sequestration mechanism is provided by MCP, as this is the only pump that leads to the production of refractory DOC fractions. The MCP thus possesses a profound impact on global carbon cycle and global climate.

Keywords

Carbon sequestration · Microbial pump · Dissolved organic carbon (DOC) · Recalcitrant dissolved organic carbon (RDOC) · Labile dissolved organic carbon (LDOC)

S. N. Ahmad · T. A. Mir · T. Shareef · S. A. Lone (✉)
Centre of Research for Development (CORD), University of Kashmir, Srinagar, India

S. Pattnaik
Department of Microbiology, Government Medical College Srinagar, Srinagar, India

S. A. Lone, A. Malik (eds.), *Microbiomes and the Global Climate Change*,
https://doi.org/10.1007/978-981-33-4508-9_2

2.1 Introduction

Carbon is the principle element and basis of all life forms present on the face of earth. The ocean is an enormous reservoir of carbon, containing about 60 times more carbon than in the atmosphere, which is about 38,000 petagrams (Pg) of C quantitatively (Sarmiento and Gruber 2002; Solomon et al. 2007). This reservoir of carbon is dynamic and keeps on transforming from one pool to other within the ocean and with the atmosphere over a varying timescale, the marine carbon cycle. There is continuous exchange of CO_2 at the air–sea interface, about 90 gigatons (Gt) of carbon per year is exchanged in each direction (Falkowski et al. 2000). Moreover 30% CO_2 released by anthropogenic activities is absorbed by the ocean, making it a major sink of atmospheric CO_2 (IPCC 2013).

The carbon flux in the ocean takes place by various complex processes, collectively termed as ocean carbon pumps. These processes, viz. solubility pump, carbonate pump, and biological pump are responsible for exporting carbon into the interiors of ocean. The solubility pump and carbonate pump generate two different pools of carbon in the ocean, the dissolved inorganic carbon (DIC) and particulate organic carbon (POC). However, studies have shown that sequestration of carbon via the carbonate pump requires release of CO_2 and which is why this pump is now termed as carbonate counter pump. The biological component of the ocean carbon pumps, the biological carbon pump (BCP) produces organic carbon pool of the ocean, the dissolved organic carbon (DOC), and particulate organic carbon (POC). The DOC and POC are operationally defined and categorized based on the parameter of size. All the carbon compounds having size less than 0.2 μm represent the DOC fraction, while the carbon compounds surpassing that baseline make up the POC (Carlson et al. 2002).

The POC is primarily formed as the phytoplanktonic biomass, which subsequently circulates through different trophic levels in the marine food web (Gehlen et al. 2006; Wassmann 1998). 50% POC is converted into DOC and the production of DOC involves many mechanisms like viral lysis (rupturing of bacterial cells because of viral infection), sloppy feeding of phytoplanktons by metazoans, excretion of protists and metazoans contains DOC, action of microbial ectohydrolases which convert POC into DOC, etc. (Anderson and Tang 2010; Suttle 2007; Nagata et al. 2000). It must be pointed out that primary source of DOC production varies with ecological frame besides spatially and temporally.

The majority of DOC is used by the heterotrophic prokaryotes and is thus respired back to the atmosphere but a fraction is converted into unusable and undegradable forms termed as recalcitrant DOC. This recalcitrant DOC forms a huge reservoir of carbon in the ocean, scientists and researchers are continuously trying to find and understand the processes and phenomena that convert the carbon into these unusable forms. Recently new concept of microbial carbon pump (MCP) has been proposed by N. Jiao to address the scenario (Jiao et al. 2014).

2.2 Understanding DOC Fractions

The ocean stores a large reservoir of carbon, quantitatively 662 Pg C, in the form of DOC (Hansell et al. 2009). Different fractions of this DOC pool show diversity in terms of their lifetime, i.e. some of the fractions of DOC are very short-lived, while other fractions have a lifetime of about thousands of years. There has been huge ambiguity in the scientific world while defining these fractions of DOC, but a beautiful concept has been propounded by Hansell which categorizes these DOC fractions based on their lifetime or reactivities (Hansell 2013). It must be pointed out that in case of DOC the lifetime is defined in terms of "e folding lifetime" rather than half lifetime. The "e folding lifetime" is given by the time in which DOC fraction decreases to 1/e of its initial value. The value of e is 2.71828 and is termed as Napierian constant (Jiao et al. 2014; Hansell 2013).

The DOC has been categorized into five fractions, based on their lifetime, viz. labile DOC (LDOC), semi-labile DOC (SLDOC), semi-refractory DOC (SRDOC), refractory DOC (RDOC), and ultra-refractory DOC (URDOC) (Hansell 2013).

LDOC is very short-lived (hours–days) and is the major DOC fraction produced at the surface of ocean. This fraction supports the microbial loop in the euphotic zone and is respired back to the atmosphere by the heterotrophic prokaryotes. Thus, this fraction has no role in carbon sequestration. SLDOC has a lifetime of approximately 1.5 years and is primarily exported to upper mesopelagic zone (Hansell 2013). This fraction supports the microbial loop in both euphotic and upper mesopelagic zone and has a very small role as far as carbon sequestration is concerned. SRDOC plays an important role in carbon sequestration as this fraction has a lifetime of around 20 years (Hansell 2013). This fraction reaches further depths and is exported to mesopelagic zone. RDOC and URDOC possess an approximate lifetime of 16,000 years and 40,000 years, respectively, and are thus major contributors to carbon sequestration (Hansell 2013). These are the least understood DOC fractions.

Many authors use the term recalcitrant DOC, which is defined as the DOC fraction that is resistant or reluctant to rapid degradation. Thus, DOC can be broadly classified into labile DOC and recalcitrant DOC, the recalcitrant DOC accumulates all the sub-fractions, i.e. (SLDOC, SRDOC, RDOC, URDOC). Few authors propose a new dimension of recalcitrance and categorize recalcitrant DOC into RDOCt and RDOCc based on environmental context and concentration, respectively (Jiao et al. 2014). The former fraction is defined to be recalcitrant because the microbes present in a particular environment are not able to degrade it making it recalcitrant in that particular context. The latter is recalcitrant because of the concentration, if the concentration is very low, i.e. below a particular threshold, the microbes are not able to use it.

2.3 Classical Ocean Carbon Pumps

Three major carbon pumps were previously believed to sequester carbon into the interiors of ocean and thus play an active role in carbon cycle. The carbon pumps are solubility pump, biological carbon pump, and carbonate counter pump.

2.3.1 Solubility Pump

It consists of the physio-chemical processes that involve the exchange of carbon between atmosphere and ocean and its transport into the interiors of ocean. The solubility of CO_2 in seawater is the physical parameter that governs the exchange of CO_2 between the atmosphere and the ocean. It is given by Henry's law, i.e. $S = kP$, where S is the solubility of gas in liquid, P is the pressure of gas in the atmosphere, k is the solubility constant (Carlson et al. 2001). The physical properties of seawater like salinity, temperature, etc. determine the solubility of CO_2. The solubility of CO_2 is higher in colder water and lesser in warmer water (Carlson et al. 2001).

At the air–sea interface, the CO_2 is transferred into the oceanic surface waters by the process of diffusion, which is driven by the concentration gradient. The dissolved CO_2 combines with H_2O to produce carbonic acid which subsequently dissociates into proton and bicarbonate ion.

$$H_2O + CO_{2(g)} \leftrightarrow H_2CO_3 \leftrightarrow H^+ + HCO_3{}^-$$

$$HCO_3{}^- \leftrightarrow H^+ + CO_3{}^{2-}$$

The inorganic compounds mentioned in the above reactions collectively form a pool of carbon termed as DIC. The export of DIC into the interiors of oceans is carried by many processes like thermohaline circulation and it is these reactions that act as buffering system and maintain the pH of ocean (Carlson et al. 2001).

2.3.2 Biological Carbon Pump

This carbon pump can be simply defined as the sequestration of carbon into the interiors of ocean which is driven by biological processes.(Sigman and Haug 2006) The phytoplanktons present in the euphotic zone utilize light energy to carry out the process of photosynthesis, which results in the fixation of dissolved inorganic carbon into reduced organic compounds. They uptake nutrients like nitrogen, phosphorus, and many trace elements besides dissolved inorganic carbon to produce reduced organic compounds. This foundational step from where all the food chains start is termed as primary production. The organic compounds formed, the POC and DOC, are used by the heterotrophs as their energy source, leading to the remineralization of CO_2, by the counter process of respiration.

Almost all of the primary production is counterbalanced by respiration; however, a very small fraction is not remineralized and is thus stored as POC and DOC forms (Carlson et al. 2001). The export of POC into the interiors of ocean occurs both passively and actively. The former involves sinking of POC, while the latter involves the vertical migration of zooplanktons, etc. Unlike photosynthesis the remineralization occurs throughout the water column and thus the organic compounds that cannot be used at the surface levels are used by the heterotrophs at different vertical levels in the ocean. This downward export ensures that remineralization occurs at deeper levels in the ocean, which has a huge significance in terms of maintaining the undersaturation of DIC in surface waters. This process ensures continuous CO_2 influx into the ocean from the atmosphere. The CO_2 is released back into the atmosphere on timescale of months to years. The POC fraction that crosses greater depths is excluded from the cycling processes and is stored for thousands of years. Approximately 10 Pg C y^{-1} is exported from the surface waters but only less than 1% is stored in the abyssal sediment (Carlson et al. 2001).

2.3.3 Carbonate Pump

An additional dimension of biological pump involves the formation of calcareous skeletal moieties by phytoplanktons (like coccolithophores) and animals (like foraminifera and pteropods). The formation of calcareous materials or the process of calcification involves reaction of calcium with bicarbonate ions which leads to the production of calcium carbonate, a form of PIC. This calcium carbonate forms hard coverings of organisms present in the ocean. The calcium carbonate being dense sinks rapidly and is exported into the interiors of the ocean. On the way down the calcium carbonate is not remineralized; however, after reaching carbonate compensation depth it is dissolved completely.

This pump is also termed as carbonate counter pump, as the reaction that involves the production of calcium carbonate is also accompanied by the formation CO_2 as a by-product.

$$Ca^{2+} + 2HCO_3^{-} \leftrightarrow CaCO_3 + CO_2 + H_2O$$

Thus, during this process carbon is sequestered in the form of calcium carbonate but at the same time CO_2 is also released back.

2.4 Microbial Carbon Pump (MCP)

Scientific world is at the beginning to unfold the mysteries about the activity of microorganisms in diverse environments. The unexplored microbes are performing their job silently in different domains of the biosphere. Recently, a novel idea of microbial carbon pump (MCP) was introduced by Jiao Nianzhi and co-workers. The concept of MCP refers to the processes driven by microbes which involve the

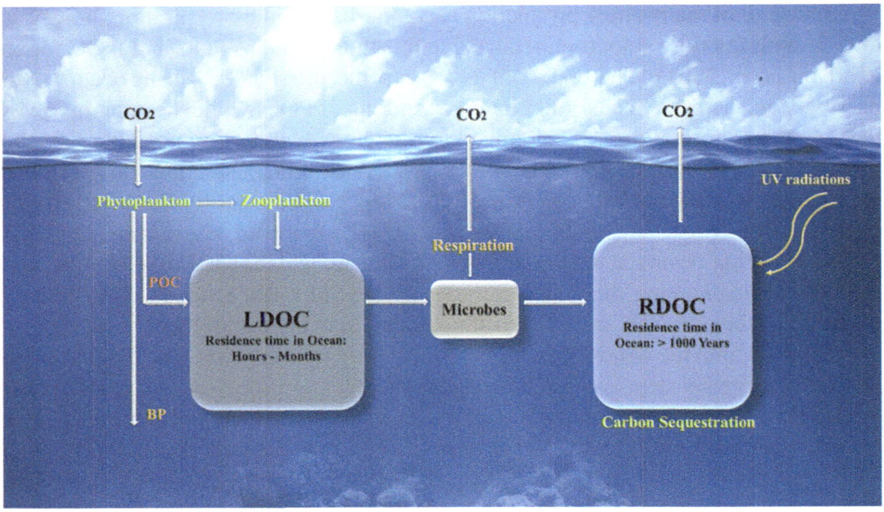

Fig. 2.1 Diagrammatic representation of microbial carbon pump. The left portion shows carbon sequestration through biological pump and the right portion shows carbon sequestration in the form of RDOC mediated by MCP. (Image modified from: the Microbial carbon pump; from genes to ecosystem by Jiao and Zheng (2011))

transformation of LDOC into RDOC. The process makes DOC highly resistant to microbial degradation, and thus it perpetuates in the oceanic environment for longer period of time (hundred to thousands of years) (Ogawa et al. 2001; Jiao et al. 2011). The microbes that are involved in this transformation process are bacteria, viruses, archaea, and phytoplankton. Traditionally, microbes were known as decomposers, but the novel role of microbes (i.e. MCP) is completely different from its traditional one. The concept of MCP was formulated within the framework of marine microbiology and describes the biochemical mechanisms that produce recalcitrant form of organic carbon. The labile form of DOC is readily available for organisms, whereas the RDOC is resistant to degradation. The microbial carbon pump increases the lifetime of DOC by transforming it to recalcitrant form, which is then sequestered into the interiors of ocean forming a large recalcitrant carbon pool (Jiao et al. 2010a) (Fig. 2.1). It is estimated that approximately 5–7% of DOC produced by marine microbes is RDOC (Gruber et al. 2006; Koch et al. 2014). The carbon sequestration in the ocean by marine microbes (via MCP) is indispensably connected to the inventory of carbon in atmosphere, and thus it has a substantial impact on global carbon cycling and its climate (Jiao et al. 2014).

The DOC pool in the ocean contains millennial compounds of organic carbon and constitutes nearly 95% of the total residual organic carbon (Ogawa and Tanoue 2003). The microbial activity changes structure of the DOC pool. The diversity and composition of microbial community depend upon the availability of DOC compounds to the microbes. For illustration, the SAR11 clade and Roseobacter clade are predominant marine bacterial groups in oligotrophic and eutrophic waters, respectively (Giovannoni et al. 2005a; Giovannoni et al. 2005b; Moran et al. 2007). The strategies for carbon assimilation are different in each clade and thus respond differently to the DOC reservoir in the ocean. The SAR11 clade constitutes nearly 25% heterotrophic bacterial community present in marine environments, whereas Roseobacter clade is another major bacterial group which constitutes nearly 15–20% of bacterial groups present in oceanic and coastal waters (Selje et al. 2004; Giovannoni and Stingl 2005; Carlson et al. 2009).

The transformation of DOC from labile to recalcitrant form is eventually carried out by the activity of functional genes present in microbes. For instance, ATP-binding cassette (ABC) transporter genes, including both importers and exporters, present in microbes trigger a mechanism that helps to mediate the transport of DOC fractions into or out of the microbial cells (Jiao and Zheng 2011). The ABC importers are particularly found in prokaryotic organisms; however, exporters are found in all kingdoms of life (Locher 2009; Rees et al. 2009). The metagenomic and metatranscriptomic analysis of microbial communities in marine environments has shown that a large number of genes associated with ABC transporters are linked with nutrient uptake and carbon metabolism (Venter et al. 2004; Poretsky et al. 2005; DeLong et al. 2006; Mou et al. 2008). The ABC transporters are considered as indicators of the potentiality of marine microbial groups or communities to utilize or generate corresponding DOC compounds (Jiao and Zheng 2011). Therefore, an approach to determine various activities of the MCP could rely upon quantification of functional genes present in the microbes.

Genes associated with various microbial processes such as extracellular hydrolysis, motility, cross-membrane transport are essential to determine the availability of DOC compounds for microbial assimilation (Jiao and Zheng 2011). The current data reveals that Roseobacter clade present in marine environments prefer to assimilate carbohydrate containing DOC compounds, whereas the SAR11 bacterial strains are more capable to assimilate amino acid or nitrogen-rich DOC compounds, because carbohydrate and amino acid related ABC transporter are widely present in these microbes, respectively (Poretsky et al. 2010).

The heterotrophic bacterial groups present in the ocean could be major producers as well as consumers of DOC compounds (Ogawa et al. 2001; Jiao et al. 2010a). They take up DOC compounds, a portion of that carbon is utilized in the formation of their biomass such as cell wall components, a part of it is released as CO_2 via respiration, and the remaining fraction of carbon is released back in the DOC pool as transformed DOC compounds (Benner et al. 1992; McCarthy et al. 1998; Ogawa et al. 2001; Benner and Herndl 2011). Microbes synthesize D-amino acids, which bind to their cell walls and make it highly resistant to degradation by extracellular hydrolytic enzymes present in the marine environment. After the death of microbes, the D-amino acids get released into the water column, over time it gets accumulated in the ocean as RDOC and thus forms a huge carbon pool (McCarthy et al. 1998; Repeta et al. 2002). Viral lysis is an another important mechanism by which organic carbon gets released from bacterial cells in the marine environment. However, majority of the components released via viral lysis are again assimilated very quickly by heterotrophic microbes, but a fraction of the components released by the lysis mechanism such as cell wall components are recalcitrant to microbial degradation and hence remains as RDOC in the environment (Suttle 2007; Brussaard et al. 2008). Therefore, these tiny creatures play a significant role in long term carbon storage in the marine environment and thus may be the key players for mitigating climate change on the planet (Hansell 2013).

2.5 Disturbances and Effects on Microbial Carbon Pump

The biological carbon sequestration is of great importance as it has been estimated that levels of CO_2 in the atmosphere without BCP would be 50% higher than the current value (Parekh et al. 2006). There are a number of different kinds of effects on sequestration flux of carbon. Two major unnatural effects of climate change on oceanic carbon sequestration are warming and acidification. To the warming environment and acidification of ocean, three general responses of the MCP have been proposed (Sarmiento and Gruber 2002). Increase in microbial activity, due to changes in seawater chemistry and introduction of large fraction of photosynthetically fixed carbon into DOC, thus the magnitude of the MCP is potentially increased in comparison to other pumps in carbon sequestration (Jiao et al. 2010a; Solomon et al. 2007). The warming of ocean would directly result in increase in ocean stratification and hence nutrient supply to the euphotic zone is decreased (oligotrophic) which ultimately results in an increase in primary production (PP) towards DOC than to POC, thus increases the part of MCP carbon sequestration (Falkowski et al. 2000; Jiao et al. 2010a). The microbial dominance and degradative capacities in an open ocean could be shifted because of overall global changes, which may have direct impact on the long-lived DOC reservoir in the ocean (Jiao et al. 2010a).

2.5.1 Warming of Ocean Waters

The increasing temperatures of oceanic waters are suggested to be directly involved in increasing the rate of regeneration of CO_2 which means the release of locked carbon increases. The temperature rise is believed to enhance the capability of bacteria to consume more planktons at shallower depths. In an experimental study called as mesocosm experiment it was shown that the effect of temperature results in reduction in POC downward transport, thus reduces sequestration or transformation of organic matter to RDOC. Ultimately it was concluded from these results that increasing ocean temperature enhances the MCP.

2.5.2 Ocean Stratification and Nutrient Supply

Ocean stratification is another physical process which greatly influences carbon sequestration in an ocean. Increase in stratification hampers supply of nutrients from deep ocean to photic zone, which in turn will have direct effect on export of

POC (Doney 2006; Capotondi et al. 2012; Passow and Carlson 2012). The decrease in nutrient supply to euphotic zone would spread the oligotrophic conditions in the sea. The MCP could get affected as maximum proportion of PP is directed to DOC (PER) and thereafter some part of that DOC is converted to RDOC (Jiao et al. 2010a).

An enhancement in the nutrient transport to coastal waters is supposed to bring about enhancement in PP, POC, and BP. But high nutrient concentrations could negatively affect MCP (Jiao et al. 2010a; Jiao et al. 2010b). Efficiency of bacterial growth, bacterial respiration, and PP would react differently to enhancing nutrient conditions. As the phytoplankton number increases with enhancing nutrient conditions, reaching a stage where number will be maximum and the PP starts to come down. This bloom in phytoplankton number will bring a supply of labile DOC for bacterial growth. The amount of oxygen present over there will be consumed through bacterial respiration and ultimately leads to hypoxia. In hypoxic and anoxic conditions, anaerobic bacteria start degrading leftover organic matter, results in generation of gases like methane and H_2S, thus excess nutrients ultimately result in lower organic carbon storage.

2.5.3 Exposure to UV Radiation

Exposure of organic matter to UV radiation is increased because of longer retention time near the ocean surface. It is thought that DOC polymers are readily cleaved by high energy ultraviolet (UV-B) or we can say that UV radiations cause photodegradation of SRDOC or RDOC. It is also thought to transform labile DOC to more recalcitrant forms. To make a net effect of SRDOC or DOC exposure to increased UV radiation needs further experimental studies. It is yet to finalize that whether there is decrease or increase in the magnitude of MCP (Legendre et al. 2015).

2.5.4 Ocean Acidification

Ocean acidification remains one of the major global issues caused due to rising concentrations of atmospheric CO_2. Microorganisms remain the prerequisite components of carbon cycle in marine ecosystem but the microorganisms and their biological processes depend upon pH of the environment. Therefore, ocean acidification directly affects microbial diversity, PP, and various gaseous emissions in an ocean. Human activities like burning of fuels, industrialization, and deforestation lead to increase in atmospheric CO_2 which ultimately results in decrease in pH of ocean. This acidification of water leads to the changes in chemistry of the ocean. With the increase of CO_2 in the atmosphere, the levels of aqueous CO_2 in the oceanic surface waters also increase on account of greater partial pressure of CO_2 in the atmosphere. The reactions of aqueous CO_2 with H_2O lead to the production of carbonic acid.

$$CO_2 + H_2O \rightarrow H_2CO_3$$

$$H_2CO_3 \rightarrow H^+ + HCO_3^-$$

Besides bicarbonate ion formation takes place by the combination of hydrogen ions and carbonate ions.

$$H^+ + CO_3^- \rightarrow HCO_3^-$$

Hence, the result of dissolving CO_2 in seawater is enhancing the concentration of aqueous CO_2, carbonic acid, hydrogen ions, and bicarbonate ions and a decrease in the carbonate ions. As a result there is increase in the concentration of H^+ ions which results in decrease in the pH of ocean water (Joint et al. 2011).

In mesocosm experiment, it was shown that more DOC is released by phytoplankton community when CO_2 concentrations are high or low pH (Riebesell et al. 2007). In natural environment, a larger proportion of carbon is fixed by photosynthesis and is directed into DOC as the acidity of ocean increases. Some fraction of this DOC is converted to refractory compounds by microbial activities and thus importance of MCP is increased (Eichinger et al. 2009).

2.5.5 Thermohaline Circulation

Thermohaline circulation is driven by temperature (thermo) and salinity (halo) (Rahmstorf 2006). In this process deep waters of ocean return to the surface of ocean. Some part of SRDOC and RDOC present at the surface of ocean water is deteriorated by incoming UV radiations (photodegradation) and this process leads to release of some part of dissolved CO_2 in water. It has not been clearly understood whether photodegradation of RDOC and release of dissolved CO_2 would be differentially affected on steady decrease in thermohaline circulation.

2.6 Conclusion

The continuous increment in the levels of CO_2 emissions into the atmosphere is furiously driving global warming and hence global climate change. The ocean carbon pumps set the basis of carbon relations between the atmosphere and hydrosphere, thus making ocean a major carbon sink. Among all the carbon pumps, MCP is the most prominent carbon pump qualitatively as well as quantitatively. Unlike physical and biological pumps, MCP is a non-sinking process and thus sequesters carbon into the ocean irrespective of its vertical level, i.e. carbon is sequestered throughout the water column. MCP may provide a future plan and path to tackle the increasing levels of CO_2 in the atmosphere and thus global climate change, as with the environmental distortions all other pumps are showing anomalies, for instance, oceanic warming impacts solubility pump and eutrophication affects BCP. With the challenges of global climate change, MCP may come to the forefront as the efficient and cheapest phenomena to reduce the level of CO_2 in atmosphere.

References

Anderson TR, Tang KW (2010) Carbon cycling and POC turnover in the mesopelagic zone of the ocean: insights from a simple model. Deep-Sea Res Pt II 57:1581–1592
Benner R, Herndl GJ (2011) Bacterially derived dissolved organic matter in the microbial carbon pump. In: Jiao N, Azam F, Sanders S (eds) Microbial carbon pump in the ocean. Science/AAAS, Washington, DC, pp 46–48. https://doi.org/10.1126/science.opms.sb0001
Benner R, Pakulski JD, McCarthy M, Hedges JI, Hatcher PG (1992) Bulk chemical characteristics of dissolved organic matter in the ocean. Science 255(5051):1561–1564
Brussaard CPD et al (2008) Global-scale processes with a nanoscale drive: the role of marine viruses. ISME J 2:575–578
Capotondi A et al (2012) Enhanced upper ocean stratification with climate change in the CMIP3 models. J Geophys Res Ocean 117(C4):C04031
Carlson CA, Bates NR, Hansell DA, Steinberg DK (2001) Carbon cycle. Elsevier, Amsterdam
Carlson CA, Giovannoni SJ, Hansell DA, Goldberg SJ, Parsons R, Otero MP, Vergin K, Wheeler BR (2002) Effect of nutrient amendments on bacterioplankton production, community structure, and DOC utilization in the northwestern Sargasso Sea. Aquat Microb Ecol 30:19–36
Carlson CA et al (2009) Seasonal dynamics of SAR11 populations in the euphotic and mesopelagic zones of the northwestern Sargasso Sea. ISME J 3:283–295
DeLong EF et al (2006) Community genomics among stratified microbial assemblages in the ocean's interior. Science 311:496–503
Doney SC (2006) The dangers of ocean acidification. Sci Am 294(3):58–65
Eichinger M et al (2009) Consumption and release of dissolved organic carbon by marine bacteria in a pulsed-substrate environment: from experiments to modelling. Aquat Microb Ecol 56:41–54
Falkowski P et al (2000) The global carbon cycle: a test of our knowledge of earth as a system. Science 290:291
Gehlen M et al (2006) Reconciling surface ocean productivity, export fluxes and sediment composition in a global biogeochemical ocean model. Biogeosciences 3:521–537
Giovannoni SJ, Stingl U (2005) Molecular diversity and ecology of microbial plankton. Nature 437:343–348
Giovannoni SJ et al (2005a) Genome streamlining in a cosmopolitan oceanic bacterium. Science 309:1242–1245
Giovannoni SJ et al (2005b) Proteorhodopsin in the ubiquitous marine bacterium SAR11. Nature 438:82–85
Gruber DF, Simjouw JP, Seitzinger SP, Taghon GL (2006) Dynamics and characterization of refractory dissolved organic matter produced by a pure bacterial culture in an experimental predator-prey system. Appl Environ Microb 72:4184–4191
Hansell DA (2013) Recalcitrant dissolved organic carbon fractions. Annu Rev Mar Sci 5:421–445
Hansell DA, Carlson CA, Repeta DJ, Schlitzer R (2009) Dissolved organic matter in the ocean: new insights stimulated by a controversy. Oceanography 22:52–61
IPCC (2013) Summary for policymakers. In: Stocker TF, Qin D, Plattner GK, Tignor M, Allen SK, Boschung J, Nauels A, Xia Y, Bex V, Midgley PM (eds) Climate change 2013: the physical science basis. Contribution of Working Group I to the fifth assessment report of the intergovernmental panel on climate change. Cambridge University Press, Cambridge, New York
Jiao N, Zheng Q (2011) The microbial carbon pump: from genes to ecosystems. Appl Environ Microbiol 77:7439–7444

Jiao N, Herndl GJ, Hansell DA, Benner R, Kattner G, Wilhelm SW, Kirchman DL, Weinbauer MG, Luo T, Chen F, Azam F (2010a) Microbial production of recalcitrant dissolved organic matter: long-term carbon storage in the global ocean. Nat Rev Microbiol 8:593–599
Jiao N et al (2010b) The microbial carbon pump: from genes to ecosystems. Appl Environ Microbiol 77(21):7439–7444
Jiao N, Herndl GJ, Hansell DA, Benner R, Kattner G, Wilhelm SW, Kirchman DL, Weinbauer MG, Luo T, Chen F, Azam F (2011) The microbial carbon pump and the oceanic recalcitrant dissolved organic matter pool. Nat Rev Microbiol 9:555–555. https://doi.org/10.1038/nrmicro2386-c5
Jiao N, Robinso C, Azam F, Thomas H, Baltar F, Dang H, Hardman-Mountford NJ, Johnson M, Kirchman DL, Koch BP, Legendre L, Li C, Liu J, Luo T, Luo Y-W, Mitra A, Romanou A, Tang K, Wang X, Zhang C, Zhang R (2014) Mechanisms of microbial carbon sequestration in the ocean – future research directions. Biogeosciences 11:5285–5306
Joint I, Doney SC, Karl DM (2011) Will ocean acidification affect marine microbes? ISME J 5:1–7
Koch BP, Kattner G, Witt M, Passow U (2014) Molecular insights into the microbial formation of marine dissolved organic matter: recalcitrant or labile? Biogeosciences 11:4173–4190. https://doi.org/10.5194/bg-11-4173-2014
Legendre L et al (2015) The microbial carbon pump concept: potential biogeochemical significance in the globally changing ocean. Prog Oceanogr 134:432–450
Locher KP (2009) Structure and mechanism of ATP-binding cassette transporters. Philos Trans R Soc B 364:239–245
McCarthy MD, Hedges JI, Benner R (1998) Major bacterial contribution to marine dissolved organic nitrogen. Science 281:231–234
Moran MA et al (2007) Ecological genomics of marine roseobacters. Appl Environ Microbiol 73:4559–4569
Mou X, Sun S, Edwards RA, Hodson RE, Moran MA (2008) Bacterial carbon processing by generalist species in the coastal ocean. Nature 451:708–711
Nagata T, Fukuda H, Fukuda R, Koike I (2000) Bacterioplankton distribution and production in deep Pacific waters: large-scale geographic variations and possible coupling with sinking particle fluxes. Limnol Oceanogr 45:426–435
Ogawa H, Tanoue E (2003) Dissolved organic matter in oceanic waters. J Oceanogr 59:129–147
Ogawa H, Amagai Y, Koike I, Kaiser K, Benner R (2001) Production of refractory dissolved organic matter by bacteria. Science 292:917–920
Parekh P et al (2006) Atmospheric carbon dioxide in a less dusty world. Geophys Res Lett 33(3): L03610
Passow U, Carlson CA (2012) The biological pump in a high CO_2 world. Mar Ecol Prog Ser 470:249–271
Poretsky RS et al (2005) Analysis of microbial gene transcripts in environmental samples. Appl Environ Microbiol 71:4121–4126
Poretsky RS, Sun S, Mou X, Moran MA (2010) Transporter genes expressed by coastal bacterioplankton in response to dissolved organic carbon. Environ Microbiol 12:616–627
Rahmstorf S (2006) Thermohaline ocean circulation. In: Elias SA (ed) Encyclopedia of quaternary sciences. Elsevier, Amsterdam
Rees DC, Johnson E, Lewinson O (2009) ABC transporters: the power to change. Nat Rev Mol Cell Biol 10:218–227
Repeta DJ, Quan TM, Aluwihare LI, Accardi A (2002) Chemical characterization of high molecular weight dissolved organic matter in fresh and marine waters. Geochim Cosmochim Acta 66:955–962
Riebesell U et al (2007) Enhanced biological carbon consumption in a high CO_2 ocean. Nature 450:545–548
Sarmiento JL, Gruber N (2002) Sinks for anthropogenic carbon. Phys Today 55:30–36
Selje N, Simon M, Brinkhoff T (2004) A newly discovered Roseobacter cluster in temperate and polar oceans. Nature 427:445–448

Sigman DM, Haug GH (2006) The biological pump in the past. In: Treatise on geochemistry, vol 6. Elsevier, Amsterdam, pp 491–528

Solomon S, Qin D, Manning M et al (2007) Climate change: the physical science. IPCC, Geneva, Switzerland

Suttle CA (2007) Marine viruses — major players in the global ecosystem. Nat Rev Microbiol 5:801–812

Venter JC et al (2004) Environmental genome shotgun sequencing of the Sargasso Sea. Science 304:66–74

Wassmann P (1998) Retention versus export food chains: processes controlling sinking loss from marine pelagic systems. Hydrobiologia 363:29–57

Climate Change Extenuation by Greenhouse Gas Quenching Microflora

3

Ruqeya Nazir and Riasa Zaffar

Abstract

The major cause of climate change is global warming, which refers to an increase in the concentration of three most important greenhouse gases such as CO_2, CH_4, and N_2O in the atmosphere. Climate change has become a hot topic because of its pervasive detrimental impacts globally. World over, the high level think tanks suggest the remedial measures to the policy makers so as to ameliorate the ramifications of climate change but till now, no suitable method has been devised which can mitigate the effects of climate change. The world is in need of a strategy that will be cost-effective and sustainable. Such techniques can now be used by harnessing the ability of microbes to sequester the greenhouse gases in an irreversible manner, thus aiding in climate change mitigation. The varied biogeochemical processes on earth are wholly and solely dependent upon the metabolic activities of microbes, without which our planet would have been packed with debris. Microbes thriving in the extreme conditions have an inherent ability to sequester CO_2, CH_4, and N_2O. Microorganisms such as Methylobacillus and Methanotrophs are important carbon recyclers as they use the greenhouse gases as their sole carbon and nitrogen sources which are consequently needed for their metabolic activities. Such bacteria are found in abundance in paddy fields, landfill sites, extreme alkaline environments and geothermal areas. There are some microbes which can convert CO_2 to $CaCO_3$ that can fetch minerals of economic value. Thus microorganisms have a great potential to fight against global

R. Nazir (✉)
Centre of Research for Development (CORD), University of Kashmir, Srinagar, India

R. Zaffar
Centre of Research for Development (CORD), University of Kashmir, Srinagar, India

Department of Environmental Science, University of Kashmir, Srinagar, India

S. A. Lone, A. Malik (eds.), *Microbiomes and the Global Climate Change*,
https://doi.org/10.1007/978-981-33-4508-9_3

warming and also for combating pollution. This chapter primarily focusses on the role of microbes in mitigating the effects of climate change.

Keywords

Bacteria · Climate change · Microorganisms · Global warming

3.1 Introduction

A thick layer of gases encapsulates the planet Earth, which work like a blanket and aid in keeping the planet warm in addition to allowing the flora and fauna to survive and sustain. The atmosphere is so pivotal in a way that in its absence, the planet would have been much colder (20–30 °C), thus rendering it inhospitable. Climate change has become a global phenomenon and its evidence can be exemplified by the increasing overall surface temperatures worldwide by a process called global warming. Even if, there is a minuscule alteration in earth's temperature, it can have disastrous consequences such as melting of glaciers and flooding. The atmosphere becomes warmed up because of the absorption of heat by the constituent gases. Consequently, the primary reason of such warming is the heat energy absorption by greenhouse gases (methane, carbon dioxide, ozone and nitrous oxide) that are reflected by the surface of Earth. The source and sink of these gases are the microbes which have an important part to play because of the recycling and transformation capability of essential elements, particularly carbon and nitrogen, which make up the cells. Of all the microorganisms, bacteria and archaea are concerned with the cyclic transition of every essential element. For example, in the process of methanogenesis, the concerned bacteria convert carbon dioxide into methane in Carbon cycle; the nitrogen-fixing bacteria such as Rhizobium fixes nitrogen, thus helping in the conversion of nitrogen in atmosphere into the form of nitrogen that can be taken up by plants in usable form. Photosynthetic algae and cyanobacteria are also involved in these cycles and form a major component of phytoplanktons. Different microorganisms (bacteria and fungi mostly) residing in soil are known for their decomposition ability as they break down complex organic matter into simpler ones, thereby releasing carbon dioxide back into the atmosphere (Panikov 1999). This chapter primarily focusses on the ability of microbes in mitigating the effects of climate change.

3.2 Soil Microbes and Climate Change

The microorganisms residing in the soil form an indispensible component for maintaining various ecological processes such as cycling of nutrients, decomposition of detritus, utilization and the making of trace gases, ultimately the biotransformation of metals (Panikov 1999). As per the IPCC report, there has been an unprecedented warming of climate systems plus the basic cause has been found out to be of anthropogenic origin. Additionally, the human activities, particularly the disposal of waste and intensive agricultural practices have paved a way for the production of greenhouse gases by microorganisms (IPCC 2007). There is an immense need to contemplate upon the role, soil microbiota has both as contributors and pacifiers with regard to climate change. Soil has been considered as a living, fertile and a substance for it is a home for varying life forms ranging from microorganisms to higher animals. When there is an increase in the temperature of Earth, the metabolic activities of microbes are enhanced, thereby expediting the biodegradation processes which further results in the release of more carbon dioxide into the environment. It has been shown experimentally that there has been an increase in soil respiration and carbon dioxide emission with the increase in temperature. The mass release of CO_2 is directly proportional to the global warming, that in turn augments the pace of microbial activity in soil (Briones et al. 2004). Nitrous oxide, another deadly greenhouse gas being 300 times more dangerous which depletes the ozone layer. Fortunately, mother nature has a substitute for combating the greenhouse effect (Sanford et al. 2012). There are various groups of microorganisms known as denitrifying bacteria which have the ability to convert nitrous oxide into harmless nitrogen. Scientists opine that other than the denitrifying bacteria, there are other microbes which have the same potential of using nitrous oxide, thereby mitigating the emissions (Willey et al. 2009). Researchers have estimated that *Prochlorococcus* and *Synechococcus* (single celled cyanobacteria) eliminate almost 10 billion tons of carbon from atmosphere every year (Willey et al. 2009). Microbes with such capabilities are found in most of the sediments and soils, implying that large microbiological masses contribute actively to nitrous oxide utilization (Willey et al. 2009).

3.3 Microbes and Global Warming

The technical and scientific world can enhance its comprehension on global environmental roles of the emissions of nitrous oxide and for enhancing the cycle models of greenhouse gases by the innovation of microbial diversification and its aid in nitrous oxide utilization (Sanford et al. 2012). This will pave a way for better understanding the ramifications of anthropogenic activities on climate change. One cannot accurately predict the global nitrous oxide emissions by analysing the microbial populations which are responsible for denitrification. Cellulose is the structural support and framework of plant cell wall, which being insoluble cannot be broken down by animals because of the lack of cellulase enzyme. Nature has solution for everything and in this perspective a specific group of animals known as ruminants are able to digest cellulose because of their four chambered stomach in which the largest compartment being the rumen is replete with millions of microorganisms which in turn are able to digest cellulose. Methanogens, living in rumen can break down complex food substance into methane, consequently, ruminants belch the gas out of digestive system. Methane is potent than carbon dioxide as it traps about 20 times more heat than it, resulting in the warming of planet

20 times more than CO_2. About 20% of global methane emanation is from farm animals as per the estimation of scientists (Panikov 1999) and they are now looking at ways of reducing the amount of methane emissions from ruminants. A vaccine has been developed in Australia which prevents the microbes present in the rumen from producing methane and it was tested on sheep and to our surprise it belched 8% less methane in a 13 h test. Currently, this vaccine is only effective against 20% of the microflora which produces methane. It is imperative for scientists to manufacture such vaccine which is effective against most of the microbes producing methane (IPCC 2007). *Methylokorus infernorum*, the methanotrophic bacteria, found in geothermal sites has the ability of consuming methane gas (11 kg/year) as the sole energy source, ultimately converting it to carbon dioxide during the digestion process. It is hoped that this bacterium can be very helpful in reducing methane emission from various sources (Jenkinson et al. 1991). Similarly other forms of bacteria known as *Methylobacillus*, found in astonishing numbers in all water ecosystems are of most important carbon recyclers of Earth. Generally, methylotrophs use greenhouse gases for fulfilling their energy needs. Recently, a microbe has been developed by Indian that can convert carbon dioxide into calcium carbonate that can fetch minerals of economic value. The enzyme that is used for such conversion can be installed in any industrial for the conversion. The discovery has been seen as a cornerstone which can go a very long way to fight global warming (Jenkinson et al. 1991).

3.4 Microbes as Carbon Sink

A cluster of oceanic microorganisms play a fundamental role in combating climate change and the task of oceans in absorbing carbon dioxide is crucial to comprehend. The photosynthetic activity taking place in the oceans by phytoplanktons is the main way that CO_2 is absorbed by the ocean. This is where the concept of food chain sets in starting from algae to the larger animals is followed by the death, sinking and decomposition of these creatures to the ocean floor. Such remains are known as particulate organic matter (POM). The organic compounds are contained within the dead creatures. Most of which is dissolved, which was not thought to be an important component of the total carbon dioxide (Anisimov et al. 1999).

3.5 Combating Global Warming Through Biofuels

Entire world is struggling to address the issues related to global warming and primarily the basis for this is industrial revolution which played a major role in the emission of greenhouse gases. This issue is such that it cannot be addressed in 1 day, rather it will take years altogether and the involvement of each and every individual, both at the local and the global level. Undoubtedly these emissions can be lowered by shifting to low carbon fuels and other sources of alternate energy. Important to mention that a hefty cluster of microorganisms have played a pivotal role to reduce these gases. The responsibility of microbes in determining the concentration of GHGs is commonly acknowledged (Singh et al. 2010). Ocean and forest ecosystems have been presumed to act as sinks for carbon and their protection through green technologies is considered to be another way of mitigation. The United Nations Environment Program (UNEP) is mainly involved in sustaining an international low carbon planet through climate change mitigation approaches. Techs such as geothermal power, hydrogen fuel cells, tidal energy, solar and wind power have been used to reduce greenhouse gas pollution (Edenhofer et al. 2012). Microorganisms such as methanotrophs already have propensity at using methane as a carbon and energy sources and moreover help to restrict the volume of GHG inside the system. There seems to be myriad studies that demonstrate microbiome can play a critical role in controlling global climate change (Singh et al. 2010). Earth's microbial ecological communities too are sensitive to man made climate change in two ways. First by manipulating the morphology of prevailing microbial species and then reshaping the makeup of microbial communities. Modifying the microbial communities provides a powerful effective method for mitigating global climate change.

3.6 Volatile Organic Carbon Mitigation and Methylotrophs

The chemical composition of the environment is changed directly by the volatile organic carbons (Cappellin et al. 2017). The intermediates of global carbon cycling such as methane and methanol are plentifully present in the earth. The bidirectional exchange of volatile organic compounds between red oaks and atmosphere was reported as per the recent research, that also emphasized the role and effectiveness of microbiota in curbing volatile components. The exchange was such that methanol in addition to the deposition in fumigation experiment, methanol was also deposited in trees and in soil (Wohlfahrt et al. 2012) Mainly, methanol generation can be linked to plant decay associated with degradation of lignin and pectin. Few facultative methanotrophs have been determined in earlier investigations to use organic compounds such as ethanol and acetate, although most of them are effective utilizing sole and various carbon compounds (Kolb 2009; Dunfield et al. 2010). Gout (2000) suggested that methanol is taken up by leaf stomata through leaf pores followed by transforming it into formaldehyde. Even though methylotrophic bacteria glossily appear on the surface of leaf, they are susceptible of using methanol by means of metabolic pathways (Subhaswaraj et al. 2017; Meena et al. 2012; Holland and Polacco 1994) or by chemical modification reactions (Elliot and McCracken 1989), while further lowering the volatile compound to certain magnitude.

3.7 Carbon Cycling and Climate Change

The global carbon perturbations are actually stated by the global carbon cycling of the natural ecosystems. Carbon is amongst the most complex elements on earth, a functional life support, and hence plays a fascinating role in weather, its reliability and human energy wealth.

The microbial transformation of damaged cells and organic matter lead to the elimination of CO_2 into the atmosphere as a consequence of their respiration and finally CO_2 is used by plants for photosynthesis. Methylotrophs are the pivotal class of bacteria which utilize greenhouse gases. Furthermore, the influence of global warming is greatly reduced (Kumar et al. 2016). As well as the various autotrophs such as vegetation and photosynthetic bacteria, methanogens utilize CO_2 for their energy demand. Heterotrophs additionally utilize organic compounds and transform them into CO_2 for their development. Several chemical transformations such as carbon dioxide fixation, methanotrophy, methanogenesis, fermentation, and anaerobic respiration maintain the stability of carbon cycling. A squad of bacteria named methylotrophs oxidize the single most significant and active greenhouse gas, i.e. methane and its equivalents (Meena et al. 2012; Kumar et al. 2015; Kumar et al. 2016; Holland and Pollacco 1994). There is the process of methylotrophic bacteria degradation and decomposition which sustains carbon cycling in the environment. Biological degradation of organic molecules results in CO_2 generation in the atmosphere (Allison et al. 2010). In conjunction, prokaryotes which include Actinomycetes, Fermicutes, Arthrobacters and Pseudomonads also have a huge role to play in perilous carbon biodegradation and its equivalents. The distinct soil microbiomes are climate change variables that enhance our insight of how articulately they adapt to climatic changes (UCAR 2011). Numerous human activities, such as habitat destruction, development of mills, automotive combustion of carbon fuels, air and water pollution, affect climate change (UCAR 2011). The above entanglements have precipitated changes in the concentration of carbon and nitrogen throughout the world. The rapid expansion of greenhouse gases as well as other atmospheric changes simultaneously affects the environment. Humanity has a considerable part to play in improving the architecture of the environment and the energy cycle which has been influenced by microbial communities for so many years (Udakis 2013).

3.8 Methylotrophs Mitigating Methane

Methane is next just to carbon dioxide, amongst the most important greenhouse gases as far as their role in climate change is concerned. Frequent discharge of methane from various sources may pose a potential temperature threat in near future. Therefore, a well-developed mechanism should be put in place to reduce the methane emissions. A wider array of mitigation strategies have also been introduced that address different human sources, which need to be customized for subsequent use. The use of methylotrophs and a thorough explanation of soil methanotrophy may be a very valuable method for resolving the natural release of methane from closed landfills and a major reduction in waste-related GHG pollution from methanotrophic processes (Hettiaratchi and del Castillo Sternenfels 2013). Methanotrophs have an incredible ability to flourish and adapt and change aerobically on CH_4 as a common source of carbon and energy, and these microorganisms play a central role in the conversion of CH_4 into organic compounds available as CO_2 to autotrophs (Large 1983). In addition, methane oxidation occurring under environmental conditions in the presence of hydroxyl radicals is the key component disintegration that occurs from a variety of photochemical reactions. A successful research was performed on methane biodegradation and consequential agglomeration of poly-β-hydroxybutyrate (PHB) using only a methanotrophic conglomerate and an independent strain, culminating in significant methane remediation outcomes. Lately, the straightforward spectrophotometric test for methane screening using microbial strains has been investigated, reviewed thoroughly with the different methods available; i.e. the methylotrophs that reduce methane, traditional gas liquid chromatographic methods, monitoring of specific enzymes, and molecular characterization/detection of MMO and mxaF encoding genes (methane monooxygenase and methanol dehydrogenase). Jhala et al. (2015) however successfully engendered the strains of bacteria, deteriorating methane by boosting soil with methane as the principal energy source and water as the basal medium. In conjunction, colorimetric assay plates established their presence in evacuated tubes containing methane and the nature of soluble methane enzyme monooxygenase (sMMO). The potential of the isolates to breakdown methane was subsequently developed by the identification of enzyme gene encoding (methane monooxygenase and ethanol dehydrogenase) along with the detailed evaluation of the isolate enzyme function (Jhala et al. 2015). A research on slurry content captured from the Herman Pit (once a mercury mine) demonstrated the key role of methanotrophic bacteria in extracting methane from the sediments under anaerobic environment. The methanogenic operation was also conducted in artificial acid environments that demonstrated the presence of acidophilic or acid-tolerant methanotrophs. Furthermore, the acid-resistant methanotrophs were conclusively proven using incubated slurries to achieve optimum operation at pH 4.5. In conjunction, certain methanotrophs were subjected to sterol and hopanoid lipid extraction (methanotrophic characteristic), and their prevalence was exacerbated with expanded use of methane sediments. A sequence enriched for pmo A gene that associated with methanotrophic members of the Gammaproteobacteria has also been seen by the genomic DNA engendered by certain supplementation cultures that oxidize the methane gas. An enrichment culture was formed under acidic conditions (pH 4.5), by methane oxidation (Baesman et al. 2015). Strengthening the amount of CO_2 in the atmosphere is yet another concern for the global scientific establishment, and consideration is now being given to tracking the role of methylotrophs in carbon dioxide prevention and management. Carbon emissions from most biomass or biomass-based waste depletion are usually not included in the national or international greenhouse gas list, as biomass linked with waste is anticipated to be produced on sustainable basis and no net carbon dioxide emissions are forecasted, as it is assumed that carbon dioxide emitted from the decay of food waste can be absorbed by next year's crop.

3.9 Methylotrophs Mitigating Methane in Paddy Fields

The disproportional balance of methane has led to the wide range of changes in the properties of climate at the physical and chemical levels. The wide fields of rice epitomizes the same as an acme of methane source. (Dubey 2005). The methane as such in the vast rice fields is biodegraded and reduced, with methanotrophic bacteria playing a major role of catalysing the same..The ample presence of water in the paddy fields activates the methanogens thereby generating methane gas and thereby leads to trapping of methane gas by methanotrophic bacteria and subsequently the methane gets converted to methanol and biomass. Methane monooxygenase (MMO) enzyme and oxygen as a triggering agent are preconditions for any methanotrophic action. This enzyme mechanism triggers to destroy the methane methanotrophs. This aerobic condition is exacerbated by the green algae prevalent over the water-filled rice paddies surface (Singh and Dubey 2012). The methylotrophic isolates with functioning enzyme systems were obtained from Gujarat wetland paddy fields, and through biochemical and molecular characterization they transformed out to be separate species of *Bacillus* and *Penibacillus*. The active enzyme mechanism indicates the existence of the genes of particulate methane monooxygenase (pmoA), encoding the gene cluster α subunits. The involvement of the gene pmoA clearly indicates the use of methane by microbiota such as *P. illinoisensis, B. Aerius, B. Subtilis* and *Rhizobium sp.*, whereas in *Methylobacterium extrorquens* the existence of mmoX gene encrypts (α subunit of the hydroxylase component). Where as in another analysis, methane effectively exploits communities such as *P. illinoisensis* and *Rhizobium sp.* Prevalence of the mxaF coding gene α subunit of methanol dehydrogenase enzyme was identified. Communities with a propensity to degrade methane were baptized from wetland rice fields which revealed the presence of methane depletion enzymes and genes within a recognized plant growth facilitating microbial community (Jhala et al. 2014). In the paddy fields there is a structured structure of certain unique methylotrophic groups with a latent propensity and capability to absorb greenhouse gas methane, culminating in aerobic conditions on the surface of the ground in paddy fields. This systemic structured film is linked to the algal colonies commonly seen in rice fields. Algal ecosystems play a key role in reducing global greenhouse gas emissions by influencing methane gas oxidation activity. In yet another trial, in the isolation of rice plants, a small quantity of algae minimized the methane emission (Wang et al. 1995). The algae on the exterior of waterlogged paddy fields also engendered the methanotrophs and halted the generation of methanogen. Peng et al. (1995) showing that, in the availability of rice, CH_4 emissions are mainly from aerenchyma. Studies thus reveal and illustrate the relationship and role of methylotrophs in the removal of greenhouse gases in the atmosphere.

3.10 Conclusions

Methylotrophic bacteria use and disrupt condensed carbon compounds such as methane, which have an unprecedented function in combating climate change. This community of bacteria is exceptional in mitigating emissions of greenhouse gases. The rice field, where enzymatic operations are facilitated by various species such as algae and methanogens, is the most suitable habitat for methanotrophs. Rice production makes a significant contribution up to 15% of global methane emissions. Methylotrophs, furthermore which degrade many methane equivalents in the atmosphere, also assist in reducing CO_2. Such conditioned response of methylotrophs thus facilitates the climate latently. Varying bacterial species with methylotrophic properties as observed in soil, water and plant habitats contribute immensely to reduction of greenhouse gases. Therefore fundamental science cannot neglect the significance and role of methylotrophs in combating climate change. Together with recent perspectives, the role of methylotrophic ecosystems in addressing climate change has led us to investigate their function in global carbon cycling and minimize the impact of greenhouse gases such as methane and implicitly carbon dioxide on the landscape.

We attempted to explain the methylotrophic community's position in the rice environment as well as climate change. Also stressed was their role in sequestration of carbon or release of climate-active metabolites as a result of climate change. Methylotrophs represent a practical microbe community that uses decreased carbon substrates that don't even contain carbon–carbon bonds. Methylotrophic bacteria include methanotrophs (which digest methane) and methylotrophs besides the methanotrophs (which eat non-methane-reduced carbon substrates). This comprehensive functional group can use different compounds like formaldehyde methanol, methylamine, dimethylamine, as well as methane. In addition to methane, this systematically integrated group will use different compounds such as formaldehyde methanol, methylamine, dimethylamine as the sole source of carbon and energy and is a persistent stakeholder in the global carbon cycle. Methanotrophs also oxidize greater than 90% of the methane contained in the soil prior to evaporating into the atmosphere (Oremland and Culbertson 1992). Methanotrophic biological oxidation of methane contributes for just approximately 5% of the global dump of atmospheric Methane (Hanson and Hanson 1996).

References

Allison SD, Wallenstein MD, Bradford MA (2010) Soil-carbon response to warming dependent on microbial physiology. Nat Geosci 3:336–340

Anisimov OA, Nelson FE, Pavlov AV (1999) Predictive scenarios of permafrost development under conditions of global climate change in the XXI century. Earth Cryol 3:15–25

Baesman SM, Miller LG, Wei JH, Cho Y, Matys ED, Summons RE, Welander PV, Oremland RS (2015) Methane oxidation and molecular characterization of methanotrophs from a former mercury mine impoundment. Microorganisms 3:290–309

Briones MJI, Poskitt J, Ostle N (2004) Influence of warming and enchytraeid activities on soil CO_2 and CH_4 fluxes. Soil Biol Biochem 36:1851–1859

Cappellin L, Alarcon AA, Herdlinger Blatt I, Sanchez J, Biasioli F, Martin ST, Loreto F, McKinney KA (2017) Field observations of volatile organic compound (VOC) exchange in red oaks. Atmos Chem Phys 17:4189–4207

Dubey SK (2005) Microbial ecology of methane emission in rice agroecosystem: a review. Appl Ecol Environ Res 3(2):1–27

Dunfield PF, Belova SE, Vorob'ev AV, Cornish SL, Dedysh SN (2010) Methylocapsa aurea sp. nov., a facultative methanotroph possessing a particulate methane monooxygenase, and emended description of the genus Methylocapsa. Int J Syst Evol Microbiol 60:2659–2664

Edenhofer O, Pichs Madruga R, Sokona Y (2012) Renewable energy sources and climate change mitigation: special report of the intergovernmental panel on climate change. IPCC, Geneva, Switzerland

Elliot AJ, McCracken DR (1989) Effect of temperature on reactions and equilibria: a pulse radiolysis study. Int J Rad Appl Instrum Part C 33(1):69–74

Gout E (2000) Metabolism of methanol in plant cells. Carbon-13 nuclear magnetic resonance studies. Plant Physiol 123(1):287–296

Hanson R, Hanson T (1996) Methanotrophic bacteria. Microbiol Rev 60(2):439–471

Hettiaratchi J, del Castillo Sternenfels U (2013) Mitigation/reduction of GHG emissions in solid/hazardous waste management. In: Climate change modeling, mitigation, and adaptation. American Society of Civil Engineers, Reston, VA, pp 600–620. https://doi.org/10.1061/9780784412718.ch22

Holland MA, Polacco JC (1994) PPFMs and other contaminants: is there more to plant physiology than just plant? Annu Rev Plant Physiol Plant Mol Biol 45:197–209

IPCC (2007) Climate change 2007: synthesis report. http://www.ipcc.ch/publications_and_data/publications_ipcc_fourth_assessment_report_synthesis_report.Htm

Jenkinson DS, Adams DE, Wild A (1991) Model estimates of CO_2 emissions from soil in response to global warming. Nature 351:304–306

Jhala YK, Rajababu VV, Shelat Harsha N, Patel HK, Patel HK, Patel KT (2014) Isolation and characterization of methane utilizing bacteria from wetland paddy ecosystem. World J Microbiol Biotechnol 30(6):1845–1860

Jhala YK, Rajababu VV, Panpatte Deepak G, Shelat Harsha N (2015) Rapid methods for isolation and screening of methane degrading bacteria. J Bioremed Biodegr 7:322

Kolb S (2009) Aerobic methanol oxidizing bacteria in soil. FEMS Microbiol Lett 300:1–10

Kumar M, Srivastava AK, Pandey AK (2015) Biocontrol activity of some potent methylotrophs isolated from Bhitarkanika mangrove sediment. Int J Curr Res Biosci Plant Biol 2(6):101–106

Kumar M, Tomar RS, Paul D, Lade H (2016) Methylotrophic bacteria in sustainable agriculture. World J Microbiol Biotechnol 32:120

Large PJ (1983) Methylotrophy and methanogenesis. In: Aspects of microbiology, vol 8. Springer, New York

Meena KK, Kumar M, Kalyuzhnaya MG, Yandigeri MS, Singh DP, Saxena AK, Arora DK (2012) Epiphytic pink pigmented methylotrophic bacteria enhance germination and seedling growth of wheat (Triticum aestivum) by producing phytohormone. Antonie Van Leeuwenhoek 101 (4):777–786

Oremland RS, Culbertson CW (1992) Importance of methane-oxidizing bacteria in the methane budget as revealed by the use of a specific inhibitor. Nature 356:421–423

Panikov NS (1999) Understanding and prediction of soil microbial community dynamics under global change. Appl Soil Ecol 11:161–176

Peng S, Ingram KT, Neue HU, Ziska LH (1995) Climate change and rice. Springer, New York, pp 81–91

Sanford RA, Wagner DD, Cu QW, Chee-Sanford J, Thomas SH, Cruz-Garcia C, Rodriguez G, Massol-Deya A, Krishnani KK, Ritalahti KM, Nissen S, Konstantinidis KT, Loffler FE (2012)

Unexpected non -denitrifier nitrous oxide reductase gene diversity and abundance in soils. Proc Natl Acad Sci 109(48):19709–19714. https://doi.org/10.1073/pnas.1211238109
Singh A, Dubey SK (2012) Temporal variations in methanogenic community structure and methane production potential of tropical rice ecosystems. Soil Biol Biochem 48:162–166
Singh BK, Bardgett RD, Smith P, Reay DS (2010) Microorganisms and climate change: terrestrial feedbacks and mitigation options. Nat Rev Microbiol 8:779–790
Subhaswaraj P, Jobina R, Parasuraman P, Siddhardha B (2017) Plant growth promoting activity of pink pigmented facultative methylotroph – Methylobacterium extorquens MM2 on Lycopersicon esculentum L. J App Biol Biotechnol 5(01):042–046
UCAR (2011) Biogeochemical cycles. Retrieved from https://spark.ucar.edu/longcontent/biogeochemicalcycle
Udakis L (2013) Microbes and climate change. Society for General Microbiology, Reading, MA
Wang Z, Crozier CR, Patrick WH Jr (1995) Methane emission in a flooded rice soil with and without algae. In: Lal R et al (eds) Soil and global change. Lewis Publishers, Boca Raton, FL, pp 245–250
Willey JM, Sherwood LM, Woolverton CJ (2009) Prescott's principles of microbiology. McGraw-Hill, New York
Wohlfahrt G, Brilli F, HöRtnagl L, Xu X, Bingemer H, Hansel A, Loreto F (2012) Carbonyl sulfide (COS) as a tracer for canopy photosynthesis, transpiration and stomatal conductance: potential and limitations. Plant Cell Environ 35(4):657–667

Role of Methanotrophs in Mitigating Global Warming

4

Farhana Masood, Saghir Ahmad, and Abdul Malik

Abstract

There is growing concern about global warming worldwide. Greenhouse gases, which absorb the heat energy reflected by the earth's surface, are the main causes of global warming. Carbon dioxide, methane, nitrous oxide, and ozone are the main greenhouse gases. Methane is about 23 times more effective as a greenhouse gas than carbon dioxide. Anthropogenic sources release methane directly or indirectly into the atmosphere account for up to one-third of the global warming currently taking place. Methanotrophic bacteria or methanotrophs may serve as a biofilter and use methane as a source of energy before it is released into the atmosphere. They have been the only recognized major biological sink for atmospheric methane and play a key role in reducing the load of methane up to 15% to the total global methane destruction. Because of its physiologically adaptable nature, methanotrophs exist in diverse habitats and present in a wide range of pH, temperature, oxygen, salinity, and radiation. In this chapter, role of methanotrophs as an effective tool in mitigating greenhouse gas emissions is reported.

Keywords

Carbon dioxide · Global warming · Greenhouse gas · Methane · Methanotrophs

F. Masood (✉) · S. Ahmad
Department of Post Harvest Engineering and Technology, Faculty of Agricultural Sciences, Aligarh Muslim University, Aligarh, India

A. Malik
Department of Agricultural Microbiology, Faculty of Agricultural Sciences, Aligarh Muslim University, Aligarh, India

S. A. Lone, A. Malik (eds.), *Microbiomes and the Global Climate Change*,
https://doi.org/10.1007/978-981-33-4508-9_4

4.1 Introduction

The atmospheric level of greenhouse gases (GHGs) like carbon dioxide (CO_2), nitrous oxide (N_2O), nitric oxide (NO), methane (CH_4), ozone (O_3), and halogenated hydrocarbons increases considerably with time due to biotic (plants, animals, microorganisms, etc.) and anthropogenic (mining, industry, transportation, deforestation, construction, settlements, etc.) activities. In the future, these gasses will continue increasing with increasing populations, crop production and changing shifting patterns of food consumption, along with increased demand for ruminant meats. Combustion of fossil fuels and industrial procedures are the primary sources contributing to GHGs. A notable contribution is also made by the agricultural sector and land use. The emission of various greenhouse gases and their role to worldwide complete emissions vary with land use. At the global level, N_2O, CO_2, and CH_4 contribute 8%, 77%, and 15% of total emissions, respectively. The agricultural sector accounts for 32% of total worldwide emissions, of which 6%, 18%, and 8% are N_2O, CO_2, and CH_4 (De la Chesnaye et al. 2006).

Greenhouse effect is the interaction between incident solar radiation and these gasses to store heat in the atmosphere and stop infrared radiation from escaping. Higher amount of these gasses results in a "radiation drive," which leads to an increase in temperature. Among them, CO_2 and water vapor dominate (some of which are formed by oxidation of CH_4 by OH radicals in the environment), that together represent approximately 85% of the retention of solar heating (Fig. 4.1).

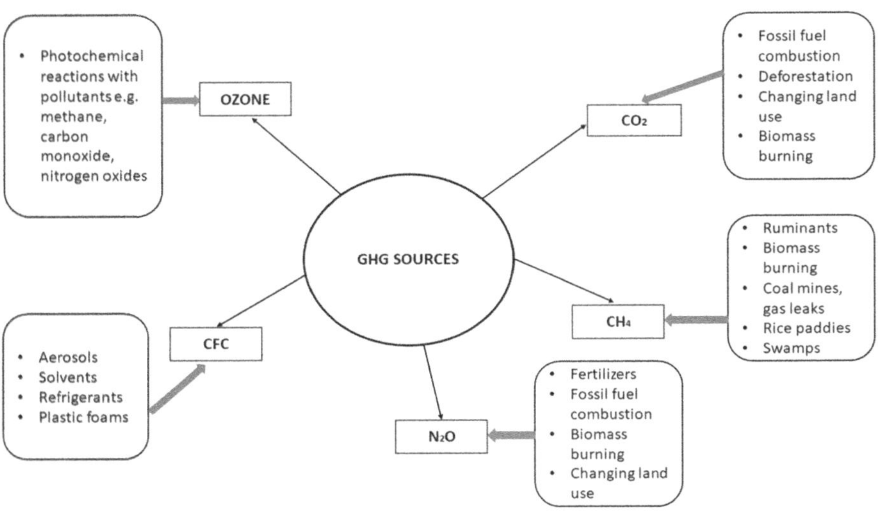

Fig. 4.1 Different sources of greenhouse gases

4.2 Methane and its Sources

Methane is produced during food digestion in ruminants (e.g., buffaloes, goats, camels) through the action of methanogenic archaea and by methanogens in landfills, anoxic sediments (oceans and lakes), wetlands, termite nests, and soils (in particular rice fields) as well as anthropogenic coal mines and natural gas. Methane is the second most prevalent greenhouse gas after CO_2, accounts for around 20% of anthropogenic radiation (Nisbet et al. 2014). Methane contributes the most to the greenhouse effect after CO_2 and is about 23 times more effective than CO_2 on a molar basis as a greenhouse gas (IPCC 2007a, b). It is estimated that microbial production accounts for 70–90% of the atmospheric CH_4 flux. Atmospheric levels of methane increased to 1800 ppb in 2008 from about 715 ppb prior to the industrial revolution (Montzka et al. 2011). CH_4's potential for global warming is 25 times that of CO_2. Slight alterations in methane levels in atmosphere will therefore make a significant contribution to future worldwide warming (Bridgham et al. 2013). Global climate change like enhanced CO_2 emissions and warming has changed the characteristics and functions of terrestrial ecosystems dramatically (Rosenzweig et al. 2007; Austin et al. 2009; Singh et al. 2010).

4.2.1 Paddy Fields

Agricultural practices also lead to important emissions of methane with rice manufacturing and livestock being the two primary sectors. Rice fields are mainly man-made wetlands and are characterized by elevated concentrations of humidity, lack of oxygen, and elevated concentrations of organic substrates and nutrients. As such, they provide perfect circumstances for methanogenic bacteria and trigger substantial methane emissions of about 40 Mt annually. Populations of methanotrophic organisms that convert methane to carbon dioxide absorb up to 90% of this methane, but the remaining 10% escape to the environment. One kilogram of rice production corresponds to 100 g of methane emission. It should be noted that the precision of methane emission estimates has greatly increased in the last decade, with current figures accounting for almost half of the previous estimates. Rice fields (Yan et al. 2009; Liu et al. 2012), representing 5–19% of worldwide CH_4 emissions, are one of the main sources of atmospheric CH_4 (IPCC 2007a, b). In reaction to high atmospheric CO_2 and high temperatures, an increase in CH_4 emissions from these sources has already been identified resulting in positive feedback from the process of worldwide warming (Allen et al. 2003; Tokida et al. 2010; van Groenigen et al. 2011). Recent proof has shown that high concentrations of CO_2 increased CH_4 emissions from rice soils by an average of 43% (van Groenigen et al. 2011); high soil temperatures (2 °C) resulted in a 42% rise in CH_4 emissions (Tokida et al. 2010). Increased CO_2 is usually considered to encourage photosynthesis, root biomass, and rice exudates (Pritchard, 2011; Okubo et al. 2014), possibly giving more CH_4 substrate (Inubushi et al. 2003).

4.2.2 Methane Hydrates

Although methane hydrates are neither a source nor a sink at the moment, they are by far the biggest methane reservoir on Earth, representing 53% of all fossil fuels on Earth. They are a crystalline solid combination of water and methane (mainly methane trapped in ice) and occur in seabed sediments and permafrost in the Arctic (Kvenvolden 1993, 1998; Kvenvolden 1988; Buffett 2004). The methane in the sediment hydrates of the oceans is trapped deep in the ocean by the high pressure, but is released above a depth of 400 m when the pressure drops. The energy industry strives to exploit these benefits and reduce these occurrences. The methane hydrate melts and releases methane gas into the atmosphere as the temperature increases (Kvenvolden 1993; Shakhova et al. 2007, 2008; Portnoy et al. 2016). Methane release from the Siberian Arctic permafrost can create an important contribution to atmospheric pollution each year (Kvenvolden 1993; Shakhova et al. 2007, 2008; Portnoy et al. 2016). There is concern that as global temperatures rise due to anthropogenic climate change that leads to Arctic permafrost melting, large amounts of methane will be released into the atmosphere, resulting in a catastrophic greenhouse effect even higher than the assumed upper estimate of 5.8 °C going beyond IPCC.

Another unanticipated source of about 4% of overall methane in the atmosphere appears to be aerobic bacterial activity in oceans and precursors like methylphosphonate (Karl et al. 2008). The substituted methane, both chlorine and bromomethane, is also released from marine algae and sources of terrestrial and aquatic bacteria into the atmosphere, contributing to global warming, which has not been quantified sufficiently to date.

4.2.3 Coal Mines

By 2008, 826,001 million tons of coal reserves were estimated globally. World coal consumption reduced in 2008 but continued to rise by 3.1% and stayed for six successive years the world fastest increasing fuel (BP 2009). Methane emissions from coal mines, measured approximately 5–30 Tg annually (Breas et al. 2001), are primarily due to methane desorption during mining, comminution, and inefficient combustion. Coal mine gas can also trigger heavy explosions in relation to the greenhouse effect. For 2004, it is estimated that methane from Chinese coal mines gets into the atmosphere. More than 80% of heavy coal mine accidents in China are caused by coal mine explosions, accounting for one-third of total national deaths (Song and Wang 2005). Therefore, the control of mine gas is important for the reduction of greenhouse gases and for the safe regulation of coal mines for energy industry (http://www.chinasafety.gov.cn).

4.3 Methanotrophs Based Mitigation of Methane

Microbial community manipulation provides a possibly efficient means of mitigating global climate change. It is essential to know their ecology and function in order to deal with useful microbial communities. Methane biocycle is more apparent than other GHG cycles because methane pathway is easy and involves advanced microorganisms. The role of methylotrophic groups in mitigating climate change recognized and their position in the global carbon cycle and in decreasing greenhouse gas environmental impacts like methane and indirect carbon dioxide is identified. Methylotrophs are a functional group of microbes that use reduced substrates of carbon comprising of no carbon–carbon bonding. Methylotrophic bacteria include methanotrophs (consuming CH_4) and methylotrophs other than methanotrophs (consuming other substrates of reduced carbon than methane). This functional group can use methanol, methylamine, dimethylamine, formaldehyde, formate, alongside methane as the sole carbon and energy source, and is often involved in the global carbon cycle (Iguchi et al. 2015; Kolb and Stacheter 2013). Methanotrophic biological oxidation of CH_4 accounts for only about 5% of the worldwide sink of atmospheric methane (Hanson and Hanson 1996). Methanotrophs can oxidize as much as 90% of the CH_4 produced in the soil prior it enters into the atmosphere (Oremland and Culbertson 1992). Previously a number of methylotrophic strains are identified as engaging effectively in climate change and reducing greenhouse gas emissions (Kappler and Nouwens 2013; Jhala et al. 2014, 2015; Oshkin et al. 2014; Baesman et al. 2015).

4.3.1 Methanotrophs

It is thought that microorganisms control global methane emissions more directly than CO_2 emissions. Microbial methanogenesis is performed by anaerobic archaea community in oceans, wetlands, rumens, and termite intestines, dominates natural emissions (~250 million tons CH_4 per year). Such biotic sources, however, are surpassed by emissions from human actions (primarily rice farming, fossil fuel extraction, landfilling, and livestock farming) (~320 million tons of methane annually), that, in addition to fossil fuel emissions, also account for emissions primarily from microorganisms.

Methanotrophic bacteria play important role in controlling high levels of methane in some of these environments. Methanotrophs are broadly categorized into two types: Type I methanotrophs or low-affinity methanotrophs primarily belong to gammaproteobacteria class, usually consume large percentage of CH_4 produced in soils prior it escapes in atmosphere. The Type I methanotrophs are present in the Methylococcaceae family that includes the genera *Methylococcus*, *Methylobacter*, *Methylomonas*, *Methylosphaera*, *Methylothermus*, *Methylosarcina*, *Methylocaldum*, *Methylosoma*, and *Methylohalobius*. Methanotrophs can also act as a net CH_4 sink for CH_4, which is already in the atmosphere. Type II methanotrophs or "high-affinity" methanotrophs belong primarily to Alphaproteobacteria class. They make an important contribution to the annual reduction of about 30 million tons of CH_4 from the atmosphere. Type II methanotrophs of the genera Methylosinus and Methylocystis are included in Methylocystaceae family. Other Type II genera (in the family Beijerinckiaceae) are *Methylocella* and *Methylocapsa*. *Methylocella silvestris* is the first studied facultative methanotroph, which can use either methane or multi-carbon compounds (Dedysh et al. 2005; Theisen et al. 2005). Thermotolerant species of *Methylococcus* and *Methylocaldum* are frequently called as Type X methanotrophs because their biochemical, physiological, and phylogenetic features are different from other Type I methanotrophs.

Our knowledge of the greenhouse gas cycle for CH_4 is more comprehensive than for CO_2 or N_2O because it involves an easy pathway and advanced microorganisms. Since large amount of methane in atmosphere is produced by microorganisms, a large share of CH_4 emissions from terrestrial ecosystems can be regulated technically by the management of microbial community's structure and process. Methanotrophic microbial oxidation of methane is just about 5% of the worldwide sink of atmospheric methane (approximately 30 million tons annually) and thus appears to be less significant. Methanotrophs, however, also oxidize about 90% of methane produced in the soil prior escaping into atmosphere. Transformation of farmland or grassland into forests leads to a significant decrease in CH_4 flux and it is obvious that both the type and frequency of methanotrophs are predictive of CH_4 flux. Future study therefore needs to concentrate on including these information and relationships to enhance CH_4 fluxes in different ecosystems. This understanding can also be used to reduce methane emissions through land use and management modifications. Methanotrophs, for instance, performed a major role in absorbing portion of the generated methane in rice cultivation. This would strengthen the control of flood frequency and duration by increasing the availability of soil oxygen. There is also considerable prospective for using methanogenesis inhibitors, like ammonium sulfate fertilizers, in controlled systems for promoting sulfate reducers growth at the cost of methanogens. Reducing methane emissions from ruminants includes approaches to improve feed quality and inhibit rumen methanogens by employing antibiotics, vaccines, etc.

Microbial use of CH_4 occurs both aerobically and anaerobically. Aerobic methanotrophs can be distinguished in three main groups: Group I (Gammaproteobacteria; also denoted as Type I and Type X) (Anthony 1982; Semrau et al. 2010), Group II (Alphaproteobacteria, also known as Type II and Type III) (Dedysh et al. 2001), and Group III (Verrucomicrobia, frequently known as Type IV) (Murrell and Jetten 2009). They occur in diverse environments and play a significant role in methane oxidation in nature. All recognized aerobic methanotrophs employ methane monooxygenase (MMO) which catalyzes first oxidation step which transforms methane into methanol (Semrau et al. 2010). Methanol is then oxidized to formaldehyde and can then be transformed to biomass or further oxidized to CO_2. Two iso-enzymes of MMO are recognized: cytoplasmic soluble MMO (sMMO), present in only a subset of methanotrophs, and membrane bound (or particulate) MMO (pMMO), that is nearly present in all known methanotrophs. Because of their distinctive microbiological and metabolic characteristics, methanotrophs are of excellent concern for their industrial application. Research on methanotrophs has enhanced considerably with the fast growth of science and technology since the first methanotroph was isolated in 1906.

4.3.2 Biodiversity of Methanotrophs

In the environment, methanotrophs tend to be ubiquitous and may be isolated from different environments where methane and oxygen are present. Methanotrophs are widely distributed in the environment, including different extreme environments, and can grow at temperatures up to 4 °C (Bowman et al. 1997) or up to 72 °C (Bodrossy et al. 1999). While the first Methanotroph, *Bacillus methanicum* was isolated back in 1906, the basis for the existing classification of these bacteria was not established until Whittenbury et al. (1970) isolated and described over 100 new methane using bacteria. Since 1990s, several extremophilic methanotrophs were discovered from extreme environments, including alkaliphiles (Khmelenina et al. 1997; Sorokin et al. 2000; Kaluzhnaya et al. 2001), acidophiles (Dedysh et al. 2000, 2002, 2005; Dunfield et al. 2007; Pol et al. 2007; Islam et al. 2008), thermophiles (Bodrossy et al. 1999; Whittenbury et al. 1970; Malashenko et al. 1975; Bodrossy et al. 1995, 1997; Tsubota et al. 2005), psychrophiles (Omelchenko et al. 1993, 1996; Bowman et al. 1997), and halophiles (Lidstrom 1988; Fuse et al. 1998; Heyer et al. 2005).

Research has shown a diverse variety of methanotrophs are found in rice fields, comprising *Methylomonas*, *Methylococcus*, *Methylobacter*, *Methylocaldum*, *Methylomicrobium*, *Methylocystis,* and *Methylosinus*. Several factors, like supply of oxygen and rice growth time, influence the diversity of methanotrophs in rice fields (Eller and Frenzel 2001). Several methanotrophs are present in landfills, like *Methylobacter*, *Methylomicrobium*, *Methylocystis*, *Methylosarcina,* and *Methylosinus*. Factors like temperature, oxygen, pH, earthworms, moisture affect the diversity of methanotrophs in landfills (Gebert et al. 2003). A number of methanotrophs have been detected in freshwaters and sediments, mainly Type I methanotrophs like *Methylobacter*, *Methylomonas*, *Methylosarcina*, *Methylosoma,* and *Methylococcus*.

4.3.3 Catalytic Properties of MMOs

Methane monooxygenases (MMOs) catalyze the first oxidation step, i.e., oxidation of methane to methanol, in methanotroph metabolism. While the methane C–H bond is strong and extremely stable, under favorable conditions MMO can break this bond (Dalton 2005). Two genetically unrelated MMOs are produced by Methanotrophs: soluble MMO (sMMO) is expressed by a subset of methanotrophs and membrane bound, particulate MMO (pMMO) is expressed in all methanotrophs with the exception of the genus *Methylocella* (Theisen et al. 2005). Indeed, cultures expressing pMMO usually display greater affinity toward methane in comparison to cells expressing sMMO. In addition, it is known that cells that use pMMO for growth show higher growth yield, indicating more effectiveness of pMMO for methane oxidation (Leak and Dalton 1986). pMMO is found in complex internal membrane structures, referred as intracytoplasmic membranes (Anthony 1982; Semrau et al. 2010). Both types of enzymes exhibit wide substrate specificities, particularly sMMO, that can oxidize alkanes, alkenes, aromatics, alicyclics, ethers, heterocycles, and ammonia. Several methanotrophs like *Methylomicrobium album* BG8 and *Methylomonas methanica* have only the pMMO, *Methylococcus capsulatus* Bath, *Methylosinus trichosporium* OB3b and some others can express both forms (Hanson and Hanson 1996) depending on the amount of copper in the medium. Expression levels are associated with intracellular concentrations of copper in species that have genes for both sMMO as well as pMMO, a mechanism called the "copper switch" where sMMO is produced at low levels of copper while pMMO expression is slightly upregulated when copper is available, sMMO expression is downregulated (Nielsen et al. 1997; Murrell et al. 2000; Kenney et al. 2016). Active sites of the enzymes reveal this metalloregulation: sMMO includes a non-heme diiron active site and pMMO contains a copper active site (Rosenzweig et al. 1993; Lieberman and Rosenzweig 2005; Balasubramanian et al. 2010). While several characteristics of the role of sMMO are known, the current knowledge of pMMO is less known.

4.4 Role of Methanotrophs in Mitigating Methane Emission

Combined with a comparatively brief atmospheric lifetime (about 10 years) compared to CO_2 (50–200 years), the elevated global warming potential of methane demonstrates that mitigation impacts can be accomplished comparatively rapidly (Lelieveld et al. 1998; Breas et al. 2001). Furthermore, the disparity between methane sources and sinks in the worldwide budget accounts for $<6\%$ of total global methane sources, that indicates that a slight reduction in methane emissions could stabilize atmospheric methane or possibly lead to substantial stabilization or reduction of concentration of methane in the atmosphere (Lelieveld et al. 1998). Unfortunately, due attention has not yet been paid to control methane emissions, and after nearly a decade of modest changes in atmospheric methane, global atmospheric methane regained growth relative to the mole fraction from early 2007 (Rigby et al. 2008). This will be an important issue for climate change, and this field of research must be given greater attention. The maximum CH_4 released into the atmosphere is around 520 Tg^{-1} of which 90% oxidizes by photochemical processes in the troposphere and 10% is removed biologically from soil and water, primarily by methanotrophs (Hanson and Hanson 1996; Breas et al. 2001). Approximately 70% of the world's methane emissions are believed to be anthropogenic (IPCC 1996). Landfill gas and coal mine fuel are two major sources of methane emissions from human activities and therefore considered as significant targets for methane mitigation.

Methanotrophs possess considerable ability for applied microbiology and bioprocess engineering, along with bioremediation of pollutants like halogenated hydrocarbons by co-metabolization by MMOs, biotransformation of various organic substrates, methane assimilation for mitigating greenhouse effects, and production of commercially important compounds (e.g., unicellular protein, astaxanthin, and poly-β-hydroxybutyrate). Therefore, for their industrial applications, engineering on methanotrophs is very essential. Methanotrophs have significant role in the global carbon cycle as they are the only biological methane sink. Methanotrophs can also use several different toxic organic compounds. Such characteristics illustrate that methanotrophs are useful in environmental biotechnology.

4.4.1 Mitigation of Methane Emissions from Landfills

Landfills around the world discharge almost 10% of CH_4 into the atmosphere (Breas et al. 2001) and are considered significant anthropogenic methane sources. In general landfill gas composition involves 30–70% (v/v) methane, 20–50% (v/v) CO_2, nitrogen, oxygen, ammonia, sulfur compounds, hydrogen, carbon monoxide, and trace quantities of various volatile organic compounds like dichloromethane, vinyl chloride, tetrachloroethylene, toluene, benzene, and xylene (Nikiema et al. 2007). Landfills have several advantages, like the ability to oxidize different concentrations of methane and toxic contaminants. Different biological systems are created to decrease methane emissions to landfills by offering optimal circumstances for microbial procedures and effective transport of landfill gas to locations where it is to be produced. Huber-Humer et al. (2008) studied the different bio-based process designs for daily use, comprising transient or long-term biocovers, passively or actively aerated biofilters, biowindows, and biotarps. Biofilters can be more practical owing to their small footprint and elevated removing capability for gases due to the growing use of gas collection systems. Latest IPCC evaluation report identifies biocovers and biofilters as important mitigation techniques that should be commercialized before 2030 (IPCC 2007a, b). Nikiema et al. (2007) reviewed effect of various parameters like different filter beds, operating conditions, inlet loads, elimination capacities, and conversions on methane biofiltration.

There are important variations in methanotrophic reaction to varying levels of methane. Some methanotrophs in upland and forest soils possess strong affinity to methane as they start uptaking methane at low levels and thus consuming methane from the atmosphere (Henckel et al. 2000; Horz et al. 2005). Some low-affinity methanotrophs do not work until the methane amount reaches 0.8–66μmol l^{-1} (Huber-Humer et al. 2008). Nitrifying bacteria also capable of oxidizing methane, but the removal rates are <5% of that of methanotrophs (Bodelier and Frenzel 1999). Some methanol oxidizing bacterial consortia are also capable of oxidizing methane only when concentration of methane is below 10% (Hughes et al. 2002).

In addition to the biological system configurations and the landfills characteristics, their waste, environmental, and operational parameters also have a major impact on the microbial methane mitigation rates. These parameters are nutrients (Nikiema et al. 2005; Albanna et al. 2007), filter bed type (Dobbie and Smith 1996; Humer and Lechner 1999; Wilshusen et al. 2004; Haubrichs and Widmann 2006), temperature (Whalen et al. 1990; Cai and Yan 1999; Visvanathan et al. 1999; Borjesson et al. 2004; Kettunen et al. 2006), moisture content (Borjesson et al. 1998; Visvanathan et al. 1999; Albanna et al. 2007), and pH (Bender and Conrad 1994; Syamsul Arif et al. 1996). Therefore, for controlling landfill methane emissions, engineering and standardization of the methanotrophic bioprocess is essential.

4.4.2 Mitigation of Methane Emissions from Coal Mines

Methane is released from operational and inactive underground and surface mines due to post-mining practices such as coal processing, storage, and transport. It has been measured that in the year 2010 global methane emissions from coal mines were approx. 584 $MMTCO_2$, representing 8% of overall global methane emissions. Coal mine gas is a complex mixture having higher methane concentrations. In addition to the greenhouse effect, coal mine gas may also cause heavy explosions. In China, more than 80% of heavy coal mine accidents are due to coal mine explosions, which account for one-third of total national deaths (Song and Wang 2005). Therefore, the control of mine gas is necessary for the mitigation of greenhouse gases and for the safe coal mining management for the energy sector (http://www.chinasafety.gov.cn).

The biological process employing methanotrophs can be more economical and effective in comparison to seam gas drainage and air ventilation that are presently the main efficient approaches for controlling coal mine gas. Biofiltration is a safe and less expensive technology because it uses microorganisms for oxidation of methane at ambient temperature to carbon dioxide and biomass. In the last decade, methane biofiltration approach is considered for treatment of effluent gases produced during landfill and animal husbandry activities, where gas flow levels are relatively low (Cloirec et al. 2001; Nikiema et al. 2005; Gebert and Gröngröft 2006; Dever et al. 2007). Methane removal studies with packaging materials like polypropylene Raschig rings (Apel 1991), glass tubes (Sly et al. 1993), mature compost (Nikiema et al. 2005), gravel (Nikiema et al. 2007; Girard et al. 2011) have shown comparatively slow conversion. A bench-scale bioreactor has been employed to study the ability of *Methylomonas methanica* for removing methane from coal mines, and 90.4% of methane was removed in a 35% methanol-air mix in 24 h (Apel 1991). At flow rates of between 0.2 and 2.4 L min^{-1} at 30 °C, a laboratory-based coal-packed biofilter has been developed and moderately methane is removed from humidified air. Oxidation of methane was catalyzed by a diverse microbial community, the most prevailing of which was recognized as *Methylocystis* by the 16S rRNA sequencing. Furthermore, *Methylosinus sporium* was also tested in parallel. The results indicate that low-cost coal packaging has a favorable potential as an effective growth surface and contains methanotrophs for methane removal (Limbri et al. 2014).

Detecting the methanotrophs source in the environment is important for basic and applied research. Using culture-independent molecular biological methods, the diversity of methanotrophs present in alkaline soil of a Chinese coal mine was investigated. A wide variety of methanotrophs along with several uncultivated methanotrophs were reported (Han et al. 2009). The occurrence of viable aerobic methanotrophs in geological coal- and lignite-bearing formations at substantial depths has been proved (Mills et al. 2010; Stępniewska et al. 2013). Such results are significant to the mining industry in particular. Methanotrophs are exceptionally efficient in removal of methane under optimal conditions, for instance, the removal of the methane released in the Gulf of Mexico during the 2010 Deepwater Horizon oil spill (Kessler et al. 2011). A strict control of global emissions of greenhouse gas and further progress in research into methanotrophs can allow these microorganisms to be used in the near future to control coal-gas emissions and explosions.

4.5 Engineered Strategies for Methane Removal

Various strategies may be used, taking into account the stimulation of methanotrophic activity for methane reduction in the atmosphere. The renowned engineered method engaged in this operation is "Biofilters or biocovers," Biocovers are usually organic substances in which the permeable materials, like compost, sewage sludge, or wood chips, cover the landfill surface. These materials are characterized by effective gas transport for methane coming out from landfills and atmospheric oxygen, as well as the ability to retain water for methanotrophic activity (Huber-Humer et al. 2008; Scheutz et al. 2009).

Knief and Dunfield (2005) isolated two methanotrophic strains, which contributes to the reduction of CH_4 over a 3-month period. However, in the atmospheric concentration of CH_4, these strains are unable to grow and require unique conditions. Surprisingly, a new strain of *Methylocystis* that can grow at 10 p.p.m.v. CH_4 has been found. Further studies of this strain indicate that this cell has two pMMO isozymes, pMMO1 and pMMO2, resulting in the finding that pMMO2 oxidation of CH_4 is more efficient than pMMO1. twenty-first century scientists are now using biotechnology to effectively improve methanotrophic bacteria, i.e., to control methane emissions, bioremediation, biobleaching, and known methanotrophic biofilters for reducing methane in landfills (Scheutz et al. 2009; Semrau et al. 2011).

Important biotechnological factors, like growth rate, yield, and genetic tractability, were reported for limited methanotrophs. Therefore, today majority of metabolic engineering attempts are focusing on well-known species like *Methylococcus capsulatus* Bath, *M. trichosporium* OB3b, and *Methylocystis* spp. Nevertheless, recent attempts in the cultivation and characterization of new methanotrophic organisms have led to the isolation of a number of strains, which can give a wider variety of biotechnology potential applications. Several number of extremely psychrophilic, thermophilic, alkaliphilic, acidophilic, and halophilic methanotrophs have been identified, thereby increasing the physiological range of aerobic methanotrophy (Kalyuzhnaya et al. 1999; Trotsenko and Khmelenina 2008; Murrell 2010). Growth factors of methanotrophs differ considerably, covering a wide range of pH (1–10), temperature (4–65 °C), and salinity (0–10%) (Trotsenko and Khmelenina 2008).

While methanotrophs have shown promising potential in methane mitigation and biodegradation of hazardous contaminants, but from a biochemical engineering point of view, still several significant limiting factors remain, like unavailability of desirable methanotrophic strains with better controlled metabolic activity, complexity in the design and control of microbial consortia, low methane and oxygen solubility in aqueous phases resulting in slow growth, competitive inhibition among growth substrate (methane) and co-metabolites. Molecular ecology research indicated that there are far more methanotrophs exist in the ecosystem than documented in culture collections. Enrichment and isolation of novel methanotrophics, especially from more extreme environments, require considerable efforts. It is also necessary to address the nature of non-cultivated methanotrophs that grow in atmospheric methane concentrations. Extremophilic methanotrophs capable of growing at very high or very low pH levels give a means to investigate the mechanisms through which such methanotrophs sustain at extremes.

In technical applications, a microorganism's big biodiversity and exploitability often demonstrate its excellent potential. The earlier results show that for their industrial application we have plenty of methanotrophs available though their diversity is overlooked. New methods are required for better understanding the molecular mechanisms of their adaptation to different stresses such as pH, temperature, salinity, and diverse chemicals. Comparative genomics and proteomics will give insights about their adaptation to extreme situations and on the growth and phylogeny of methanotrophs. However, recent developments in methanotrophic physiology lead to new possibilities for exploiting their machinery. Still there are significant gaps in the fundamental knowledge of methanotrophy. With the development of new approaches several significant breakthroughs in this field will be achieved in the coming years.

4.6 Conclusions

Global warming is seen as a major environmental issue. The increase in the consumption of fossil fuels and deforestation leads to the anthropogenic greenhouse gas emissions contributing to global warming. Methane accounts for approximately 16% of the greenhouse effect among these gases. Methanotrophs possess the remarkable potential to use methane as their only carbon and energy source. Methanotrophs are found in diverse environments and have significant role in methane mitigation. The resources are available for manipulating methanotrophs and new methanotrophic strains are now available to extend the metabolic features for industrial biotechnology. Hence, it is necessary to adopt various strategies that may have practical importance in stimulating the methane sink activities by methanotrophs to mitigate the problem of global warming due to methane.

Acknowledgement F. M. is thankful to U.G.C., New Delhi for financial assistance in the form of Postdoctoral Research Fellowship for Women.

References

Albanna M, Fernandes L, Warith M (2007) Methane oxidation in landfill cover soil; the combined effects of moisture content, nutrient addition, and cover thickness. J Environ Eng Sci 6:191–200

Allen LH, Albrecht SL, Colon-Guasp W, Covell SA, Baker JT, Pan D et al (2003) Methane emissions of rice increased by elevated carbon dioxide and temperature. J Environ Qual 32:1978–1991

Anthony C (1982) The biochemistry of methylotrophs. Academic Press, London

Apel W (1991) Use of methanotrophic bacteria in gas phase bioreactors to abate methane in coal mine atmospheres. Fuel 70:1001–1003

Austin E, Castro H, Sides K, Schadt C, Classen A (2009) Assessment of 10 years of CO_2 fumigation on soil microbial communities and function in a sweet gum plantation. Soil Biol Biochem 41:514–520

Baesman SM, Miller LG, Wei JH, Cho Y, Matys ED, Summons RE, Welander PV, Oremland RS (2015) Methane oxidation and molecular characterization of methanotrophs from a former mercury mine impoundment. Microorganisms 3:290–309

Balasubramanian R, Smith SM, Rawat S, Yatsunyk LA, Stemmler TL, Rosenzweig AC (2010) Oxidation of methane by a biological dicopper centre. Nature 465:115–119

Bender M, Conrad R (1994) Methane oxidation activity in various soils and freshwater sediments: occurrence, characteristics, vertical profiles, and distribution on grain size fractions. J Geophys Res Atmos 99(D8):16531–16540

Bodelier P, Frenzel P (1999) Contribution of methanotrophic and nitrifying bacteria to CH_4 and NH^{4+} oxidation in the rhizosphere of rice plants as determined by new methods of discrimination. Appl Environ Microbiol 65:1826–1833

Bodrossy L, Murrell JC, Dalton H et al (1995) Heat-tolerant methanotrophic bacteria from the hot water effluent of a natural gas field. Appl Environ Microbiol 61:3549–3555

Bodrossy L, Holmes EM, Holmes AJ et al (1997) Analysis of 16S rRNA and methane monooxygenase gene sequences reveals a novel group of thermotolerant and thermophilic methanotrophs, *Methylocaldum* gen. nov. Arch Microbiol 168:493–503

Bodrossy L, Kovacs KL, McDonald IR, Murrell JC (1999) A novel thermophilic methane-oxidizing *γ-Proteobacterium*. FEMS Microbiol Ecol 170:335–341

Borjesson G, Sundh I, Tunlid A et al (1998) Microbial oxidation of CH_4 at high partial pressures in an organic landfill cover soil under different moisture regimes. FEMS Microbiol Ecol 26:207–217

Borjesson G, Sundh I, Svensson B (2004) Microbial oxidation of CH4 at different temperatures in landfill cover soils. FEMS Microbiol Ecol 48:305–312

Bowman JP, McCammon SA, Skerratt JH (1997) *Methylosphaera hansonii* gen. nov., sp. nov., a psychrophilic, group I methanotroph from Antarctic marine-salinity, meromictic lakes. Microbiology 143:1451–1459

BP (2009) Statistical review of world energy. www.bp.com

Breas O, Guillou C, Reniero F, Wada E (2001) The global methane cycle: isotopes and mixing ratios, sources and sinks. Isot Environ Health Stud 37:257–379

Bridgham SD, Cadillo Quiroz H, Keller J, Zhuang Q (2013) Methane emissions from wetlands: biogeochemical, microbial, and modeling perspectives from local to global scales. Glob Chang Biol 19:1325–1346

Buffett B (2004) Global inventory of methane clathrate: sensitivity to changes in the deep ocean. Earth Planet Sci Lett 227:185–199

Cai Z, Yan X (1999) Kinetic model for methane oxidation by paddy soil as affected by temperature, moisture and N addition. Soil Biol Biochem 31:715–725

Cloirec PL, Humer P, Ramirez-Lopez EM (2001) Biotreatment of odours: control and performances of a biofilter and a bioscrubber. Water Sci Technol 44:219–226
Dalton H (2005) The Leeuwenhoek lecture 2000 the natural and unnatural history of methane-oxidizing bacteria. Philos Trans R Soc Lond Ser B Biol Sci 360:1207–1222
De la Chesnaye FC, Delhotal C, DeAngelo B, Ottinger Schaefer D, Godwin D (2006) Past, present, and future of non-CO2 gas mitigation analysis in human-induced climate change: an interdisciplinary assessment. Cambridge University Press, Cambridge
Dedysh SN, Liesack W, Khmelenina VN et al (2000) *Methylocella palustris* gen. nov., sp. nov., a new methane-oxidizing acidophilic bacterium from peat bogs, representing a novel subtype of serine-pathway methanotrophs. Int J Syst Evol Microbiol 50:955–969
Dedysh SN, Horz HP, Dunfield PF, Liesack W (2001) A novel pmoA lineage represented by the acidophilic methanotrophic bacterium *Methylocapsa acidiphila* (correction of acidophila) B2. Arch Microbiol 177:117–121
Dedysh SN, Khmelenina VN, Suzina NE et al (2002) *Methylocapsa acidiphila* gen. nov., sp. nov., a novel methane-oxidizing and dinitrogen-fixing acidophilic bacterium from Sphagnum bog. Int J Syst Evol Microbiol 52:251–261
Dedysh SN, Knief C, Dunfield PF (2005) *Methylocella* species are facultatively methanotrophic. J Bacteriol 187:4665–4670
Dever SA, Swarbrick GE, Stuetz RM (2007) Passive drainage and biofiltration of landfill gas: Australian field trial. Waste Manag 27(2):277–286
Dobbie K, Smith K (1996) Comparison of CH_4 oxidation rates in woodland, arable and set aside soils. Soil Biol Biochem 28:1357–1365
Dunfield PF, Yuryev A, Senin P et al (2007) Methane oxidation by an extremely acidophilic bacterium of the phylum *Verrucomicrobia*. Nature 450:879–882
Eller G, Frenzel P (2001) Changes in activity and community structure of methane oxidising bacteria over the growth period of rice. Appl Environ Microbiol 67:2395–2403
Fuse H, Ohta M, Takimura O et al (1998) Oxidation of trichloroethylene and dimethyl sulfide by a marine *Methylomicrobium* strain containing soluble methane monooxygenase. Biosci Biotechnol Biochem 62:1925–1931
Gebert J, Gröngröft A (2006) Performance of a passively vented field-scale biofilter for the microbial oxidation of landfill methane. Waste Manag 26:399–407
Gebert J, Groengroeft A, Miehlich G (2003) Kinetics of microbial landfill methane oxidation in biofilters. Waste Manag 23:609–619
Girard M, Ramirez AA, Buelna G, Heitz M (2011) Biofiltration of methane at low concentrations representative of the piggery industry—influence of the methane and nitrogen concentrations. Chem Eng J 168:151–158
Han B, Chen Y, Abell G (2009) Diversity and activity of methanotrophs in an alkaline Chinese coal mine soil. FEMS Microbiol Ecol 70:40–51
Hanson RS, Hanson TE (1996) Methanotrophic bacteria. Microbiol Rev 60:439–471
Haubrichs R, Widmann R (2006) Evaluation of aerated biofilter systems for microbial methane oxidation of poor landfill gas. Waste Manag 26:408–416
Henckel T, Jackel U, Schnell S, Conrad R (2000) Molecular analyses of novel methanotrophic communities in forest soil that oxidize atmospheric methane. Appl Environ Microbiol 66:1801–1808
Heyer J, Berger U, Hardt M, Dunfield PF (2005) *Methylohalobius crimeensis* gen. nov., sp. nov., a moderately halophilic, methanotrophic bacterium isolated from hypersaline lakes of Crimea. Int J Syst Evol Microbiol 55:1817–1826
Horz HP, Rich V, Avrahami S, Bohannan BJ (2005) Methane-oxidizing bacteria in a California upland grassland soil: diversity and response to simulated global change. Appl Environ Microbiol 71:2642–2652
Huber-Humer M, Gebert J, Hilger H (2008) Biotic systems to mitigate landfill methane emissions. Waste Manag Res 26:33–46

Hughes K, Daneel R, Senior E (2002) Physiological characterization of a methanol-oxidizing microbial association isolated from landfill final covering soil. S Afr J Sci 98:434–436
Humer M, Lechner P (1999) Alternative approach to the elimination of greenhouse gases from old landfills. Waste Manag Res 17:443–452
Iguchi H, Yurimoto H, Sakai Y (2015) Interactions of methylotrophs with plants and other heterotrophic bacteria. Microorganisms 3(2):137–151
Inubushi K, Cheng W, Aonuma S, Hoque MM, Kobayashi K, Miura S et al (2003) Effects of free-air CO_2 enrichment (FACE) on CH_4 emission from a rice paddy field. Glob Chang Biol 9:1458–1464
IPCC (1996) Climate change 1995: the science of climate change. In: Houghton JT, Filho LGM, Callander BA, Harris N, Kattenberg A, Maskell K (eds) Contribution of working group I to the second assessment report of the intergovernmental panel on climate change. Cambridge University Press, Cambridge, 572 pp
IPCC (2007a) Climate change 2007- the physical science basis: working group I contribution to the fourth assessment report of the intergovermental panel on climate change. Cambridge University Press, Cambridge
IPCC (2007b) Fourth assessment report: climate change 2007. http://www.ipcc.ch
Islam T, Jensen S, Reigstad LJ et al (2008) Methane oxidation at 55 degrees C and pH 2 by a thermoacidophilic bacterium belonging to the *Verrucomicrobia* phylum. Proc Natl Acad Sci U S A 105:300–304
Jhala YK, Rajababu VV, Shelat Harsha N, Patel HK, Patel HK, Patel KT (2014) Isolation and characterization of methane utilizing bacteria from wetland paddy ecosystem. World J Microbiol Biotechnol 30(6):1845–1860
Jhala YK, Rajababu VV, Panpatte Deepak G, Shelat Harsha N (2015) Rapid methods for isolation and screening of methane degrading bacteria. J Bioremed Biodegr 7:322
Kaluzhnaya M, Khmelenina V, Eshinimaev B et al (2001) Taxonomic characterization of new alkaliphilic and alkali tolerant methanotrophs from soda lakes of the southeastern Transbaikal region and description of *Methylomicrobium buryatense* sp. nov. Syst Appl Microbiol 24:166–176
Kalyuzhnaya MG, Khmelenina VN, Lysenko AM, Suzina NE, Trotsenko YA (1999) New methanotrophic isolates from soda lakes of the southern Transbaikal region. Mikrobiologiia (Russian) 68:689–697
Kappler U, Nouwens AS (2013) Metabolic adaptation and trophic strategies of soil bacteria C1-metabolism and sulfur chemolithotrophy in *Starkeya novella*. Front Microbiol 4:1–12
Karl DM, Beversdorf L, Björkman KM, Church MJ, Martinez A, DeLong EF (2008) Aerobic production of methane by the sea. Nat Geosci 1:473–478
Kenney GE, Sadek M, Rosenzweig AC (2016) Copper-responsive gene expression in the methanotroph Methylosinus trichosporium OB3b. Metallomics 8(9):931–940
Kessler JD, Valentine DL, Redmond MC, Du M, Chan EW, Mendes SD, Quiroz EW, Villanueva CJ, Shusta SS, Werra LM, Yvon-Lewis SA (2011) A persistent oxygen anomaly reveals the fate of spilled methane in the deep Gulf of Mexico. Science 331:312–315
Kettunen RH, Einola JKM, Rintala JA (2006) Landfill methane oxidation in engineered soil columns at low temperature. Water Air Soil Pollut 177:313–334
Khmelenina VN, Kalyuzhnaya MG, Starostina NG et al (1997) Isolation and characterization of halotolerant alkaliphilic methanotrophic bacteria from Tuva soda lakes. Curr Microbiol 35:257–261
Kolb SP, Stacheter A (2013) Prerequisites for amplicon pyrosequencing of microbial methanol utilizers in the environment. Front Microbiol 4:268
Knief C, Dunfield PF (2005) Response and adaptation of different methanotrophic bacteria to low methane mixing ratios. Environ Microbiol 7:1307–1317
Kvenvolden KA (1988) Methane hydrate—a major reservoir of carbon in the shallow geosphere? Chem Geol 71:41–51
Kvenvolden K (1993) Methane hydrates and global climate. Glob Biogeochem Cycles 3:221–229

Kvenvolden KA (1998) A primer on the geological occurrence of gas hydrate. Geol Soc Lond Spec Publ 137:9–30

Leak DJ, Dalton H (1986) Growth yields of methanotrophs. Appl Microbiol Biotechnol 23:470–476

Lelieveld J, Crutzen PJ, Dentener FJ (1998) Changing concentration, lifetime and climate forcing of atmospheric methane. Tellus B 50:128–150

Lidstrom ME (1988) Isolation and characterization of marine methanotrophs. Antonie Van Leeuwenhoek 54:189–199

Lieberman RL, Rosenzweig AC (2005) Crystal structure of a membrane-bound metalloenzyme that catalyses the biological oxidation of methane. Nature 434:177–182

Limbri H, Gunawan C, Thomas T, Smith A, Scott J et al (2014) Coal-packed methane biofilter for mitigation of green house gas emissions from coal mine ventilation air. PLoS One 9(4):e94641

Liu GC, Tokida T, Matsunami T, Nakamura H, Okada M, Sameshima R et al (2012) Microbial community composition controls the effects of climate change on methane emission from rice paddies. Environ Microbiol Rep 4:648–654

Malashenko YR, Romanovskaya VA, Bogachenko VN, Shved AD (1975) Thermophilic and thermotolerant methane assimilating bacteria. Mikrobiologiya 44:855–862

Mills CT, Amano Y, Slater GF, Dias RF, Iwatsuki T, Mandernack KW (2010) Microbial carbon cycling in oligotrophic regional aquifers near the Tono Uranium Mine, Japan as inferred from 14C values of in situ phospholipid fatty acids and carbon sources. Geochim Cosmochim Acta 74:3785–3805

Montzka SA, Dlugokencky EJ, Butler JH (2011) Non-CO_2 greenhouse gases and climate change. Nature 476:43–50

Murrell JC (2010) Genomics of *Methylococcus capsulatus* (Bath). In: Timmis KN (ed) Handbook of hydrocarbon and lipid microbiology, vol 36. Springer, Heidelberg, pp 1328–1333

Murrell JC, Jetten MSM (2009) The microbial methane cycle. Environ Microbiol Rep 1:279–284

Murrell JC, McDonald IR, Gilbert B (2000) Regulation of expression of methane monooxygenases by copper ions. Trends Microbiol 8:221–225

Nielsen AK, Gerdes K, Murrell JC (1997) Copper-dependent reciprocal transcriptional regulation of methane monooxygenase genes in Methylococcus capsulatus and Methylosinus trichosporium. Mol Microbiol 25:399–409

Nikiema J, Bibeau L, Lavoie J et al (2005) Biofiltration of methane: an experimental study. Chem Eng J 113:111–117

Nikiema J, Brzezinski R, Heitz M (2007) Elimination of methane generated from landfills by biofiltration: a review. Rev Environ Sci Biotechnol 6:261–284

Nisbet EG, Dlugokencky EJ, Bousquet P (2014) Methane on the rise-again. Science 343:493–495

Okubo T, Tokida T, Ikeda S, Bao Z, Tago K, Hayatsu M et al (2014) Effects of elevated carbon dioxide, elevated temperature, and rice growth stage on the community structure of rice root–associated bacteria. Microbes Environ 29:184–190

Omelchenko MV, Vasilieva LV, Zavarzin GA (1993) Psychrophilic methanotroph from tundra soil. Curr Microbiol 27:255–259

Omelchenko MB, Vasilieva LV, Zavarzin GA et al (1996) A novel psychrophilic methanotroph of the genus *Methylobacter*. Mikrobiologiya 65:339–343

Oremland RS, Culbertson CW (1992) Importance of methane-oxidizing bacteria in the methane budget as revealed by the use of a specific inhibitor. Nature 356:421–423

Oshkin IY, Beck DAC, Lamb AE, Tchesnokova V, Benuska G, Benuska G, McTaggart TL, Kalyuzhnaya MG, Dedysh SN, Lidstrom ME, Chistoserdova L (2014) Methane-fed microbial microcosms show differential community dynamics and pinpoint taxa involved in communal response. ISME J 9(5):1119–1129

Pol A, Heijmans K, Harhangi HR et al (2007) Methanotrophy below pH 1 by a new *Verrucomicrobia* species. Nature 450:874–878

Portnoy A, Vadakkepuliyambatta S, Mienert J, Hubbard A (2016) Ice-sheet-driven methane storage and release in the Arctic. Nat Commun 7:10314

Pritchard S (2011) Soil organisms and global climate change. Plant Pathol 60:82–99
Rigby M, Prinn RG, Fraser PJ et al (2008) Renewed growth of atmospheric methane. Geophys Res Lett 35:1–6
Rosenzweig C, Casassa G, Karoly DJ, Imeson A, Liu C, Menzel A et al (2007) Assessment of observed changes and responses in natural and managed systems. In: Parry ML, Canziani OF, Palutikof JP, van der Linden PJ, Hanson CE (eds) Proceedings of the contribution of working group II to the fourth assessment report of the intergovernmental panel on climate change: climate change 2007: impacts, adaptation and vulnerability. Cambridge University Press, Cambridge, pp 79–131
Rosenzweig AC, Frederick CA, Lippard SJ (1993) Crystal structure of a bacterial non-haem iron hydroxylase that catalyses the biological oxidation of methane. Nature 366:537–543
Scheutz C, Kjeldsen P, Bogner JE, De Visscher A, Gebert J et al (2009) Microbial methane oxidation processes and technologies for mitigation of landfill gas emissions. Waste Manag Res 27:409–455
Semrau JD, DiSpirito AA, Yoon S (2010) Methanotrophs and copper. FEMS Microbiol Rev 34:496–531
Semrau JD, DiSpirito AA, Vuilleumier S (2011) Facultative methanotrophy: false leads, true results, and suggestions for future research. FEMS Microbiol Lett 323:1–12
Shakhova N, Seemiletov I, Salyuk A, Kosmach D, Bel'cheva N (2007) Methane release on the Arctic East Siberian shelf. Geophys Res Abstr 9:01071
Shakhova N, Seemiletov I, Salyuk A, Kosmach D (2008) Anomalies of methane in the atmosphere over the East Siberian shelf: is there any sign of methane leakage from shallow shelf hydrates? Geophys Res Abstr 10:01526
Singh BK, Bardgett RD, Smith P, Reay DS (2010) Microorganisms and climate change: terrestrial feedbacks and mitigation options. Nat Rev Microbiol 8:779–790
Sly L, Bryant L, Cox J, Anderson J (1993) Development of a biofilter for the removal of methane from coal mine ventilation atmospheres. Appl Microbiol Biotechnol 39:400–404
Song YM, Wang T (2005) Current status and counter-measures for gas control in Chinese coal mines. China Coalbed Methane 2:3–6
Sorokin DY, Jones BE, Kuenen JG (2000) An obligate methylotrophic, methane-oxidizing *Methylomicrobium* species from a highly alkaline environment. Extremophiles 4:145–155
Stępniewska Z, Pytlak A, Kuźniar A (2013) Methanotrophic activity in carboniferous coal bed rocks. Int J Coal Geol 106:1–10
Syamsul Arif M, Houwen F, Verstraete W (1996) Agricultural factors affecting methane oxidation in arable soil. Biol Fertil Soils 21:95–102
Theisen AR, Ali MH, Radajewski S et al (2005) Regulation of methane oxidation in the facultative methanotroph *Methylocella silvestris* BL2. Mol Microbiol 58:682–692
Tokida T, Fumoto T, Cheng W, Matsunami T, Adachi M, Katayanagi N et al (2010) Effects of free-air CO_2 enrichment (FACE) and soil warming on CH_4 emission from a rice paddy field: impact assessment and stoichiometric evaluation. Biogeosciences 7:2639–2653
Trotsenko YA, Khmelenina VN (2008) Extremophilic methanotrophs. ONTI PNTs RAN, Pushchino, p 206
Tsubota J, Eshinimaev B, Khmelenina VN, Trotsenko YA (2005) *Methylothermus thermalis* gen. nov., sp. nov., a novel moderately thermophilic obligate methanotroph from a hot spring in Japan. Int J Syst Evol Microbiol 55:1877–1884
Van Groenigen KJ, Osenberg CW, Hungate BA (2011) Increased soil emissions of potent green house gases under increased atmospheric CO_2. Nature 475:214–216
Visvanathan C, Pokhrel D, Cheimchaisri W et al (1999) Methanotrophic activities in tropical landfill cover soils: effects of temperature, moisture content and methane concentration. Waste Manag Res 17:313–323
Whalen SC, Reeburgh WS, Sandbeck KA (1990) Rapid methane oxidation in a landfill cover soil. Appl Environ Microbiol 56:3405–3411

Whittenbury R, Phillips KC, Wilkinson JF (1970) Enrichment, isolation and some properties of methane-utilizing bacteria. J Gen Microbiol 61:205–218

Wilshusen J, Hettiaratchi J, Stein V (2004) Long-term behavior of passively aerated compost methanotrophic biofilter columns. Waste Manag 24:643–653

Yan X, Akiyama H, Yagi K, Akimoto H (2009) Global estimations of the inventory and mitigation potential of methane emissions from rice cultivation conducted using the 2006 intergovernmental panel on climate change guidelines. Glob Biogeochem Cycles 23:1–15

Paradigm Ecological Shift and Succession in Microbiomes: A Climatic Advent

5

Avijit Tarafdar, Devashish R. Chobe, and Mamta Sharma

Abstract

Surge in the global mean annual temperature a resultant of the significant rise in CO_2 concentration and other greenhouse gases have become a serious concern internationally due to their unfavourable impacts on the physical and biological entities. Throughout Earth's history, microbiomes are known to play important roles in changing climate and are in turn being continuously changed by the climate. Certain recent studies are even pointing towards changes within the Earth's diverse microbiome as the causality of the warming Earth. Amongst all the factors, climate change has taken over as the most important determinants of the alterations in diversity, distribution and abundance of species in both artificially managed as well as natural terrestrial and marine ecosystems. Understanding microbial ecology of different ecosystems on earth, whether terrestrial or aquatic, is essential for assessing the relationship between various biogeochemical cycles with climate, its responses and feedbacks. But the microbial communities are very complex in their organization and function and are easily influenced by the changes in climate. Several studies have revealed the extent to which the shifts in microbial species interactions due to changing climatic scenario lead to alterations in biodiversity and function of both terrestrial and aqueous ecosystems. In fact, the interactions of soil microbes within their community and as well as rhizoplanes and root-rhizosphere of plants can directly or indirectly reform landscape patterns and abundance, diversity and composition of flora and fauna. Events such as ecological succession are also of prime importance in this context, as acute climatic changes could result in faster succession of a microbial community by another within a particular ecosystem.

A. Tarafdar · D. R. Chobe · M. Sharma (✉)
Legumes Pathology, International Crops Research Institute for the Semi-Arid Tropics, Patancheru, Hyderabad, Telangana, India
e-mail: mamta.sharma@cgiar.org

S. A. Lone, A. Malik (eds.), *Microbiomes and the Global Climate Change*,
https://doi.org/10.1007/978-981-33-4508-9_5

Keywords

Microbial community · Climate change · Econiches · Ecological succession · Spatial turnover · Temporal turnover

5.1 Introduction

Ecological succession of a microbial community is defined as the change in the species abundance and microbial population within an ecological niche over the time. The microbial community in an ecosystem is shaped by the environmental factors in nature, and the ecological succession of the microbial community in the particular space occurs when these environmental factors are modified. For example, climatic elements like rainfall, moisture and temperature have a direct impact on the biology of freshwater fungi responsible for the shifts in various important components of heterotrophic food webs in woodland streams. Similarly, ocean acidification rising temperature and seawater threaten the survival of marine ecology involving marine microbial diversity, coral physiology, etc. Microbiomes are the pioneers in mineralization of litter carbon and conversion of complex, dead, organic matter into simpler biomass, due to which even a minimal change in climatic elements could indirectly deliver a large scale impact on agriculture, fresh water and marine ecology. Microbiomes respond, adapt, and evolve to their surroundings at much higher rates than any other higher organisms. This property allows scientists to correlate and derive scalable assessments in a little span of time on the impacts of changing climatic factors on the nature of microbiomes. This may in turn be used to understand and forecast imminent effects on all forms of life in changing climatic scenario. Hence, the chapter puts some efforts in explaining the effects of changing climate on the diverse microbiomes under different ecosystem.

Microbes are the main decomposers of organic materials existing in environment. Changes in microbial community or microbial succession are occurred due to their adaptation in various habitats. Due to global change, the species distribution is also altering and thus interactions among organisms within the specific ecological niche. In a niche, microorganisms are living with thousands of other different species. Of them few are beneficial to each other and helping them to live in community, whereas some are pathogenic and have effects on other species. But some microbes have no effect to the other microbes within the community either positive or negative. Since in natural conditions the microbial community is made up with several microorganisms with having different characters and dispersal ability, it can be assumed that the organisms present within the same econiches will react in dissimilar way under changing climatic conditions. The microorganisms take a major role in regulating nutrient transformation in a macroecological system and allow the co-existence of higher organism. When changes come to the microorganisms or their surrounding environment at large spatial scale due to the climate change, the patterns of abundance, distribution and diversity of the species are altered within the niche where it belongs, and not only a single community, rather than whole ecosystem could be affected. Thus, the species are altered permanently; and the gradual microbial succession is taken place. In this chapter, we provide information about succession of different microbial community existing in different natural habitat. We elucidated in this chapter that how microbes are being stimulated with changing climatic factors and discussed the real consequences of changes in microbial interactions which are the main functional composition in a frame of microbial ecosystem.

5.2 Responses of Soil Microbial Community Under Changing Multiple Climatic Factors

Recycling of nutrients and carbon (C) in ecosystems soil microbial communities have a big role to play, where all activities are influenced by two main external factors (Hector et al. 2010), abiotic components, e.g. quantity and quality of the litter in the soil, soil moisture conditions, soil temperature, temperature of the open environment, humidity, CO_2 concentration and even other gaseous components of the open air; and biotic factors, e.g. microbiome of the soil, plant diversity and activity of the other higher organisms. Sudden or gradual but stable changes in surrounding atmospheric or climatic conditions always have influential role on both abiotic and biotic components of an ecosystems directly and indirectly by modification in their responses in ecosystems (Fig. 5.1). Again, the co-responses of the ecosystem from the opposite side to the environment as feedback may also be regulated by microbial populations present in the soil (Bardgett et al. 2008). It has

Fig. 5.1 Unusual precipitation results either low soil moisture stress (drought) or off time high soil moisture stress (water lodging/flood)

been shown in some previous research that changes in climatic factors, e.g. raising temperature and elevated CO_2 would act together to alter the overall process and common properties of the ecosystem. But the core process of the ecosystem that known to be regulated by the general responses generated from the microbial communities is very unpredictable, and it is often unanswered how the composition and abundance of microbial communities associate with the upshot of ecosystem processes and climatic perturbations. Till date, the research conducted related to climate change had intended on microscale responses to climate change, e.g. much of the research to date related to climate change generally focused on the effect and response at microscale level such as alteration of compositions in the plant community (Bakkenes et al. 2002; Lloret et al. 2009), plant growth (Norby et al. 2001, 2004) and coarse scale soil processes (Garten et al. 2008; Franklin et al. 2009), and interestingly all of which are indirectly effects of microbial processes. In response to single climatic factor, some gross parameters of microbial process like the basic microbial community profile, microbial biomass, and enzymatic activity may also be changed. In a study, the direct and interactive effects of single or different combinations of several climatic factors on fungal and bacterial communities (composition and abundance) present in soil were observed in an old-field ecosystem, where different environments were created on the basis of changing climatic scenarios by altering and simulating the multi-factor of the climate. The environments of old-field ecosystem were stimulated and exposed it to different atmospheric CO_2 concentration (ambient and elevated), rainfall precipitation (dry and wet) and temperature (ambient and elevated up to +3 °C). The old-field ecosystem was left for interaction and alteration in abundance and community structure of soil microorganisms, e.g. fungus and bacteria. In that study, it was found that (a) in warmed conditions, fungal abundance increased; (b) bacterial abundance decreased under ambient atmospheric CO_2 with warmed conditions; (c) bacterial abundance increased under elevated CO_2 with warmed conditions; (d) the distribution of the fungal and bacterial spp. in phylogenetic cluster was varied among treatments; (e) the composition of fungal community changed with the alteration in precipitation; (f) the ratio of Proteobacteria and Acidobacteria was altered with the change in rainfall precipitation. The results of the study indicated that change in climatic factors and their interactive effects in turn to changes in overall abundance of soil microorganism, i.e. fungal and bacterial population in an ecosystem. But changes in precipitation, the overall effect is too high to establish a stable alteration in community composition.

5.3 Development and Evolution of Microbial Ecosystems and Its Gradual Succession

The development of ecosystem in an area is usually called ecological succession. So under climatic change scenario the major questionable area related to evolution of ecosystems or ecological succession is (a) the parameters on which stability of the community depends even after disturbance of the environment; (b) for community

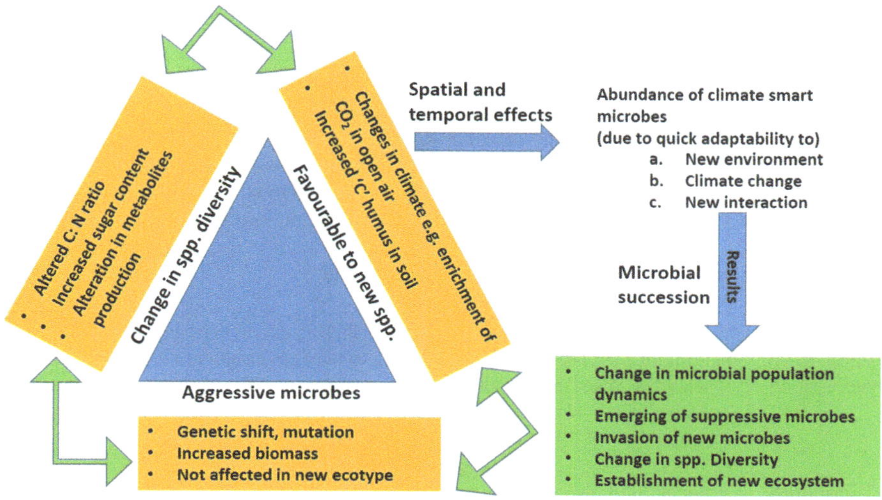

Fig. 5.2 The spatial and temporal effects of climate change on establishment of new or altered ecosystem

dynamics, what are the driving forces; (c) what is the relation between ecosystem functions and community functions. The true microbial succession is happened when a better adapted and best fitted microbial population is stabilized in an ecological niche with an altered environment and climatic conditions by replacing previous stabled any microbial population. The development of the microbial ecosystem in a particular geographical area is brought by a slow but a continuing and steady changes within the main composition of a microbial and climax characteristic (Fig. 5.2). The changes in community composition of a microbial ecological niche are seasonal and multiyear dynamic process which is cyclic and random because of fluctuations. On the contrary, the microbial ecological succession is an irreversible, unidirectional and orderly process. Usually, the succession in a microbial community started by intense change within abiotic environmental conditions, e.g. desertification, heavy rain and flooding, cooling or warming climate, volcano eruption, forest fire, etc. We can easily differentiate between a primary and a secondary succession. The primary succession assumes the development and establishment of initial microbial community within a land that bared originally, e.g. suddenly released rock, magma rock of lava-stream or sand dune where the ecosystem starts from zero. But the process and establishment of secondary microbial succession are much faster than a primary microbial succession. Generally the secondary microbial succession is occurred when the soil is already been improvised naturally by the heavy deposition of organic matter. Again, the succession of a microbial community is a continuous and concomitantly process with the evolution of the entire microbial ecosystem as the microbial microenvironment affects by the orderly and gradual replacement of higher organisms, especially plant and species. The ecologists view the whole mechanisms of succession in two competing

Table 5.1 Evolution of microbial community and role of holistic and meristic paradigm as a driving force

Components	Paradigm	
	Holistic	Meristic
Community arrangement	Persistent, specifically determined by resilient interspecific links	Non-persistent, be influenced by emigration and immigration, local conditions and past history
Boundary ecosystem	Definite and clearly expressed	Indefinite and not clearly expressed
Risk in forecast	High	Low
Pioneer stages of succession	Inappropriate abiotic atmosphere for starting of life. The ecology begins when stress resistant species start to colonize which advance the environment for other species	Begins due to ecological vacuum, where environment is suitable but little saturation of organisms within the community. Opportunistic species colonize first without any severe stress and competition
Transitory short-live community	Edification improves abiotic environment for the community; the productivity increases. The organisms are not emphasized although not rejected by the reverse effect of environment	The organisms are selectively chosen by the strong environmental effects. Progressive increase of competition for resources. Edification improves abiotic environment for the community; the productivity increases. The organisms are not emphasized although not rejected by the reverse effect of environment
Climax community	Symbiotic association within the maximum organisms helps community to be stabilized	In similar econiches, the maximum microorganisms are interacted each other competitively, but the community become stabilized

paradigm: holistic and meristic (Table 5.1). In holistic paradigm concept, the first biotic component (e.g. microbial organisms) of econiches acts as a kind of superorganisms. In the beginning of ecology in a bare area, the superoganisms are stress resistant and start to colonize in an inappropriate abiotic atmosphere and make a favourable ecosystem to grow other organisms. This paradigm has steady structure with non-selective and strong deterministic interactions among the different econiches, similar to interaction between cells and specialized tissues within the higher organisms (Table 5.2). According to this concept the previous microbial population amend the habitat and physical state which improve the living conditions for successive organisms. The meristic paradigm approaches give freedom to some independent organisms for making biotic interaction with the other organisms living within the same econiches. As the species are independent and community is open, the organisms can freely enter or leave the community by immigration or emigration. The replacement of the species occurs when two species compete each other within the same econiche (Fig. 5.3).

Table 5.2 Type of ecological interactions within the microbial community and its effect on population growth

Interactions	Group I organisms	Group II organisms
Competition	Negative effect; compete with group II organisms	Negative effect; compete with group I organisms
Mutualism	Positive effect; helping to survival of group II organisms	Positive effect; helping to survival of group I organisms
Parasitism/ predation	Positive effect; dominant over Group II organisms	Negative effect; dominated by group I organisms
Commensalism	No effect	Positive effect; benefited from group I organisms
Amensalism	No effect	Negative effect; inhibition of survival by group I organisms
Neutralism	No effect	No effect

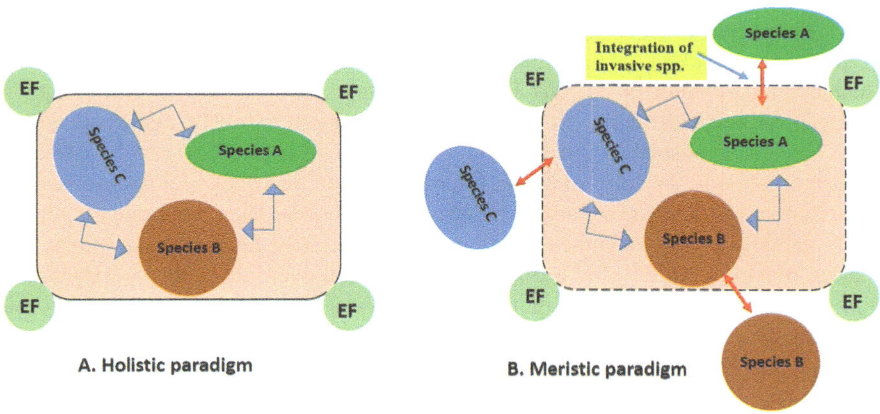

Fig. 5.3 Establishment of stable microbial community in an econiche through a holistic or meristic paradigms succession. "EF" is denoting environmental factors

5.4 Role of Microbes in Global Warming and Recycling the Essential Elements

The grazing animals like cattle, sheep, camel, goats, etc. have a special chamber in stomach called rumen. This part is swarming with billions and billions of microorganisms like moulds, yeasts, bacteria and protozoa. Those microbes are essential for digesting cellulose that is consumed by animals. Animals cannot produce cellulose degrading enzymes, so they absorb the cellulose directly. The microbes break it down the complex food into simpler organic substances. On the other hand, some methane loving archaea bacteria live in the rumen, which are specialized in producing methane by breaking down the animal's food. The methane, a very powerful greenhouse gas then is out from their digestive system. The methane can trap about 20 times more heat than CO_2. It means that the methane can warm the atmosphere 20 times more than same volume of CO_2 gas can. But the question of worry is that the cattle and farm animals are producing around 20% of total global methane. Nowadays, scientist are concerned about to stop or decrease the methanogenesis inside the rumen of grazing animals, but that results in the animal being smaller. Some researchers are finding certain additives for their diet which can reduce emission of methane. The soil microbes have also a big role in changing climate. Soil is the immense collection of a range of diversified microbes. In changing climate scenario, it is predicted that the emission of CO_2 will be increased and carbon-based materials in the soil will be broken down very swiftly because of changing activity of microbes due to climate heats up. If this happens then global temperature will increase because release of CO_2 in the environment will be more as a waste product due to increased microbial activity and their respiration.

Microbes play an important role to recycle and transform the essential elements, e.g. organic carbon and nitrogen as either generators or users of these vital gases in the atmosphere. Since beginning of the life lower organisms like bacteria and archaea is recycling of all the essential elements. During methanogenesis, methanogenic microbes convert CO_2 to methane through carbon cycle. Again, nitrogen-fixing bacteria fix nitrogen by the process of nitrogen cycle, by which they convert atmospheric nitrogen into biological nitrogen that can be used by higher organisms. In spite of those microbes, other microbes, e.g. cyanobacteria and photosynthetic algae are also involved in these cycles. The taxonomically higher algae capable to photosynthesis and cyanobacteria are the major component of marine plankton which play a major role in carbon cycle through photosynthesis and the basis of food chain in the marine ecosystem. On the other side, soil microorganisms, e.g. fungi and bacteria are the main decomposer of organic matter and they break down the complex organic matter to simple compound and bring back CO_2 into the atmosphere.

But the question is that the microbes can help to solve the planet from global warming. Photosynthetic microorganisms have big power to convert atmospheric CO_2 into O_2. Some single-celled blue green algae (BGA), e.g. *Prochlorococcus* and *Synechococcus* are very small in size but most abundant BGA the ocean. Near about 10×10^7 cells of *Prochlorococcus* are present in 1 litre of marine water. It was estimated that *Synechococcus* and *Prochlorococcus* can eliminate about 10×10^9 tons of carbon from the atmosphere every year; i.e. almost 66.66% of the total C fixation succeeds in the sea water. If we are able to use power of such microbial organisms, the level of greenhouse gases including CO_2 could reduce from the atmosphere and eventually could slow down the global climate change.

5.5 Acceleration of the Spatial Turnover of Soil Microbial Communities Under Elevated CO_2

Elevated CO_2 considerably affects the function, composition, dynamics, interaction and diversity of soil microbes. The impacts of elevated CO_2 on geographic distribution of microbial community at regionally or globally are little known. The research on climate change showed that elevated CO_2 shifted the structure, interaction and composition of microbial community present in soil (Deng et al. 2012; He et al. 2012a, b; Van Groenigen et al. 2014). Generally elevated CO_2 increases the total biomass of plants and productivity leading more deposition of organic carbon in soil (He et al. 2014), the soil microenvironment (Van Groenigen et al. 2014; Xiong et al. 2015), and the microbial growth and activity (Kelley et al. 2011; Van Groenigen et al. 2014) but these responses seemed to be largely inconstant among the different econiches. So many studies were carried out, but effect of elevated CO_2 on geographic distribution like soil microorganisms and their association with different habitats and associated ecosystems is still unclear.

The distance–decay relationship (DDR) is a model to understand the mechanisms help in forming community assembly and biodiversity across a range of organisms in different geographical patterns at different spatial scales (Hanson et al. 2012). This model is shown the relationship between decline patterns of community similarity and increasing distance of geographical location. According to hypothesis, with the increase of geographic distance the similarity within microbial community will be declined, and the changes of DDR would be greatly determined by the niche assembly theory. The niche assembly theory predicts the stability of the community biodiversity and how a certain number of microbial species coexists in a close proximity by segregating themselves to a specific econiche (Webb et al. 2002). For predicting the response of microbial communities in a particular econiches to future climate change, understanding the mechanisms of maintaining biodiversity is needed. In some study it has been proved that the turnover rate of a soil microbial community at a spatial scale under elevated CO_2 is higher. Where the soil C: N ratio could be prime contributor to DRR, C: N ratios have a close relation with microbial growth and total biomass (Fierer et al. 2009). In point of fact, environmental selection has a major role in structuring the microbial community, where four evolutionary and/or ecological processes, i.e. selection, mutation drift and dispersal would construct biogeography of microbial ecosystem (Hanson et al. 2012).

Plant–Microbe Interaction and the Temporal Turnover of Soil Microbial Community.

The number of species replaced or eliminated per unit of time is called species temporal turnover (Hatosy et al. 2013). Another common model for describing species temporal turnover is the time–decay relationship, where the changes in number of the species in a community similarity over time are used to explain (Nekola and White 1999). The rates of temporal turnover depend on local environmental factor, ecosystem types (Korhonen et al. 2010), disturbances (Svensson et al. 2009) and temporal scale (Hatosy et al. 2013). The temporal turnover of animals and plants is well documented, but temporal turnover of microbial world is yet to be known. Only few researches have been conducted on microbial temporal turnover for a short period, i.e. only few months, which were mainly based on microbial growth under a controlled engineered system. Recently a study on temporal beta-diversity of three bacterial communities from three different marine environments showed that different factors are involved in appearance of dissimilarity in turnover at different time scales (Hatosy et al. 2013). The variations in temporal turnover of soil microbial community with changing climatic scenario and related mechanisms remain largely unknown.

The soil microbes can remain active potentially or dormant for a long period within an econiche and those microbes have a role in shaping of soil microbial ecosystem. As, the soil microbial populations can evolve within the short period of time scale, it can shift the plant–microbe interactions quickly (Chave 2013); for example, in a micro-ecosystem like plant rhizosphere, the microbial communities change their composition time to time throughout the plant's growing season (Cregger et al. 2012; Kumar et al. 2016). The changes in microbial communities due to seasonal variation are more imperative than ecosystem C-fluxes during inter-annual variations. In a study, it was observed that background microbial variation was constantly higher when precipitation is unchanged since few years then altered precipitation (Gutknecht et al. 2012). In a community, microbial acclimation and adaptation are often alluding to elucidation that climate manipulation for the shorter period has no influence at long time scales (Bradford et al. 2008). Rapid adaptation of microbial communities may allow the modern plant species which are resilient and up-to-date to new climate, as the longer generation times of host plant lag behind the dynamic microbial communities (Lau and Lennon 2012). Evolution may swamp long-term impacts of climate and background microbial community variations may overcome climate treatments on seasonal timescales.

5.6 Few Example of Microbial Community Succession Under Changing Environment Stimulated Through Soil Transplant

The transplantation of soils from one geoclimatic zone to another geoclimatic zone gives an effective method to examine the response of microbial organisms in an econiche due to alteration of multiple climatic factors under changing climatic scenario (Sun et al. 2014; Zhao et al. 2014). It is reported that the soil transplantation has an impact on composition of microbial communities, but till now it is debatable. For instances, transplantation of soil from cooling area to warming area resulted in a change of microbial biodiversity and loss in total biomass. These functional and structural changes in microbial community in warmer condition indicated that exposure in higher temperature for few years could alter the soil microbiome (Vanhala et al. 2011). The microbial composition, biomass, respiration and enzymatic activities found to be decreased, after transplanting of soils under oak canopies to an open environment of grassland with lower soil water content and higher temperature (Waldrop and Firestone 2006a, b). The bacterial and fungal community structures of a forest soil were found to be significantly changed after 2 years of incubation subsequently transplanting the soil to meadows (Kageyama et al. 2006). Similarly, transplanting the black soil to warmer areas and red soil to colder regions (Zhao et al. 2014) caused huge alterations in the activities of microbial populations.

5.7 Influence of Climate Change in Shifting Plant Diseases from Minor to Major

To develop any disease in plants, infection, establishment and colonization of pathogen into the host tissue are the most important steps. Again, the infection of pathogen to host is completely depended on environmental factors (Madgwick et al. 2011; Sharma et al. 2015, 2019). Environmental stress plays an important role on community interaction between plants and pathogens (Chobe et al. 2016; Tarafdar et al. 2017, 2018). If any of three factors (host × pathogen × environment) in disease 'Δ' is altered, the disease progression would change. Elevated temperature and CO_2, ozone depletion, high or low humidity all are events of imperious deviances in climatic components. Alteration in above climatic components plays important roles in expression and pathosystems. The diversity of pathogens, e.g. nematode, protozoa, phytoplasma, mycoplasma, fungi, oomycetes, bacteria, virus and viroids (Tarafdar et al. 2012; Singh et al. 2013; Biswas et al. 2014; Padaria et al. 2016) infecting crops is overwhelming and becoming more pathogenic by continuous expanding themselves through mutation, adaptation, evolution and dissemination of new strains/variants/pathotypes/races (Table 5.3). Recently emerged fungi, *Puccinia striiformis* and *P. graminis* are among the utmost destructive and swift spreading pathogens. For instances, other late blight genotypes of potato have been displaced and rapidly emigrant by a new lineage of *P. infestans*. Another example,

Table 5.3 Effect of elevated CO_2 on pathogenic microbes causing plant diseases

No effect	Positive effect (increase)	Negative effect (decrease)
• Sudden death syndrome (*F. virguliforme*) in soybean. • *Phytophthora* root rot (*P. parasitica*) in tomato	• *Fusarium* rot (*F. verticillioides*) in maize. • *Fusarium* wilt (*F. oxysporum* f. sp. *lactucae*) in lettuce. • *Phytophthora* blight in pigeonpea by *P. cajani*, chilli pepper by *P. capsici* and potato by *P. infestans*. • Crown rot (*F. pseudograminearum*) in wheat. • Brown spot (*Septoria glycines*) in soybean. • Sheath blight (*Rhizoctonia solani*) and blast (*Pyricularia oryzae*) in rice	• Downy mildew (*Peronospora manshurica*) in soy bean. • Anthracnose (*Colletotrichum acutatum*) and two bacterial diseases, viz. bacterial wilt (*Ralstonia solanacearum*) and bacterial spot (*Xanthomonas campestris* pv. *vesicatoria*) in chilli paper. • Powdery mildew (*Erysiphe graminis*) in barley and wheat. • Rust (*Puccinia sparganioides* Ellis & Barth) in C3 sedge

Fusarium head blight, a damaging disease of wheat at anthesis period is emerging again in the USA for warm and wet condition. As pathogen aetiology and growth are often dependent on soil moisture, the availability of water, the infection development and colonization establishment to plants can greatly affect either positively or negatively (Ghosh et al. 2016, 2017; Tarafdar et al. 2017, 2018). For instance, under low soil moisture conditions, the susceptibility of host plants to bacterial pathogen can increase (Mohr and Cahill 2003). Similarly, low soil moisture stress significantly increases the incidence and severity of dry root rot disease (*Rhizoctonia bataticola*) in chickpea (Tarafdar et al. 2018) but on contrary lowering the collar rot disease (*Sclerotium rolfsii*) in similar condition. Although till now very limited information is available about the effect of elevated CO_2 in developing plant diseases and their pathogens, whether the incidence and/or severity of the diseases are remaining unaffected or it may increase, or decrease. The studies of Sharma et al. (2015) and Ferrocino et al. (2013) showed that the elevated CO_2 enhanced the colonization of *Phytophthora cajani* (*Phytophthora* blight) in pigeon pea and *Fusarium oxysporum* f. sp. *lactucae* (*Fusarium* wilt) in lettuce roots, respectively. On the other hand, the study of Thompson and Drake supported the reduction of disease severity in wheat by *Erysiphe graminis* (powdery mildew) severity and in C3 sedge (*Scirpus olneyi*) by *Puccinia sparganioides* (Ellis & Barth) Grey (rust). Thus the real interest in our study to decipher how elevated CO_2 level would influence the chickpea–Foc interactions which may help to better understand the change in *Fusarium* wilt disease dynamics and aetiology in chickpea under elevated CO_2. Again sheath blight (*Rhizoctonia solani*) and blast (*Pyricularia oryzae*) in rice, and higher threats of late blight (*Phytophthora infestans*) in potato are increasing day to day due to elevated temperature and CO_2.

5.8 Effect of Climate Change on Marine Microbial Succession

The microbial ecosphere is frequently left out from the dialogues in changing climate scenario. But microbial world is one of the acute contributors in the biogeochemical cycles, e.g. carbon and nitrogen (Fig. 5.3). So, the attention is needed to be given on microbial responses to global warming. About 93% of total CO_2 in world is accumulated in oceans, out of this 90 billion tons of CO_2 cycles per year. The mechanism of carbon cycle in the ocean is mostly dominated by microbial species existing in it. Earth's climate is regulated by the oceans so the small change in oceans is responsible for the change in earth's climate. On the other way, the ocean is increasing temperature of the climate by continuous depositing and distributing heat around the globe and transferring heat within the atmosphere. The exterior part of the stratum warms very swiftly, when the atmosphere heated up and so does ocean. The acidification within the ocean water is a natural phenomenon. Although acidification and warming are different singularity, the elevated CO_2 is increasing acidification in ocean by raising temperature of the surrounding atmosphere of ocean. Both the phenomenon are interacting and becoming detrimental for the marine ecosystem and associated microbial community. The CO_2 level of the ocean surface and the atmosphere was constant since millions of years. From past few centuries due to human interference, e.g. land-use practice and burning fossil fuels, the level of CO_2 significantly augmented and reached to serious concern level. As an effect, the ocean absorbs about 29% of more carbon as compare to normal. This extra CO_2 has substantial effects on water of ocean, make it 30% more acidic. Every drop of seawater contains thousands upon thousands of small and single-celled microorganisms which are the part of ocean ecosystem.

Since past century, the temperature of coastal water has been upsurge, and estimated to be increased by 4–8 °F in the twenty-first century (US EPA 2015). The ocean ecosystems are dependent on so many properties of the water, which are vital and critical. But those properties are being now affected when the fluctuation of the water temperature occurs. As a result, the dramatic changes in the ocean life happen (NASA 2015). These environmental situations are likely to be favourable for some microbial species which are adapted to that condition. This is specifically understood in the case of two species of common oceanic nitrogen fixers, viz. *Trichodesmium* and *Crocosphaera* (Michaels et al. 2001).

Microbiomes retort, alter and evolve to their ecosystem at much higher rates than any other higher creatures. This property permits scientists to associate and originate scalable assessments in a short period of time on the effects of changing climatic factors on the nature of microbiomes. This may in turn be used to recognize and predict the future repercussions of climate change on all forms of biodiversity. Microbiomes are the forerunners in mineralization of litter carbon and alteration of complex dead organic matter into simpler biomass, due to which even a trifling change in climatic elements could secondarily deliver a large scale impression on marine ecology. 70% of earth surface is covered by marine biomass including mangroves, coral reefs, costal estuaries and the open ocean. During photosynthesis the phototropic microorganisms use sun in the top 200 m of the water column, whereas marine ecosystem in deeper zones uses organic and inorganic compounds for energy. In addition to sunlight, the accessibility of other energy forms and water temperature stimulates the composition of marine communities. In marine environments, microbial prime production provides significantly to CO_2 sequestration.

Marine microorganisms recycle nutrients to use them in CO_2 release process and ultimately in marine food web. In accumulation, CO_2 accretion is prominent to its acidification. The expansion of oxygen-depleted zones has also been ascended due to stratification of oceans. Alterations like this are likely to have impacts on the microbial food pyramid and ultimately on the biogeochemical cycles (Walsh et al. 2016). Although distinct from sure, the repercussions of global warming on marine microbial populations were projected in several studies.

Similarly, bacterio-plankton plays a key function in the marine food pyramid. These microorganisms are accountable to an activity known such as re-mineralization. Complex organic matter breaks down to CO_2 and vital inorganic nutrients. So this answer that ocean does not stink but at some places it does. Stinky organic matter and by-products of organic matter decay are the product of slow re-mineralization of microscopic organisms from which plenty of nutrients are brimmed up from some highly fruitful coastal locations. While on the huge clear surface of water of the ocean, preliminary production of these organic matters is closely attached to re-mineralization.

Bacterio-plankton obscuring and restraining energy and carbon transfer up the food chain rapidly guzzle organic matter directly released from phytoplankton. Plenty of carbon goes into organization of new bacterio-plankton cells, and rest of reverted to water and atmosphere as CO_2. Hence to generate the energy essential for endurance and reproduction the cells 'burn' the carbon. At the same time by regenerating the nutrients required to sustain primary production bacterio-plankton plays an acute role.

In addition, the ocean-bacteria grown on phytoplanktons and break them into molecules such as lipids and fats. This may become airborne to link aerosols as the ocean agitates. Ultimately the further process is cloud condensation where the further progress into a context where the clouds made by the conversion of moisture droplets. The total volume of oceanic microbes can significantly be altered by the alterations of climatic elements, and this alteration can directly disturb the amount of partially water soluble organic materials such as lipids (Iacurci 2015). Change in climatic element can also be the major range in extension of detrimental microbes and causes biodiversity. These facts can be evaluated through parasite *Perkinsus marinus* has responsible for causing huge oyster deaths.

Same as in coral reefs are the most distinct and parsimoniously important ecosystem in the marine ecology as well as on earth. They are in danger due to change in various climate related marine ecosystem like raising sea level, ocean acidification and raising ocean temperature. Coral reefs require $CaCl_3$ to build frames, but the ocean acidification is driving its availability downwards. Near about 14.2% great barrier coral reefs has been lost due to calcification.

A significance of change in physiognomies, death rates and uniform allele rate of recurrence is all dependent on degree of change in climate is that the development and resilience of whole populations. If the population size changes, it interfaces among the species may also disturbed and eventually regimen shift may occur with undergoing rapid and irrevocable transition from one phase to another in biodiversity. For example, propagation of micro algae can be led by underlying damage and death to coral reefs due to bleaching and storms. As corals are dwelling for many marine microbial species loss of coral will lead to change in biodiversity and amount of such species and it may be responsible for extension of reef specialties. In addition to it change in reef specialties may adversely affect the fishes and invertebrates and may have large scale implications on fisheries, tourism and other human uses of reefs.

Covering 65% of earth surface benthic deep-sea ecosystems are the major ecosystems on the earth. A critical share of global biogeochemical cycles is played by microorganisms dwelling in these huge biomes at all water depths which has a significant share of biomass and biosphere. As benthic deep-sea ecosystem is not much explored till now they received a less attention. Hence the shift in benthic deep-sea biota remains mysterious. But the deep-sea ecosystem directly or indirectly affecting by the global climate changes (Gooday et al. 2008; Goineau and Gooday 2017). This predominantly applies to deep-sea microorganisms like micro eukaryotes their viruses, bacteria and archaea. Unexplored information on the response of deep-sea microbes to these alterations limits our ability and study to forecast the enormousness of the repercussion of global climate changes on the deep-sea microbial biodiversity. These repercussions may lead to regime shifts, shifting of deep-ocean life amenities like carbon sequestration capacity and nutrient regeneration, which is mostly dependent on microbes (Danovaro et al. 2017). The maintainance of the ocean ecosystems depends mostly on the functioning and distribution of benthic deep-sea microorganisms. The effect on the functioning of the ocean interior is a main concern to better forecast the potential implications at global scale.

There are various effects of human interference, agriculture, rivers and local environmental factors, including light and soil type affects the existing complex network for microbial exchange significantly, that occur with other microorganism. These exchanges facilitate how microbes retort to change in climatic elements and affect climate change and how climate change in turn affects microbial responses.

In conclusion, studies are going on in commencement for demonstration that marine microorganisms can be useful as watchdogs to assess the effect of climate change on marine ecosystem. Hence there is an acute necessity in lasting phase sequence biotic assessment of marine ecosystem. We may exactly foresee the change? But still the microbial ocean leftover is cryptic.

Importance of microorganisms in influencing the greenhouse gases concentration in atmosphere is worthy, but till now it is not properly recognized and praised by the scientists. Keeping in mind the observed suspicions, it is obvious that considering the straight and secondary repercussions of climate change on the microorganisms and their consistent lasting and temporary reactions can capture the major influences of these microbes. So, for variable climate change, hitched appropriately microorganism could be the potential natural resources. Conversely, proper attention is needed to stop to be turn as most incipient accelerator to hitch. Now it is a time to study this facet, to act and understand the mechanism more exactly. Hence use it appropriately to draw some solutions (Fig. 5.4).

5.9 Conclusion

Ecological succession of a microbial community provides conceptual framework in which many ecological topics of concern can be examined. In future, the microbial succession could be the major indicator or a central concept in ecology that would provide the information that makes critical contribution to understand the environmental changes as well as environmental challenges. Use of modern tools like mathematical modelling, meta-analysis of multi-site data and expert systems could be the medium to better understanding of microbial succession due to climate change, biodiversity loss, restoration and species invasion. Indirect influence of climate change on microbial populations can be facilitated via plants which may be stronger and durable than direct impact, as the community compositions and microbial activities can be influenced by the climate configuration. The novel

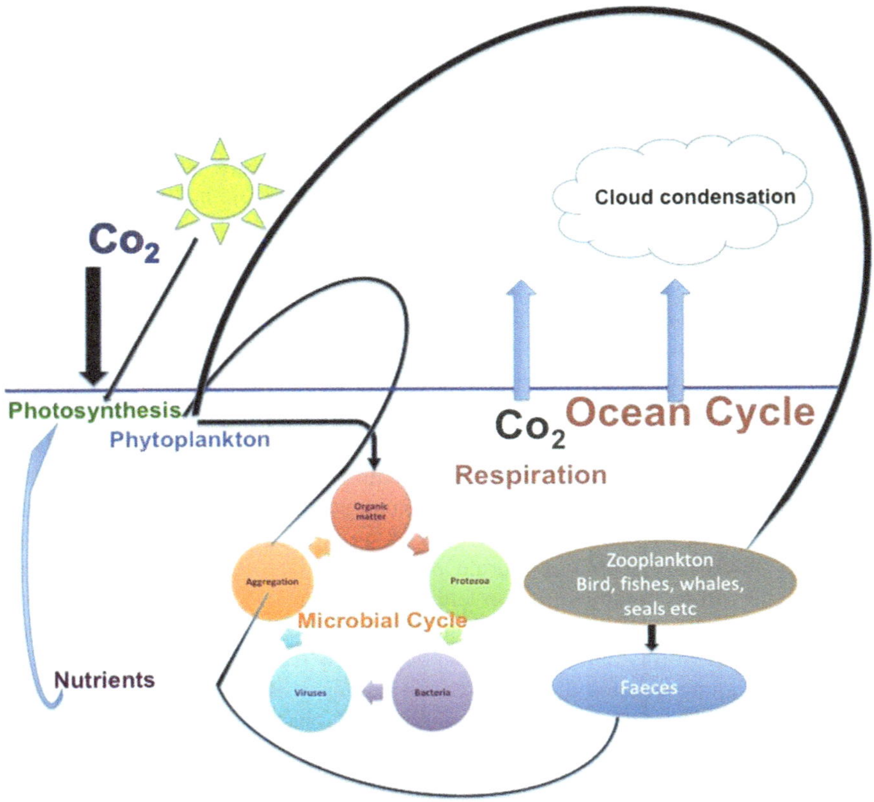

Fig. 5.4 The role of climatic factors on marine ecosystem

technologies will be essential to gather knowledge on microbe-centric information to identify the climate sensitive taxa whose response leads to shift in population dynamics of microbial community and alter in microbial activities. So, for predicting future ecosystem under changing climate scenario, we need best approaches and model for quantifying, observing and scaling the microbes–microbes interactions.

References

Bakkenes M, Alkemade JRM, Ihle F, Leemans R, Latour JB (2002) Assessing effects of forecasted climate change on the diversity and distribution of European higher plants for 2050. Glob Chang Biol 8:390–407

Bardgett RD, Freeman C, Ostle NJ (2008) Microbial contributions to climate change through carbon cycle feedbacks. ISME J 2:805–814

Biswas KK, Tarafdar A, Sharma SK, Singh JK, Biswas K, Diwedi S (2014) Status of disease incidence and distribution of *Citrus tristeza virus* in Indian citrus growing geographical zones. Indian J Agric Sci 84:184–189

Bradford MA, Davies CA, Frey SD, Maddox TR, Melillo JM, Mohan JE, Reynolds JF, Treseder KK, Wallenstein MD (2008) Thermal adaptation of soil microbial respiration to elevated temperature. Ecol Lett 11:1316–1327

Chave J (2013) The problem of pattern and scale in ecology: what have we learned in 20 years? Ecol Lett 16:4–16

Chobe DR, Gupta O, Pawar M (2016) Radiation induced mutation for resistance against races/ pathotypes of *Fusarium oxysporum* f.sp *ciceris* in chickpea (*Cicer arietinum* L.). Indian Phytopathol 69:260–265

Cregger MA, Schadt CW, McDowell NG, Pockman WT, Classen AT (2012) Response of the soil microbial community to changes in precipitation in a semiarid ecosystem. Appl Environ Microbiol 78:8587–8594

Danovaro R, Rastelli E, Corinaldesi C, Tangherlini M, Dell'Anno A (2017) Marine archaea and archaeal viruses under global change. F1000 Res 6:1241

Deng Y, He Z, Xu M, Qin Y, Van Nostrand JD, Wu L, Roe BA, Wiley G, Hobbie SE, Reich PB, Zhou J (2012) Elevated carbon dioxide alters the structure of soil microbial communities. Appl Environ Microbiol 78:2991–2995

Ferrocino I, Chitarra W, Pugliese M, Gilardi G, Gullino M, Garibaldi A (2013) Effect of elevated atmospheric CO_2 and temperature on disease severity of *Fusarium oxysporum* f.sp. *lactucae* on lettuce plants. Appl Soil Ecol 72:1–6

Fierer N, Strickland MS, Liptzin D, Bradford MA, Cleveland CC (2009) Global patterns in belowground communities. Ecol Lett 12:1238–1249

Franklin O, McMurtrie RE, Iversen CM, Crous KY, Finzi AC, Tissue DT, Ellsworth DS, Oren R, Norby RJ (2009) Forest fine-root production and nitrogen use under elevated CO_2: contrasting responses in evergreen and deciduous trees explained by a common principle. Glob Chang Biol 15:132–144

Garten CT, Classen AT, Norby RJ, Brice DJ, Weltzin JF, Souza L (2008) Role of N_2-fixation in constructed old-field communities under different regimes of [CO_2], temperature, and water availability. Ecosystems 11:125–137

Ghosh R, Tarafdar A, Sharma M (2016) Rapid detection of Fusarium oxysporum f. Sp ciceris from disease infested chickpea fields by loop-mediated isothermal amplification. Indian Phytopathol 69(4s):47–50

Ghosh R, Tarafdar A, Sharma M (2017) Rapid and sensitive diagnoses of dry root rot pathogen of chickpea (*Rhizoctonia bataticola* (Taub.) Butler) using loop-mediated isothermal amplification assay. Sci Rep 7:42737

Goineau A, Gooday AJ (2017) Novel benthic foraminifera are abundant and diverse in an area of the abyssal equatorial Pacific licensed for polymetallic nodule exploration. Sci Rep 7:45288

Gooday AJ, Nomaki H, Kitazato H (2008) Modern deep-sea benthic foraminifera: a brief review of their biodiversity and trophic diversity. In: Austin WEN, James RH (eds) Biogeochemical controls on palaeoceanographic environmental proxies, vol 303. Geological Society of London, London, pp 97–119

Gutknecht JL, Field CB, Balser TC (2012) Microbial communities and their responses to simulated global change fluctuate greatly over multiple years. Glob Chang Biol 18:2256–2269

Hanson CA, Fuhrman JA, Horner-Devine MC, Martiny JBH (2012) Beyond biogeographic patterns: processes shaping the microbial landscape. Nat Rev Microbiol 10:497–506

Hatosy SM, Martiny JBH, Sachdeva R, Steele J, Fuhrman JA, Martiny AC (2013) Beta diversity of marine bacteria depends on temporal scale. Ecology 94:1898–1904

He Z, Piceno Y, Deng Y, Xu M, Lu Z, Desantis T, Andersen G, Hobbie SE, Reich PB, Zhou J (2012a) The phylogenetic composition and structure of soil microbial communities shifts in response to elevated carbon dioxide. ISME J 6:259–272

He Z, Van Nostrand JD, Zhou J (2012b) Applications of functional gene microarrays for profiling microbial communities. Curr Opin Biotechnol 23:460–466

He Z, Xiong J, Kent AD, Deng Y, Xue K, Wang G, Wu L, Van Nostrand JD, Zhou J (2014) Distinct responses of soil microbial communities to elevated CO_2 and O^{-3} in a soybean agro-ecosystem. ISME J 8:714–726

Iacurci J (2015) Ocean microbes: how they may directly impact climate change. http://www.natureworldnews.com/articles/14761/20150519/oceanmicrobeshowtheymaydirectlyimpactclimatechange.htm

Kageyama SA, Posavatz NP, Jones SS, Waterstripe KE, Bottomley PJ, Cromack K Jr, Myrold DD (2006) Responses of soil bacterial and fungal communities to reciprocal transfers of soil between adjacent coniferous forest and meadow vegetation in the Cascade Mountains of Oregon. Plant Soil 289:35–45

Kelley AM, Fay PA, Polley HW, Gill RA, Jackson RB (2011) Atmospheric CO_2 and soil extracellular enzyme activity: a meta-analysis and CO_2 gradient experiment. Ecosphere 2:96

Korhonen JJ, Soininen J, Hillebrand H (2010) A quantitative analysis of temporal turnover in aquatic species assemblages across ecosystems. Ecology 91:508–517

Kumar R, Pankaj VP, Tarafdar A, Biswas K, Kumar S (2016) Soil microbes and their interaction with plants. In: Mitra R, Barman A (eds) Plant pathogen interaction: recent trends. Sharma Publications, New Delhi, pp 1–46

Lau JA, Lennon JT (2012) Rapid responses of soil microorganisms improve plant fitness in novel environments. Proc Natl Acad Sci USA 109:14058–14062

Lloret F, Penuelas J, Prieto P, Llorens L, Estiarte M (2009) Plant community changes induced by experimental climate change: seedling and adult species composition. Perspect Plant Ecol Evol Syst 11:53–63

Madgwick J, West J, White R, Semenov M, Townsend J, Turner J, Fitt BL (2011) Impacts of climate change on wheat anthesis and *Fusarium* ear blight in the UK. Eur J Plant Pathol 130:117–131

Michaels AF, Karl DM, Capone DG (2001) Element stoichiometry, new production and nitrogen fixation. Oceanography 14:68–77

Mohr P, Cahill D (2003) Abscisic acid influences the susceptibility of Arabidopsis thaliana to *Pseudomonas syringae* pv. Tomato and *Peronospora parasitica*. Funct Plant Biol 30:461–469

NASA (2015) Responding to climate change. http://climate.nasa.gov/solutions/adaptation-mitigation

Nekola JC, White PS (1999) The distance decay of similarity in biogeography and ecology. J Biogeogr 26:867–878

Norby RJ, Cotrufo MF, Ineson P, O'Neill EG, Canadell JG (2001) Elevated CO_2, litter chemistry, and decomposition: a synthesis. Oecologia 127:153–165

Norby RJ, Ledford J, Reilly CD, Miller NE, and. O'Neill EG. (2004) Fine-root production dominates response of a deciduous forest to atmospheric CO_2 enrichment. Proc Natl Acad Sci USA 101:9689–9693

Padaria JC, Tarafdar A, Raipuria R, Lone SA, Gahlot P, Shakil NA et al (2016) Identification of phenazine-1-carboxylic acid gene (phc CD) from *Bacillus pumilus* MTCC7615 and its role in antagonism against *Rhizoctonia solani*. J Basic Microbiol 56:1–10

Sharma M, Ghosh R, Tarafdar A, Telangre R (2015) An efficient method for zoospore production, infection and real-time quantification of Phytophthora cajani causing Phytophthora blight disease in pigeonpea under elevated atmospheric CO_2. BMC Plant Biol 15:90

Sharma M, Ghosh R, Tarafdar A, Rathore A, Chobe DR, Kumar AV, Gaur PM, Samineni S, Gupta O, Singh NP, Saxena DR, Saifulla M, Pithia MS, Ghante PH, Mahalinga DM, Upadhyay JB, Harer PN (2019) Exploring the genetic cipher of chickpea *(Cicer arietinum* L.) through identification and multi-environment validation of resistant sources against *Fusarium* wilt (*Fusarium oxysporum* f. sp. *ciceris*). Front Sustain Food Syst 3:78

Singh JK, Tarafdar A, Sharma SK, Biswas KK (2013) Evidence of recombinant *Citrus tristeza virus* isolate occurring in acid lime cv. Pant lemon orchard in Uttarakhand terai region of northern Himalaya in India. Indian J Virol 24:35–41

Sun B, Wang F, Jiang Y, Li Y, Dong Z, Li Z, Zhang X (2014) A long- term field experiment of soil transplantation demonstrating the role of contemporary geographic separation in shaping soil microbial community structure. Ecol Evol 4:1073–1087

Svensson JR, Lindegarth M, Pavia H (2009) Equal rates of disturbance cause different patterns of diversity. Ecology 90:496–505

Tarafdar A, Ghosh PD, Biswas KK (2012) *In planta* distribution, accumulation, movement and persistence of *Citrus tristeza virus* in citrus host. Indian Phytopathol 65:184–188

Tarafdar A, Rani TS, Chandran USS, Ghosh R, Sharma M (2017) Impact of moisture stress on collar rot (Sclerotium rolfsii Sacc.) development in chickpea (Cicer arietinum L.). in: Proceedings of ISMPP International Conference on "Plant health for human welfare." Organized by Indian Society of Mycology and Plant Pathology, (Jaipur), 119

Tarafdar A, Rani TS, Chandran USS, Ghosh R, Chobe D, Sharma M (2018) Exploring combined effect of abiotic (soil moisture) and biotic (*Sclerotium rolfsii* Sacc) stress on collar rot development in chickpea. Front Plant Sci 9:1154

U.S. EPA (2015) Greenhouse gas emissions and fuel efficiency standards for medium- and heavy-duty engines and vehicles—phase 2. Fed Regist

Van Groenigen KJ, Qi X, Osenberg CW, Luo Y, Hungate BA (2014) Faster decomposition under increased atmospheric CO_2 limits soil carbon storage. Science 344:508–509

Vanhala P, Karhu K, Tuomi M, Björklöf K, Fritze H, Hyvärinen H, Liski J (2011) Transplantation of organic surface horizons of boreal soils into warmer regions alters microbiology but not the temperature sensitivity of decomposition. Glob Chang Biol 17:538–550

Waldrop MP, Firestone MK (2006a) Response of microbial community composition and function to soil climate change. Microb Ecol 52:716–724

Waldrop MP, Firestone MK (2006b) Seasonal dynamics of microbial community composition and function in oak canopy and open grassland soils. Microb Ecol 52:470–479

Walsh EA, Kirkpatrick JB, Rutherford SD, Smith DC, Sogin M, D'Hondt S (2016) Bacterial diversity and community composition from seasurface to subseafloor. ISME J 10:979–989

Webb CO, Ackerly DD, Mcpeek MA, Donoghue MJ (2002) Phylogenies and community ecology. Annu Rev Ecol Syst 33:475–505

Xiong J, He Z, Shi S, Kent A, Deng Y, Wu L, Van Nostrand JD, Zhou J (2015) Elevated CO_2 shifts the functional structure and metabolic potentials of soil microbial communities in a C4 agroecosystem. Sci Rep 5:9316

Zhao M, Xue K, Wang F, Liu S, Bai S, Sun B, Zhou J, Yang Y (2014) Microbial mediation of biogeochemical cycles revealed by simulation of global changes with soil transplant and cropping. ISME J 8:2045–2055

Exploring the Diversity of Marine Microbiome in Response to Changes in the Environment

6

Sana Khan and Abdul Malik

Abstract

For any environment, the total microorganisms along with their collective genetic material constitute a "Microbiome." The microorganisms being ubiquitous in nature inhabit the marine environment, i.e., the ocean, the seas, estuaries, bays, and their coasts as free entities or live in association with the marine animals (fish, sponges, corals, sea grass and algae, etc.). In the marine environment, coral reefs represent one of the most diverse ecosystems, having immense biodiversity and thus contributing to the primary productivity in the greater extent. Over last few decades, the researchers have explored much of the marine microbiology using new technologies and tools, but vast majority of the marine microbiome still remained to be uncovered, whose potential would definitely play an important role in future marine science. Up to the recent time, the number of marine microbes has reached to billions microbes per liter of seawater. Owing to their extreme diversity and versatility in terms of their metabolic activities, the marine microbes play an important role for supporting the marine food web. Generally, these microbes derive energy either through photosynthesis or through chemosynthesis. The marine microbiomes are responsible for more than half of the all primary production occurring on Earth. Moreover, these microorganisms contribute significantly to the biogeochemical cycling of the nutrients, oxygen production, and degradation of organic matters. In this chapter, the diversity and ecology of the marine microorganisms (Bacteria, Archaea, and Eukarya) and viruses are discussed along with their response to change in environmental factors like temperature, pressure, nutrient, oxygen and sunlight availability, and alterations in ocean stratification.

S. Khan (✉) · A. Malik
Department of Agricultural Microbiology, Faculty of Agricultural Sciences, Aligarh Muslim University, Aligarh, Uttar Pradesh, India

S. A. Lone, A. Malik (eds.), *Microbiomes and the Global Climate Change*,
https://doi.org/10.1007/978-981-33-4508-9_6

Keywords

Algal bloom · Biogeochemical cycling · Climate change · Coral reef microbiomes · Marine microbiome · *Pelagibacter* · *Prochlorococcus* · Stratification

6.1 Introduction

Seas and oceans are the largest biospheres on earth, covering more than 70% of its surface and hosting majority of its biomass (Whitman et al. 1998). Since all life forms originated from microbes in the ocean, these microbes dominate the living biomass after billions of years of evolution. Microorganisms are the most important entities in marine environments, representing about 90% of the biomass, and are responsible for about 98% of primary marine productivity (Alvarez-Yela et al. 2019). Marine microbes play a fundamental role in maintaining structure and function of marine ecosystem, as these microorganisms are the main source of mass and energy for all other life in the ocean and also sustain all life on earth by generating much of the oxygen in the planet. Marine microorganisms have the potential to mitigate the effects of climate change since they are the major processors of the world's greenhouse gases. Marine microbes play a central role in biogeochemical cycling of carbon, nitrogen, and sulfur and are potential source of biotechnologically important compounds (Tinta et al. 2019). Owing to the remarkable diversity and clear importance of the marine microbes, uncovering oceanic microbial taxa remains a fundamental challenge in microbial ecology (Gajigan et al. 2018). Major limitation in studying the diversity of marine microbial communities includes that vast majority (about 90–99%) of the microorganisms cannot be grown under standard laboratory conditions (Delmont et al. 2011). To overcome this limitation, molecular biology and sequencing technologies (metabarcoding and metagenomics) have been developed (Streit and Schmitz 2004; Pavan-Kumar et al. 2015), that are continuously enriching our knowledge about phylogenetic and functional diversity of the marine microbes (Debnath et al. 2007; Dionisi et al. 2012; Glöckner et al. 2012). With the advent of several new tools and technologies (such as deep sea sampling, metagenomics, next generation sequencing, etc.), there is possible opportunities for unlocking the potential of the marine microbiome in future (de la Calle 2017).

A coral reef is an important ecosystem found beneath the surface of the water and characterized by corals which form reefs. Often called sea rainforests, shallow coral reefs form one of the most diverse and productive ecosystems on Earth. Although they occupy less than 0.1% of the world's ocean's surface, coral reefs provide food, shelter, and habitat to about 25% of known marine species (Spalding and Grenfell 1997; Spalding et al. 2001; Mulhall 2009), including crustaceans, echinoderms, fish, molluscs, sponges, tunicates, worms, and other cnidarians (Hoover 2007). They offer various important economic services like coastal protection, fisheries, and tourism (Bourne et al. 2016; Silveira et al. 2017; Torda et al. 2017) (Fig. 6.1a).

Coral reefs are microbially driven ecosystems where microbes play a fundamental role in maintaining holobiont health and resilience of reef ecosystem under environmental perturbation (Vanwonterghem and Webster 2020).

6.2 Marine Microbiome

The "marine microbiome" is the diverse community of all microorganisms found in oceans and other aquatic environments. Marine microbiome controls the health of life on earth (de la Calle 2017). A drop of seawater may contain up to a million microbes with estimated number of 3.5×10^{30} microbes on the oceanic subsurface. In marine environments, microbes either live as free communities in the water column and benthonic substrates or in symbiotic association with other oceanic macroorganisms like macro-algae, corals, and sponges (Glöckner et al. 2012; Grossart et al. 2013). The symbiotic interactions between the oceanic microbes and host organisms could be harmful or beneficial depending on its microbial taxonomic structure and functionality (Konopka 2009; Kellogg et al. 2014). Any environmental factors or host characteristics (such as age, diseases, and physiological state) change the composition of associated microbial communities that ultimately influences the nutrient uptake, light absorption, and pathogen interactions of the hosts (Webster et al. 2008; Lachnit et al. 2011; Wahl et al. 2012; Zhang et al. 2015; Lawler et al. 2016).

Coral reefs consist a complicated network of free living and host associated microbial communities with strong benthic-pelagic coupling (Lesser 2006; Bourne and Webster 2013). In order to flourish in oligotrophic waters, these communities help them in efficient capture, retention, and recycling of nutrients and trace elements (de Goeij et al. 2013; Cardini et al. 2014).

Corals obtain a small portion of their nutrients through heterotrophic feeding (predation of zooplankton), and mainly depend on their microbiome for the proficient acquisition and recycling of nutrients in the sea waters (Bourne et al. 2016). In addition, the coral-associated microbes also provide essential amino acids and co-factors (B vitamins) to the coral host and algal endosymbionts (Robbins et al. 2019; Matthews et al. 2020). Moreover, the native microbiome also protects the coral hosts from pathogens by colonizing the coral surface, by competing for nutrients, space, and the synthesis of antimicrobial compounds (Peixoto et al. 2017).

6.3 Marine Microorganisms (Bacteria, Archaea, and Eukarya) and Viruses

Microorganisms are the most abundant and diverse residents of marine environment. These microbes are the key players in maintaining marine ecosystems health owing to their integral contribution to biogeochemical cycles and other biological processes

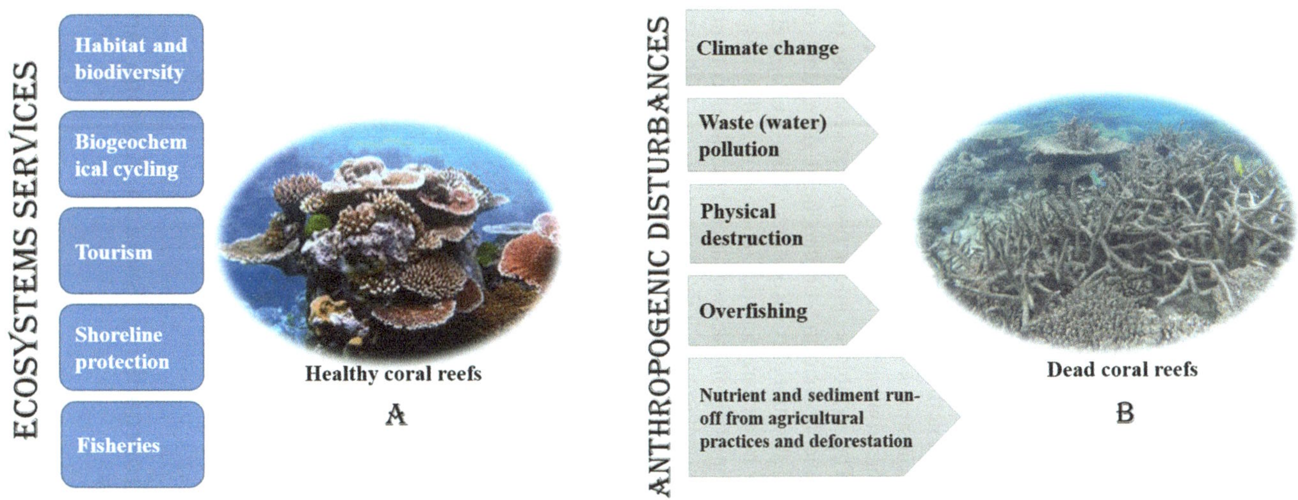

Fig. 6.1 (**a**) Overview of important ecosystem services provided by healthy coral reefs and (**b**) impact of anthropogenic perturbations that lead to decline of coral reefs

(carbon, nitrogen, sulfur cycling, etc.) (Caron 2005; Sogin et al. 2006). Marine microorganisms include archaea, bacteria, fungi, protists, and viruses which collectively comprise millions of cells in each milliliter of the oceanic water (Eakins and Sharman 2010).

Marine microorganisms acquire specific properties owing to extreme marine environmental conditions in which they live such as alkaline or acidic water, high or low temperature, high pressure and limited substrate in the deep sea water (Baharum et al. 2010).

Over 100 species of bacteria can be found in just a single drop of sea water. Their size may vary from the smallest, i.e., one-hundredth of a millimeter to the largest, i.e., three-quarters of a millimeter (found in ocean sediments off the coast of Namibia). Marine bacteria are well adapted to their environmental condition in the ocean. Bacteria living close to the water's surface obtain energy through photosynthesis like cyanobacteria. In contrast, bacteria living at deeper depths where there is no sunlight, acquire unique adaptations to obtain energy through different chemical reactions and this leads to greater bacterial diversity at water depth. Some marine bacteria feed either on other bacteria (like *Bdellovibrio*) or on dying phytoplankton. Among marine bacteria, *Prochlorococcus*, (a cyanobacterium) and *Pelagibacter* are of particular interest. The ocean surface (rich in sunlight) is home to some of the world's biggest photosynthesizers. *Prochlorococcus*, being one of the most abundant photosynthesizer on the planet, is responsible for producing 20% of the O_2 in the atmosphere. *Pelagibacter* accounts for about 25% of all the microbes in the water column. This bacterium plays significant role in cleaning the ocean as it feeds on organic matter, dissolved in the ocean water.

Archaea constitute about 40% of the marine microbes. In addition to living in extremely hot, acidic, or low oxygen environments, archaea are found in both freshwater and saltwater environments.

Viruses are the most abundant entities in the oceanic water, with estimated number of more than 10^{30} viruses in the ocean. Since viruses living in the ocean generally infect specific hosts, the number of viruses fluctuates with change in the host bacterial communities. It is found that in the sunlit portion of the water column, viral infection rates are comparatively higher.

Although protists are one of the least studied microbes in the ocean, they can have significant impacts on ecosystems. Some protists are voracious predators that impose check on the number of bacteria in the ocean. A type of protist, a dinoflagellate, live in symbiotic association with corals.

In the marine environment, although fungi are comparatively hard to find, they play important roles in the marine ecosystem (recycling nutrients). Generally, marine fungi live in association with the decomposition of plant materials or as parasites within marine plants, algae, and animals.

6.4 Importance of Marine Microbes

6.4.1 Biogeochemical Cycling of the Nutrients

Marine microbes are of critical importance for maintaining environmental and human health. They are key players to biogeochemical cycles, fluxes, and processes which occur in marine environment; and thus these microorganisms are very crucial for the functioning of marine ecosystems. Marine phototrophic microbes produce most of the global oxygen that is indispensable for all life on Earth.

6.4.2 Degradation of Organic Matters

Marine microorganisms are responsible for the degradation of organic compounds in the ocean, thus maintain the balance between free and fixed carbon dioxide (CO_2). Due to their extreme abundance and diversity, these microbes produce and release carbon products (CO_2 and CH_4) that regulate the Earth's climate.

6.4.3 Source of Novel Bioactive Compounds

Marine microorganisms are potential source of biotechnologically important enzymes and compounds that are used in industrial and medical sectors. Polymer degrading enzymes and robust enzymes have been isolated from extremophilic marine microbes and are being successfully used in several industries like laundry and food processing industries (Antranikian 2007; Kennedy et al. 2008). Marine bacteria are also a source of biosurfactants and (extracellular) polymeric substances that are used as bio-adhesives, biodegradable plastics, dyes, drag reducing coatings on ship hulls, sunscreens, underwater surface coatings, etc. (Munn 2011). Moreover several bioactive compounds have been isolated from marine microbes and tested for their biomedical potential (anticancer and anti-inflammatory drugs, antibacterial, antifungal and antiviral agents, drug delivery agents, etc.) (Fenical 2006; Newman and Hill 2006; Williams 2009; Choudhary et al. 2017).

6.4.4 Tackling Pollution

Owing to their incredibly diverse genes, marine microorganisms are being studied and explored for their bioremediation activities. It has been found that these microbes have significant potential to break down the environmental pollutants, e.g., oil, Br-containing pesticides and flame retardants. Several marine bacteria have natural tendency to breakdown and remove oil in the ocean. Ocean plastic pollution is another issue of concern. Marine bacteria have been found to grow and feed on plastics (Yoshida et al. 2016). Moreover, marine microbes are being investigated as a source of new sustainable bioplastics (Narancic and O'Connor

2017). Various marine bacteria act as indicators of effluent discharges and find application in heavy metal bioremediation (Ramaiah et al. 2007; Selvin et al. 2009; Dash and Das 2016).

6.4.5 Maintenance of Marine Food Chain and Food Web

Marine microorganisms occupy the critical bottom trophic level in ocean foodwebs through continuously supplying the seafood products. The microscopic remnants of dead organisms and their waste products are fed by marine microbes, which in turn are consumed by tiny creatures called nanoflagellates. Then, the slightly larger creatures called ciliates feed on those nanoflagellates. These ciliates are the food source of copepods (bug-like crustaceans) and other zooplankton, which are further engulfed by small fish.

6.5 Environmental Factors Affecting the Marine Microbiome

Environmental pollution and climate change have significant hazards to the corals, sponges, and other marine organisms, which in turn, may introduce risks to human health and economic growth. Coral reefs are facing unprecedented losses on local and global scales leading to its degradation. Under future climate change conditions, they are expected to be among the most adversely affected ecosystems. Localized reef degradation is caused by declining water quality, disease, overfishing, physical destruction, pollution, and outbreaks of coral predating fish (De'ath et al. 2012; Pawlik et al. 2016; Hoegh-Guldberg et al. 2017) whereas climate change (i.e., elevated temperature, oceanic acidification) is responsible for reef degradation at a global scale (Hoegh-Guldberg et al. 2007).

In 1988, the first mass coral bleaching event was recorded (Cziesielski et al. 2019), and prominent coral reef ecosystems such as the Great Barrier Reef have lost about 50% of shallow-water corals in the past 4 years alone (Hoegh-Guldberg et al. 2019).

Climate change may cause a destabilization of the coral natural microbiome, resulting in a dysbiosis that will in turn cause the emergence of opportunistic and potentially pathogenic microorganisms, resulting in an increased incidence of disease, bleaching and eventually host mortality (Littman et al. 2011; van Oppen and Blackall 2019) (Fig. 6.1b).

6.5.1 Global Climate Change

Due to anthropogenic activities, there is an increase of greenhouse gases in the atmosphere that in turn results in increasing the average global temperature. This increase in global temperature will also cause a rise in oceanic temperature that pose significant effects on the marine microbiome and other oceanic life it supports.

Metabolic activity and growth rate of marine microorganisms are differentially affected by temperature. One of the important consequences of increased water temperature results in the occurrence of harmful "algal bloom," where microorganisms grow rapidly in large quantities, producing toxins and deplete vital nutrients in the water. Thus the algal bloom presents significant hazards to aquatic life, seabirds, and humans.

Due to the increased global temperature, the polar icecaps are melting and massive volumes of freshwater are released in the oceans that results in sea level rise. This sudden release of water will change the salinity of the receiving water bodies, i.e., oceans, seas, etc., which will affect marine microorganisms causing dramatic shifts in the microbial community composition (Gao et al. 2012; Thomas et al. 2012).

Moreover, high temperature of the ocean affects the availability of certain nutrients and lowers the amount of dissolved oxygen in the water. This may affect the growth of some microorganisms particularly, planktons, which are an important food source for other oceanic life (Hutchins and Fu 2017).

Marine environment is also affected by the anthropogenic emissions of CO_2 to the atmosphere that results in ocean acidification. About one-third of the CO_2 emitted into the atmosphere by combustion of fossil fuel enters the ocean surface (Orr et al. 2005). CO_2 dissolution in seawater disturbs the "equilibrium of the seawater carbonate buffer system" thus lowering its pH. However the effects of low pH on the growth and activity of marine microorganisms are not fully understood.

In case of coral reef ecosystems, it has been reported that due to higher ocean temperature and/or acidification, there is a microbial community shift, i.e., from beneficial bacterial taxa (Endozoicomonas) to opportunistic and pathogenic groups (Alteromonadaceae and Vibrionaceae) that will result in elevated incidence of disease (Bourne et al. 2016; O'Brien et al. 2016; McDevitt-Irwin et al. 2017).

Another impact of high CO_2 is alteration of ocean stratification (i.e., a physical barrier/density driven structure in a water column in which colder and/or more saline water underlies warmer and/or less saline water). The intensified "stratification" process leads to lowering of vertical fluxes of critical deep-water nutrients, thus affecting the growth of plankton in the surface ocean (Hutchins and Fu 2017).

6.5.2 Environmental Pollution

Inadequately planned and poorly managed agriculture and aquaculture practices are generating nutrient-abundant wastes being dumped into aquatic environments. It may cause eutrophication, algal bloom events, and oxygen-depleted "dead zones."

The effect of oil pollution on marine microbial community is not fully understood. Some marine bacteria have natural tendency to breakdown the oil in the ocean. Discharge of large amount of oil in the ocean results in growth of their numbers, which in turn, may cause reduction in overall marine microbial biodiversity (Yang et al. 2016).

6.6 Conclusion

Marine microbiome represents the totality of all microbes and viruses in the ocean/seas and in any related environment (the seafloor and marine animals/plants). The diversity of microbial life remains largely unidentified, and may represent a secret treasure for human society. Moreover, in today's ocean, the abundance, distribution, diversity, interactions, and functions of marine microorganisms are either directly or indirectly affected by the climate change and anthropogenic disturbances, resulting in altered nutrient cycling, loss of microbial diversity and biomass, local extinction, and community shifts. Although coral reefs support enormous biodiversity and provide critical ecosystem services, they suffer rapid degradation on a local and global scale. Therefore, a detailed understanding of the mechanisms involved in coral bleaching and resilience is required, coupled with concerted global action to reduce carbon emissions to prevent further decline in coral reefs.

References

Alvarez-Yela AC, Mosquera-Rendón J, Noreña A, Cristancho Ardila MA, López-Alvarez DC (2019) Microbial diversity exploration of marine hosts at Serrana Bank, a coral atoll of the Seaflower biosphere reserve. Front Mar Sci 6:338

Antranikian G (2007) Industrial relevance of thermophiles and their enzymes. In: Robb F, Grogan D (eds) Thermophiles: biology and technology at high temperatures. GRC Press, Boca Raton, FL, pp 113–189

Baharum SN, Beng EK, Mokhtar MAA (2010) Marine microorganisms: potential application and challenges. J Biol Sci 10(6):555–564

Bourne DG, Webster NS (2013) Coral reef bacterial communities. In: Rosenberg E, DeLong EF, Lory S, Stackebrandt E, Thompson F (eds) The prokaryotes. Springer, Berlin, pp 163–187

Bourne DG, Morrow KM, Webster NS (2016) Insights into the coral microbiome: underpinning the health and resilience of reef ecosystems. Annu Rev Microbiol 70:317–340

Cardini U, Bednarz VN, Foster RA, Wild C (2014) Benthic N2 fixation in coral reefs and the potential effects of human-induced environmental change. Ecol Evol 4(9):1706–1727

Caron DA (2005) Marine microbial ecology in a molecular world?: what does the future hold? Scentia Mar 69(Suppl):97–110. https://doi.org/10.3989/scimar.2005.69s197

Choudhary A, Naughton LM, Montánchez I, Dobson ADW, Rai DK (2017) Current status and future prospects of marine natural products (MNPs) as antimicrobials. Mar Drugs 15:272

Cziesielski MJ, Schmidt-Roach S, Aranda M (2019) The past, present, and future of coral heat stress studies. Ecol Evol 9(17):10055–10066

Dash HR, Das S (2016) Diversity, community structure, and bioremediation potential of mercury-resistant marine bacteria of estuarine and coastal environments of Odisha, India. Environ Sci Pollut Res 23:6960–6971

De Goeij JM, Van Oevelen D, Vermeij MJ, Osinga R, Middelburg JJ, De Goeij AF, Admiraal W (2013) Surviving in a marine desert: the sponge loop retains resources within coral reefs. Science 342(6154):108–110

de la Calle F (2017) Marine microbiome as source of natural products. Microb Biotechnol 10 (6):1293

De'ath G, Fabricius KE, Sweatman H, Puotinen M (2012) The 27–year decline of coral cover on the great barrier reef and its causes. Proc Natl Acad Sci 109(44):17995–17999

Debnath M, Paul AK, Bisen PS (2007) Natural bioactive compounds and biotechnological potential of marine bacteria. Curr Pharm Biotechnol 8:253–260. https://doi.org/10.2174/138920107782109976

Delmont TO, Robe P, Clark I, Simonet P, Vogel TM (2011) Metagenomic comparison of direct and indirect soil DNA extraction approaches. J Microbiol Methods 86(3):397–400

Dionisi H, Lozada M, Oliviera NL (2012) Bioprospection of marine microorganisms: biotechnological applications and methods. Rev Argent Microbiol 44:49–60. https://doi.org/10.1590/S0325-75412012000100010

Eakins BW, Sharman GF (2010) Volumes of the World's oceans from ETOPO1. National Geophysical Data Center, Boulder, CO

Fenical W (2006) Marine pharmaceuticals: past, present and future. Oceanography 19:112–119

Gajigan AP, Yñiguez AT, Villanoy CL, Jacinto GS, Conaco C (2018) Diversity and community structure of marine microbes around the benham rise underwater plateau, northeastern Philippines. PeerJ 6:e4781

Gao K, Helbling EW, Häder DP, Hutchins DA (2012) Responses of marine primary producers to interactions between ocean acidification, solar radiation, and warming. Mar Ecol Prog Ser 470:167–189

Glöckner FO, Stal L, Sandaa RA, Gasol JM, O'Gara F, Hernandez F et al (2012) Marine microbial diversity and its role in ecosystem functioning and environmental change. In: McDonough N, Calewaert J-B (eds) Marine board position Paper 17. European Scientific Foundation, Belgium, p 84. https://doi.org/10.13140/RG.2.1.5138.6400

Grossart HP, Riemann L, Tang KW (2013) Molecular and functional ecology of aquatic microbial symbionts. Front Microbiol 4:59. https://doi.org/10.3389/fmicb.2013.00059

Hoegh-Guldberg O, Mumby PJ, Hooten AJ, Steneck RS, Greenfield P, Gomez E, Harvell CD, Sale PF, Edwards AJ, Caldeira K, Knowlton N (2007) Coral reefs under rapid climate change and ocean acidification. Science 318(5857):1737–1742

Hoegh-Guldberg O, Poloczanska ES, Skirving W, Dove S (2017) Coral reef ecosystems under climate change and ocean acidification. Front Mar Sci 4:158

Hoegh-Guldberg O, Jacob D, Taylor M, Bolaños TG, Bindi M, Brown S, Camilloni IA, Diedhiou A, Djalante R, Ebi K, Engelbrecht F (2019) The human imperative of stabilizing global climate change at 1.5 C. Science 365(6459):1–13

Hoover J (2007) Hawaii's sea creatures. Mutual Publishing, Honolulu, HI

Hutchins DA, Fu F (2017) Microorganisms and ocean global change. Nat Microbiol 2(6):17058

Kellogg CA, Piceno YM, Tom LM, DeSantis TZ, Gray MA, Andersen GL (2014) Comparing bacterial community composition of healthy and dark spot-affected Siderastrea siderea in Florida and the Caribbean. PLoS One 9:e108767. https://doi.org/10.1371/journal.pone.0108767

Kennedy J, Marchesi JR, Dobson ADW (2008) Marine metagenomics: strategies for discovery of novel enzymes with biotechnological applications from marine environments. Microb Cell Fact 7:27

Konopka A (2009) What is microbial community ecology? ISME J 3:1223–1230. https://doi.org/10.1038/ismej.2009.88

Lachnit T, Meske D, Wahl M, Harder T, Schmitz R (2011) Epibacterial community patterns on marine macroalgae are host-specific but temporally variable. Environ Microbiol 13:655–665. https://doi.org/10.1111/j.1462-2920.2010.02371.x

Lawler SN, Kellogg CA, France SC, Clostio RW, Brooke D, Ross SW (2016) Coral-associated bacterial diversity is conserved across two deep-sea anthothela species. Front Microbiol 7:458. https://doi.org/10.3389/fmicb.2016.00458

Lesser MP (2006) Benthic–pelagic coupling on coral reefs: feeding and growth of Caribbean sponges. J Exp Mar Biol Ecol 328(2):277–288

Littman R, Willis BL, Bourne DG (2011) Metagenomic analysis of the coral holobiont during a natural bleaching event on the great barrier reef. Environ Microbiol Rep 3(6):651–660

Matthews JL, Raina JB, Kahlke T, Seymour JR, Van Oppen MJ, Suggett DJ (2020) Symbiodiniaceae-bacteria interactions: rethinking metabolite exchange in reef-building corals as multi-partner metabolic networks. Environ Microbiol 22(5):1675–1687

McDevitt-Irwin JM, Baum JK, Garren M, Vega Thurber RL (2017) Responses of coral-associated bacterial communities to local and global stressors. Front Mar Sci 4:262

Mulhall M (2009) Saving rainforests of the sea: an analysis of international efforts to conserve coral reefs. Duke Environ Law Policy Forum 19:321–351

Munn CB (2011) Marine microbiology: ecology and applications. CRC Press, Abingdon, UK

Narancic T, O'Connor KE (2017) Microbial biotechnology addressing the plastic waste disaster. Microb Biotechnol 10(5):1232–1235

Newman DJ, Hill RT (2006) New drugs from marine microbes: the tide is running. J Ind Microbiol Biotechnol 33:539–544

O'Brien PA, Morrow KM, Willis BL, Bourne DG (2016) Implications of ocean acidification for marine microorganisms from the free-living to the host-associated. Front Mar Sci 3:47

Orr JC, Fabry VJ, Aumont O, Bopp L, Doney SC, Feely RA, Gnanadesikan A, Gruber N, Ishida A, Joos F, Key RM (2005) Anthropogenic Ocean acidification over the twenty-first century and its impact on calcifying organisms. Nature 437(7059):681–686

Pavan-Kumar A, Gireesh-Babu P, Lakra WS (2015) DNA metabarcoding: a new approach for rapid biodiversity assessment. J Cell Sci Mol Biol 2:111

Pawlik JR, Burkepile DE, Thurber RV (2016) A vicious circle? altered carbon and nutrient cycling may explain the low resilience of Caribbean coral reefs. Bioscience 66(6):470–476

Peixoto RS, Rosado PM, Leite DCDA, Rosado AS, Bourne DG (2017) Beneficial microorganisms for corals (BMC): proposed mechanisms for coral health and resilience. Front Microbiol 8:341

Ramaiah NV, Rodrigues V, Alwares E, Rodrigues C, Baksh R, Jayan S, Mohandass C (2007) Sewage-pollution indicator bacteria. In: Shetye SR, Dileep Kumar M, Shankar D (eds) The Mandovi and Zuari estuaries. National Institute of Oceanography, Goa, pp 115–120

Robbins SJ, Singleton CM, Chan CX, Messer LF, Geers AU, Ying H, Baker A, Bell SC, Morrow KM, Ragan MA, Miller DJ (2019) A genomic view of the reef-building coral Porites lutea and its microbial symbionts. Nat Microbiol 4(12):2090–2100

Selvin J, Priya SS, Kiran GS, Thangavelu T, Bai NS (2009) Sponge-associated marine bacteria as indicators of heavy metal pollution. Microbiol Res 164:352–363

Silveira CB, Cavalcanti GS, Walter JM, Silva-Lima AW, Dinsdale EA, Bourne DG, Thompson CC, Thompson FL (2017) Microbial processes driving coral reef organic carbon flow. FEMS Microbiol Rev 41(4):575–595

Sogin ML, Morrison HG, Huber JA, Welch DM, Huse SM, Neal PR et al (2006) Microbial diversity in the deep sea and the underexplored "rare biosphere". Proc Natl Acad Sci U S A 103:12115–12120. https://doi.org/10.1073/pnas.0605127103

Spalding MD, Grenfell AM (1997) New estimates of global and regional coral reef areas. Coral Reefs 16(4):225–230

Spalding MD, Ravilious C, Green EP (2001) World atlas of coral reefs. University of California Press, Berkeley, CA

Streit WR, Schmitz RA (2004) Metagenomics - the key to the uncultured microbes. Curr Opin Microbiol 7:492–498. https://doi.org/10.1016/j.mib.2004.08.002

Thomas MK, Kremer CT, Klausmeier CA, Litchman E (2012) A global pattern of thermal adaptation in marine phytoplankton. Science 338(6110):1085–1088

Tinta T, Kogovšek T, Klun K, Malej A, Herndl GJ, Turk V (2019) Jellyfish-associated microbiome in the marine environment: exploring its biotechnological potential. Mar Drugs 17(2):94

Torda G, Donelson JM, Aranda M, Barshis DJ, Bay L, Berumen ML, Bourne DG, Cantin N, Foret S, Matz M, Miller DJ (2017) Rapid adaptive responses to climate change in corals. Nat Clim Chang 7(9):627–636

van Oppen MJ, Blackall LL (2019) Coral microbiome dynamics, functions and design in a changing world. Nat Rev Microbiol 17(9):557–567

Vanwonterghem I, Webster NS (2020) Coral reef microorganisms in a changing climate. Science 23(4):100972

Wahl M, Goecke F, Labes A, Dobretsov S, Weinberger F (2012) The second skin: ecological role of epibiotic biofilms on marine organisms. Front Microbiol 3:292. https://doi.org/10.3389/fmicb.2012.00292

Webster NS, Xavier JR, Freckelton M, Motti CA, Cobb R (2008) Shifts in microbial and chemical patterns within the marine sponge Aplysina aerophoba during a disease outbreak. Environ Microbiol 10:3366–3376. https://doi.org/10.1111/j.1462-2920.2008.01734.x

Williams PG (2009) Panning for chemical gold: marine bacteria as a source of new therapeutics. Trends Biotechnol 27:45–52

Yang S et al (2016) Hydrocarbon degraders establish at the costs of microbial richness, abundance and keystone taxa after crude oil contamination in permafrost environments. Sci Rep 6:37473

Yoshida S et al (2016) A bacterium that degrades and assimilates poly(ethylene terephthalate). Science 351(6278):1196–1199

Zhang YY, Ling J, Yang QS, Wang YS, Sun CC, Sun HY et al (2015) The diversity of coral associated bacteria and the environmental factors affect their community variation. Ecotoxicology 24:1467–1477. https://doi.org/10.1007/s10646-015-1454-1454

Polar Microbes as Climate-Resilient Pathways for Mitigation of Climate Change

7

Shah Ishfaq, Baba Uqab, Jeelani Gousia, Ruqeya Nazir, and B. A. Ganai

Abstract

Changing climate challenges economic and sustainable growth. IPCC in all its reports has highlighted that climate change has grown its roots and needs immediate attention. For natural environments, documentation of observable impacts from climate change is strongest and much more extensive. Several effects have also been related to climate change on human environments, with a significant or small effect to climate change distinguishable from other factors. Changes in climate have effects throughout all continents and around the world in both ecological and human processes. Soil microbes shape the foundation of the Earth ecological model by regulating biogeochemical processing of essential resources, such as carbon (C) and nitrogen (N). Through part of these biogeochemical processes, bacteria generate and absorb heat-trapping pollutants, such as carbon dioxide (CO_2), methane (CH_4), and nitrous oxide (N_2O). Regardless of the large carbon deposits sequestered in soils of Arctic, the carbon cycles in these regions do need to be inferred if the effects of global warming are to be correctly measured. Microbial cells inside the permafrost are frozen, hence preserved. While permafrost comprises just about 9% of total land mass on earth, it is assumed that it produces about 25–50% of the soil organic carbon on earth. As it is discharged atmospherically in the form of methane or carbon dioxide, polar carbon has been described as a possible symptom and engine of global warming. Global warming and other human impacts reveal formally frozen ecosystems to increase temperatures which are likely to lead higher performance of accumulated organic material and nutrients. Only small amounts of warming can have significant effects on the structure and function of microbial community in polar soils,

S. Ishfaq · B. Uqab (✉)
Department of Environmental Science, University of Kashmir, Srinagar, India

J. Gousia · R. Nazir · B. A. Ganai
Centre of Research for Development, University of Kashmir, Srinagar, India

S. A. Lone, A. Malik (eds.), *Microbiomes and the Global Climate Change*,
https://doi.org/10.1007/978-981-33-4508-9_7

which eventually shifts the competitive balance between taxa and can lead to processes that promote the release of carbon deposits as greenhouse gases. Within this book chapter we will concentrate on the factors within polar soils known to affect microbial abundance and community structure. Further it will also focus on physiology and functioning of the polar microbes and their potential in mitigating the climate change.

Keywords

Biogeochemical cycles · Climate change · Permafrost · Polar microbiota · Soil microbes

7.1 Introduction

Anthropogenic activities are estimated to be responsible for causing 1 °C increase in global temperature above pre-industrial levels and are likely to further reach 1.5 °C between 2030 and 2050 that will impact the sea water levels and changes in climate (IPCC 2018). Ultimately these will affect the carbon budgets in ecosystems. Soil microbes play an important role in determining these changes. Global warming increases the heterotrophic microbial activity which in turn increases the CO_2 efflux to the atmosphere and affects the leaching of dissolved organic carbon (Davidson and Janssens 2006). Studies in subalpine forests of Colorado have shown that reduction in snow cover suppresses the soil respiration due to soil–microbe interactions that are sensitive to changes in temperature (Monson et al. 2006). Soil microbes have an indirect role in climate change by influencing the photosynthesis and respiration. In addition, the changes in precipitation patterns have also affected the vegetation distribution on local as well as global scale. For example, changes in vegetation distribution have been reported in tropical rainforests (Engelbrecht et al. 2007), Canadian tundra (which has been replaced by boreal forests) (Danby and Hik 2007), and African savannas (Sankaran et al. 2005). Such changes lead to alteration in carbon feedback system and ultimately affect the microbes and their metabolism. Moreover, nitrogen deposition, species invasion and changes in land use affect the microbial diversity directly or indirectly. N enrichment affects the extracellular enzymes that help in decomposition of organic matter. Although there are several challenges in understanding the impact of climate change on microbes and vice versa, but progress can be achieved through long term multifactor experiments in the specific biomes by collaborating the microbial ecology, carbon feedbacks and whole ecosystem flux measures.

7.2 Polar Regions and Climate Change

Polar regions are experiencing a shift in their climate that has led to the potential changes in their environment. Research in past few decades has enabled to understand the polar environments and its change in depth. The research has led to a number of discoveries like halogen chemistry playing a role in reducing lower tropospherical ozone and mercury (Simpson et al. 2007; Steffen et al. 2008), snowpack lit by sun being highly photo reactive (Grannas et al. 2007) and biology playing a vital role in these processes. In order to study these polar processes, sea ice plays a pivotal role. Sea ice shows interaction with solar radiation, provides habitat for animals, mammals and microbes, and is a key component in elemental cycles in polar regions. It also enhances the chemical processes by providing saline conditions for self-cleansing of environment. Sea ice limits production as well as consumption of biogenic gases (Loose et al. 2011) that largely depend on the temperature fluctuations, i.e. the transport of gases through ice pores is low in cold and gradually increases in springtime which is warm at poles. Temperature is a determining factor for the processes in polar environments. However, due to the rapid shift in global anthropogenic activities, there has been an increase in the concentration of greenhouse gases (GHGs) that resulted in increase in the global temperature in general and poles in particular. This shift in anthropogenic activities has led to a number of changes in polar environments.

7.3 Recent Environmental Changes in Polar Regions

7.3.1 Changes in Atmospheric Circulation

Both the polar regions have recorded a shift in their atmospheric circulation patterns. For example, in the Arctic, the Aleutian Low-one of the characteristic feature of atmospheric circulation in northern hemisphere- shifted to more poleward in winters due to increase in GHGs and aerosols of sulphur (Fyfe 2006). In Antarctic, the Southern Annular Mode (SAM) has shifted to positive phase due to cooling of stratosphere as result of ozone hole during the springtime which otherwise is not the case. This shift in SAM has resulted in colder temperatures and increase in coastal sea ice. However this cooling has resulted in overall increase in surface temperatures.

7.3.2 Changes in Temperature and Ocean Circulation

Due to the changes in global atmosphere and increasing anthropogenic impact on global environment, the temperature at poles has seen deviation from the normal patterns. For instance, in the first decade of twenty-first century the colder regions have seen an increase which has resulted in loss of sea ice. The central arctic has experienced 5 °C increase in temperature (Turner and Overland 2009). Similarly the west Antarctic has seen 0.1 °C increase in temperature on per decade trend and at certain places like Vernadsky which is warming at 5 °C during the winter. The increase in the air temperatures has resulted in increase in river discharge into the ocean especially in Arctic Ocean thus refreshing the waters. In addition the Arctic waters are sensitive to warm Atlantic currents resulting in warming of Arctic Ocean.

Similarly the Antarctic Circumpolar current has warmed rapidly due to various processes that govern the global climate and ocean circulation.

7.3.3 Changes in Sea Ice and Ice Sheets

It is difficult to predict any change in ice sheet cover, however, sea ice in Arctic has decreased while in Antarctic it has increased. There has been an increase in Green land Ice sheet accumulation because most of the ice sheet lies at higher elevations (Box et al. 2006). In Antarctic region, due to negligible climatic changes in recent decades, there has not been any significant loss in ice sheets but at certain regions ice sheets have shown some loss (Turner and Overland 2009).

7.3.4 Changes in Microbial Interactions

Polar environments are characterised by high productivity and hence act as carbon sink thus play an important role in climate system of the world (Arrigo et al. 2008; Bates and Mathis 2009). Despite harsh conditions, polar environment is rich in microbial biota that governs nutrient cycles (Cavicchioli 2015; Pedros-Alio et al. 2015). Microbial biota and their interaction deeply depend on the changes in the physical processes. For example, the rise in temperature in ocean water enhances microbial growth that is coupled with phytoplanktons in marine waters. Also changes in water flux greatly affect the microbial world in polar regions.

7.4 Microbial Diversity in Polar Regions

The polar regions being far apart geographically make it an opportunity for study of factors shaping the marine microbial studies. The microbial life forms at poles are adapted to extreme conditions and are referred to as extremophiles or psychrophils. These include marine planktons, diatoms, fungi, Cyanobacteria, bacteria, etc. Both the polar regions despite having similar external physical forces are dissimilar in various aspects like freshwater supplies, land mass cover, etc. However, few microbial communities common to both regions are the bacteria like *Shewanella frigidimarina* (Staley and Gosink 1999) and *Polaribacter irgensii* (Zeng et al. 2010); Foraminiferan *Neogloboquadrina pachyderma* (Darling et al. 2007) and the haptophyte *Phaeocystis* (Verity et al. 2007). In addition some planktonic *Archaea* were found similar between the two poles. Some of the microscopic organisms are discussed as follows:

7.4.1 Diatoms

Diatoms are single cell algae, the cell wall which consists of transparent opaline silica and are found in marine waters, freshwaters and soils. Diatoms have also been studied in extreme environments like that of poles. These inhabit the waters and sea ice in the polar regions and are responsible for most of the primary productivity. Study of diatoms in polar environments dates back to 1873 during the *HMS challenger* expedition from 1873 to 1876 (Castracane 1886) and voyage by *S. Y. Belgica* during 1897–1899 (van Heurck 1909). The dominant species of diatoms found in polar oceans include *Fragilariopsis kerguelensis*, *Thalassiosira Antarctica*, *Corethron criophilum*, *T. Pseudonana*, *Actinocyclus karstenii*, etc. (Reinsch 1890). Diatoms are the primary component of the food chain in polar regions and act as food to fish, whales, penguins, etc. During spring season, diatoms produce vast blooms to such extent that can be remotely seen from the satellite imaginary.

7.4.2 Snow Algae

Microalgae species belong to genera *Chlamydomonas*, *Chloromonas*, *Raphidonema* and *Ankistrodesmus*. Actively grow on snow ice when sunlight penetrates the snow during spring and summer. These microalgal groups have the tendency to colour the snow pink, green, yellow or red. Microalgae-*Chlamydomonas nivalis* has been found to show tolerance to UV radiation by accumulating astaxanthin and producing flavonoids (Duval et al. 2000; Hoham and Ling 2000).

7.4.3 Cyanobacteria

Cyanobacteria or blue-green algae are the most dominant photosynthetic prokaryotic microbes present in polar environments. However, their presence in marine waters of poles is rare, and they are present in all freshwaters of Antarctica (Seckbach and Oren 2007). Along with the diatoms, Cyanobacteria form major prokaryotic phototrophic microorganisms in polar regions. These two also have the ability to biodegrade several compounds, for example, cyanobacterium *Phormidium foveolarum* along with the diatom *Achnanthes minutissima* can degrade naphthenic hydrocarbons (Antić et al. 2006). Due to the bioremediation capabilities, these possess an important role in cleansing the polar environments.

7.4.4 Other Microbes

Similar to the temperate and tropical microbial diversity, polar regions have also shown a high diversity of *Crenarchaeota*, *Gamma-proteobacteria*, *Alphaproteobacteria* and *Flavobacteria* (Kirchman et al. 2010; Wilkins et al. 2013). Other bacterial groups include Epsilon-*proteobacteria* (Murray and

Grzymski 2007), *Actinobacteria* (Jamieson et al. 2012), endolithic microbes including Cyanobacteria, algae and fungi (Omelon 2008). The microbial diversity varies throughout the latitude, i.e. from equator to poles, however certain groups are adapted to a wide variety of stresses and can be found throughout the latitudinal gradient.

7.5 Adaptations in Low Temperatures

Microorganisms thrive in cold environments due to certain adaptations to their habitat (Collins and Margesin 2019). These include a wide range of adaptations in cellular components in terms of structure and function (Siddiqui et al. 2013).

7.5.1 Adaptations in Cell Envelope and Cell Membrane

Cell envelope serves as an important component having structural and regulatory role in a cell. Its function includes providing shape, support, regulate movement of substances, nutrient uptake, etc. The low temperature leads to reduced permeability, diffusion rates and mobility of embedded proteins, etc. (Collins and Margesin 2019). Similarly the cell membrane has modifications in fatty acid composition of lipid bilayer (Siddiqui et al. 2013), i.e. production of branched chain fatty acids (He et al. 2015). Further, there has been a change in genomic content responsible for degradation of membrane rigidifying molecule that may result in reduction in membrane rigidity at lower temperatures.

7.5.2 Adaptations in Membrane Pigments

Pigments, mainly carotenoids play role in membrane fluidity. At lower temperature, it has been reported that these pigments tend to increase (Pandey et al. 2018). This increase in turn increases the cell rigidity, plays role in photoprotection and acts as antioxidants (Pandey et al. 2018).

7.5.3 Adaptations in Cell Wall

Studies related to cold adaptations in cell wall are limited. However, few studies have revealed an upregulation in genes responsible for biosynthesis of peptidoglycans and outer membrane components and thick peptidoglycan layer in cold adapted microbes (Mykytczuk et al. 2013). In the outer membrane layer of cell wall, the lipopolysaccharide structure is of shortened length at low temperatures (Corsaro et al. 2017) with increased fatty acid content and upregulated outer membrane proteins (DeMaayer et al. 2014).

7.5.4 Antifreeze Proteins

Antifreeze proteins bind to ice and inhibit its growth and recrystallisation (Voets 2017). These AFPs lead to curvatures of the ice face that are unfavourable for the ice growth which in turn leads to thermal hysteresis (Voets 2017). In addition the AFPs help in providing integrity to cell structure and play role in enhancement in access to nutrients and oxygen.

7.5.5 Compatible Solutes

These are the low molecular mass, non-toxic organic osmolytes like trehalose, glycerol, sarcosine, etc. These solutes restore the osmotic balance and prevent water loss and cell shrinkage due to freezing (Goordial et al. 2016). These also help in neutralising free radicals and prevent protein aggregation leading to enhanced protein folding and stabilisation of proteins and membranes at low temperatures.

7.5.6 Other Adaptations

Other adaptations in polar microbes involve the production of biosurfactants, enzymes, chaperones, etc. Biosurfactants act as osmolytes and show ice recrystallisation inhibitor (IRI) role. Enzymes act as catalysts for operation of various activities at lower temperature. Chaperones help in protein folding and prevent protein aggregation and stabilise the RNA and DNA. They also maintain translation and transcription processes as well as play role in DNA replication. Under sub-zero condition all these components have been studied for their upregulation or enhanced production. In addition to these, various microbes have been studied for reduced metabolic activity at lower temperature by synthesis of cold shock proteins, chaperones, etc.

7.6 Role of Microbes in Mitigation of Climate Change

Role of polar microbes in climate change mitigation has not been extensively studied. The main causing factor for climate change is greenhouse gases (GHGs) like carbon dioxide, methane, etc. So far as carbon dioxide is concerned, microorganisms contribute to its exchange through the process of decomposition. The microbes sequester carbon in soil as well as in marine ecosystems. Various proteobacteria, acidobacteria and actinobacteria act as copiotrophic organisms (Singh et al. 2010). Proteobacteria genera like *Methylococcus, Methylocaldum, Methylobacter* are also studied for role in methanotrophy. These microbes use methane as a sole energy source and are capable of growth in extreme environments (Dunfield 2009).

7.7 Conclusion

For natural systems, proof of observable impacts from climate change is best and most detailed. Some effects have also been related to climate change on human environments, with a significant or minor effect of climate change distinguishable from other causes. Changes in climate have shown effects across all continents and around the oceans on both natural and human systems. Soil microbes form the backbone of the Earth's ecological processes by regulating biogeochemical cycling of essential elements such as carbon (C) and nitrogen (N). Global warming increases the heterotrophic microbial activity which in turn increases the CO_2 efflux to the atmosphere and affects the leaching of dissolved organic carbon due to the rapid shift in global anthropogenic activities, there has been an increase in the concentration of greenhouse gases (GHGs) that resulted in increase in the global temperature in general and poles in particular. This shift in temperature has led to a number of changes in polar environments. Microorganisms make a major difference to pollution sequestration, this is evident as more and more CO_2 is taken by microbes than terrestrial herbs. For that function, environmental shifts impact the microbial photosynthesis and resulting conservation of set data carbon in every setting is of great significance to the world. Addressing unique problems concerning microorganisms targeted laboratory studies of model microorganisms will be needed. Mesocosm and field studies in situ are especially relevant to obtain insight into community-level reactions to specific environmental conditions. Efficient testing approach includes responsible decision-making, including awareness from various fields related to the maritime field (e.g. practical oceanography) and terrestrial (e.g. geochemistry) biomes.

References

Antić MP, Jovancicevic B, Vrvić MM, Schwarzbauer J (2006) Petroleum pollutant degradation by surface water microorganisms. Environ Sci Pollution Res 13:287–360

Arrigo KR, Van Dijken G, Long M (2008) Coastal southern ocean: a strong anthropogenic CO_2 sink. Geophys Res Lett 35:L21602. https://doi.org/10.1029/2008GL035624

Bates N, Mathis J (2009) The Arctic Ocean marine carbon cycle: evaluation of air-sea CO_2 exchanges, ocean acidification impacts and potential feedbacks. Biogeosciences 6:2433–2459. https://doi.org/10.5194/bg-6-2433-2009

Box JE, Bromwich DH, Veenhuis BA, Bai LS, Stroeve JC, Rogers JC, Steffen K, Haran T, Wang SH (2006) Greenland ice sheet surface mass balance variability (1988–2004) from calibrated polar MM5 output. J Clim 19:2783–2800

Castracane AFA (1886) Report on the Diatomaceae collected by the H. M. S. challenger during the years 1873-76. In reports of the scientific voyage of H. M. S. Challenger Botany 2:1–178

Cavicchioli R (2015) Microbial ecology of Antarctic aquatic systems. Nat Rev Microbiol 13:691–706. https://doi.org/10.1038/nrmicro3549

Collins T, Margesin R (2019) Psychrophilic lifestyles: mechanisms of adaptation and biotechnological tools. Appl Microbiol Biotechnol 103(7):2857–2871

Corsaro MM, Casillo A, Parrilli E, Tutino ML (2017) Molecular structure of lipopolysaccharides of cold-adapted bacteria. In: Margesin R (ed) Psychrophiles: from biodiversity to biotechnology, 2nd edn. Springer, Cham, pp 285–303

Danby RK, Hik DS (2007) Variability, contingency and rapid change in recent subarctic alpine tree line dynamics. J Ecol 95:352–363
Darling KF, Kucera M, Wade CM (2007) Global molecular phylogeography reveals persistent Arctic circumpolar isolation in a marine planktonic protist. Proc Natl Acad Sci U S A 104:5002–5007
Davidson EA, Janssens IA (2006) Temperature sensitivity of soil carbon decomposition and feedbacks to climate change. Nature 440:165–173
DeMaayer P, Anderson D, Cary C, Cowan DA (2014) Some like it cold: understanding the survival strategies of psychrophiles. EMBO Rep 15(5):508–517. https://doi.org/10.1002/embr.201338170
Dunfield PF (2009) Methanotrophy in extreme environments. In: Encyclopedia of life sciences (ELS). Wiley, Chichester. https://doi.org/10.1002/9780470015902.a0021897
Duval B, Shetty K, Thomas WH (2000) Phenolic compounds antioxidant properties in the snow alga Chlamydomonas nivalis after exposure to UV light. J Appl Phycol 11:559–566
Engelbrecht BM, Comita LS, Condit R, Kursar TA, TyreeMT TBL et al (2007) Drought sensitivity shapes species distribution patterns in tropical forests. Nature 447:80–82
Fyfe JC (2006) Southern Ocean warming due to human influence. Geophys Res Lett 33:L19701. https://doi.org/10.1029/2006GL027247
Goordial J, Davila A, Lacelle D, Pollard W, Marinova MM, Greer CW et al (2016) Nearing the cold-arid limits of microbial life in permafrost of an upper dry valley, Antarctica. ISME J 10 (7):1613–1624
Grannas AM et al (2007) An overview of snow photochemistry: evidence, mechanisms and impacts. Atmos Chem Phys 7:4329–4373. https://doi.org/10.5194/acp-7-4329-2007
He J, Yang Z, Hu B, Ji X, Wei Y, Lin L, Zhang Q (2015) Correlation of polyunsaturated fatty acids with the cold adaptation of Rhodotorula glutinis. Yeast 32(11):683–690. https://doi.org/10.1002/yea.3095
Hoham RW, Ling HU (2000) Snow algae. The effects of chemical and physical factors on their life cycles and populations. In: Seckbach J (ed) Journey to diverse microbial worlds. Kluwer Academic Publishers, Dordrecht, pp 131–145
IPCC (2018) Summary for policymakers. In: Allen M, Babiker M, Chen Y, de Coninck H, Connors S, et al (eds) Global warming of 1.5 °C: an IPCC special report on the impacts of global warming of 1.5 °C above pre-industrial levels and related global greenhouse gas emission pathways, in the context of strengthening the global response to the threat of climate change, sustainable development, and efforts to eradicate poverty. World Meteorological Organization, Geneva, pp. 1–32
Jamieson RE, Rogers AD, Billett DS, Smale DA, Pearce DA (2012) Patterns of marine bacterioplankton biodiversity in the surface waters of the Scotia Arc, Southern Ocean. FEMS Microbiol Ecol 80(2):452–468
Kirchman DL, Cottrell MT, Lovejoy C (2010) The structure of bacterial communities in the western Arctic Ocean as revealed by pyrosequencing of 16S rRNA genes. Environ Microbiol 12 (5):1132–1143
Loose B, Miller L, Elliot S, Papakyriakou T (2011) Sea ice biogeochemistry and material transport across the frozen interface. Oceanography 24:202–218. https://doi.org/10.5670/oceanog.2011.72
Monson RK, Lipson DL, Burns SP, Turnipseed AA, Delany AC, Williams MW et al (2006) Winter soil respiration controlled by climate and microbial community composition. Nature 439:711–714
Murray AE, Grzymski JJ (2007) Diversity and genomics of Antarctic marine micro-organisms. Philos Trans R Soc B: Biol Sci 362(1488):2259–2271
Mykytczuk NC, Foote SJ, Omelon CR, Southam G, Greer CW, Whyte LG (2013) Bacterial growth at− 15 C; molecular insights from the permafrost bacterium Planococcus halocryophilus Or1. ISME J 7(6):1211–1226

Omelon CR (2008) Endolithic microbial communities in polar desert habitats. Geomicrobiol J 25 (7–8):404–414. https://doi.org/10.1080/01490450802403057
Pandey N, Jain R, Pandey A, Tamta S (2018) Optimisation and characterisation of the orange pigment produced by a cold adapted strain of Penicillium sp. (GBPI_P155) isolated from mountain ecosystem. Mycology 9(2):81–92. https://doi.org/10.1080/21501203.2017.1423127
Pedros-Alio C, Potvin M, Lovejoy C (2015) Diversity of planktonic microorganisms in the Arctic Ocean. Prog Oceanogr 139:233–243. https://doi.org/10.1016/j.pocean.2015.07.009
Reinsch PF (1890) Die Süsswasseralgenflora von Süd-Georgien. In: Neumeyer G (ed) Die Deutschen Expeditionen und ihre Ergebnisse, vol 1882-01883. Biodiversity Heritage Library, pp 329–365
Sankaran M, Hanan NP, Scholes RJ, Ratnam J, Augustine DJ, Cade BS et al (2005) Determinants of woody cover in African savannas. Nature 438:846–849
Seckbach J, Oren A (2007) Oxygenic photosynthetic microorganisms in extreme environments: possibilities and limitations. In: Seckbach J (ed) Algae and cyanobacteria in extreme environments. Springer, Cham, pp 5–25
Siddiqui KS, Williams TJ, Wilkins D, Yau S, Allen MA, Brown MV et al (2013) Psychrophiles. Annu Rev Earth Planet Sci 41:87–115
Simpson WR et al (2007) Halogens and their role in polar boundary-layer ozone depletion. Atmos Chem Phys 7:4375–4418. https://doi.org/10.5194/acp-7-4375-2007
Singh BK, Bardgett RD, Smith P, Reay DS (2010) Microorganisms and climate change: terrestrial feedbacks and mitigation options. Nat Rev Microbiol 8:780–790
Staley JT, Gosink JJ (1999) Poles apart: biodiversity and biogeography of sea ice bacteria. Annu Rev Microbiol 53:189–215
Steffen A et al (2008) A synthesis of atmospheric mercury depletion event chemistry in the atmosphere and snow. Atmos Chem Phys 8:1445–1482. https://doi.org/10.5194/acp-8-1445-2008
Turner J, Overland J (2009) Contrasting climate change in the two polar regions. Polar Res 28 (2):146–164. https://doi.org/10.1111/j.1751-8369.2009.00128.x
Van Heurck H (1909) Diatomées. In: Expédition Antarctic Belge, Résultats du Voyage du S. Y. Belgica en 1897–1899. Botanique, vol 6. Antwerp, Buschmamn, pp 1–126
Verity PG et al (2007) Current understanding of Phaeocystis ecology and biogeochemistry, and perspectives for future research. Biogeochemistry 83:311–330
Voets IK (2017) From ice-binding proteins to bio-inspired antifreeze materials. Soft Matter 13 (28):4808–4823. https://doi.org/10.1039/c6sm02867e
Wilkins D, Yau S, Williams TJ, Allen MA, Brown MV, DeMaere MZ et al (2013) Key microbial drivers in Antarctic aquatic environments. FEMS Microbiol Rev 37(3):303–335
Zeng Y, Zheng T, Yu Y, Chen B, He J (2010) Relationships between Arctic and Antarctic Shewanella strains evaluated by a poly- phasic taxonomic approach. Polar Biol 33:531–541

Part II

Climate Change and Pathogens

Climate Change and Population Health

8

Mariya Amin Qurieshi and Mehbooba Rasool

Abstract

Human health and climate have a deep rooted co-existence since the inception of mankind on mother earth. An understanding between the two has been established since times immemorial with the great work of scientists and scholars. Human health and disease have made great strides in the present day world, while more emphasis on causal relationships of various diseases with various factors both internal as well as external is being studied. No conclusive data or little evidence suggesting climate change as responsible in the changing patterns of disease is available. But extreme changes in temperatures like cold and heat waves coupled with the increased or decreased precipitation levels causing famines and floods and the presence of aeroallergens with increased air pollution levels cause a brunt on human health on an individual level. Vector borne diseases are directly affected with the change in temperature and precipitation levels. Water related diseases arising due to poor quality, quantity of water with minimal personal hygiene share a complex association. Loss of human lives, health infrastructure, and public properties by the calamities like storms, cyclones, hurricanes which we witnessed in Asia, America in recent times are a forecast for the future world to tackle this situation now or never. On the other hand, vulnerable population of low to middle income countries suffer the greatest brunt of climate change related health burden. Thus mitigation of health related impact of climate change should be tailor made keeping in view the regional health threats. Effective strategies need to be devised at all levels of healthcare delivery system.

M. A. Qurieshi (✉) · M. Rasool
Department of Community Medicine, Government Medical College, Srinagar, Jammu and Kashmir, India

S. A. Lone, A. Malik (eds.), *Microbiomes and the Global Climate Change*,
https://doi.org/10.1007/978-981-33-4508-9_8

Keywords

Climate change · Carbon dioxide · Direct and indirect effects · Population health · Mitigation

8.1 Introduction

Change is the essence of life and is inevitable. Change is what made life possible on this mother Earth. Change can be seen both in positive as well as negative directions. Climate change has evolved from times immemorial. It brought so many boons and banes in every aspect of life, be it the extinction of some species or origin of others. It led mankind for betterment from ice age towards civilization. Health of the population too has been affected with this change in climate. Climate change broadly affects the environmental and social dimensions of health and well-being of people in particular and populations in general. However, in last few decades the changes have been quite remarkable owing to changing population size and extent of economic activities, which has started creating imbalance in the ecosystem per se.

It is very important to understand what has led to climate change. The surface temperature of earth is a result of combination of natural processes that take place inside and around the earth's atmosphere which include solar activity, volcanic activity, dust and aerosol in the atmosphere, reflection of some energy back by clouds, ice and snow. Far more important than all these is greenhouse gases (GHGs) which are produced as a result of human activity (anthropogenic greenhouse gases). The greenhouse gases trap solar energy and create a blanket for earth which prevents heat loss from earth's surface. This action of greenhouse gases can be exemplified by comparing it with the effect observed inside a car parked in sun with window panes up. Among greenhouse gases, carbon dioxide (CO_2) is the most abundant gas. In nature, the concentration of carbon dioxide is maintained by going through atmosphere to oceans and from oceans it gets impounded in rocks which ultimately reach back to atmosphere after being thrown out by volcanic eruption. What is so astonishing is the fact that over the last half of the century, there has been remarkable escalation of atmospheric CO_2 as a result of human activities which include overuse of energy sources and deforestation. The concentration of CO_2 in the atmosphere has risen gradually from about 280 parts per million (ppm) in pre-industrial times to 400 ppm in the year 2014. In addition to carbon dioxide, methane and nitrous oxide are two other shorter-lived but more potent greenhouse whose concentration has also increased acutely in the last 50 years (Woodward and Macmillan 2015).

Though the reported rise in global temperature has been 0.6 °C predominantly from 1970 onwards, the Intergovernmental panel on Climate change (IPCC) have forecasted that the atmospheric carbon dioxide will double its pre-industrial level between the years 2050 and 2100, which will shift the global temperature upwards by 2–2.5 °C by the end of this century (Soloman et al. 2007).

Environmental and Social Scientists see climate change detrimental to mother Earth, so does health care providers. Hippocrates, known as father of Medicine made valuable contribution in our understanding about role of environmental determinants on health of population. The Aphorisms of Hippocrates is direct evidence to it, which he has penned down in the introductory portion of treatise on "Airs, Waters, Places" and this is considered as the first ground work knowledge on the environmental health determinants. It begins as saying "*Whoever would study medicine aright must learn of the following subjects. Firstly, he must consider the effect of each season of the year and the differences between them. Secondly, he must study the warm and the cold winds, both those that are common to every country and those peculiar to a particular locality. Lastly, the effect of water on the health must not be forgotten. When, therefore, a physician comes to a district previously unknown to him, he should consider both its situation and its aspect to the winds. Similarly, the nature of the water supply must be considered. Then think of the soil, whether it be bare and waterless or thickly covered with vegetation and well watered, whether in a hollow and stifling, or exposed and cold. Lastly consider the life of the inhabitants themselves, are they heavy drinkers and eaters and consequently unable to stand fatigue or, being fond of work and exercise, eat wisely but drink sparely*"(Lloyd 1978). It is important to appreciate the observations made more than 2000 years ago about various components of environment including climate, and their effect on human health.

If we go back in history, human population has been witness to the extreme changes in temperature which led to flooding in 1980 in Sudan with consequent diarrheal disease outbreak. Similarly, in 1998 floods in one of the states of India led to massive Cholera Epidemic due to 01Eltor, Ogawa. Outbreak of Leptospirosis in Brazil (1983, 1988 and 1996), India (1999), and Thailand (2000) was also attributed to climate change (WHO 2002). In Burma (2008), Cyclone Nargis caused approximately 138,000 deaths and US $10 billion damage; in the USA, Hurricane Katrina in 2005 was responsible for nearly thousand deaths (Fritz et al. 2009) (Brunkard et al. 2008). Other example like occurrence of acute illness that happened after eating sea-foods contaminated with *Vibrio parahaemolyticus* in North America which was attributed to rising temperature of the farm beyond a critical level that was considered sensitive for the survival of organism. Being a temperature sensitive organism, rising temperature favored its survival in local shell-fish which was consumed by the local populace (McLaughlin et al. 2005). Researchers have pointed out direct connection of climate change to the human deaths. Be it the wild fires in the USA which ravaged into larger swaths of its country land or the 2014 floods in Kashmir, India which marooned the Valley for more than a month. In 2003 European Heat wave resulted in nearly 70,000 deaths in 12 countries and climate change increased the heat related deaths by 70% in Paris and 20% in London. Hot days are common and summers tend to be longer in some regions where there was a reverse trend.

In order to understand the population health and impact of any event or risk factor on health one should have methods to measure it. Traditionally health and disease have been measured in terms of mortality rates. However, the major drawback of this indicator is that it ignores the parameters that compromise the function and quality of life without causing end of life event like accidents causing disability or mental ill health. Hence a new metric, Disability adjusted life years (DALYs) which are a composite index of quantifying years of life lost due to premature death and years lived with disability were introduced (WHO 2017).

Global Climate change is being incriminated for about 1,50,000 deaths and 5.5 million DALYs lost per year with highest burden of climate change per million populations in the regions of Sub-Saharan Africa despite the fact that their contribution towards global climate change has been very minimal. Recent estimates given by World Health Organization reveal that globally climate change will cause 2,50,000 additional deaths per year between 2030 and 2050 and direct financial loss to health is estimated to be USD 2 to 4 billion per year by 2030 (Haines and Berlin 2014; WHO 2018).

The health impacts of climate change differ with population because of variation in resources which determine their vulnerability. It has been observed that overcrowding, food insecurity, local environmental degradation, conflicts, socio economic and ecological conditions play a crucial role in determining the impact of climate change. Hence the climate change does not only act as a risk multiplier but itself contributes to the risk for ecosystem (Woodward and Macmillan 2015).

Climate Change and population health are indispensible to each other. Climate Change can have direct and indirect effects on population health which depends upon various demographic groups having an altogether change in their health patterns than others (Table 8.1). Today we can see climate change as a public health emergency and can be predicted for future generations too. Direct changes are an outcome of extreme weather and indirect changes are secondary to changes in the range of vector population. Events like increased frequency, duration and intensity of heat waves, draughts, floods, storms, wild fires which mankind is witness to today have led to death, injury, disease and displacement of thousands of people. In this chapter, impact of climate change on population health will be considered as a consequence of following alterations (McMichael et al. 1996): (Table 8.1)

(a) Direct Changes

- Exposure to thermal extremes
- Intensity of other extreme weather events

(b) Indirect Changes

- Effect on vector activity
- Altered ecology of waterborne and food-borne agents
- Alteration in food productivity
- Population displacement

(c) Direct and Indirect Effects of Increased Ultraviolet Levels (McMichael et al. 1996)

Table 8.1 Direct and indirect pathways of climate change and their influence on human health

I. Temperature and weather changes	Consequence of climate change	Health outcome
	Direct pathway	
	Exposure to thermal exposure	Heat and cold related illness Death
	Alteration in other weather events	Death Injuries Psychological disorders Damage to public health infrastructure
	Indirect pathway	
	Effect on range and activity of vectors	↓↑incidence of vector borne disease
	Alteration in infective agents	↓↑incidence of diarrheal disease
	Alteration in crop production	Malnutrition Impairment of child growth and development
	Rise in sea level→ Population displacement Damage to infrastructure	↑ Infectious disease risk ↑Psychological disorder
	Biological impact of air pollution Pollens and spores	Respiratory disorders Deaths Asthma and allergic disorders
	Effect on economy, infrastructure and resource supply→sociodemographic dislocation	↑ Mental ill health ↑ Nutritional impairment ↑ Infectious disease Civil strife
	Direct pathway	
II. Stratospheric ozone depletion		Skin cancer Cataract Immune suppression → ↑susceptibility to infections
	Indirect pathway	
	Impaired productivity of agriculture → ↓food production → hunger	Malnutrition Impaired growth and development
	Impaired productivity of aquatic systems → ↓ population of predators of insect vectors	Vector borne infectious disease

8.2 Consequences of Climate change

8.2.1 Exposure to Thermal Extremes and Other Weather Events

Health impact of thermal extremes encompasses effect of both hot and cold environmental temperature which includes heat stress, heat exhaustion, and heat stroke. The most vulnerable population groups like children, elderly, women, and people who move out to earn livelihood and work in external environmental conditions for at least 6–8 h are at greater risk of facing heat related morbidity and mortality. Excessive daily heat exposures decrease work efficiency and interfere with daily activities. The impact of extremes of cold shows marked seasonal variation with 10–25% excess cold related death rate in winter season. As compared to heat related mortality which are mostly attributed to direct impact of heat in the form of heat exhaustion and heat stroke, cold related deaths are mostly due to exacerbation of underlying morbidity like cardiovascular disease or chronic obstructive pulmonary disease (COPD). The increased mortality has been attributed to cold induced tendency for blood to clot (Keatinge et al. 1989).

Increased precipitation levels have led to increasing entry of human and animals into waterways and drinking sources, thus there will be an increase in the frequency and magnitude of common disasters like flood, storm, and drought. These increase the chances of injuries as well as disease outbreak (Infectious), mental trauma, and reduction of work productivity (McMichael et al. 1996).

8.2.2 Biological Impact of Air Pollution, Pollens, and Spores

It has been hypothesized that due to increased atmospheric temperature the production of secondary air pollutants is accelerated. On inhalations of such chemicals the respiratory cells are acutely affected thereby impairing lung function (Romieu et al. 1990). Seasonal allergic disorders like hay fever and asthma are also impacted due to alteration in the production of plant aeroallergens which are considered to be sensitive to climate change (Emberlin 1994).

8.2.3 Effect Due to Change in Range and Activity of Vectors

The indirect effects of change in global climatic conditions are widespread and diverse. Human health effects are interactive and complex with the changing climatic conditions of a region which is primarily linked to alteration in various biological organisms and the processes linked with their transmission to human host. Temperature, humidity, precipitation and wind, influence biting rate, egg production, vector distribution, and passive dispersal of flying insects and consequent survival benefit to the affected vector species. Dispersion of insect vectors could possibly pave way for creation of new species which might pave way to new hosts previously immune to the parent vector. Population displacement as will be observed during floods and droughts would expose human population to vectors which were otherwise maintaining their lifecycle in wildlife and did not cause any significant disease in humans. At the same time one needs to consider existing vector population dynamics before making predictions about the impact of climate change on distribution of vector borne disease (McMichael et al. 1996).

Malaria affects about 50% of the world's population which is projected to increase to 60% and with more frequent epidemics (Martin and Lefebure 1995). In 2017 local transmission of malaria and Chikungunya was reported in Italy, an area declared malaria free in 1970 by WHO (CDC 2017a). Exact cause was uncertain but it showed a vulnerability to the changes presented by climate change. Scenario based modeling studies have predicted increase in prevalence of malaria in response to rise to global mean temperature in many parts of the world by the year 2100. The climate change will have potential effect in the transmission of malarial parasite favoring its spread by both latitude and longitude in tropical areas. Areas which previously exhibited stable transmission of *Plasmodium falciparum* in particular will become more unstable with high mortality and morbidity among the newly exposed population (CDC 2017b).

Dengue, like malaria is transmitted by vector mosquito (*Aedes aegypti and A. albopictus*) and exhibits similar dynamics with temperature being the key predictor for transmission of infection. Diseases like tick borne encephalitis and Lyme disease also have been found to be influenced by climate change though the exact mechanism of how climate change will affect pattern of transmission of these infection is uncertain. However, it has been said that increase in temperature to a certain level and precipitation are more hospitable for fleas and ticks also. On the one hand, high temperature is suitable for multiplication of tick population and on the other side, it causes reduction in predators of deer mice (McMichael et al. 1996).

Arboviral diseases, most of which are mosquito borne are predicted to be changed throughout the globe due to the effect of changing climate. Resurgence of one of the worst outbreak of West Nile fever in 2012 in America has been attributed to change in climatic conditions. Similarly, Ross River virus disease, a major Australian arboviral infection is greatly influenced by alteration in rainfall pattern (Young et al. 2014).

Onchocerciasis, transmitted by blackflies, occurring mostly in Africa is also a climate sensitive vector. Simulation studies have predicted that blackfly population could increase by 25% which can increase the possibility of risk of onchocerciasis in previously naïve population as these vectors move to newer areas creating new habitats. Climate change can make animal reservoirs and hosts to migrate from one area to another thus bringing them closer to each other and increasing exposure to populations with lower immune status (Hales et al. 1999) (McMichael et al. 1996).

Change in climatic conditions would also affect the life cycle of Schistosoma which enter human host by penetrating through the skin. It does so through two mechanisms. Firstly, in response to hot climate there will be likely spreading out of irrigation system with subsequent increase in host snail population, secondly, high temperature would arbitrate transmission of infective stage throughout the year which would otherwise lose their infectivity in winter seasons.

Leptospirosis, a spirochaetal disease transmitted through rodents could also increase due to expected increase in rodent population in warmer temperate areas. Similarly, rabies virus infection which is transmitted through bite of blood sucking bats in America is anticipated to be increased as climate changes and brings more favorable temperature for this biting mammal.

Population of pests like flies and cockroaches which transmit diseases by carrying infectious agent on their body or proboscis may increase because of higher temperature making difficult to control these pests and the diseases transmitted by them (McMichael et al. 1996).

8.2.4 Effect of Alteration in Infective Agents

Increase in land temperature will impact the pathogenic proliferation and also affect the rate of transmission of various infectious agents which are primarily transmitted through direct route (food, water). With expected scarcity of clean and safe water, the incidence of diarrheal disease caused by *Salmonella*, *Shigella, Campylobacter, Rotavirus,* and protozoans (*Giardia lamblia*, *Entamoeba histolytica*) are likely to increase. Diarrheal diseases exhibit seasonal trends intensifying during summer season and is compounded by poverty, population displacement, overcrowding, poor sanitation, and hygiene. These conditions will tend to exacerbate the occurrence of diarrheal diseases caused by *Vibrio cholera* in particular which is a climate sensitive infectious disease agent. Disease models used to study transmission of Cholera have shown that alteration in temperature and precipitation with increase in sea levels is expected to increase frequency of cholera infection in future. Similarly, *Salmonella* causing food poisoning proliferate rapidly at higher temperature and change in eating pattern (consuming foods frequently contaminated with *Salmonella*) due to change in outdoor temperatures might also affect exposure (D'Souza et al. 2004).

Looking at the current scenario of some infectious diseases like recent shift in Malaria in Sub-Saharan Africa, tick borne encephalitis in Sweden and Schistosomiasis in Eastern China due to change in regional temperature and alteration in seasonal and inter-annual pattern of certain enteric pathogens like Cholera and Salmonella is a stronger indicator reflecting impact of climate change on the dynamics of infectious disease (Woodward and Macmillan 2015).

8.2.5 Effect of Alteration in Crop Production

Climatic change threatens mankind with food shortage as crop yields are going to be directly affected by droughts and famines thus leading to food shortage and poverty. However, populations which are already food deficit or are directly related to the food production process, even a slight reduction in the yield predispose them to malnutrition (under nutrition). Crop yields are in addition to temperature sensitive to change in soil moisture and activity of pollinating insects. Though scientists have foreseen initial increased production of certain crop variety in response to increased carbon dioxide levels and soil moisture, however, with altered precipitation flooding and soil erosion will be precipitated which will have strong negative influence. The world is already suffering from under nutrition with regional disparities, be it the vast deserts of Africa or Central America where scores of children are facing different forms of undernourishment in the form of wasting, stunting, and underweight. In India, empowered action group states which are classified on the basis of poor health indicators (Madhya Pradesh, Orissa, Bihar, Uttar Pradesh, West Bengal, and Rajasthan) are badly hit by under nutrition. The under nutrition leads to impaired childhood development which occurs in first 10 crucial years of life thus effecting all stages of an individual's development. It has been estimated that people who would be at extra risk of hunger due to alteration in climate by 2060 would be 5–35% more than is expected in the absence of climate change (Woodward and Macmillan 2015).

8.2.6 Effect of Extreme Weather Events

Extreme weather events like floods, droughts, forest fires, and landslides have increased in frequency over past few decades with tropical Asia and tropical America considered being high risk zones. Impact of extreme weather condition on population health is greatly influenced by social, economic, and political factors (Fig. 8.1). For example, impoverished populations who have limited accessibility to health services are at higher risk of facing health inequality when faced with consequences of extreme weather events like floods. Extreme weather events are likely to induce population migration on large scale as is observed following floods

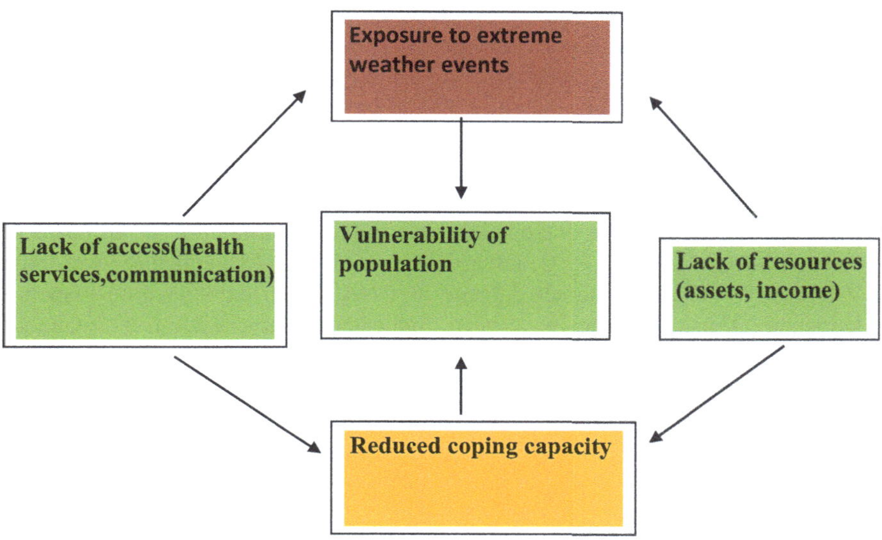

Fig. 8.1 Schematic illustration depicting factors determining vulnerability of population

which would affect the risk of infectious disease outbreaks. Heat waves coupled with air pollution lead to higher burden of respiratory diseases as well as non-communicable diseases. Burning of fossil fuels with higher temperatures exacerbates chronic lung conditions like asthma and allergic disorders. The clinical condition of patients with heart disease, diabetes deteriorates if exposed to excessive heat levels. The mental well-being of a person depends directly upon the economic status. With large spells of drought conditions farmers out of frustration attempt suicides, this is a growing concern in various states of India where farmer deaths are occurring day in and day out (McMichael et al. 1996; Woodward and Macmillan 2015).

8.2.7 Effect of Stratospheric Ozone Depletion

Climate change and stratospheric ozone depletion are closely related entities. With the accumulation of GHGs, there is enhanced effect of radioactive elements on climate and destruction of ozone by chlorine radicals increases ultraviolet radiation entry into lower atmosphere. In nature, stratospheric ozone is responsible for absorption of enormous amount of incoming solar radiation which is a mixture of visible, infrared, and ultraviolet rays. Ultraviolet rays are further subdivided into ultraviolet A, ultraviolet B, and ultraviolet C rays. Stratospheric ozone absorbs all of UV-C radiations, nearly three forth UV-B but only a small amount of UV-A radiation. In addition to ozone, some amount of ultraviolet rays is absorbed in the troposphere layer of the earth's atmosphere by suspended particulate matter, dust, and clouds. Recent modeling estimates show concordance with the observations made in certain countries regarding increased ground levels of UVR in relation to ozone depletion. The percentage increase in UVR reaching the earth varies exponentially with the percentage decline in stratospheric ozone depletion. The increase in ground level of ultraviolet radiation has a varying impact on health of human beings which has both positive and negative consequences. For instance, UV-B is needed by human body to synthesize vitamin D which is an essential fat soluble vitamin and has a role in maintaining calcium level in the human body. However, sustained exposure to UV radiations has been found to be harmful for all forms of life on earth (Quaite et al. 1992) (Blaustein et al. 1994). At molecular level, both UV-B and UV-C induce direct damage to DNA which exhibits dose–response relationship. These radiations have been implicated in the causation of cutaneous malignant melanoma and non-melanocytic skin cancer (Fig. 8.2). Non-melanotic skin cancers are of two types: basal cell carcinoma and squamous cell carcinoma. Incidence of skin cancer is increasing rapidly among fair-skinned populations and who spend most of the time in outdoor activities. The incidence of these cancers has doubled over the past 25 years in Canada. It has been estimated that at least 80% of all melanomas are caused by exposure to sunlight, particularly in highly exposed populations with fair skin, like Australians of Caucasian origin. Ultraviolet radiations have been classified as Group 2A carcinogen by International agency

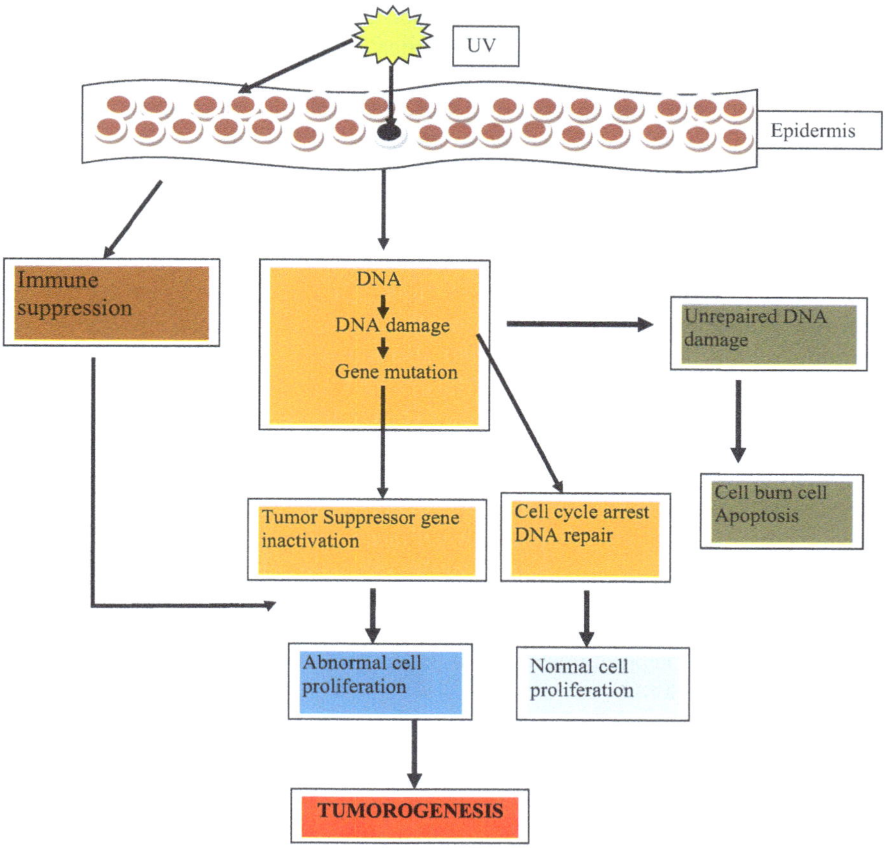

Fig. 8.2 Pathway depicting mechanism of non-melanoma skin cancer by ultraviolet radiation. (Source: IARC Monographs on the evaluation of carcinogenic risks to humans 1992)

for research on Cancers (IARC) which is for being a probable carcinogen to humans (IARC 1992).

In addition to this effect, sufficient amount of evidence exists on the role of ultraviolet radiations in the suppression of cell mediated immunity and ocular effects (photoconjunctivitis and acute photokeratitis) (McMichael et al. 1996).

8.3 Future Trends

Mathematical modeling, an important epidemiological tool is very commonly used framework not only to understand the complex biological or population-based process but also to develop a more simple and easily understandable sequence of events and future projection and consequent impact which helps to define processes clearly and identify the most important components of a system. It is used to explain the real world problem in terms of mathematical equations. It is employed in many areas of public health research and practice. For example, this approach has been widely used to understand infectious disease epidemiology and is now being used by epidemiologists to elucidate the complex phenomenon of climate change as well as to depict future trend and consequent impact of climate change of ecosystem (Woodward and Macmillan 2015).

In addition to current estimates, Global Burden of Disease project 2002 included estimates of deaths and disability adjusted life years attributable to climate change in 2030 (Fig. 8.3). In the case of diarrheal disease, it was assumed that the relation between temperature and incidence would not change, and that vulnerability to warming-related diarrheal disease would also remain as it is currently. Based on outputs from the MARA (Mapping Malaria Risk in Africa) project it has been projected that within 50 years, malaria could be established in the densely populated and presently malaria-free highlands of Zimbabwe. Similarly, in China, it has been estimated that additional 8% of land area will become suitable for snail that harbor the parasite responsible for Schistosomiasis (Zhou and Yang 2008). Additionally, a study modeled the current relation between mean annual temperature and hospital admissions for nephrolithiasis (stones in the urinary tract) in the USA. Based on future projections on mean annual temperatures the potential increase in hospital admissions and associated costs was calculated. Nearly two million additional lifetime cases of nephrolithiasis have been projected to attribute to climate (Brikowski et al. 2008).

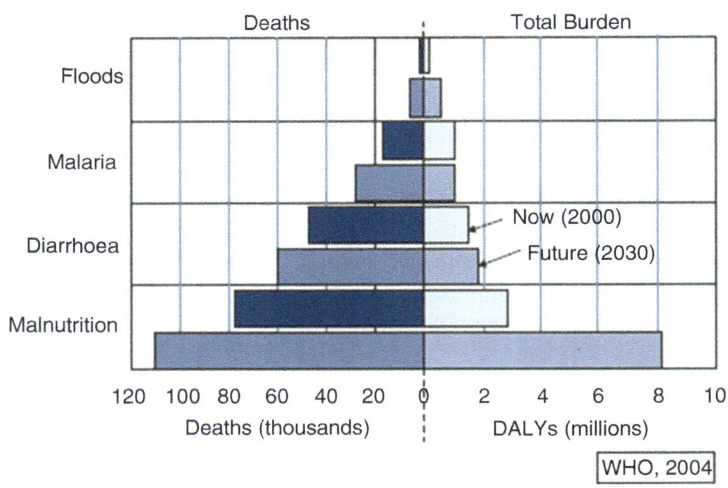

Fig. 8.3 Deaths and DALYs attributable to Climate Change, 2000 and 2030 Selected conditions in developing countries. Source: data from World Health Organization, The Global Burden of Disease 2004 Update, World Health Organization, Geneva, Switzerland, Copyright© (World Health Organization 2008)

With expected increase in world population from about 7 billion presently to around 9.5 billion by 2050, and projected decline in mortality which is more pronounced in high income group countries it has been estimated that the overall burden of disease will be reduced by about 30% by 2030 (WHO 2008). With continued socioeconomic development, the vulnerability of populations to climate change is expected to continue due to increased greenhouse emissions. With country like India which is going through enormous social and economic transition has not made much progress in health indicators (Subramanyam et al. 2011).

Although projection estimates created by mathematical modeling is far from complete, however it provides a basis for likely impact of climate change thereby enabling nations to prepare and respond in time.

8.4 Strategies to Minimize the Health Risk

As described in the earlier section, climate change acts as a risk multiplier with populations having poor health indicators being the most sufferers. This was also reflected by World Bank, in its 2010 World Development Report considering baseline health status as the single most important determinant of both future losses, and the cost of adaptation (World Bank 2009). Hence it becomes imperative to take immediate action to combat ill effects of global climate change to ensure protection and promotion of health of the people with focused approach towards vulnerable population. As we know climate change has more adverse effects than benefits to population health, but the health benefits are overshadowed by far impacting adverse outcomes as being observed presently and may get worse in future. Climate change mitigation strategies should be devised to reduce its ill impact on population health for better future as well as present. With life expectancy to grow further in low income countries, lively environment is needed for future generation to combat the harmful effects of global climate change. To achieve this, the existing health systems need to be strengthened for concrete coping capacity. As observed climate change will act primarily by intensifying existing loopholes in the socio-ecological environment of people, hence the existing health sectors have to develop adaptive strategy and provide strong foundation to safeguard health and reduce health impact of climate change. The strategies need to extend across different allied sectors and should include the following components:

1. Health risk communication to population with special focuses on susceptible ones.
2. Establishment of event specific early warning signals.
3. Better urban planning, climate-proof housing and improved farming practices.
4. Disaster preparedness to cope up with increase in acute demands on health system.
5. Expanded infectious disease control programmes (development of candidate vaccines and stock piling, integrated vector management, case identification and treatment).

6. Better disease and risk factor surveillance network.
7. Development of appropriate Manpower (recruitment, training for capacity building).

Risk communication involves the process of identifying the key message chosen by experts in the field which is relevant and beneficial to the public. Effective risk communication is quite essential in managing risk and enables people to make right decision at right time. This will prevent damage and injuries due to impending risk which is climate change in this context and encourage people to safeguard their life. Risk communication plays pivotal role in risk management.

Surveillance plays key role in protection and promotion of environmental health. It involves systemic collection, analysis, and interpretation of data specific to the health event in a population and application of that information in the prevention and control of that factor. The ultimate aim of surveillance (disease, risk factor) is protection and promotion of health of the population. Setting up of strong risk factor and disease surveillance network will help in monitoring the trends in illness due to specific exposure (for example, exposure to extreme heat and heat stroke), and identifying opportunities to prevent and control effects of climate change.

Another key area for mitigation strategies is to identify and protect the susceptible. Susceptible population poses special challenge for any intervention. It is of utmost importance to identify the susceptible in the population because their needs are different due to enhanced vulnerability to any health risk. This population includes children, pregnant women, people who live in poverty, people with chronic diseases and those who work in high risk areas (extremes of temperature). Children by virtue of their poor judgment cannot avoid exposure to environmental pollutants in and around the home and need adult supervision. Further children in poor resource settings are further exposed to poor drinking water, overcrowding, lack of sanitation, burning of fossil fuels within the households which has an additive to their risk (Woodward and Macmillan 2015).

In addition to adaption strategy it is equally pertinent to initiate mitigation strategy to reduce the emission of greenhouse gases with utmost priority. Limiting global warming to 2 °C by the end of the century has been the consensus target for policymakers and scientists across the globe. Since energy sector emissions contribute to more than one-third greenhouse gas emissions currently, low carbon source in the form of nuclear and other renewable energy needs to become main source of energy supply by the end of year 2050. This will not only reduce impact of greenhouse gases and air pollution but will have an additional benefit of ensuring energy security.

At household level use of clean fossil fuel, biogas, or ethanol for cooking purpose and space heating will reduce emission of black carbon emission which is a short lived climate pollutant thereby reducing health risks.

Transport sector, despite being the fastest growing sector in terms of energy demand a substantial scope exists to reduce the energy demand by 40% and consequent CO_2 emissions by 15–40% by 2050. Transport mitigation measures include modal shift of motorized traffic to dedicated bus rapid transit (BRT), complemented by dedicated walking and cycling infrastructure. Compact urban planning and neighborhoods with safe walking/cycling routes to reduce trip length and carbon intensity of journeys and improved vehicle and engine efficiencies along with uptake of low carbon fuel sources will achieve population health benefits.

Mitigation measures to reduce building emissions which has been projected to increase by 50–150% by year 2050 should include climate-adapted building design to reduce the need for heating in temperate countries and air conditioning in warm countries. Secondly, selective use of day lighting and natural ventilation can be adapted to reduce cooling requirements. Low carbon heating and cooling systems, and energy efficient building appliances should be used to improve thermal comfort.

With increasing burden of non-communicable diseases like diabetes mellitus, cardiovascular diseases, cancers, and chronic respiratory diseases, the health systems capacity to respond to staggering demand for complex health care services that increasingly rely on diagnostic and treatment options that are highly technology dependent, is posing great challenge. Hence it is important to devise mitigation measures at the health system level in order to reduce the carbon-footprint without compromising on patient care. Even though there is sparse evidence available on greenhouse gas emissions by the health system. However, some estimates from certain health sectors of UK and USA suggest that 3–8% of GHG emissions comes from the health sector of these countries. As health sector has a central role in protecting and promoting population health, it has to play a key role and lead by example in implementing mitigation measures in order to avoid contributing to the devastating health impacts of climate change. This can be achieved by adopting clean and more reliable energy sources and climate resilient infrastructure. Building of Climate smart health facilities especially in high income settings can save 30–50% of energy in the long run. This can be achieved through modification of parameters of physical environment like use of mixed mode and natural alone ventilations, use of passive cooling and heating through landscaping and other building modifications procurement of efficient and sustainable health care commodities like cleaning agents and other medical supplies and pharmaceuticals will not only reduce greenhouse gas emissions but will also reduce occupational hazards.

Now, it is for all countries to devise their national policies wherein such measures are integrated. It needs lot of groundwork as the countries have to go for cost benefit analysis as well as health impact assessment of mitigation measures in their own settings for rational policymaking. The successful deployment of mitigation policies needs strong political commitment.

Another key area of concern is the need to ensure that policy measures also influence behavioral changes at the individual level which is conductive to more sustainable living which include shifting from over consumptions of foods which form risk factors for major non-communicable diseases like diabetes, cardiovascular diseases, obesity, and certain cancers like colorectal and breast cancer (WHO 2015).

8.5 What is Being Done at Country and International Level?

The United Nations Framework Convention on Climate Change is a central forum for intergovernmental negotiations on climate change. Under the Convention the member states have committed to "employ appropriate methods, for example impact assessment . . . with a view to minimizing adverse effects on the economy, on public health and on the quality of the environment, to projects or measures undertaken by them to mitigate or adapt to climate change" (WHO 2015).

At United Nations Climate Change summit (UNCC) 2018, the country heads expressed their pledge to unite the world to agree on the next steps towards achieving climate goals. A key part of the process was to limit global warming to 1.5 °C which requires concerted efforts by all sections of the society. Multiple efforts are being taken at international forum to combat global effects of climate change. The countries exchange their ideas and share their experiences and steps taken at national and regional levels and express their commitment towards mother earth which is not only home to the current generation but the future generations to come. The Katowice climate package is recent in this path wherein more than 20,000 individuals participated which included local people, heads of state, media representatives, and members of various organizations including United Nations who are working towards combating climate change. They provided operational framework for climate action and guidance for countries to track greenhouse gas emissions and evaluating efforts at the national and international level. The framework sets year 2050 for net zero emission of GHGs. The countries are encouraged to continue their efforts towards combating climate change as the past emissions would continue to change the climate which will have adverse effect on underprivileged and vulnerable communities of the world. To address loss and damage due to climate change, the countries are encouraged to report on efforts to enhance actions that overcome such effects and this will be assessed every 5 years at the time of global stocktake of progress. In addition, every year, regional climate week takes place at Nairobi, Singapore, and Montevideo to inspire countries to take greater action towards reducing their greenhouse gas emissions.

Through Clean Development mechanism, under Kyoto Protocol, in which 140 countries which include 36 least developed countries, contributed towards reduction or avoidance of 2 billion tonnes of CO_2 through sales of credits known as certified emission reductions (CERs) (which represent one tonne of carbon dioxide (CO_2) equivalent). This was achieved by planting 152 million trees, use of natural resources in 25% of projects, installation of one million efficient cook stoves. In order to achieve long-term low greenhouse gas emission, development strategies have been devised globally with support from various organizations and local governments.

Involvement of communities, through various platforms where they can share their views, ideas and efforts towards response to the challenges posed by climate change, has opened further space for exchange of ideas and engagement of people from all sections of the communities (United Nations 2018).

Human race has always adapted socially, culturally to the ecosystem in which they lived. Migration has been one of the primary steps in case where climate change made impossible for man to live in; but keeping in view today's world scenario increasing urban settlements for economic or social reasons one will find hard to migrate. The strategies thus devised should be made keeping in view the regional, cultural, social determinants. Carbon dioxide and other greenhouse gas emission is a prime concern where global economies fight with each other on that account. The Paris agreement in 2016 set the targets for greenhouse gas emissions. To solve this situation, they have to take a call and find long-lasting solutions for its mitigation otherwise the world will not be an environment of sustenance in next 100 years for man to live and endure.

Reaching out to those people who are in absolute poverty and depend upon fossil fuels as a primary source of household energy will have to be given a better option in the form of renewable sources of energy like solar energy.

A disaster management policy with their implementation is the need of the day especially in low income countries where disasters take a heavy toll on the already scarce resources. Buildings should be designed in such a manner that produces a healthy indoor environment for living as well as working in extreme temperatures and yet reduces the energy demand. As vector borne diseases are influenced by climate change, preventive programmes focusing on development of effective vaccines, mosquito control, food hygiene, sanitation and nutritional supplementation to vulnerable population need to be effectively devised. We can responsibly take action with sensible steps towards safeguarding health of the population. Energy efficient public transport and use of bicycles and walking for short daily trips by the masses will lessen the use of energy. Better land use planning for building purposes can reduce flooding, restriction of building permissions in flood prone areas with relocating areas which experience frequent flooding.

Beyond this, policy decision at national and international level is immensely critical which will in a longer time horizon help in averting climate change. The healthcare providers have opportunity as well as responsibility to make this happen at individual level which will ultimately culminate into an immense change at global level.

References

Blaustein AR, Hoffman PD, Hokit DG et al (1994) UV repair and resistance to solar UV-B in amphibian eggs: a link to population declines. Proc Natl Acad Sci U S A 91(5):1791–1795. https://doi.org/10.1073/pnas.91.5.1791

Brikowski TH, Lotan Y, Pearle MS (2008) Climate related increase in the prevalence of urolithiasis in the United States. Proc Natl Acad Sci U S A 105(28):9841–9846. https://doi.org/10.1073/pnas.0709652105

Brunkard J, Namulanda G, Ratard R (2008) Hurricane Katrina deaths, Louisiana, 2005. Disaster Med Public Health Prep 2(4):215–223

Centers for Disease Control and Prevention (2017a) Malaria in italy. https://www.cdc.gov/malaria/new_info/2017/Italy_2017.html. Accessed 11 Sept 2019

Centers for Disease Control and Prevention (2017b) Chikungunya in Italy. https://wwwnccdcgov/travel/notices/watch/chikungunya-italy. Accessed 11 Sept 2019
D'Souza RM, Becker NG, Hall G et al (2004) Does ambient temperature affect foodborne disease? Epidemiology 15(1):86–89
Emberlin J (1994) The effect of patterns in climate and pollen abundance on allergy. Allergy 49:15–20
Fritz HM, Blount CD, Thwin S, Thu MK, Chan N (2009) Cyclone Nargis storm surge in Myanmar. Nat Geosci 2(7):448–449
Haines A, Berlin A (2014) Global health. In: Manson's tropical diseases, 23rd edn. Elsevier, Amsterdam, pp 9–15
Hales S, Weinstein P, Sovares Y, Woodward A (1999) El Nino & the dynamics of vector borne disease transmission. Environ Health Perspect 107(2):99–102
International Agency for Research on Cancers, Ultraviolet and solar radiation (1992) IARC monographs on the evaluation of carcinogenic risks to humans. IARC Press Lyon, France
Keatinge WR, Coleshaw SRK, Holmes J (1989) Changes in seasonal mortalities with improvement in home heating in England and Wales from 1963–1984. Int J Biometeorol 33:71–76
Lloyd GER (1978) Hippocratic writings. In: Harmondsworth. Penguin classics, London
Martin PH, Lefebure M (1995) Malaria and climate: sensitivity of malaria potential transmission to climate. Ambio 24(4):200–207
McLaughlin J, Paola D, Bopp C (2005) Outbreak of Vibrio parahaemolyticus gastroenteritis associated with Alaskan oysters. N Engl J Med 353(14):1463–1470
McMichael AJ, Haines A, Slooff R et al (1996) Climate change and human health. An assessment prepared by task group of World Health Organization, World Meteorological Organization and United Nations Environment Programme. World Health Organisation, Geneva
Quaite FE, Sutherland BM, Sutherland JC (1992) Action spectrum for DNA damage in alfa alfa lowers predicted impact of ozone depletion. Nature 358:576–578
Romieu I, Weitzenfeld H, Finkelman J (1990) Urban air pollution in Latin America and Caribbean health perspectives. World Health Stat Q 23(2):153–167
Solomon S, Qin D, Manning M, Marquis M et al (2007) Climate change: the physical science basis. Contribution of working group I to the fourth assessment report of the intergovernmental panel on climate change. Summary for policymakers. Cambridge University Press, New York
Subramanyam MA, Kawachi I, Berkman LF et al (2011) Is economic growth associated with reduction in child undernutrition in India ? PLoS Med 8(3):e1000424
Technical Briefing for the World Health Organisation (2015) Conference on Health and Climate. Promoting health while mitigating climate change. 27–29 August, Geneva. https://www.who.int/globalchange/publications/briefing-health-mitigating/en/. Accessed 15 Nov 2019
United Nations Framework Convention on Climate Change (2018) United Nations Climate change Annual report. http://unfccc.int/news/un-climate-change-launches-first-ever-report. Accessed 15 Nov 2019
Woodward A, Macmillan A (2015) Environment and climate change. In: Oxford textbook of global health, 6th edn. Oxford University Press, Oxford, UK, pp 201–215
World Bank (2009) World development report. Development and climate change. Washington, DC
World Health Organisation (2002) Reducing risks, promoting healthy life. World Health Report, Geneva. http://www.who.int/whr/2002/en/
World Health Organisation (2017) WHO methods and data sources for global burden of disease estimates 2000–2015. World Health Organisation, Geneva
World Health Organisation (2018) Climate change and health. http://www.who.int/news-room/fact-sheets/detail/climate-change-and-health. Accessed 20 Mar 2020
World Health Organization (2008) The global burden of disease geneva
Young PR, Lisa FP, Hall RA et al (2014) Arboviral infections. In: Manson's tropical diseases, 23rd edn. Elseiver, Cham, pp 129–161
Zhou X, Yang G (2008) Potential impact of climate change on schistosomiasis transmission in China. Am J Trop Med Hygiene 78(2):188–194

Impact of Climate Change on the Incidence and Transfer of Food- and Water-Borne Diseases

9

Mohammad Anas, Mohammad Asim Sami, Zarreena Siddiqui, Khalida Khatoon, Mohammad Tarique Zeyad, and Abdul Malik

Abstract

Presently, climate change is considered as a serious global environmental challenge influencing growth and survival of pathogens of many food- and water-borne diseases as well as their transmission pathways. Climate change is the major factor to increase the chances of discomfort by alternation in weather, high temperatures, variation in rainfall, and deficit of water. Heavy rainfall will raise the risk of more waterborne illnesses especially in such localities where water drainage system is poorly developed leading to water stagnation. Changes in the climatic condition can increase the burden of disease. Fundamentally all the calamity of climate change like higher temperature, heavy rainfall can have adverse effect on disease development. And there are documented evidences that these changes affect food security and food safety. Environmental changes can possibly impact the world's natural framework. Developing countries like India with constrained assets are required to confront a large group of health impacts because of environmental change, including food- and water-borne disease, for example, cholera, typhoid, shigellosis, and food poisoning. This chapter summarizes how environmental changes impact the proliferation and transfer of food- and water-borne pathogens.

Keywords

Climate change · Food safety · Pathogens · Waterborne disease · Cholera · Shigellosis · Proliferation

M. Anas (✉) · M. A. Sami · Z. Siddiqui · K. Khatoon · M. T. Zeyad · A. Malik
Department of Agricultural Microbiology, Faculty of Agricultural Sciences, Aligarh Muslim University, Aligarh, India

S. A. Lone, A. Malik (eds.), *Microbiomes and the Global Climate Change*,
https://doi.org/10.1007/978-981-33-4508-9_9

9.1 Introduction

Climate change is a continuing trend with potentially unpredictable, serious, and permanent human and ecological implications (IPCC, UNDP 2007–2008, WHO 2018a). A significant number of extreme weather conditions on the globe comprising flooding in southern India, forest fires in western United States (US), and heatwaves through most of the northern hemisphere can contribute to changes in threats to human health and multiply existing health issues (Lake et al. 2012). Climate change has the potential to influence the planet's biological systems including influencing proliferation and survival of pathogens as well as their routes of transmission (FAO 2008). Higher ambient temperatures raise the cycles of proliferation of food and waterborne pathogens; extended seasons can increase food handling errors. Besides the infectious agents, changing water flows will affect the transportation and existence of chemicals, for example, heavy metals and organic compounds in the surroundings. Reductions in the accessibility of fresh water due to climate change in water stressed regions will have important consequences on water quality and disproportionately affecting children and new born. Reduced water resources will also have complications for healthy storage, production, and processing of food. Climate change related diseases are already projected to account for 4.6% of all environmental threats. It was evaluated that around 2.4% of all global diarrhoea outbreaks, 6% of malaria outbreaks in some developing nations, and 7% of dengue fever in some industrial nations were attributed to climate change in the year 2000 (WHO 2015b). Together, the estimates show that deaths due to climate change is 0.3%, whereas the related disease risk was 0.4% (Department for Environment, Food and Rural Affairs, 2013). The WHO (2015a) presumed that there will be 250,000 deaths annually around 2030 and 2050 due to climate change; 38,000 from heat exposure in the elderly, 48,000 from diarrhoea, 60,000 from malaria, and 95,000 from childhood malnutrition.

9.2 Climate Change a Global Concern

Climate change may be the most important environmental issue facing mankind in the coming century. There may be direct effect on diseases and conditions that can lead to severe temperature-related diseases or mortality, and many other indirect health impact like, diseases linked to ingestion of contaminated potable water, food or vector borne and zoonotic diseases, and lack of food and water-related health conditions. Geographical, atmospheric alterations are expected to influence the vectors of disease and have an effect on its transmission, for example, leishmaniasis, Lyme borreliosis, malaria (in endemic areas), dengue, and tick-borne encephalitis, etc. Floods and rising temperature are the causes of increased water pollution and consequent increase in water and food pathogens. Changing climate is also likely to have effect on aeroallergen dispersal, particularly that of pollens, resulting in changes in the dispersal of allergic diseases (Climate change related event and transmission of diseases are elaborated in Fig. 9.1). Changing habits, food

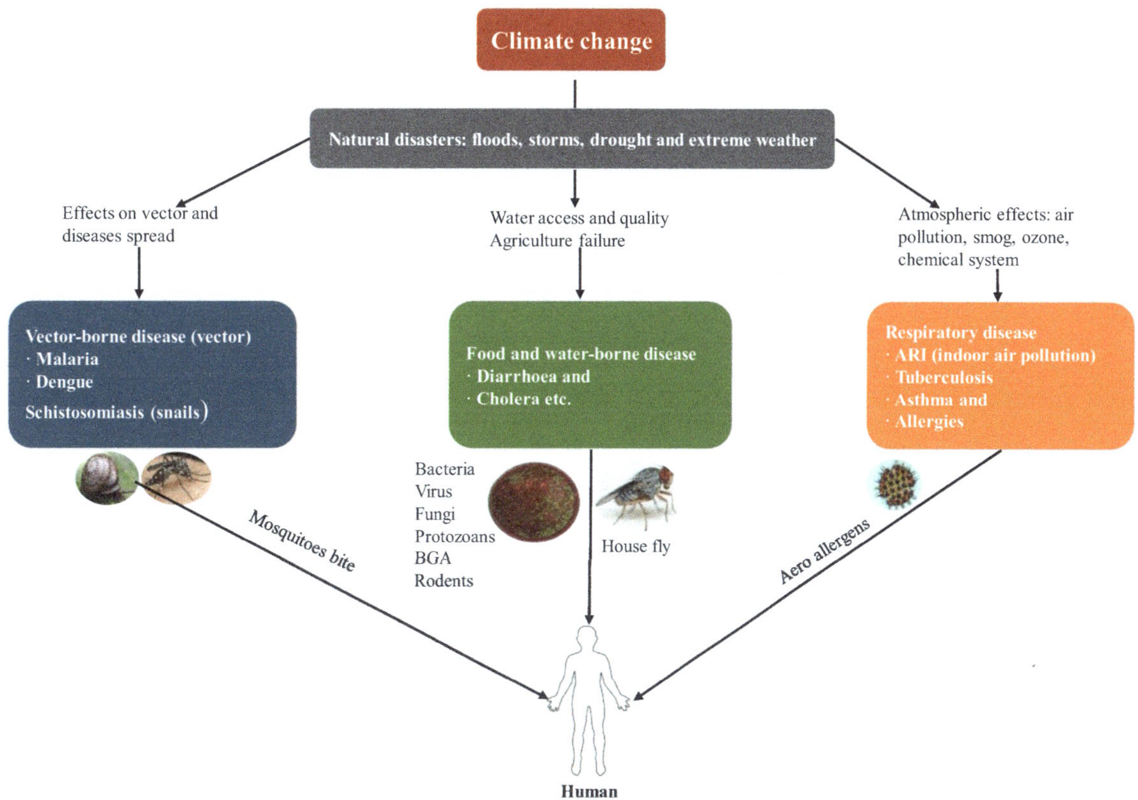

Fig. 9.1 Climate change related events and their impacts on transmission of disease to human

fabrication, urban modernization, and changing climate increase the risk of zoonotic expansion (McMichael et al. 2006). In a recent NASA report, which is the first to thoroughly investigate the effect of the major global climate phenomenon on public health, the 2015–2016 El Niño happening affected the weather conditions in a way that caused local disease outbreaks around the globe (https://www.nasa.gov/feature/goddard/2019/2015-2016-el-nino-triggered-disease-outbreaks-across-globe/). An increases in precipitation, temperatures, changing vegetation had encouraged disease transmission environment during the years 2015–2016, leading to an increase in reported incidents of Cholera in Tanzania, Hantavirus and plague in New Mexico and Colorado, and dengue fever in Southeast Asia and Brazil, respectively. Based on monthly epidemic reports in Colorado and New Mexico from 2002 to 2016, confirmed plague case occurred in 2015, whereas Hantavirus cases peaked in year 2016. The reason for the rise of both potentially fatal diseases was an uptick in precipitation caused by El Niño and mild temperatures across the Southwest of America, which facilitated vegetative growth of Hantavirus by providing sufficient fodder for rodents. A subsequent explosive increase in rodent population puts them in more regular contiguity with humans, mostly due to faecal or urine contamination, which contract the potentially fatal disease. The plague-bearing fleas were also proliferating as were the rodent hosts. In Tanzania on the continent of East Africa, the total confirmed cases of Cholera in year 2015 and 2016 were the second and third highest in the 18-year stretch since 2000–2017, respectively. Cholera is a potentially fatal bacterial infection of the small intestines that spreads through contamination of food and water. In Southeast Asia and Brazil dengue fever increased during the El Niño. Between 2000 and 2017 in Brazil, the greatest number of reported case of potentially fatal mosquito transmitted infection was in 2015 (Anyamba et al. 2019).

9.3 Incidence of Diseases in Relation to the Climate Change

9.3.1 Food Handling and Security

The Food and Agriculture Organization (FAO) defines food security as a situation as it stands if everyone has at all times physical, social, and economic access to appropriate, safe, and healthy foods that meet their dietary requirements and food preferences for an active and healthy lifestyle (FAO 2008). This definition includes four key food supply arrays: (a) availability (b) access (c) stability, and (d) usage. The first aspect concerns the availability of sufficient food, i.e. the agribusiness system's overall potential to fulfil food requirement. The sub-dimensions include the agro-climatic basics of plant and farmland productivity (Schmidhuber and Tubiello 2007) and the entire spectrum of socio-economic and cultural factors that determine how and where farmers respond to the businesses. Climate change has broad consequences for agriculture and for the production of food. It affects production of food directly through alterations in agro-ecological conditions and indirectly through productivity and distribution of wealth, and, therefore, demand for agricultural products (Izrael et al. 2007). Climate change will also impact individuals' ability to use nutrition efficiently by increasing food safety environments and raising the host, air, and foodborne disease environment. Climate change and food security are mainly concerned that altering weather situations which has the potential to trigger a vicious cycle in which infectious diseases lead to or cause hunger compounds, enabling the affected populations to become more infectious. As a result, labour productivity would significantly decline, unemployment and even mortality would increase. Principally all forms of climate change, including heavy rainfall, high temperatures and drought, influence the food safety and lead to diseases. Greater rainfall will pose high risk of occurrence of waterborne diseases, particularly where conventional water management systems are poor (Schmidhuber and Tubiello 2007). Likewise, flooding impacts will be most strong in environmentally degraded zones, and where there is a lack of proper public infrastructure, specially sanitation and hygiene. It will also increase the number of people exposed to waterborne diseases, thereby reducing their ability to consume food effectively. The global and regional effect of climate changes on food security is that it affects farm food systems in all communities, comprising exporters and importers, and those at subsistence level. There are also many other impacts which need to be addressed through this sector, such as increasing soil erosion and land degradation, changes in water quality, loss of biodiversity, widespread and more severe pest and disease outbreaks. To illustrate the effect of climate change at a more global scale, World Health Organization released data on the expected effect on human health as of 2004. It is easy to see from these statistical data that climate change has adversely affected the developing regions of the world in comparison to the developed regions. This disparity contrasts sharply with the difference of greenhouse gas emissions that is almost exclusively due to developed countries and fast growing economies. The WHO studies also focus on forecasts of the future global load of diseases which may result from climate change. It is estimated that 10% more diarrhoeal disease would occur by 2030 as compared to that would have occurred without the change in climate and considered that it will by and large affect young children's health. As estimated, with increase in global average temperature by 2–3 °C, the risk of malaria in population is expected to increase by 3–5 billion (World Health Organization 2004).

9.3.2 Foodborne Diseases

World Health Organization (World Health Organization 2018a) describes foodborne diseases as, the disease caused by toxic or infectious agents acquired through the ingested food. On the global basis, WHO reports that foodborne disease is a significant public health issue and in 2010 nearly 600 million Cases of foodborne diseases and 420,000 related deaths were reported (WHO 2015b; Lake and Barker 2018). A number of articles emphasize that climate change will affect the incidence of foodborne, waterborne and particularly diarrhoeal diseases globally (Lake and Barker 2018; Levy et al. 2018; Schijven et al. 2013). In the African region, 91 million episodes of disease and 137,000 deaths are caused each year by foodborne diseases (World Health Organization 2018b).

Climate change leads to increased microbial contamination (e.g. bacteria, virus, and other pathogens) of food and water by changing the mechanisms of survival and transmission through altering meteorological conditions, such as temperature and humidity. In 2010 it was reported that bacteria, viruses, and parasitic protozoa induced more than 2 billion diseases, contributing 31 million disability-adjusted life years (DALYs). 29% of these diseases were confirmed to have spread via infected food (Springmann et al. 2016). It has been estimated that, every year, 760,000 illness and 24,000 deaths (Approx.) have been reported due to the severe nuisance of *V. cholera*. The occurrence and distribution of these diseases are mainly associated with ingestion/consumption of contaminated food and water (Drinking water and some foodborne microbial pathogens are given in Table 9.1). In this regard, climate changes due to these disease causing organisms have been designated and the major cause of these threatening problems is global expansion of algal blooms as these are the major factors for contamination of water bodies (Springmann et al. 2016). Fisheries and aquaculture contamination procedures, such as oyster beds, have been correlated with heavy rainfall incidents (Doyle and Schoeni 1984), sea water heating has already contributed to an increase in contaminants historically confined to warmer waters. In particular, oyster-related *V. parahaemolyticus* outbreaks have been described in earlier too cold regions to support pathogen survival (Drake et al. 2007). Change in climatic conditions such as temperature and humidity and crop water stress affect the occurrence of mycotoxins releasing fungi that grow on different crops (Paterson and Lima 2011). This is particularly challenging in developing countries to protect crops and consumers from toxins because of poor food safety regulations. In recent years, the afla mycotoxins present in wheat had severe health consequences in Kenya (Daniel et al. 2011). Temperatures affect the handling and processing of food at the end of the product supply chain, increase the risk of consuming contaminated food, and put greater emphasis on proper handling and storage of food. Warming temperatures impact traditional food processing and storage (Ford and Hamner 2015) methods as well as supply in some Arctic regions. An increase in temperature, or the length of episodes of high temperature, in some geographical areas, may provide better conditions for *Salmonella* spp. to multiply in foodstuffs. As quoted by WHO in the 2017 report on protecting health in Europe from climate change, salmonellosis cases increased by 5–10% for every 1 °C rise in temperatures per week when ambient temperatures exceed 5 °C. In the same document, citing a study in Kazakhstan, the rate of salmonellosis rose by 5.5 million with a rise of 1 °C in the average monthly temperature. A number of foodborne pathogenic bacteria which cause the main concern in the climate change context are described:

9.3.2.1 Campylobacteriosis

Campylobacteriosis is the most frequently reported foodborne gastrointestinal disease caused by the thermophilic *Campylobacter* spp., with an increasing threat to public health. Campylobacter-related clinical complications include Guillain-Barre syndrome, which in 20 percent of the cases requires intensive care and may be fatal (WHO 2013). *Campylobacter* shows a high seasonal variability which indicates that

Table 9.1 Drinking water and some foodborne microbial pathogens

Organisms	Diseases	Transmission	Climatic condition
Protozoa			
Giardia duodenalis	Giardiasis	Oral faecal distribution through drinking water or recreational water	While moisture and relatively humid
Cryptosporidium parvum	Cryptosporidiosis	-do-	-do-
Cyclospora cayetanensis	Cyclosporiasis	Faecal-oral distribution through Drinking water	Influenced by several of them such as rainfall, temperature, and humidity.
Entamoeba histolytica	Amoebiasis	Oral faecal distribution through drinking water	Humid environment and moderate temperature
Toxoplasma gondii	Toxoplasmosis	Feline animals contaminated drinking water	Moderate temperatures and high humidity favour survival
Free-living amoebae	Amoebic meningoencephalitis		-do-
Bacteria			
Vibrio cholerae	Cholera	Drinking water or water contaminated food	Monsoons can influence the concentration in water
Salmonella spp.	Salmonellosis	Sporadic outbreaks with drinking water	Warmer ambient temperatures that favour the growth
Salmonella typhi	Typhoid	Drinking water	-do-
Shigella spp.	Shigellosis (bacillary dysentery)	Both drinking and recreational Water	Moderate to high temperature with humid environment
Campylobacter spp.	Campylobacteriosis	Both drinking water and food	Relatively high temperature
Enterotoxigenic *E. coli*		Drinking water	Increases in ambient temperature and precipitation
Enterohaemorrhagic *E. coli*		Drinking and recreational Water contact	--do-
Yersinia spp.	Yersiniosis	Drinking water and contaminated food	Optimum temperature is 28–29 °C but can survive—2 to 42 °C
Francisella tularensis	Tularaemia	Drinking water	It can survive for several weeks at low temperatures but best grow at 35–37 °C

(continued)

Table 9.1 (continued)

Organisms	Diseases	Transmission	Climatic condition
Helicobacter pylori		Drinking water	*H. pylori* is capable of survival in different types of aquatic environments under an array of physical variables
Mycobacteria spp. not *M. Tuberculosis*	Varies	Potable water systems in hospitals, some recreation	Optimal growth temperature lies in the range 40–45 °C
Viruses			
Hepatitis A and hepatitis E viruses	Viral hepatitis	Drinking and recreational Water contact and food	The HEV virus is very resistant and remains intact under environmental conditions, allowing its transmission
Enteroviruses	Various, including poliomyelitis	Drinking and recreational Water contact	
Various, esp. Norwalk-like Viruses	Viral gastroenteritis	Drinking and recreational water contact	Survival measured under various temperatures (4 °C, 15 °C, 25 °C, and 40 °C) and relative humidity 50% and 70%
Helminths			
Schistosoma spp.	Schistosomiasis	Touch with free swimming cercariae contaminated surface water	Survive between 14–32 °C and maximum risk of infection at 6–18 °C
Dracunculus medinensis	Dracunculiasis	Drinking water	Temperatures between 25 and 30 °C are best for larval development in fresh water

climate change may affect it. There is concrete evidence that weather and climate play an important role in transmitting it to humans. Uncertainties about the exact processes through which weather affects disease however, make it difficult to evaluate the probable effect of climate change because *Campylobacter*, unlike many other bacteria, does not multiply outside the intestine. The most frequently reported factor for *Campylobacter* transmission is positive correlation to temperature (Kovats et al. 2005a; Naumova et al. 2007; Lake et al. 2009). Recent studies have indicated that the risk of *Campylobacter* is positively linked with average weekly temperatures and incidence increases by 2–5% per degree Celsius increase of temperature, even though the strength of association is not consistent (Kovats et al. 2005b; Bi et al. 2008; Fleury et al. 2006). In 2018, the number of reported cases of human campylobacteriosis was 246,571, corresponding to a reporting rate of 64.1 per 100,000 populations in the European Union (EU) and Poultry meat (both fresh and broiler) was one of the largest recognized source of *Campylobacter* (EFSA and ECDC 2019). The colonization of *Campylobacter* in poultry increases rapidly with an increase of temperature although their transmission processes are complex for a large number of cases (Tam et al. 2006; Nichols et al. 2012). In context of UK health outcomes, infection with *Campylobacter* leads to around 500,000 case and 80,000 general practice consultations per year (Tam et al. 2012).

9.3.2.2 Salmonellosis

Salmonellosis (caused by *Salmonella* spp.) is the second largest foodborne human disease. *Salmonella* causes diarrhoea, fever, and abdominal cramps, usually 1–3 days following the initial infection. Symptoms normally last 4–6 days, however in some cases, persons need to be hospitalized. *Salmonella* can be transmitted through the intake of raw or undercooked poultry or ovine (major source of several species of *Salmonella*). *Salmonella enteritidis*, for example, is frequently associated with eggs, while *Salmonella typhimurium* is associated with a broader range of foods (Lake et al. 2009). In 2018 the incidence of the diseases was recorded with an EU reporting rate of 20.1 cases per 100,000 populations (91,857 confirmed cases of salmonellosis in humans), at the same level as in 2017, the biggest contributors of these outbreaks were eggs after that fresh poultry and pork (EFSA and ECDC 2019). There is a strong biological understanding of processes involved in proliferation of salmonella in food stored at ambient temperature (D'Souza et al. 2004). Higher ambient temperatures were correlated with 5–10% higher salmonella infections for each degree of weekly temperature increase (for ambient temperatures above 5 °C). Approximately one-third of *Salmonella* transmission (the fraction of population) traced due to temperature effect in England, Germany, the Netherlands, Spain, Switzerland, and the Czech Republic. Therefore, the studies showed that *Salmonella* is sensitive to climate, and its infections are more common in hot weather (Kovats et al. 2004; Naumova et al. 2007).

9.3.2.3 Listeriosis

Listeriosis is a bacterial disease caused by *Listeria monocytogenes* with a high mortality rates in the western countries (Adak et al. 2002). Recent large-scale outbreaks, such as the epidemic in South Africa from 2017 to 2018, have reinvigorated the risk for widespread disease. South Africa's outbreak has been the largest Listeria outbreak to date, with over 1000 laboratory-confirmed cases and even more than 200 deaths (NICD 2019).

Since 2016, European nations data collecting information on infections with *L. monocytogenes* recorded 2555 confirmed cases of listeriosis in the EU/European Economic Area (EEA), with the highest levels of infants below 1 year of age (1.3 per 100,000 population) and older people over 64 years of age (1.6 per 100,000) (Listeriosis 2019).

In 2016, 375 cases were registered in France which has a comprehensive national surveillance programme (Listeriosis 2019). In 2014 the United States also reported 675 confirmed cases (Food Net 2019). Listeriosis is most frequently a sporadic outbreak, as seen in a study from the United States Foodborne Diseases Active Surveillance Network (Food Net 2019). Due to its high case fatality rate listeria has emerged as major foodborne disease.

Over the last decade, several food items such as unpasteurized milk and dairy products, soft cheese, ready-to-eat cooked sausages and sliced, refrigerated smoked seafood, pasta and meat products have been reported for outbreaks and transmission of *L. monocytogenes*. A ready-to-eat meat product called polony in South Africa was identified as outbreak source of *L. monocytogenes* and the strains was also identified in the manufacturer processing environment of the implicated product (Allerberger and Wagner 2010; McCollum et al. 2013; WHO 2019).

Under a sheet of extracellular polysaccharide matrix (named as Biofilms) these microbes protects themselves from stress and antimicrobial agents. *L. monocytogenes* biofilms are of specific concern because the biofilm formed on food processing plants make it difficult to eliminate *L. monocytogenes*. For example, in a study in Gauteng, microbes were isolated from stainless steel surfaces of a food processing plant although it has been cleaned and disinfected by applying a variety of cleaning methods (Lambrechts et al. 2014). The large environmental impact associated with global climate change significantly alters foodborne disease epidemiology, including that of *L. monocytogenes* (Hellberg and Chu 2016). Although *Listeria* spp. are globally present in nature but several features of the epidemiology and the microbe's characteristics make it particularly sensitive to climate (Chersich et al. 2018). For example, increases in ambient temperatures and high summer temperatures are associated with listeriosis, as with many other diarrhoeal pathogens (Goulet et al. 2006; EFSA and ECDC 2018). Extreme hot weather, which becomes more prevalent with climate change, increases the replication cycles of *L. monocytogenes* and may cause breakdowns in food cooling chains, with rapid increments in the number of bacteria in food products (Miettinen et al. 1999). However, besides of rising temperatures, altered rain fall patterns and prolonged dry seasons can influence the transmission of *Listeria*. *Listeria* dispersal is also affected by the changes in precipitation patterns caused by large-scale alteration of the climate. In short bursts of 5–10 min, rainfall favours the dispersal of *Listeria* and other pathogens from the soil to crops, whereas longer downpours have wash out effect (Hellberg and Chu 2016). Unlike fresh produce, runoff water can contaminate fish farm as this water flows through the gills of the fish *L. monocytogenes* present in water can infect fishes and through food chain it may be transmitted to the other organisms (Miettinen and Wirtanen 2006).

9.3.2.4 *Bacillus cereus*

Bacillus cereus is categorized as Gram-positive, spore-forming, widespread environmental bacterium that can decay organic matter (Stenfors Arnesen et al. 2008) and spores are extremely resistant to unfavourable environmental conditions and can commonly spread to food stuffs due to contamination. Such spores are highly sticky and hydrophobic, enabling them to survive in food producing areas, e.g. *B. cereus* spores can be attached to cattle udders during weeding, can enter in the dairy production site, and cause persistent contamination of the end products (Montville and Matthews 2012). On the basis of strain type and toxins production *B. cereus* causes two forms of diseases, i.e. (a) diarrhoeal and (b) emetic. Diarrhoeal form has been related with a range of food items including meat, vegetables, and dairy products, whereas starchy foods such as rice and pasta are primarily associated with the emetic form. Most of the *B. cereus* strains are motile and exhibit salinity tolerance at high level (7.5%) which commonly falls under the category of facultative anaerobe. Likewise, many *B. cereus* strains are mesophilic in nature and 35–40 °C is an optimum temperature for their growth (Stenfors Arnesen et al. 2008; FDA 2012). However, it has been observed that *B. weihenstephanensis* which is a psychrotrophic strain can grow at 7 °C but not at 43 °C. The variability of temperature tolerance shown by *B. cereus* strains and closely related species has been recognized as a possible mechanism for the adaptation of this microorganism to climate change (Carlin et al. 2010). Seasonal variability also has an impact on the incidence of *B. cereus*, for instance, a study conducted by Bartoszewicz et al. (2008) in Poland showed that seasonal variation was important for *B. cereus* and *B. weihenstephanensis* in both raw and refined milk, with the highest and lowest levels, respectively, in the spring and summer and autumn and winter seasons. Though this may have been because of lower exposure of cattle to the bacteria in winter.

9.3.2.5 *Clostridium*

This bacterium is Gram-positive, anaerobic, spore-forming, and widespread in nature (FDA 2012). There are two species of *Clostridium* (a) *C. botulinum* and (b) *C. perfringens* which are commonly associated with foodborne illness. Spores of these species exhibit high heat resistance and can live in adverse conditions. The toxin produced by *C. botulinum* causes botulism (a serious form of food intoxication) which may cause paralysis and mortality. Canned foods (low-acid, having pH 4.6) and Vacuum-packed (VP) fish products are more prone to the growth of *C. botulinum* and thereby causing botulism (Montville and Matthews 2012). Some strains of *C. perfringens* also produce enterotoxins which cause a range of diseases such as gastroenteritis, enteritis necroticans in human, and enterotoxaemia in livestock (Sayeed et al. 2008; FDA 2012). *C. perfringens* enterotoxin (CPE) is known to be leading cause of food poisoning and antibiotic-related diarrhoea. Due to its heat resistant, survivability in adverse condition, and persistence in animal and human faeces *Clostridium* is also considered as indicator of faecal contamination (Brynestad and Granum 2002). A number of studies observing the influence of precipitation on microbial dispersion have focused on indicator microorganisms, for example, *C. perfringens*, faecal coliforms, and *E. coli* to show the chance of faecal contamination. Over-all a positive association has been reported between rainfall and these indicator microorganisms in estuaries and the rivers (Cooley et al. 2007; Wilkes et al. 2009; Kim et al. 2013; Hofstra et al. 2013).

9.3.2.6 *Staphylococcus*

Staphylococcus aureus is a non-spore-former, coagulase-positive, Gram-positive, and facultative anaerobic bacterium commonly found in outdoor environment (i.e. air, water, soil, and dust) and in human and animals as natural microflora, predominantly present on skin, hair, and nasal passage (Le Loir et al. 2003; Madigan et al. 2012; FDA 2012). Through contamination, *S. aureus* can enter in food chain and may cause a number of human illnesses. Staphylococcal food poisoning (SFP) is known to be one of the world's most serious foodborne diseases caused by intake of highly heat-stable enterotoxins produced in foods (Jablonski 1997). Some other strains of *Staphylococcus* spp. such as *S. intermedius* sporadically produce enterotoxins. Food items like dairy, bakery, meat and poultry products are more prone to *S.aureus* contamination and intoxication therefore, require considerable handling during the processing. The mesophilic growth condition (optimal temperature 35 °C), osmotolerant nature, survival in dry state for long period, and ability to proliferate in low water activity food items as 0.83–0.86 make the *S. aureus* to survive in adverse climatic conditions (Montville and Matthews 2012; Anas et al. 2019).

9.3.2.7 *Escherichia coli*

E. coli (a natural gut microbiota of human and other animals) is a collection of Gram-negative, motile, and facultative anaerobes with some pathogenic strains causing mild to severe diarrhoeal infection (Montville and Matthews 2012). The diarrhoeagenic *E. coli* strains comprise: enteropathogenic, enterotoxigenic, enteroinvasive, diffusely adhering, enteroaggregative, and enterohaemorrhagic; these are largely transmitted through faecal-oral route. Over all, these diarrhoeagenic *E. coli* strains were evaluated to cause yearly two lakh case of foodborne disease in the United States (Scallan et al. 2011; Hellberg and Chu 2016). Enterohaemorrhagic *E. coli* (EHEC) strains produce a cytotoxic factor known as Verotoxins or Shiga toxins (Stx), infections caused by these strains may, therefore, lead to severe problems such as haemorrhagic colitis, kidney disorder, purpura thrombotic thrombocytopenia (TTP), and death. *E. coli* O157:H7 is the main serotype of EHEC which globally accounts for about 75% human infections with very low projected contagious dose of 10–100 cells (FDA 2012). *E. coli* O157:H7 normally escalate best at approximately 37 °C; yet, optimal growth temperature differs and depends on strain. At declined temperatures (e.g. $\leq$10 °C) the growth of *E. coli* O157:H7 ceased while at freezing temperature it may survive for a longer period of time (Doyle and Schoeni 1984; Gonthier et al. 2001). Unusually *E. coli* O157:H7 also exhibits acid tolerant nature as revealed by some other foodborne pathogens, and proliferating as low as at pH 4.0 to 4.5 (Montville and Matthews 2012). Remarkably, a single type of environmental stress given at sub lethal doses showed an increment in the overall resistance of *E. coli* O157:H7 compared with the other kinds of stressors, thus boosting its longevity in these environments (Franz and Van Bruggen 2008). Outbreaks of foodborne diseases caused by *E. coli* O157:H7 is primarily related with the ingestion of partially cooked meat products, fruits, vegetables, unpasteurized juices, etc. (FDA 2012). Faecal contaminants from domestic livestock and wild life may contaminate fresh produce in the pre-harvest environment. For example, outbreak analysis of *E. coli* O157:H7 in spinach during 2006 matches to the outbreak species that were identified in wild swine faeces, cattle faeces, and soil or sediment samples taken nearby spinach field (Jay et al. 2007; Mandrell 2009). Once in the field *E. Coli* O157:H7 has survived for a long time, can potentially contaminate crops, and may be moved to many other areas, such as groundwater and reservoirs (Fremaux et al. 2008; van Elsas et al. 2011). Meteorological events such as heavy precipitation resulted in to more flooding and run-off plays a crucial role in the transmission of *E. coli* O157:H7. Sewage overflows due to heavy rainfall increase the water turbidity resulting in re-suspension of pathogens from sediments (Hofstra 2011). In a study Cooley et al. (2007) found that at month of peak flooding there was fivefold increment in the level of generic *E. coli*, although, in 85% of testing samples occurrence of *E. coli* O157:H7 has grown from undetectable levels. Though, most foodborne pathogens are likely to be affected adversely by drought because of water necessity for their survival and growth. A major outbreak of *E. Coli* O157:H7 occurred in southern Africa when heavy rainfall and surface water pollution accompanied 3 months of long drought (Effler et al. 2001).

9.3.3 Waterborne Diseases

Waterborne diseases are conditions that developed by pathogenic microorganisms or their toxins that are transmitted in water. The diseases can be spread through drinking water or consuming food exposed to polluted water and also through bathing and cleaning. A major proportion of the global population do not have access to pure and hygienic water needed for drinking or required for basic necessities. Approximately 1.1 billion people throughout the world lack access to improved water supplies and 2.3 billion lack adequate sanitation (World Health Organization 2017). As per the WHO, waterborne diseases contribute to an approximate 3.6% of total DALY (disability-adjusted life year) global burden of disease and cause around 1.5 million annual fatalities. The WHO reports also concluded that 58% of this burden, or 842,000 deaths each year, are due to a lack of sufficient supply of drinking water, sanitation, and hygiene (World Health Organization 2018a). The lack of proper hygiene, sanitation, and facilities still remain two of the major challenges in the global fight against waterborne diseases (Ford and Hamner 2015). Additionally, there are rising trends of certain diseases particularly those associated with contaminated water include viral (Hepatitis A and Hepatitis E), bacterial (Salmonellosis, Shigellosis, cholera, and alimentary toxic infections, etc.), and protozoan (*Giardia lamblia*) (WHO 2015b). These waterborne diseases mainly transfer through the contaminated potable or recreational water (David et al. 2014; Murphy et al. 2014). The transmission and the survival of these infectious agents in nature are greatly affected by increased extreme weather frequencies of incidents such as increased precipitation, heavy rainfall flash floods and droughts (Thomas et al. 2006). The impact of climate change on the hydrological cycle is expected to result in more floods and droughts that increase the incidence and transfer of waterborne diseases (Costello et al. 2009). Excess water in flooding areas could affect the wastewater system and contribute to leakage of waste; this may affect water quality and quantity. Heavy rain is frequently linked to contamination, leaching pathogens such as *Salmonella, E. coli, Campylobacter* spp., and *Cryptosporidium* into the water supply and contaminating the water bodies (Schuster et al. 2005; Dorner et al. 2006; Patz et al. 2005). Prolonged hot weather increases the temperature of many water bodies which may lead to extend the frequency of cyanobacterial blooms which contain injurious toxins (Paerl and Huisman, 2009). Similarly, the incidence of diseases (well growing in warm water) such as cholera, Legionnaire, and cryptococcosis has increased and can be detected as a result of climate change (McMichael et al. 2006; Greer et al. 2008; Raffa et al. 2012). Besides waterborne diseases human health has also been affected by exposure to chemical contaminants, for example, pesticides, biocides, and pharmaceutical (Noyes et al. 2009; Pangare and Idris 2012; Zeyad et al. 2019). Urbanization patterns and domestic use of chemical products (e.g. chemicals, detergents, disinfectants) will also lead to increase in pollutant discharges and expanding the plant and human diseases (Evans et al. 2012). Climatically-induced hydrological shifts would alter the manner in which pollutants are introduced into the aquatic environment and the capacity for rivers and streams to dilute. Increases in extreme climate happenings like droughts and floods will change the transmission of contaminants (Manuel 2006). Changes in irrigation practices in agricultural areas, such as wastewater reuse, may transfer pollutants and infectious agents from aquatic bodies to farmlands (Tirado et al. 2010). The resulting increased severity and frequency of extreme weather conditions will be influenced by changes in microbial communities and interactions between species (Walker 2018). Some recognized waterborne pathogens in the reference of climate change are as follows:

9.3.3.1 *Vibrio* Spp.

Vibrio spp. comprise naturally occurring, motile, Gram-negative, and comma-shaped facultatively anaerobic bacteria commonly found in shore estuaries and marine water (Montville and Matthews 2012). A number of Vibrio spp., for example, *V. cholerae*, *V. parahaemolyticus*, and *V. vulnificus* have been found to cause human diseases, while the majority of these species are non-pathogenic. *V. vulnificus* and *V. parahaemolyticus* are able to grow in saline water, while *V. cholerae* can grow in both fresh as well as in saline water (Reidl and Klose 2002; Su and Liu 2007; Horseman and Surani 2011). The optimal growth temperature for these microorganisms fall around 37 °C and is more frequently found through hot months of the year. Seasonal variability of vibrio infections was observed in the United States, with a rise in late spring/early summer months at the top in August and a steady fall between September and October (CDC 2012). Gastroenteritis, wound infection, and septicaemia (in some case) are the primary symptoms shown by infectious form of *Vibrio* spp. in human (Montville and Matthews 2012). In the United States, the frequency of *V. parahaemolyticus* (a leading cause of shellfish-related gastroenteritis) infections has risen by 76% relative to the levels in 2011 to 1996–98 (CDC 2012).

Infections caused by *V. vulnificus* are sporadic but have a high fatality rate, with an estimate of ~50 cases of infection with 45 hospitalizations and 16 fatalities in the United States each year (Scallan et al. 2011; CDC 2012). *V. cholerae* (O1/O139) occasionally causes cholera in developed nations while it is frequently related with waterborne diseases in underdeveloped nations where healthy and hygienic drinking water supply systems are scarce. Growth and transmission of *Vibrio* spp. are also influenced by various climatic conditions like precipitation, salinity, temperatures, and wind speed. In a study Johnson et al. (2010), Yamazaki and Esiobu (2012) demonstrated that the degrees of prevalence of infectious *Vibrio* spp. such as *V. vulnificus*, *V. carchariae*, *V. parahaemolyticus*, and *V. alginolyticus* in coastal water are positively linked with turbidity and antecedent rainfall for 2 days. Likewise, in another study Miller et al. (2006) reported non-significant and opposite relationship among antecedent rainfall and the degrees of *V. cholerae* and *V. alginolyticus* simultaneously in mussels. Although previous study done by Lipp et al. (2001) showed that there was non-significant association between the levels of *V. vulnificus* and antecedent precipitation in the same environment. Such findings revealed a dynamic link between the *Vibrio spp.* and the rainfall in crustacean and seawater which may be affected by ambient temperature and salinity. The climate change is expected to promote calamities like desertification and storm which results in to an increment in the levels of global dust. These natural disasters can contribute to the dispersal and durability of certain foodborne microorganisms specially spore-formers. Despite the microbial transport, even dust can carry supplements that support growth of microorganisms. For example, a field study accomplished by Westrich et al. (2016) in Florida Keys revealed that the growth of total *Vibrio* spp. increased with the addition of dust to seawater. In addition, the same study was carried out in synthetic iron-diminished seawater with *V. alginolyticus* and *V. cholerae*, showing that the growth of these bacteria increases by adding this dust to the medium. Another aspect related to climate change is a rise of surface temperatures of seawater with a mean of 0.13 °C per year over the last decade. This rise in surface temperatures of sea water is assumed to remain through the twenty-first century, with estimates of 0.6–2.0 °C warming of top 100-metre sea water (Pearce et al. 2014). This trend is relating, in part for the reason that *Vibrio* spp., which dwell in coastal estuaries and marine water, have been linked with warmer temperatures. For instance, several research revealed a connection between temperature and *V. cholerae* infections was associated with El Nin˜o-related ocean warming and higher sea surface temperature incidents in South America and South-eastern Africa, respectively (Hashizume et al. 2011; Mishra et al. 2012). Also, the rise in surface temperature of sea has been found to be associated in a positive way with the abundance of many different *Vibrio* spp. such as *V. vulnificus, V. parahaemolyticus,* and *V. alginolyticus* across the globe (Böer et al. 2013; Cantet et al. 2013). For instance, in a study Alam et al. (2003) observed a significant correlation between climate-induced sea surface warming and also with the population of *Vibrio* spp. in North Sea, and Vezzulli et al. (2012) reported an increased incidence of *V. parahaemolyticus* all through the hot weather at along the coast of Japan. In a study Johnson et al. (2012) reported that the concentration of both *V. parahaemolyticus and V. vulnificus* at across the East, West, and Gulf Coasts of the United State was positively correlated with sea surface temperature. These finding indicate that climate-induced environmental changes can influence the incidence and transfer of *Vibrio* spp.

9.4 Risk and Mitigation Approach

Climate change will affect all populations but few are at high risk than others. Particularly those individuals residing in developing countries, small island, and other coastal areas, megacities, polar and mountainous areas are primarily more vulnerable. Children particularly inhabiting in an economically feeble nations are more prone to the consequential health threats and will be exposed to further health results. This also have a serious impact on adults and on immunocompromised people. Those regions having poor health infrastructure as is commonly found in developing countries will be the least capable of managing the climate change induced health effect due to limited awareness, improper planning, and delayed response (Patz et al. 2005). A feasibility review of the WHO to make standards appeals to the involvement of policymakers in the primary stages, as the best initial precautionary measures would be worthwhile if organizational objectives are not clearly delineated and addressed from the outset. In 2009, the World Health Assembly (WHA) validated a latest WHO work plan on climate change and health as described in Fig. 9.2.

9.5 Conclusion

As discussed in the chapter, that climate change may have considerable consequences on incidence and transfer of food and waterborne diseases but these interactions are complex and dependent on various factors. Consequently, there is considerable uncertainty in published literature as to which foodborne pathogens will be most affected, what the specific effect will be, and what time scales could change. The impacts of temperature and precipitation were investigated in multiple studies and variable effect was shown, relying collectively on geographical locations and many other climate variables. While, impacts of other different climatic factors likewise dust and wind incidents on foodborne pathogens are very well known, but it can be considered extremely important as it can rise in frequency and severity over the decades to come. This change of climate has an impact on incidence and transfer of food and waterborne pathogens from animal host but the process is complicated and little explained. There are various ways by which livestock and agricultural crops can be infected with pathogens. So, careful maintenance of aquatic bodies, wildlife intrusion, application of manure, and maintenance of soil coupled with a knowledge of what effect will climate change have on these components will help in minimizing the adverse impact of the climate change. Marine food hygiene and waterborne diseases are another big issue given forecasts. Climate change has an impact on crop production which would include mitigation policies to maintain food safety and security. In this area of study, much remains to be acknowledged and further study are necessitated to evaluate the impact of climatic factors on the foodborne pathogens. These interactions are easy to understand, so they could be applied to construct or reinforce frameworks of prediction that fits for a holistic approach to mitigate the threats posed by the climate change in food safety. Also,

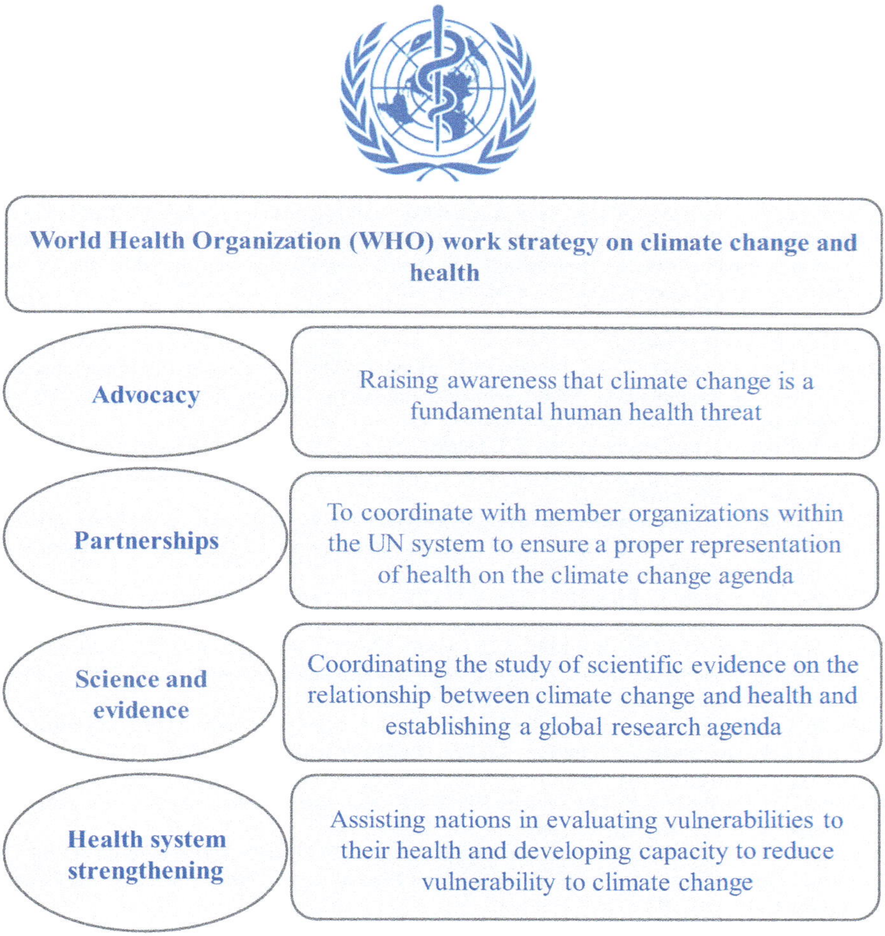

Fig. 9.2 Main points established in WHO work plan on the climate change and health

innovative approaches will allow scientists from different fields to work together to tackle the complex interactions between ecosystems, food safety, and climate.

References

Adak GK, Long SM, O'Brien SJ (2002) Trends in indigenous foodborne disease and deaths, England and Wales: 1992 to 2000. Gut 51:832–841

Alam MJ, Miyoshi SI, Shinoda S (2003) Studies on pathogenic Vibrio parahaemolyticus during a warm weather season in the Seto Inland Sea, Japan. Environ Microbiol 5(8):706–710

Allerberger F, Wagner M (2010) Listeriosis: a resurgent foodborne infection. Clin Microbiol Infect 16(1):16–23

Anas M, Ahmad S, Malik A (2019) Microbial escalation in meat and meat products and its consequences. In: Health and safety aspects of food processing technologies. Springer, Cham, pp 29–49

Anyamba A, Chretien JP, Britch SC, Soebiyanto RP, Small JL, Jepsen R et al (2019) Global disease outbreaks associated with the 2015–2016 El Niño event. Sci Rep 9(1):1930

Bartoszewicz M, Hansen BM, Swiecicka I (2008) The members of the Bacillus cereus group are commonly present contaminants of fresh and heat-treated milk. Food Microbiol 25(4):588–596

Bi P, Cameron AS, Zhang Y, Parton KA (2008) Weather and notified campylobacter infections in temperate and sub-tropical regions of Australia: an ecological study. J Infect 57(4):317–323

Böer SI, Heinemeyer EA, Luden K, Erler R, Gerdts G, Janssen F, Brennholt N (2013) Temporal and spatial distribution patterns of potentially pathogenic Vibrio spp. at recreational beaches of the German North Sea. Microbial Ecol 65(4):1052–1067

Brynestad S, Granum PE (2002) Clostridium perfringens and foodborne infections. Int J Food Microbiol 74(3):195–202

Cantet F, Hervio-Heath D, Caro A, Le Mennec C, Monteil C, Quéméré C et al (2013) Quantification of Vibrio parahaemolyticus, Vibrio vulnificus and Vibrio cholerae in French Mediterranean coastal lagoons. Res Microbiol 164(8):867–874

Carlin F, Brillard J, Broussolle V, Clavel T, Duport C, Jobin M et al (2010) Adaptation of Bacillus cereus, a ubiquitous worldwide-distributed foodborne pathogen, to a changing environment. Food Res Int 43(7):1885–1894

Centers for Disease Control and Prevention (2012) Foodborne diseases active surveillance network (FoodNet): FoodNet surveillance report for 2011 (final report). US Department of Health and Human Services, CDC, Atlanta, GA

Chersich MF, Scorgie F, Rees H, Wright CY (2018) How climate change can fuel listeriosis outbreaks in South Africa. S Afr Med J 108(6):453–454

Cooley M, Carychao D, Crawford-Miksza L, Jay MT, Myers C, Rose C et al (2007) Incidence and tracking of Escherichia coli O157: H7 in a major produce production region in California. PloS One 2(11):e1159

Costello A, Abbas M, Allen A, Ball S, Bell S, Bellamy R et al (2009) Managing the health effects of climate change: lancet and University College London Institute for Global Health Commission. Lancet 373(9676):1693–1733

D'Souza RM, Becker NG, Hall G, Moodie KB (2004) Does ambient temperature affect foodborne disease? Epidemiology 15(1):86–92

Daniel JH, Lewis LW, Redwood YA, Kieszak S, Breiman RF, Flanders WD et al (2011) Comprehensive assessment of maize aflatoxin levels in Eastern Kenya, 2005–2007. Environ Health Perspect 119(12):1794–1799

David JM, Ravel A, Nesbitt A, Pintar K, Pollari F (2014) Assessing multiple foodborne, waterborne and environmental exposures of healthy people to potential enteric pathogen sources: effect of age, gender, season, and recall period. Epidemiol Infect 142(1):28–39

Dorner SM, Anderson WB, Slawson RM, Kouwen N, Huck PM (2006) Hydrologic modeling of pathogen fate and transport. Environ Sci Technol 40(15):4746–4753

Doyle MP, Schoeni JL (1984) Survival and growth characteristics of Escherichia coli associated with hemorrhagic colitis. Appl Environ Microbiol 48(4):855–856

Drake SL, DePaola A, Jaykus LA (2007) An overview of Vibrio vulnificus and Vibrio parahaemolyticus. Compr Rev Food Sci Food Saf 6(4):120–144

Effler E, Isaäcson M, Arntzen L, Heenan R, Canter P, Barrett T et al (2001) Factors contributing to the emergence of Escherichia coli O157 in Africa. Emerg Infect Dis 7(5):812

European Food Safety Authority and European Centre for Disease Prevention and Control (EFSA and ECDC) (2018) The European Union summary report on trends and sources of zoonoses, zoonotic agents and food-borne outbreaks in 2017. EFSA J 16(12):e05500

European Food Safety Authority and European Centre for Disease Prevention and Control (EFSA and ECDC) (2019) The European Union one health 2018 Zoonoses Report. EFSA J 17(12): e05926

Evans AE, Hanjra MA, Jiang Y, Qadir M, Drechsel P (2012) Water quality: assessment of the current situation in Asia. Int J Water Res Dev 28(2):195–216

FAO (2008) Climate change: implications for food safety. Food and Agriculture Organisation, Rome, Italy

FDA (2012) Bad Bug: Handbook of foodborne pathogenic microorganisms and natural toxins. CreateSpace Independent Publishing Platform, Scotts Valley, CA

Fleury M, Charron DF, Holt JD, Allen OB, Maarouf AR (2006) A time series analysis of the relationship of ambient temperature and common bacterial enteric infections in two Canadian provinces. Int J Biometeorol 50(6):385–391

Foodnet (2019) Centers for disease control and prevention. https://www.cdc.gov/foodnet/index.html. Accessed 21 Dec 2018

Ford TE, Hamner S (2015) A perspective on the global pandemic of waterborne disease. Microb Ecol 76:2–8

Franz E, van Bruggen AH (2008) Ecology of E. coli O157: H7 and Salmonella enterica in the primary vegetable production chain. Crit Rev Microbiol 34(3–4):143–161

Fremaux B, Prigent-Combaret C, Vernozy-Rozand C (2008) Long-term survival of Shiga toxin-producing Escherichia coli in cattle effluents and environment: an updated review. Vet Microbiol 132(1–2):1–18

Gonthier A, Guérin-Faublée V, Tilly B, Delignette-Muller ML (2001) Optimal growth temperature of O157 and non-O157 *Escherichia coli* strains. Lett Appl Microbiol 33(5):352–356

Goulet V, Jacquet C, Martin P, Vaillant V, Laurent E, De Valk H (2006) Surveillance of human listeriosis in France, 2001–2003. Euro Surveill 11(6):79–81

Greer A, Ng V, Fisman D (2008) Climate change and infectious diseases in North America: the road ahead. CMAJ 178(6):715–722

Hashizume M, Faruque ASG, Terao T, Yunus M, Streatfield K, Yamamoto T, Moji K (2011) The Indian Ocean dipole and cholera incidence in Bangladesh: a time-series analysis. Environ Health Perspect 119(2):239–244

Hellberg RS, Chu E (2016) Effect of climate change on the persistence and dispersal of foodborne bacterial pathogens in the outdoor environment: a review. Crit Rev Microbiol 42(4):548–572

Hofstra N (2011) Quantifying the impact of climate change on enteric waterborne pathogen concentrations in surface water. Curr Opin Environ Sustain 3(6):471–479

Hofstra N, Vermeulen LC, Wondmagegn BY (2013, April) Climate change impacts on faecal indicator and waterborne pathogen concentrations and disease. In: EGU General Assembly Conference Abstract, EGU2013-8413

Horseman MA, Surani S (2011) A comprehensive review of Vibrio vulnificus: an important cause of severe sepsis and skin and soft-tissue infection. Int J Infect Dis 15(3):e157–e166

Izrael YA, Semenov SM, Anisimov OA, Anokhin YA, Velichko AA, Revich BA, Shiklomanov IA (2007) The fourth assessment report of the intergovernmental panel on climate change: working group II contribution. Russ Meteorol Hydrol 32(9):551–556

Jablonski LM (1997) Staphylococcus aureus. Food Microbiol: Fundam Front:353–375

Jay M, Cooley M, Carychao D et al (2007) Escherichia coli O157:H7 in feral swine near spinach fields and cattle, Central California coast. Emerg Infect Dis 13:1908–1911

Johnson CN, Flowers AR, Noriea NF, Zimmerman AM, Bowers JC, DePaola A, Grimes DJ (2010) Relationships between environmental factors and pathogenic vibrios in the northern Gulf of Mexico. Appl Environ Microbiol 76(21):7076–7084

Johnson CN, Bowers JC, Griffitt KJ, Molina V, Clostio RW, Pei S et al (2012) Ecology of Vibrio parahaemolyticus and Vibrio vulnificus in the coastal and estuarine waters of Louisiana, Maryland, Mississippi, and Washington (United States). Appl Environ Microbiol 78 (20):7249–7257

Kim JY, Lee H, Lee JE, Chung MS, Ko GP (2013) Identification of human and animal fecal contamination after rainfall in the Han River, Korea. Microbes Environ 28(2):187–194

Kovats RS, Edwards SJ, Hajat S, Armstrong BG, Ebi KL, Menne B (2004) The effect of temperature on food poisoning: a time-series analysis of salmonellosis in ten European countries. Epidemiol Infect 132(3):443–453

Kovats RS, Edwards SJ, Charron D, Cowden J, D'Souza RM, Ebi KL et al (2005a) Climate variability and campylobacter infection: an international study. Int J Biometeorol 49 (4):207–214

Kovats RS, Edwards SJ, Charron D, Cowden J, D'Souza RM, Ebi KL, Gauci C et al (2005b) Climate variability and campylobacter infection: an international study. Int J Biometeorol 49 (4):207–214

Lake IR, Barker GC (2018) Climate change, foodborne pathogens and illness in higher-income countries. Curr Environ Health Rep 5(1):187–196

Lake IR, Gillespie IA, Bentham G, Nichols GL, Lane C, Adak GK, Threlfall EJ (2009) A re-evaluation of the impact of temperature and climate change on foodborne illness. Epidemiol Infect 137(11):1538–1547

Lake IR, Hooper L, Abdelhamid A, Bentham G, Boxall AB, Draper A et al (2012) Climate change and food security: health impacts in developed countries. Environ Health Perspect 120 (11):1520–1526

Lambrechts AA, Human IS, Doughari JH, Lues JFR (2014) Efficacy of low-pressure foam cleaning compared to conventional cleaning methods in the removal of bacteria from surfaces associated with convenience food. Afr Health Sci 14(3):585–592

Le Loir Y, Baron F, Gautier M (2003) Staphylococcus aureus and food poisoning. Genet Mol Res 2 (1):63–76

Levy K, Smith SM, Carlton EJ (2018) Climate change impacts on waterborne diseases: moving toward designing interventions. Curr Environ Health Rep 5(2):272–282

Lipp EK, Rodriguez-Palacios C, Rose JB (2001) Occurrence and distribution of the human pathogen Vibrio vulnificus in a subtropical Gulf of Mexico estuary. In: The ecology and etiology of newly emerging marine diseases. Springer, Dordrecht, pp 165–173

Listeriosis (2019) European centre for disease prevention and control. http://ecdc.europa.eu/en/healthtopics/listeriosis/annual:surveillance-data/. Accessed 21 Dec 2018

Madigan, M., Martinko, J., Stahl, D., & Clark, D. (2012). Brock biology of microorganisms (San Francisco, USA:(13 eth) Pearson education)

Mandrell RE (2009) Enteric human pathogens associated with fresh produce: sources, transport, and ecology. In: Microbial safety of fresh produce. Wiley, Hoboken, NJ, pp 5–41

Manuel M (2006) In Katrina's wake. Environ Health Perspect 114:A32–A39

McCollum JT, Cronquist AB, Silk BJ, Jackson KA, O'Connor KA, Cosgrove S et al (2013) Multistate outbreak of listeriosis associated with cantaloupe. N Engl J Med 369:944–953

McMichael AJ, Woodruff RE, Hales S (2006) Climate change and human health: present and future risks. Lancet 367(9513):859–869

Miettinen H, Wirtanen G (2006) Ecology of Listeria spp. in a fish farm and molecular typing of Listeria monocytogenes from fish farming and processing companies. Int J Food Microbiol 112 (2):138–146

Miettinen MK, Siitonen A, Heiskanen P, Haajanen H, Björkroth KJ, Korkeala HJ (1999) Molecular epidemiology of an outbreak of febrile gastroenteritis caused by *Listeria monocytogenes* in cold-smoked rainbow trout. J Clin Microbiol 37(7):2358–2360

Miller WA, Miller MA, Gardner IA, Atwill ER, Byrne BA, Jang S et al (2006) Salmonella spp., Vibrio spp., Clostridium perfringens, and Plesiomonas shigelloides in marine and freshwater invertebrates from coastal California ecosystems. Microb Ecol 52(2):198–206

Mishra A, Taneja N, Sharma M (2012) Environmental and epidemiological surveillance of Vibrio cholerae in a cholera-endemic region in India with freshwater environs. J Appl Microbiol 112 (1):225–237

Montville TJ, Matthews KR (2012) Physiology, growth, and inhibition of microbes in foods. Food Microbiol: Fundam Front:1–18

Murphy HM, Pintar KD, McBean EA, Thomas MK (2014) A systematic review of waterborne disease burden methodologies from developed countries. J Water Health 12(4):634–655

Naumova EN, Jagai JS, Matyas B, DeMaria A, MacNeill IB, Griffiths JK (2007) Seasonality in six enterically transmitted diseases and ambient temperature. Epidemiol Infect 135(2):281–292

NICD (2019) National Institute for Communicable Diseases (NICD), South Africa. http://www.nicd.ac.za/wp-content/uploads/2018/08/An-update-on-the-outbreak-of-Listeria-monocytogenes-South-Africa.pdf. Accessed 7 Jan 2019

Nichols GL, Richardson JF, Sheppard SK, Lane C, Sarran C (2012) Campylobacter epidemiology: a descriptive study reviewing 1 million case in England and Wales between 1989 and 2011. BMJ Open 2(4):e001179

Noyes PD, McElwee MK, Miller HD, Clark BW, Van Tiem LA, Walcott KC et al (2009) The toxicology of climate change: environmental contaminants in a warming world. Environ Int 35 (6):971–986

Paerl HW, Huisman J (2009) Climate change: a catalyst for global expansion of harmful cyanobacterial blooms. Environ Microbiol Rep 1:27–37. https://doi.org/10.1111/j.1758-2229.2008.00004

Pangare G, Idris L (2012) Water and health security. In: Bigas H, Morris T, Sandford B, Adeel Z (eds) The global water crisis: addressing an urgent security issue (Series ed: Axworthy TS)

Paterson RRM, Lima N (2011) Further mycotoxin effects from climate change. Food Res Int 44 (9):2555–2566

Patz JA, Campbell-Lendrum D, Holloway T, Foley JA (2005) Impact of regional climate change on human health. Nature 438(7066):310

Pearce W, Holmberg K, Hellsten I, Nerlich B (2014) Climate change on Twitter: topics, communities and conversations about the 2013 IPCC Working Group 1 report. PLoS One 9 (4):e94785

Raffa RB, Eltoukhy NS, Raffa KF (2012) Implications of climate change (global warming) for the healthcare system. J Clin Pharm Ther 37(5):502–504

Reidl J, Klose KE (2002) Vibrio cholerae and cholera: out of the water and into the host. FEMS Microbiol Rev 26(2):125–139

Sayeed S, Uzal FA, Fisher DJ, Saputo J, Vidal JE, Chen Y et al (2008) Beta toxin is essential for the intestinal virulence of *Clostridium perfringens* type C disease isolate CN3685 in a rabbit ileal loop model. Mol Microbiol 67(1):15–30

Scallan E, Hoekstra RM, Angulo FJ, Tauxe RV, Widdowson MA, Roy SL et al (2011) Foodborne illness acquired in the United States-major pathogens. Emerg Infect Dis 17(1):7

Schijven J, Bouwknegt M, de Roda Husman AM, Rutjes S, Sudre B, Suk JE, Semenza JC (2013) A decision support tool to compare waterborne and foodborne infection and/or illness risks associated with climate change. Risk Anal 33(12):2154–2167

Schmidhuber J, Tubiello FN (2007) Global food security under climate change. Proc Natl Acad Sci 104(50):19703–19708

Schuster CJ, Aramini JJ, Ellis AG, Marshall BJ, Robertson WJ, Medeiros DT, Charron DF (2005) Infectious disease outbreaks related to drinking water in Canada, 1974–2001. Can J Public Health 96(4):254–258

Springmann M, Mason-D'Croz D, Robinson S, Garnett T, Godfray HCJ, Gollin D et al (2016) Global and regional health effect of future food production under climate change: a modelling study. Lancet 387(10031):1937–1946

Stenfors Arnesen LP, Fagerlund A, Granum PE (2008) From soil to gut: Bacillus cereus and its food poisoning toxins. FEMS Microbiol Rev 32(4):579–606

Su YC, Liu C (2007) Vibrio parahaemolyticus: a concern of seafood safety. Food Microbiol 24 (6):549–558

Tam CC, Rodrigues LC, O'brien SJ, Hajat S (2006) Temperature dependence of reported campylobacter infection in England, 1989–1999. Epidemiol Infect 134(1):119–125

Tam CC, Rodrigues LC, Viviani L, Dodds JP, Evans MR, Hunter PR et al (2012) Longitudinal study of infectious intestinal disease in the UK (IID2 study): incidence in the community and presenting to general practice. Gut 61(1):69–77

Thomas KM, Charron DF, Waltner-Toews D, Schuster C, Maarouf AR, Holt JD (2006) A role of high impact weather events in waterborne disease outbreaks in Canada, 1975–2001. Int J Environ Health Res 16(03):167–180

Tirado MC, Clarke R, Jaykus LA, McQuatters-Gollop A, Frank JM (2010) Climate change and food safety: a review. Food Res Int 43(7):1745–1765

Van Elsas JD, Semenov AV, Costa R, Trevors JT (2011) Survival of Escherichia coli in the environment: fundamental and public health aspects. ISME J 5(2):173

Vezzulli L, Brettar I, Pezzati E, Reid PC, Colwell RR, Höfle MG, Pruzzo C (2012) Long-term effects of ocean warming on the prokaryotic community: evidence from the vibrios. ISME J 6 (1):21–30

Walker JT (2018) The influence of climate change on waterborne disease and legionella: a review. Perspect Public Health 138(5):282–286

Westrich JR, Ebling AM, Landing WM, Joyner JL, Kemp KM, Griffin DW, Lipp EK (2016) Saharan dust nutrients promote Vibrio bloom formation in marine surface waters. Proc Natl Acad Sci 113(21):5964–5969

WHO (2015a) Climate change and health, fact sheet no. 266, September 2015, WHO. http://www.who.int/mediacentre/factsheets/fs266/en/. Accessed 14 Oct 2015

WHO (2015b). Waterborne diseases – MDGs-SDGs 2015. https://www.who.int/gho/publications/mdgs-sdgs/MDGs-SDGs2015_chapter5_snapshot_waterborne_diseases.pdf?ua=1

Wilkes G, Edge T, Gannon V, Jokinen C, Lyautey E, Medeiros D et al (2009) Seasonal relationships among indicator bacteria, pathogenic bacteria, Cryptosporidium oocysts, Giardia cysts, and hydrological indices for surface waters within an agricultural landscape. Water Res 43 (8):2209–2223

World Health Organization (2004) World Health Report, Geneva. https://www.who.int/globalchange/publications/climchange.pdf

World Health Organisation (2013) The global view of Campylobacteriosis; report of an expert consultation in Utrecht, Netherlands. World Health Organisation, Geneva

World Health Organization (2017) Progress on drinking water, sanitation and hygiene: 2017 update and SDG baselines. World Health Organization, Geneva

World Health Organization (2018a). Health topics. Foodborne diseases. Accessed 06 Mar 2018. https://www.who.int/health-topics/foodborne-diseases#tab=tab_1

World Health Organization (2018b) WHO methods and data sources for global burden of disease estimates 2000–2016. Global Health Estimates Technical Paper WHO/HIS/IER/GHE/2018.4. WHO, Geneva

World Health Organization (2019) Disease outbreak news, March 28, 2018. https://www.who.int/csr/don/28-march-2018-listeriosis. Accessed 7 Jan 2019

Yamazaki K, Esiobu N (2012) Environmental predictors of pathogenic Vibrios in South Florida coastal waters. Open Epidemiol J 5(1)

Zeyad MT, Kumar M, Malik A (2019) Mutagenicity, genotoxicity and oxidative stress induced by pesticide industry wastewater using bacterial and plant bioassays. Biotechnol Rep 24:e00389

Climate Change: Any Dangers from Antimicrobial Resistant Bacteria? 10

Charles W. Knapp, Ronald Turner, Emmanuel Salifu, Sadia Khan, Mark Stillings, and Rebecca Tonner

Abstract

Antimicrobial resistance represents a threat whereby micro-organisms (particularly bacteria) become resistant to antibiotics and disinfectants, thus complicating our ability to treat and prevent infections. This is often developed by sub-inhibitory exposures to such drugs and/or disinfectants, but it can be caused by (often negative) changes in the environment as a stress–response mechanism to environmental conditions, such as temperature, salinity, metals (potentially toxic elements), and organic pollutants.

Here, we explored possible mechanisms by which climate change could either directly (by changes in temperature and/or precipitation) or indirectly—such as shifts in human populations, disease vectors, agriculture, water availability, glaciation, and hydrology— affect the development or transfer of antimicrobial resistance in bacteria.

Keywords

Antibiotic resistance · Pollution · Pathogens · Agriculture · Water

Climate change is altering the earth's atmosphere in several ways, being expected to generate altered air temperatures and lead to more extreme changes in precipitation across the globe. The worldwide average surface temperature has increased 0.74 °C over the last century (Rosenzweig et al. 2007). These effects are not anticipated to be

---

C. W. Knapp (✉) · R. Turner · E. Salifu · M. Stillings · R. Tonner
Department of Civil and Environmental Engineering, University of Strathclyde, Glasgow, UK
e-mail: charles.knapp@strath.ac.uk

S. Khan
Department of Environmental Engineering, NED University of Engineering and Technology, Karachi, Pakistan

S. A. Lone, A. Malik (eds.), *Microbiomes and the Global Climate Change*,
https://doi.org/10.1007/978-981-33-4508-9_10

geographically uniform, and have generated multiple changes, including altered frequencies of extreme weather conditions in different parts of the world. The most significant differences in air temperature and climate classification are predicted to occur at more northern latitudes. In comparison, precipitation increases are present at both extreme northern latitudes and also sporadically across regions of the equator, including parts of India, Saudi Arabia, and East Africa (Beck et al. 2018). These changes in climate will translate to altered conditions relevant to microbial growth, at the scale of both individual microorganisms and at the community level. These changes will be relevant to all forms of microbial life, including bacteria, fungi, and viruses.

Climate change will also impact human health, increasing the prevalence of heat related diseases and leading to altered migration patterns. There is limited discussion in the literature regarding the mechanisms through which climate change will alter the development or dissemination of antimicrobial resistance (AMR) in the environment. However, it has been mentioned that global warming contributes to antibiotic resistance (AR) in the environment (e.g., MacFadden et al. 2018). Rather than attempt to list all previous articles that have been published on the subject, we will instead discuss current issues involving AMR and how they may be impacted by climate change.

Both climate change and antimicrobial resistance have been recognised as interdisciplinary problems related to the "tragedy of the commons" (Singer et al. 2016)—in which individuals behave according to one's self interests, but contrary to the common good of others. A major misconception exists that the development of AMR stems from bacterial exposure to antibiotics or synthetic antibacterial compounds; the extent of this can be argued. It has, however, become more evident that anthropogenic pressures accelerate the presence of AMR (Forsberg et al. 2014), but it can also be presumed that inherent changes in natural landscapes could contribute as well—although these changes may occur at a slower pace and their contribution may be more difficult to elucidate.

There is strong evidence demonstrating that pollution, even at relatively low concentrations, can produce shifts in bacterial communities (the "microbiome"). Resistance traits within a microbial community (i.e., the "resistome") propagate in the presence of a natural selection factor, which ultimately creates resistant populations of bacteria. Once a resistance trait is selected, host bacteria can transfer genes between individuals (horizontal gene transfer) via transferrable genetic elements (e.g., plasmids) creating an enhanced resistome (Wright 2007, 2010). Thus resistance can spread among microorganisms once a gene enters a system, including target organisms that are pathogenic (Dantas and Sommer 2012).

Microorganisms can develop resistance to many environmental factors, including heavy metals, disinfectants, nutrient conditions, and temperature. However, resistance to antibiotics is perhaps that of greatest contemporary concern in society. It has consequences which impede the efficacy of infection control and treatment—with health-related, economic, and societal consequences. It is commonly believed that this type of resistance is only driven by the excessive use of sub-inhibitory concentrations of antibiotics; again, the extent can be argued. While this mechanism remains the major contributor to antibiotic resistance, it has been recently observed that pressures from environmental pollutants can also provide a level of selection pressure.

Bacterial communities respond genetically to pollutants via evolved mechanisms for their self-protection (Alonso et al. 2001). For instance, metals (or potentially toxic elements) in the environment may contribute to microbial cellular health as a micronutrient, while at elevated levels generating a stress response and resistance development (Beaber et al. 2004; Ashbolt et al. 2013). An example of this response has been observed in the development of copper-resistance mechanisms in bacteria isolated from soil with a high-metal content (Berendonk et al. 2015). Other metals, such as chromium and mercury, have limited biochemical merit and cause oxidative stress. Unfortunately, defence-associated metal resistance genes are often closely associated with those responsible for AMR as both reside on mobile genetic elements.

This co-selection of metal resistance (or other resistance mechanisms, e.g., those against toxic organics or disinfectants) and antibiotic resistance can occur by one of two processes: co-resistance or cross-resistance (Baker-Austin et al. 2006). Co-resistance occurs when selection of one phenotype simultaneously selects for other genes located on the same genetic element. On the other hand, cross-resistance occurs when the antibiotic and metal have similar biochemical pathways or targets in the cell. Consequently, when a resistance response is triggered, the cell defence becomes effective against both metal and antibiotic toxicants. As such, the environment acts both as a reservoir of resistance traits and a bioreactor containing chemical stressors, providing opportunities for genetic exchange. There is, therefore, significant potential for these traits to disseminate to clinically relevant pathogens.

The relationship between metal tolerance (or resistance) and AR has been known; reviews can be found in the literature (e.g., Baker-Austin et al. 2006; Martinez 2009; Seiler and Berendonk 2012; Perry and Wright 2013). However, microorganisms of clinical significance have been found resistant to multiple antimicrobials, including metals, by susceptibility assays (e.g., Marques et al. 1979; Dhakephalkar and Chopade 1994; Bass et al. 1999; Ghosh et al. 2000; Guo et al. 2014), suggesting that a link between genetic traits exists.

In the environment, relationships between metals and AR proliferation were first noticed in highly contaminated areas, including outflows of insufficiently treated wastewater and biosolids (Graham et al. 2011; Knapp et al. 2012; Su et al. 2015), agricultural wastes (Ji et al. 2012; Zhu et al. 2013; Li et al. 2015), industrially contaminated areas (Stepanauskas et al. 2005; Wright et al. 2006; Graham et al. 2011; Knapp et al. 2012; Abella et al. 2015; Hu et al. 2016; Rodgers et al. 2018, 2019, 2020), and direct application via metal exposure experiments (Berg et al. 2005, 2010; Stepanauskas et al. 2006; Knapp et al. 2011). AR presence in these cases has often been indicative of exposure to elevated metal pollution.

Less information exists on less-impacted soils and basal (e.g., pristine natural) levels, but evidence suggests that risks still exist. The geochemical nature of soils, collected in the 1970s for a national archive, had correlations with various antibiotic resistance genes found among their bacterial communities—including those for tetracyclines, beta lactams, and erythromycin (Knapp et al. 2011). Similar patterns observed in residential soils, all have sub-regulatory levels of metals (Knapp et al. 2017). In some cases, the metal content may not directly correlate with AR genes, but increase the mechanisms (e.g., integrases) by which genes may become horizontally transferred to other bacteria—thus increasing the risk via frequency of potential exchange.

Besides metals, some persistent organic compounds will co-select for antimicrobial (including antibiotic) resistance genes, including pesticides (e.g., Anjum and Krakat 2016), polychlorinated biphenyls (Guidice et al. 2013), and disinfectants (Tandukar et al. 2013; Khan et al. 2016a, b, 2017; Zhang et al. 2016). Additionally, polycyclic aromatic hydrocarbons (PAHs) are ubiquitously found environmental pollutants that have the capability to enhance the abundance of AMR in microbial communities. PAHs are organic compounds with multiple aromatic rings fused in a single compound. PAHs can either be phytogenic—originating from plant materials, petrogenic—from petroleum products or sources, or pyrogenic—resulting from combustion processes. Innate with their potential sources are the distribution and abundance of compounds, and their associated toxicities.

The association of PAHs with AMR has been found in lake-sediment pollution (Máthé et al. 2012), coastal areas (Wang et al. 2017), and soils (Chen et al. 2017). Areas with PAH contamination have been reported to have increased prevalence of genes encoding efflux pumps, detoxifying transporters (Chen et al. 2017), and also conjugative transfer via class I integrons (e.g., Wang et al. 2017).

Since metals and certain PAHs (e.g.) are also widespread in the environment and do not immediately degrade, these compounds can potentially provide a long-term selection pressure. Resistance development is not only due to the presence of high concentration of antimicrobials; rather low concentrations of contaminants, which cause cellular stress to microorganisms, could also provide the selective pressure.

Climate change and AMR represent contemporary topics of discussion. In a recent meeting among the members of Centre for Water, Environment, Sustainability and Public Health (a University of Strathclyde, Department of Civil & Environmental Engineering research group), various cross-disciplinary researchers had a discussion about the possible impacts of climate change on the patterns of antimicrobial resistance. Some of their thoughts are outlined as:

1. Pathogen; their fate, prevalence and gene transfer.
2. Agriculture-related changes.
3. Water distribution and quality.
4. Melting glaciers and permafrost thaws.
5. Hydrological changes and legacy pollution.

Here, we present our thoughts.

10.1 Pathogens

10.1.1 Pathogen Prevalence

At the community scale, changes in precipitation levels and temperatures can be expected to alter the spatial distribution of pathogen vectors. These vectors may be biological (e.g., flies) or non-biological (e.g., dust, fomites) (Nordor et al. 2018).

For example, climate alterations have the potential to alter the range of disease-spreading insects such as the mosquito *Anopheles*, a vector for *Plasmodium*—an obligate parasite of the mosquito. Many *Plasmodium* species cause malaria, with *Plasmodium falciparum* putatively linked to carcinogenesis (Lehrer 2010; Nordor et al. 2018). Elevated CO_2 and changes in air temperature are climate effects with the greatest potential to alter the distribution of malaria via changes in disease vector distribution. Temperature changes will have an indirect impact on soil moisture dynamics, providing a further mechanism to alter the life cycle of malarial vectors. The presence of stagnant water due to increased precipitation rates would also increase the breeding area for the *Anopheles* vector. Therefore, climate change will alter the life cycle of the *Anopheles* vector, along the life cycle of *Plasmodium* itself (Dhimal et al. 2015; Le et al. 2019).

Changes in distribution of malarial areas are anticipated due to climate change, leading to higher malarial burdens (Campbell-Lendrum et al. 2015). Moreover, climate change is expected to impact a range of other vector-borne pathogens in a similar manner. Examples include dengue fever, west Nile virus, tularemia, rabies, chikungunya, and Lyme disease (Green et al. 2008). The principal vector of Lyme disease—the arthropod *Ixodes scapularis*—is anticipated to experience a 213% increase in suitable habitat area within the North American continent by 2080. This effect is anticipated to be driven largely by expansion of viable breeding territory into Canada due to alterations in temperature caused by greenhouse gas emissions and sulphate aerosols (Brownstein et al. 2005).

Climate change will generate changes in daily life, such as increased use of air conditioning systems due to elevated indoor temperatures (Vardoulakis et al. 2015). It has been recognised in the past that indoor environments can harbour microbial communities which vary with geographic location. For example, temperate climate zones harbour far more diverse fungal communities than tropical climate zones (Amend et al. 2010). Significant expansions in the temperate "humid continental climate" (Dfb) are anticipated by 2080, particularly across Russia and parts of India. This would suggest that regions with significant human populations may be exposed to fungal communities of increased diversity due to climate change, with potential impacts on human health.

Heating, ventilation, and air conditioning systems as well as settled dust and outdoor air are well recognised as sources of indoor airborne microorganisms (Prussin and Marr 2015). Airborne fungal particles are of particular concern, as a cause of allergies and even cancers. Climate change may lead to significant effects on global fungal populations, with the spread of thermo-tolerant species being anticipated (Garcia-Solache and Casadevall 2010). These effects may be amplified in urban areas (McLean et al. 2005). The reduction in the thermal gradient between environment and mammalian hosts generated by climate change is anticipated to increase the range of potential fungal diseases (Garcia-Solache and Casadevall 2010). Any increased use of domestic air circulation systems due to climate change—for either heating or cooling—may become linked to alterations in aerobiology, with subsequent potential impacts on human health (Fernstrom and Goldblatt 2013).

Specific geographic sites for increased fungal diseases have been identified in North America, with the increase in dry summers and increased precipitation late in the year enhancing the spread of *Blastomyces dermatitidis* and *Coccidioides imitis*, both of which are associated with varied health effects, including bone disease (Green et al. 2008).

Within the USA, a 2–4% rise in occurrence of antibiotic resistant bacteria belonging to the genera *Escherichia*, *Klebsiella*, and *Staphylococcus* has been associated with regions of elevated local temperature versus cooler areas (MacFadden et al. 2018). There is a clear link between increased temperature and population density, with increased prevalence of antimicrobial resistance being linked to these elevated temperatures (MacFadden et al. 2018). These effects were observed to be consistent across almost all classes of antibiotics and pathogens, and this association has grown stronger over time. While the selective pressure of antibiotic use is considered to be the largest contributor to the emergence of antibiotic resistance, it has become apparent that climate-change associated factors such as increases in air temperature (Beck et al. 2018) and increased urban density (Güneralp et al. 2017) have potential to increase the prevalence of antimicrobial resistance in many bacterial species relevant to human health.

The negative effects of antimicrobial resistance are ever-increasing due to the uncontrolled use of antimicrobials for the treatment of various infectious diseases. Changing patterns in pathogen distribution are expected to expand the geographic range and intensity of antimicrobial use. For example, the expansion in the viable area of *Ixodes scapularis* into Canada is likely to increase the use of doxycycline and ceftriaxone to combat the causative pathogenic bacterium *Borrelia*. While the emergence of antibiotic resistance has not been observed in *Borrelia* (Sharma et al. 2015), the increased use of doxycycline and other antibiotics such as amoxicillin and even macrolide antibiotics may be associated with increased resistance of other species present in the environment, as observed with *Streptococcus pneumonia* (Karcic et al. 2015). Similar effects may be anticipated due to increased use of antifungal treatments in response to elevated levels of fungal diseases linked to climate change (Fernstrom and Goldblatt 2013; Prussin and Marr 2015; Vardoulakis et al. 2015), leading to altered geographic distribution of antifungal resistance.

10.1.2 Gene Transfer

HGT (horizontal gene transfer) is a process through which organisms can exchange genetic information without reproduction. This process allows the exchange of functional genetic material such as virulence, xenobiotic metabolism, and antibiotic resistance genes (ARG). This transfer can occur even at the domain level—such as between bacteria and archaea—and occurs in many different environments, including soil, seawater, and within the human body (Aminov 2011). HGT is linked to emergence of some microbial traits harmful to human health and wellbeing, including antibiotic resistance.

Ecology and environmental conditions have been identified to significantly influence the gene transfer among microorganisms (Fuchsman et al. 2017). Indeed, it is thought that ecology is even more significant in influence than phylogeny (Smillie et al. 2011). This has been evidenced by the high rates of gene transfer between bacteria and archaea in environments, which are conducive to the growth of species from each of these domains. This effect is particularly pronounced in the extremophilic bacteria and archaea, such as thermophiles and halophiles.

Changes in climate are linked with increased salinity in locations such as the Arctic Ocean (Greene and Pershing, 2007), as well as coastal estuaries and aquifers (Titus 1989). These effects may expand the geographic distribution of growth sites suitable for extremophiles, with a concomitant increase in horizontal gene transfer between highly varied species of bacteria and archaea.

In soils, temperature, moisture levels, pH, and soil type have all been associated with changes in rates of conjugal plasmid transfer (Aminov 2011). Changes in soil temperature have been associated with climate change, dependent upon geographic location. For example, warming trends have been observed in the Tibetan Plateau (Fang et al. 2019), Alaska and Siberia (Oelke and Zhang 2004), and Russia (Zhang et al. 2001). There is also potential for climate-change associated soil acidification in a number of environments (Rengel 2011). Associations between climate change and soil moisture are complex, being influenced by changes in precipitation and, therefore, potentially broader climate classification changes (Seneviratne et al. 2010; Beck et al. 2018).

Increasing pH from 4.5 to 8.5 is linked with a general increase in gene transfer rate at 22 °C, with the effect becoming less pronounced at 37 °C (Rochelle et al. 1989). Similarly, increasing temperature from 22 °C to 29 °C is linked to a decrease in gene transfer rates in *Agrobacterium tumefaciens* (Dillen et al. 1997).

In summary, climate change can be expected to increase the levels of HGT in the environment through a range of mechanisms, such as increased sea and soil temperatures, soil acidification, and changes in ocean salinity. Altered microbial community compositions due to climate change effects may also bring previously disparate microbial species into closer contact. Increased levels of HGT have been directly linked to the generation and spread of antibiotic resistance in the environment (Ventola 2015). Therefore, climate-change effects have significant potential to increase the spread of antibiotic resistance genes.

10.2 Agriculture

The occurrence and prevalence of antimicrobial- and drug-resistant organisms and genes have been associated with the selective pressures on the usage of antibiotics in both clinical and agricultural sectors (Levy 1997). Agricultural practices have been considered a major culprit for the increase in antibiotic-resistant strains of bacteria (Silbergeld et al. 2008; Kennedy 2013; John 2014). For instance, in the USA alone, about 13 million kg, or 80% of all antibiotics, has been associated with livestock production annually (Hollis and Ahmed 2013). On a global scale, Van Boeckel et al. (2015) reported that 45–172 mg of antimicrobial agents is consumed annually per kg of animal (e.g., cattle, chicken, and pigs) produced.

10.2.1 AMA and AMR Pathways in Agriculture

Agriculture has been recognised to have many reservoirs/hotspots and pathways involved in the dissemination of AMA (antimicrobial agents) and AMR within the agricultural environment and the food processing industry (Thanner et al. 2016). The general transmission route or pathway is typically from animal (livestock) to animal-derived manure to soil to water and to sediments, with human (farmers) and animals as hosts. Several authors have identified hot spots of ARG (antimicrobial resistance genes) and ARB (antimicrobial resistant bacteria) to include the digestive tract of farm animals and humans, manure, wastewater treatment plants, and the soil or rhizosphere (Zhu et al. 2013; Wu et al. 2014; Woolhouse et al. 2015; Thanner et al. 2016; Manyi-Loh et al. 2018). Other possible water-related hotspots for ARG and ARB include: rivers and lakes, sediments, biofilms in aqua-cultural systems, irrigation systems, slaughterhouses, as well as on plant surfaces (Thanner, et al. 2016). Soil and water tend to function as mixing media or sinks for the mobile genetic elements (MGEs), as well as sources of AMR (Riesenfeld et al. 2004; Xiao et al. 2016).

Antibiotics such as tetracyclines, sulphonamides, lincosamides, aminoglycosides, macrolides, β-lactams, and pleuromutilins with potentially adverse effects are administered to livestock globally for treatment or therapeutic purposes, for sub-therapeutic purposes for growth promotion, as prophylaxis and/or metaphylaxis (Mellon et al. 2001; Apata 2009; Landers et al. 2012; Abou-Raya et al. 2013; Finley et al. 2013; Pham Kim et al. 2013; De Briyne et al. 2014; Baynes et al. 2016; Manyi-Loh et al. 2018). This has led to increase in resistant bacteria within the gut flora of livestock and the deposition of antibiotic residues in products such as meat/muscles/fat, milk, eggs, liver, and kidney available for human consumption (Witte 1998; Aarestrup 1999; Manyi-Loh et al. 2018). Further, animal excretions (urine and faeces) containing antibiotic residues, ARBs, and ARGs may be released into the agricultural environment as raw manure often used for soil fertilisation (Acar and Moulin 2006).

The dissemination and fate of antimicrobials within the agricultural sector depend on:

1. the pattern of administration of the antimicrobials, the metabolic activities and transformation processes within the animal, the potential of releasing the compounds into the environment through animal excrements (European Medicines Agency 2018);
2. their physicochemical properties such as shape, size, molecular structure, solubility, and water repellence (Thanner et al. 2016; European Medicines Agency 2018);
3. environmental effects such as changes in climatic conditions, soil types, erosion, hydrology—surface and groundwater (Carlson and Mabury 2006; Davis et al. 2006; Kuchta et al. 2009; Park and Huwe 2016; Pan and Chu 2017; European Medicines Agency 2018);
4. environmental fate, including sorption properties (Tolls 2001; Lin and Gan 2011; Thanner et al. 2016), abiotic or biotic transformation/degradation processes (Thiele-bruhn and Peters 2007; Reichel et al. 2013; Cui et al. 2014; Manzetti and Ghisi 2014; Duan et al. 2017); and,
5. uptake by crops or pasture (Kumar et al. 2005; Dolliver et al. 2007; Kuchta et al. 2009; Carter et al. 2014).

Soil acts as a natural reservoir for AMA and ARBs since a wide range of known and unknown (or uncultured) AMR determinants have been detected in non-fertilised soils (Riesenfeld et al. 2004; Maron et al. 2013; Marti et al. 2013; You and Silbergeld 2014; Xiao et al. 2016). However, antimicrobial agents are mostly introduced to soil via fertilisation with raw manure collected from animal sources treated with antimicrobials, bringing metabolites of AMA and microbes with ARG existing on mobile genetic elements (MGE) (Hamscher et al. 2005; Binh et al. 2008; Zhu et al. 2013; Wolters et al. 2015; Thanner et al. 2016). Additionally, AMAs are added through the use of irrigation with wastewater and treatment of crop diseases with antibiotics (Finley et al. 2013; Oluyege et al. 2015). Depending on soil temperature, pH, nutrients, and oxygen concentration as well as microbial diversity, ARG may be transferred to soil microbes through HGT (i.e., via conjugation, transduction, and transformation mechanisms), resulting in enrichment of taxa and decline in the population of organisms significant for soil quality (van Elsas and Bailey 2002; Aminov 2011; Perry and Wright 2013; Ding et al. 2014; Fahrenfeld et al. 2014; Jechalke et al. 2014; Forsberg et al. 2015). According to the European Medicines Agency (2018), there may be a positive correlation between the factors that promote the persistence of a compound in soil and factors that enhance their potential to select for resistance within the same microbiome. It has been reported that an increase in soil moisture content from 15% to 25% led to a decrease in the half-life of an antibiotic (sulfadimethoxine) from 10.4 to 4.9 days (Wang et al. 2006), while a similar increase in temperature resulted in the degradation of norfloxacin (Yang et al. 2012). In contrast, other studies have found that warmer temperatures (30–45 °C) could enhance the antimicrobial effects and antibiotic resistance of common pathogens and bacterial biofilms (Rachid et al. 2000; Hajdu et al. 2010; MacFadden et al. 2018).

The aquatic environment also serves as source and sink of antibiotic-resistant bacteria and resistance genes (Biyela et al. 2004; Baquero et al. 2008). Also, antimicrobial agents (or mobile resistant genes) in raw manure can be translocated via the action of rainfall runoff and/or soil erosion from the surface of fertilised soils to nearby surface or groundwater bodies, depending on the properties of the AMAs, the soil, and hydrological effects (Pollard and Morra 2018). Although studies on the mobility and transport of antimicrobials in the environment are scarce, the European Medicines Agency (2018) suggests that these behaviours will vary considerably for different antibiotics.

Humans, especially those involved in agriculture (e.g., farmers) typically become hosts to commensal organisms and resistant pathogens via ingestion. Farmers and their families can transmit AMAs to livestock through contact, since the human skin and digestive tract serve as reservoir for some of these pathogens (Acar and Moulin 2006). However, this route of transmission depends on several factors including geographical location or region, farming system adopted, size of farm, hygiene, ethno-cultural and religious practices (Lozano et al. 2016).

10.2.2 Influence of Climate Change on Agriculture

Climate change will directly impact agricultural climate resources (such as precipitation and temperature) and cause changes in livestock productivity, crop/plant community composition and productivity, soil microbial communities, and the general agro-related ecosystem, as well as surface and groundwater hydrology (Kim et al. 2010; Bardgett et al. 2013). It has been predicted that climate change will result in further reduced agricultural productivity in regions with hitherto low agricultural productivity levels (Harvey et al. 2014). Some studies report that changes in the amount and quality of feed as well as increased heat stress may result in a 25% decrease in animal production (van de Steeg and Tibbo 2012; Lewis, et al. 2018) with higher temperatures leading to lower daily milk yield (Verner 2012).

10.2.3 Possible Changes Induced by Climate Change on AMR in Agriculture

Thanner et al. (2016) provide a comprehensive list of knowledge gaps in AMR in agriculture, justifying the difficulty to reach valid conclusions on the prediction of key factors, like climate change, on AMR prevalence in the agricultural sector. The discussions in the aforementioned sections also show that for the few available studies, there are contradictions in findings regarding the behaviour of AMA/ARG under variable climatic conditions. It has been suggested that global-scale surveillance programmes utilising comparable test methods may be necessary to provide data on areas like the occurrence/abundance, transfer pathways, and factors influencing the selection, spread, and persistence of AMAs/AMR in the different compartments of agriculture, and the environment in general (Thanner et al. 2016;

Manyi-Loh et al. 2018). This will contribute towards better understanding, prediction, and management of AMR in agriculture.

Nevertheless, the socio-economic trends induced by climate change are expected to influence land use patterns (Karmakar et al. 2016) and general agricultural practices, with possible indirect impacts on AMR. The Food and Agriculture Organization of the United Nations (2016) predicts that climate shocks may lead to economic downturn for farmers, especially small-scale farmers, as productive capital or assets may be sold to absorb the shock. In several countries in the Near East and North Africa region, extreme droughts and climatic events could result in crop failure and lead to food insecurity, malnutrition, famine, and starvation (Elasha 2010; WFP and ODI 2015). These factors coupled with the growing population imply that there would be greater demands for food (including animal proteins) across the world, thereby placing pressure on agricultural production (Padol et al. 2015). The tendency to use antibiotics to promote animal growth will become almost inevitable especially in developing countries. Currently, the factors influencing the usage of antibiotics vary considerably across regions and countries (World Health Organisation 2012). These factors include: land use, sources of contaminated water, animal demography as well as national and international trade and policies (World Health Organisation 2012). Antimicrobials that have been banned in several countries, including developed countries, are still being used in developing countries (Moyane et al. 2013; Adebowale and Adeyemo 2016). It has also been predicted that to meet increasing consumer demands, the use of antibiotics in Brazil, Russia, India, China, and South Africa (known as the BRICS countries) is expected to double (Van Boeckel et al. 2015). China is reported to be the highest producer and user of antibiotics for humans and animals (Manyi-Loh et al. 2018). A study by Lee et al. (2001) has predicted that, in the future, a significant percentage of animal-derived proteins packaged as food for human consumption would have received some sort of chemotherapeutic or a prophylactic agent.

10.3 Water Distribution and Quality

10.3.1 Surface Waters

Drought and flood are two extreme conditions, resulting from climate change that can dramatically affect the water quality, having significant effects on concentration of metals, organic pollutants, and pathogenic bacteria (Xia et al. 2015). The increase in the world average temperature could have greater impact on world's water resources and decrease the availability of freshwater resources. There are many direct and indirect effects that could have lasting impacts: (1) re-distribution of water and its availability as a source of drinking water; (2) changes to the water quality as driven by increased biochemical activities; (3) changes in the fate and presence of pathogens likely involved in the AMR process; and (4) impacts on treatability of water, all of which could impact the risks associated with AMR.

Climate change is a major contributing factor in the disruption of water cycle and water supply systems (de Oliveira et al. 2015), and these strongly relate to elevated temperature (Xia et al. 2015). Increased temperatures accelerate evaporation and transpiration leading to water loss from soil and plants (Bates et al. 2008). Many water-supply systems face water scarcity and reduced water quality, and climate change is considered as one of the reasons for this shortage of water (Delpla et al. 2009).

Climate change deteriorates water quality (Xia et al. 2015) by different biochemical processes (Delpla et al. 2009). As we know, microorganisms support the existence of life on the earth, and climate changes depend on the response of the microorganisms to the environment (Cavicchioli et al. 2019), including their key role in different nutrient cycles (Azam and Malfatti 2007). These biochemical and microbiological reactions are temperature dependent (Whitehead et al. 2009); for example, higher temperatures reduce the concentration of dissolved oxygen in water. Anoxia, as we already know in eutrophic systems, leads to further changes in water quality, such as the further release of nutrients and metals. This cascades to increased rates of nutrient assimilation and biomass production—including algal blooms and anoxic dead-zones (de Oliveira et al. 2015). However, long-lasting results may not be clear; while immediate-nutrient effects are observable, studies show that climate change (i.e., increasing temperatures) may decrease nutrient concentrations in surface river waters (Alam et al. 2013; Hosseini et al. 2017), caused by cumulative assimilation by biota.

Anthropogenic activities tend to have a direct effect on water pollution while climate change can have indirect effects on decrease in water quality in supply system (Bates et al. 2008). Water temperatures have dramatic impacts on its treatability for drinking water and may require further treatments to maintain the potability. Increased water temperatures have the unfortunate consequence of requiring higher dosages of chemical disinfectants (e.g., chlorine), and they also result in increased formation of disinfectant by-products after disinfectant treatment—e.g., the formation of chloroform from chloramine-treated water after 3 days at higher temperature (Yang et al. 2007). This not only affects the potability of drinking water, but also the increased disinfectant chemical loadings can impact the microbiome.

10.3.2 Water Distribution

The obvious, but highly unfortunate consequence is that alterations in water availability will lead to the re-distribution of populations, but it will also encourage the use of waters with reduced quality to meet demand. Thus requiring additional treatment and/or distribution of water.

Water distribution systems are complex systems. They often represent environments with low concentrations of antimicrobials, e.g., chemical disinfectants, to maintain the level of assurance that the water remains potable. Chemical disinfectants, like the antibiotic therapies, are designed to exhibit either biocidal or biostatic effects on the bacteria. At elevated concentrations, they may be quite effective. But, disinfectant concentrations and efficacy decline with temperature and deterioration of water quality (e.g., chlorine demand). At some point along the distribution system, or in time, the concentrations could decline to "sub-lethal levels". Instead of inhibition, we could see population selection. Any surviving bacteria would have undergone selective pressures for resistance traits that contribute to their survival. This could be resistance traits against the disinfectant, but also other chemicals present, including those associated with poor water quality—e.g., metals.

Even if bacteria could be removed at the point of treatment, distribution systems still have the risk to become colonised by pathogens having antimicrobial resistance due to limitations in treatment technologies and leakage; this becomes a mechanism and source, by which resistant infections to human beings could result (Buse et al. 2012). This could be further exacerbated by the prolonged storage of water in (e.g.) building cisterns (Falkinham et al. 2015), where disinfection efficacy declines and the opportunity for contamination increases. Thus, water supply systems could represent an unsuspecting source of AMR, which could become worsen by declining availability, deteriorating water conditions, and challenged treatment efficacy. As such, there is a risk of spreading of antibiotic resistant bacteria to human populations via insufficiently treated drinking water (West et al. 2011).

A number of antibiotic resistant genes including *tet*, *sul*, *van*, and *amp* have been reported being transported via water bodies, and have been found in treatment plants and distribution systems (Schwartz et al. 2003; Pruden et al. 2006; Prado et al. 2008; Czekalski et al. 2012; Guo et al. 2014; Bergeron et al. 2015; Adesoji et al. 2016). The bacteria with antimicrobial resistance are not easy to treat (Levy 2002) and pose a potential health risk for the immune-compromised population (Lee et al. 2010). Due to their impacts on human health, antibiotic resistance has been announced as global health emergency (Sanderson et al. 2018), and there is an urgent need to minimise the environmental pollution due to antibiotics (Singer et al. 2016). Although the use of antibiotics has been reduced in different parts of the world, the problem is not diminishing as their resistance is widely spread among different environments. World Health Organization has started a global action plan to eradicate the problem through public awareness and sustainable investment (WHO 2015; Singer et al. 2016). Distribution systems and the complex interactions of source water availability, conditions, and treatment exemplify the challenges ahead.

10.4 Melting Glaciers and Permafrost Thaws

Warming climate in glacial areas causes huge physical ($\Delta\sigma > 2000$ MPa), chemical, and biological stress to surface and near surface environments. Ice sheets had once wide coverage during the Pleistocene (~2.5 Ma to ~12 ka) reaching ~40 °N in North America. Currently, glaciers and continental ice-sheets cover ~10% of the land; permafrost currently covers ~23.9% (Zhang et al. 2008) of exposed Northern Hemisphere land surface. The glacial cycle produces large vertical and flexural stress changes to the lithosphere (Lemieux et al. 2008). Glacial retreat results in hydromechanical changes: isostasy, permafrost melting, ice sheet un-loading, groundwater flow/chemistry, and pore fluid pressure. Deglaciation removes vertical, flexural, and hydraulic loads from the lithosphere. Lithospheric rebound occurs, pore fluid pressures change (Neuzil 2012) in response to the changes in stress. High-pressure heads during glacial advance cause groundwater to migrate from the surface to depth (Lemieux et al. 2008). When the hydrostatic pressure of ice sheet is removed, exfiltration of deep saline brines occurs. Fresh glacial groundwater becomes input to the groundwater system.

Groundwater chemical conditions and subsurface flow pathways change during the glacial cycle, having the potential to mobilise and transport: brines, heavy metals, and organics at the surface. Studies of the Laurentide ice sheet, during the Wisconsin and Pleistocene glaciations in North America show brine formation and migration causes changes to ground and surface water chemistry across a glacial cycle. Brine rich in sodium chloride formed from sea water during cold glacial periods under permafrost through a process of cryogenic brine formation. Cryogenic brines form when water freezes (Starinsky and Katz 2003), solutes are not stored in the ice and concentrate in the remaining water. Cryogenic brines form alongside permafrost (McIntosh et al. 2011) in the near surface. During static glacial periods brines and mineralised halite remain stable in the subsurface. Warm periods result in permafrost thawing and glacial melt; this results in the up-welling of saline brines (Starinsky and Katz 2003), which can then reach the surface and soil zone. Deglaciation has the largest chemical impact on groundwater: saline brines are driven to shallower depths. Melt water infiltrates and dissolves halite minerals (deposited during permafrost formation in glacial periods). Infiltrating melt water triggers microbial mediated methanogenesis (McIntosh and Walter 2005). The effect of glacial cycles results in large swings in the surface and soil water chemical conditions.

Conditions in the permafrost differ; Siberian permafrost has high methane and negative redox potential (Willerslev et al. 2004) indicating a harsh anaerobic condition. Antarctic permafrost has high redox potential and alkaline pH compared with neutral pH in Siberian sites. Thawing of permafrost is documented to release CO_2 as well as methane into the atmosphere (Hodgkins et al. 2014). However, during the thawing process CO_2 and CH_4 gases will also get dissolved into the melt water. This in turn will lead to lowering of pH in thawed permafrost. As permafrost thawing rate increases over time (Haldorsen et al. 2010), melt water surrounding patches of permafrost promote further thawing. Release of organic matter and methanogenesis CH_4/CO_2 production will increase exponentially as permafrost thawing rate increases. Increasingly rapid changes to the soil environment will cause stresses to the microbiome as a consequence of the changing climate. Previously trapped and accumulated organic matter in permafrost (Turetsky et al. 2000) is released into the water system during thawing. The increase in organic matter (OM) input from melting permafrost supplies the OM source used in methanogenesis. OM input from melting permafrost includes primary and secondary metabolites which are biologically available and can be directly up taken by plants and microbes.

Melting permafrost, in response to warming climate, releases major ions making them available in the soil zone. Release of major ions increases their concentration in fresh surface water environment affecting aquatic ecosystems. Both major ions (Ca^{2+}, Mg^{2+}, SO_4^{2-}, NO_3^-) and trace elements (e.g., Ni, Hg, Pb, Cr, As) increase with increased melting of permafrost. Release of toxic trace metals into freshwater environments can have wide reaching impacts on the ecosystem. Increased nitrate, phosphate, and sulphate are made available in melting permafrost. High nutrient input and rapid fluctuations stress the aquatic ecosystem. Rapid changes in toxic trace elements and nutrients into the environment will cause shifts in the microbiome, and the trace organics and metal may impact the resistome—or presence of the resistance genes (Colombo et al. 2018).

There are evidences for the release of heavy metals and the consequential presence of AMR (particularly metal). For example, strains of bacteria from permafrost soil samples in China show Pb resistance and Pb^{2+} biosorption (Li et al. 2017). Environmental strains of *Acinetobacter lwoffii* from permafrost contain genes for heavy metal resistance (including Cr, As, Hg, Cu, Ni, Co, Zn); the chromosomal resistance genes were homologous to plasmid genes found in *A. lwoffii* (Mindlin et al. 2016). Mercury resistant bacteria were discovered in permafrost near Kolyma, Canada (Petrova et al. 2002).

While it is expected to have associated resistance traits to elevated metal concentrations, there have been evidences of expanded spectrums. Viable bacteria found in Siberian permafrost showed antibiotic and heavy metal resistance. Resistance is thought to be developed because of the harsh conditions (constant freezing, poor nutrient availability, inorganic and organic inhibitors) the strong selection pressure (Zhang et al. 2013).

Microbes trapped in permafrost have been shown to have antibiotic resistance genes similar to modern day bacteria (D'Costa et al. 2011). Thawing ground ice and sediments can release ancient resistant bacteria to the surface environment. There is little evidence to show that increased organic matter is transported to aquatic environments, it instead likely remains in the soil and saturated zones.

10.5 Hydrological Changes and Legacy Pollution

Increased industrial activity in society is an inevitable and on-going process. With a growing demand for goods and services, pristine and rural landscapes have become transformed into urban centres with increased population densities, factories, and manufacturing. Other areas become exploited for natural resources (e.g., mining) and exploration and/or generation of energy (e.g., oil and gas industries). Many European and North American countries have experienced this Industrial Revolution during the eighteenth century; many other countries throughout the world have recently undergone, or undergoing, their own versions of revolution. Whilst development of industries and the technology became a period of societal advancement, it was also inevitably the era of pollution, where by-products of anthropogenic activities and industry were released into the surrounding environment (Krishna and Govil 2007; Reza et al. 2010) and became entrained into the soil and sediment layers as legacy pollutants.

These legacy pollutants include a variety of potentially toxic elements (e.g., Pb, Cu, Cr, Co, Ni, V, and Zn), polycyclic aromatic hydrocarbons (Rodgers et al. 2019), and bacteria which have been released into sediments due to anthropogenic pollution (Krishna and Govil 2007). Often, these contaminants that remain in the environment are rather recalcitrant, meaning either they do not degrade or do so very slowly. Evidences have been found in recent studies, for instance, (Rodgers et al. 2018, 2019, 2020) in the Clyde estuary (UK) highlight that a higher legacy of anthropogenic activity can be found in sediments from historical industrialisation. PCA was used in this study to distinguish between natural and anthropogenic contaminants to identify sites with high industrial depositions. Results indicate high levels of heavy metals and PAHs present near heavily used industrial sites of the estuary, several of which being pollution "hotspots", indicating pollutants have accumulated over time. Sediment quality were greatly diminished due to these inputs, and alterations in alkalinity and tides have had little to no impact on the concentrations of PTEs present (Rodgers et al. 2019).

Climate change can result in changes in the hydrological cycles, leading to consequential changes in precipitation, water flow, and ultimately erosion of soils or scouring of sediments (Krishna and Govil 2007). Pollutants, for example, in the sediment layers could increasingly become re-entrained into advective flows, becoming re-distributed downstream creating an additional pressures on environmental and public health.

There is not much evidence of risk with antimicrobial resistance resulting from changes in hydrology, but the extent of their presence in the environment and their relation to pollution has become increasingly known over the last few decades. The relationship between environmental conditions and the development and dissemination do apply; the risks could be hidden. We can postulate some scenarios where this may become problematic:

1. Polluted soils and sediments become a bioreactor for bacteria, subsequently enriching resistance traits. Post-industrial sediments, as mentioned previously, are carriers of adsorbed chemical pollutants that provide additional stress to the microorganisms. The prolonged degradation and retarded release of the pollutant could provide a concentration gradient of exposure, which could provide a wide range of stress conditions, and varied bacterial responses. Populations in situ can either become selected for resistance traits (via chemical toxicity), genetically modified (via increased mutation and genetic selection), or experience increased horizontal genetic transfer.
2. Entombed in polluted sediments, the bacteria could be co-present at the time. When reviewing the period of industrial revolution (in the UK), it was also the period of poor sanitation and wastewater treatment. Sewage and bacteria were discharged into major waterways that were also major industrial centre. In this case, pathogens associated with human or agricultural disease are present in elevated numbers, and could likewise experience the pollution stress. They then become stored, remaining dormant, in sediment layers with their fully developed resistome.

Resulting from either scenario, polluted sediments, harbouring highly multi-drug resistant and potentially pathogenic micro-organisms, become re-introduced into flowing water (Allen et al. 1977). This creates a biological risk to agricultural, aqua-cultural, and human health downstream, as many industries are located on rivers and estuaries due to their necessity for a water source.

The best evidence for this is a recent examination of the Clyde estuary where researchers found chemical pollutant profiles in sediment (30–50 cm depths) that corresponded to past pollutant events, especially in terms of zinc, copper, and arsenic (Rodgers et al. 2020). Examining the extracted DNA from the microbial communities obtained from slices of the sediment cores, there was a corresponding elevation of antibiotic resistance genes for: aminoglycosides, beta-lactams, FCA (fluoroquinolone, quinolone, chloramphenicol, florfenicol and amphenicol resistance genes), MLSB (macrolide, lincosamide, streptogramin B), tetracycline, vancomycin, sulphonamide, and efflux pumps; additionally, genes associated with transposons and integrons, which represent genetic transfer elements, were also increased (Rodgers et al. 2018). Additionally, Tonner et al. (2017) successfully recovered viable isolates from the deep sediment layers, consisting mostly of aquaculture pathogens: *Pseudomonas* spp. and *Aeromonas* spp. resistant to several heavy metals (e.g., Zn, Cu, Cr) and antibiotics (e.g., many beta lactams).

Hydrology in this context refers to the nature of waterways (i.e., rivers, channels, tributaries, and estuaries) and their natural flow pattern or regime near areas of industrial legacy. Many of these areas, which have the continuous stress of variation in salinity and pH's, have been exposed to a myriad of pollutant effluents affecting, one way or another, the presence of bacterial pathogens and the character of their resistome.

Unfortunately, variations in the flow of waterways can become influenced by alteration of the surrounding environment which can be triggered by climate change. The alterations can disrupt soils and sediments and cause potentially pathogenic bacteria to be re-introduced into in-stream processes. This leads to a decrease in the quality of the water in nearby sources such as estuaries subsequently posing a potential threat to human health. In summation, it is important to evaluate the health risks of post-industrial soils and waterways and their potential risk to human life. The hazards of bacteria, which have developed from industrial legacies, are yet to be fully understood.

10.6 Summary

Climate change represents a shift in the balance, and the outcomes and links are complex—and rather difficult to predict. It can only be best speculated based on the indirect results by the changes in the factors that may contribute to the development and dissemination of resistance traits; changes in the chemical composition of the environment that may create additional stress on the microorganisms (e.g., metals and polycyclic aromatic hydrocarbons); changes in the character of the soils and water; changes in human behaviour (e.g., agriculture) and distribution; or the distribution of pathogens. One common theme that may mitigate the impact of AMR would be (and currently is) the reduction of environmental stress. Further research is needed to understand the microbiome, the resistome, and the impact shifting environmental (e.g., chemical and temperature) stressors—these approaches could be applied to agricultural wastes and soils, wastewater treatment, pathogens discharged into the environment, and the evolution and dissemination upon exposure to ecological stress.

References

Aarestrup FM (1999) Association between the consumption of antimicrobial agents in animal husbandry and the occurrence of resistant bacteria among food animals. Int J Antimicrob Ag 12(4):279–285

Abella J, Fahy A, Duran R, Cagnon C (2015) Integron diversity in bacterial communities of freshwater sediments at different contamination levels. FEMS Microbiol Ecol 91:fiv140

Abou-Raya SH et al (2013) Effect of ordinary cooking procedures on tetracycline residues in chicken meat. J Food Drug Analysis 21(1):80–86. https://doi.org/10.6227/jfda.2013210110

Acar JF, Moulin G (2006) Antimicrobial resistance at farm level. Rev Sci Tech 25(2):775–792

Adebowale O, Adeyemo OK (2016) Antibiotic use and practices in commercial poultry laying hens in Ogun state Nigeria. Rev Elev Med Vet Pays Trop 69(1):41–45. https://doi.org/10.19182/remvt.31170

Adesoji AT, Ogunjobi AA, Olatoye IO (2016) Characterization of integrons and sulfonamide resistance genes among bacteria from drinking water distribution systems in southwestern Nigeria. Chemother 62(1):34–42. https://doi.org/10.1159/000446150

Alam A, Badruzzaman ABM, Ali MA (2013) Assessing effect of climate change on the water quality of the Sitalakhya river using WASP model. J Civ Eng 41:21–30

Allen DA, Austin B, Colwell RR (1977) Antibiotic resistance patterns of metal-tolerant bacteria isolated from an estuary. Antimicrob Ag Chemother 12(4):545–547. https://doi.org/10.1128/aac.12.4.545

Alonso A, Sánchez P, Martínez JL (2001) Environmental selection of antibiotic resistance genes: minireview. Env Microbiol 3(1):1–9. https://doi.org/10.1046/j.1462-2920.2001.00161.x

Amend AS et al (2010) Indoor fungal composition is geographically patterned and more diverse in temperate zones than in the tropics. Proc Nat Acad Sci 107(31):13748–13753. https://doi.org/10.1073/pnas.1000454107

Aminov RI (2011) Horizontal gene exchange in environmental microbiota. Front Microb 2:158. https://doi.org/10.3389/fmicb.2011.00158

Anjum R, Krakat N (2016) Detection of multiple resistance, biofilm formation and conjugative transfer of *Bacillus cereus* from contaminated soils. Curr Microbiol 72(3):321–328

Apata D (2009) Antibiotic resistance in poultry. Int J Poultry Sci 8(4):404–408

Ashbolt NJ, Amézquita A, Backhaus T, Borriello P, Brandt KK, Collignon P, Coors A, Finley R, Gaze WH, Heberer T, Lawrence JR, Larsson DG, McEwen SA, Ryan JJ, Schönfeld J, Silley P, Snape JR, Van den Eede C, Topp E (2013) Human health risk assessment (HHRA) for environmental development and transfer of antibiotic resistance. Environ Health Perspect 121:993–1001. https://doi.org/10.1289/ehp.1206316

Azam F, Malfatti F (2007) Microbial structuring of marine ecosystems. Nature Rev Microbiol 5:782–791

Baker-Austin C, Wright MS, Stepanauskas R, McArthur JV (2006) Co-selection of antibiotic and metal resistance. Tr Microbiol 14(4):176–182. https://doi.org/10.1016/j.tim.2006.02.006

Baquero F, Martinez J-L, Canton R (2008) Antibiotics and antibiotic resistance in water environments. Curr Op Biotechnol 19(3):260–265. https://doi.org/10.1016/j.copbio.2008.05.006

Bardgett RD et al (2013) Hierarchical responses of plant–soil interactions to climate change: consequences for the global carbon cycle. J Ecol 101(2):334–343. https://doi.org/10.1111/1365-2745.12043

Bass L, Liebert CA, Lee MD, Summers AO, White DG, Thayer SG, Maurer JJ (1999) Incidence and characterization of integrons, genetic elements mediating multiple-drug resistance, in avian *Escherichia coli*. Antimicrob Ag Chemother 43(12):2925–2929. https://doi.org/10.1128/AAC.43.12.2925

Bates BC, Kundzewicz ZW, Wu S, Palutikof JP (2008) Climate change and water. Technical paper of the intergovernmental panel on climate change. IPCC Secretariat, Geneva, Switzerland, p 210

Baynes RE et al (2016) Health concerns and management of select veterinary drug residues. Food Chem Toxicol 88:112–122. https://doi.org/10.1016/j.fct.2015.12.020

Beaber JW, Hochhut B, Waldor MK (2004) SOS response promotes horizontal dissemination of antibiotic resistance genes. Nature 427:72–74. https://doi.org/10.1038/nature02241

Beck HE, Zimmermann NE, McVicar TR, Vergopolan N, Berg A, Wood EF (2018) Present and future köppen-Geiger climate classification maps at 1-km resolution. Sci Data 5:180214. https://doi.org/10.1038/sdata.2018.214

Berendonk TU, Manaia CM, Merlin C, Fatta-Kassinos D, Cytryn E, Walsh F, Bürgmann H, Sørum H, Norström M, Pons M-N, Kruenzinger N, Huovinen P, Stefani S, Schwartz T, Kisand V, Baquero F, Martinez JL (2015) Tackling antibiotic resistance: the environmental framework. Nat Rev Microbiol 13:310–317

Berg J, Tom-Peterson A, Nybroe O (2005) Copper amendment of agricultural soil selects for bacterial antibiotic resistance in the field. Lett Appl Microbiol 40(2):146–151. https://doi.org/10.1111/j.1472-765X.2004.01650.x

Berg J, Thorsen MK, Holm PE, Jensen J, Nybroe O, Brandt KK (2010) Cu exposure under field conditions coselects for antibiotic resistance as determined by a novel cultivation-independent bacterial community tolerance assay. Environ Sci Technol 44:8724–8728

Bergeron S, Boopathy R, Nathaniel R, Corbin A, LaFleur G (2015) Presence of antibiotic resistant bacteria and antibiotic resistance genes in raw source water and treated drinking water. Int Biodeter Biodegrad 102:370–374. https://doi.org/10.1016/j.ibiod.2015.04.017

Binh CTT et al (2008) Piggery manure used for soil fertilization is a reservoir for transferable antibiotic resistance plasmids. FEMS Microbiol Ecol 66(1):25–37. https://doi.org/10.1111/j.1574-6941.2008.00526.x

Biyela PT, Lin J, Bezuidenhout CC (2004) The role of aquatic ecosystems as reservoirs of antibiotic resistant bacteria and antibiotic resistance genes. Water Sci Technol 50(1):45–50

Brownstein JS, Holford TR, Fish D (2005) Effect of climate change on Lyme disease risk in North America. EcoHealth 2(1):38–46. https://doi.org/10.1007/s10393-004-0139-x

Buse HY, Schoen ME, Ashbolt NJ (2012) Legionellae in engineered systems and use of quantitative microbial risk assessment to predict exposure. Water Res 46(4):921–933. https://doi.org/10.1016/j.watres.2011.12.022

Campbell-Lendrum D, Manga L, Bagayoko M, Sommerfeld J (2015) Climate change and vector-borne diseases: what are the implications for public health research and policy? Philos Trans R Soc B Biol Sci 370:1–8. https://doi.org/10.1098/rstb.2013.0552

Carlson JC, Mabury SA (2006) Dissipation kinetics and mobility of chlortetracycline, tylosin, and monensin in an agricultural soil in Northumberland County, Ontario, Canada. Environ Toxicol Chem 25(1):1–10

Carter LJ et al (2014) Fate and uptake of Pharmaceuticals in Soil–Plant Systems. J Ag Food Chem 62(4):816–825. https://doi.org/10.1021/jf404282y

Cavicchioli R, Ripple WJ, Timmis KN, Azam F, Bakken LR, Baylis M, Behrenfeld MJ, Boetius A, Boyd PW, Classen AT, Crowth TW, Danovaro R, Foreman CM, Huisman J, Hutchins DA, Jansson JK, Karl DM, Koskella B, Welch DBM, Martiny JBH, Moran MA, Orphan VJ, Reay

DS, Remais JV, Rich VI, Singh BK, Stein LY, Stewart FJ, Sullivan MB, van Oppen MJH, Weaver SC, Webb EA, Webster NS (2019) Scientists' warning to humanity: microorganisms and climate change. Nat Rev Microbiol 17:569–586. https://doi.org/10.1038/s41579-019-0222-5

Chen B, He R, Yuan K, Chen E, Lin L, Chen X, Sha S, Zhong J, Lin L, Yang L, Yang Y, Wang X, Zou S, Luan T (2017) Polycyclic aromatic hydrocarbons (PAH) enriching antibiotic resistance genes (ARGs) in the soils. Environ Pollut 220(B):1005–1013. https://doi.org/10.1016/j.envpol.2016.11.047

Colombo N et al (2018) Review: impacts of permafrost degradation on inorganic chemistry of surface fresh water. Glob Planet Chang 162:69–83. https://doi.org/10.1016/J.GLOPLACHA.2017.11.017

Cui H et al (2014) Influence of ciprofloxacin on microbial community structure and function in soils. Biol Fert Soils 50(6):939–947. https://doi.org/10.1007/s00374-014-0914-y

Czekalski N, Berthold T, Caucci S, Egli A, Burgmann H (2012) Increased levels of multiresistant bacteria and resistance genes after wastewater treatment and their dissemination into Lake Geneva, Switzerland. Front Microbiol 3(106):1–18. https://doi.org/10.3389/fmicb.2012.00106

D'Costa VM et al (2011) Antibiotic resistance is ancient. Nature 477(7365):457–461. https://doi.org/10.1038/nature10388

Dantas G, Sommer MOA (2012) Context matters – the complex interplay between resistome genotypes and resistance phenotypes. Curr Op Microbiol 15(5):577–582. https://doi.org/10.1016/j.mib.2012.07.004

Davis JG et al (2006) Antibiotic transport via runoff and soil loss. J Environ Qual 35(6):2250–2260. https://doi.org/10.2134/jeq2005.0348

De Briyne N et al (2014) Antibiotics used most commonly to treat animals in Europe. Vet Rec 175 (13):325. https://doi.org/10.1136/vr.102462

de Oliveira RP, Matos JS, Monteiro AJ (2015) Managing the urban water cycle in a changing environment. Water Utility J 9:3

Delpla I, Jung VA, Baures E, Clement M, Thomas O (2009) Impacts of climate change on surface water quality in relation to drinking water production. Environ Int 35(8):1225–1233. https://doi.org/10.1016/j.envint.2009.07.001

Dhakephalkar PK, Chopade BA (1994) High levels of multiple metal resistance and its correlation to antibiotic resistance in environmental isolates of *Acinetobacter*. Biometals 7(1):67–74

Dhimal M, Ahrens B, Kuch U (2015) Climate change and spatiotemporal distributions of vector-borne diseases in Nepal - a systematic synthesis of literature. PLoS One 10(6):1–31. https://doi.org/10.1371/journal.pone.0129869

Dillen W et al (1997) The effect of temperature on *Agrobacterium tumefaciens*-mediated gene transfer to plants. Plant J 12(6):1459–1463

Ding G-C et al (2014) Dynamics of soil bacterial communities in response to repeated application of manure containing sulfadiazine. PLoS One 9(3):1–10. https://doi.org/10.1371/journal.pone.0092958

Dolliver H, Kumar K, Gupta S (2007) Sulfamethazine uptake by plants from manure-amended soil. J Environ Qual 36(4):1224–1230. https://doi.org/10.2134/jeq2006.0266

Duan M et al (2017) Effects of biochar on reducing the abundance of oxytetracycline, antibiotic resistance genes, and human pathogenic bacteria in soil and lettuce. Environ Poll 224:787–795. https://doi.org/10.1016/j.envpol.2017.01.021

Elasha BO (2010) Mapping of climate change threats and human development impacts in the Arab region. Regional Bureau of Arab States. https://research.fit.edu/media/site-specific/researchfitedu/coast-climate-adaptation-library/middle-east/regional%2D%2D-middle-east/Elasha.-2010.-Mapping-of-CC-Threats%2D%2DHuman-Development-Impacts-In-The-Arab-Region.pdf

European Medicines Agency (2018) Reflection paper on antimicrobial resistance in the environment: considerations for current and future risk assessment of veterinary medicinal products.

Reflection paper on antimicrobial resistance in the environment: considerations for current and future

Fahrenfeld N et al (2014) Effect of manure application on abundance of antibiotic resistance genes and their attenuation rates in soil: field-scale mass balance approach. Environ Sci Technol 48 (5):2643–2650. https://doi.org/10.1021/es404988k

Falkinham JO et al (2015) Common features of opportunistic premise plumbing pathogens. Int J Environ Res Public Health 12:4533–4545. https://doi.org/10.3390/ijerph120504533

Fang X, Luo S, Lyu S (2019) Observed soil temperature trends associated with climate change in the Tibetan plateau, 1960–2014. Theoret Appl Climatol 135(1–2):169–181. https://doi.org/10.1007/s00704-017-2337-9

Fernstrom A, Goldblatt M (2013) Aerobiology and its role in the transmission of infectious diseases. J Path 2013:493960. https://doi.org/10.1155/2013/493960

Finley RL et al (2013) The scourge of antibiotic resistance: the important role of the environment. Clin Infect Dis 57(5):704–710. https://doi.org/10.1093/cid/cit355

Forsberg KJ, Patel S, Gibson MK, Lauber CL, Knight R, Fierer N, Dantas G (2014) Bacterial phylogeny structures soil resistomes across habitats. Nature 509:612–616. https://doi.org/10.1038/nature13377

Forsberg KJ, Patel S, Wencewicz TA, Dantas G (2015) The tetracycline destructases: a novel family of tetracycline-inactivating enzymes. Chem Biol 22(7):888–897. https://doi.org/10.1016/j.chembiol.2015.05.017

Fuchsman CA et al (2017) Effect of the environment on horizontal gene transfer between bacteria and archaea. Peer J 5:e3865. https://doi.org/10.7717/peerj.3865

Garcia-Solache MA, Casadevall A (2010) Global warming will bring new fungal diseases for mammals. mBio 1(1):1–3. https://doi.org/10.1128/mbio.00061-10

Ghosh A, Singh A, Ramteke PW, Singh VP (2000) Characterization of large plasmids encoding resistance to toxic heavy metals in *Salmonella abortus equi*. Biochm Biophys Res Comm 272 (1):6–11. https://doi.org/10.1006/bbrc.2000.2727

Graham DW, Olivares-Rieumont S, Knapp CW, Lima L, Werner D, Bowen E (2011) Antibiotic resistance gene abundances associated with waste discharges to the Almendares river near Havana, Cuba. Environ Sci Technol 45(2):418–424. https://doi.org/10.1021/es102473z

Green A, Ng V, Fisman D (2008) Climate change and infectious diseases in North America: the road ahead. Can Med Assoc J 178(6):715–722. https://doi.org/10.1503/cmaj.081326

Greene CH, Pershing AJ (2007) Climate drives sea change. Science 315(5815):1084–1085. https://doi.org/10.1126/science.1136495

Guidice AL, Casella P, Bruni V, Michaud L (2013) Response of bacterial isolates from Antarctic shallow sediments towards heavy metals, antibiotics and polychlorinated biphenyls. Ecotoxicology 22(2):240–250

Güneralp B et al (2017) Global scenarios of urban density and its impacts on building energy use through 2050. Proc Natl Acad Sci 114(34):8945–8950. https://doi.org/10.1073/pnas.1606035114

Guo X, Li J, Yang F, Yang J, Yin D (2014) Prevalence of sulfonamide and tetracycline resistance genes in drinking water treatment plants in the Yangtze river delta, China. Sci Total Environ 493:626–631. https://doi.org/10.1016/j.scitotenv.2014.06.035

Hajdu S et al (2010) Increased temperature enhances the antimicrobial effects of daptomycin, vancomycin, tigecycline, fosfomycin, and cefamandole on staphylococcal biofilms. Antimicrobiol Ag Chemother 54(10):4078–4084. https://doi.org/10.1128/AAC.00275-10

Haldorsen S et al (2010) Sensitivity to long-term climate change of subpermafrost groundwater systems in Svalbard. Quat Res 73(2):393–402. https://doi.org/10.1016/j.yqres.2009.11.002

Hamscher G et al (2005) Different behavior of tetracyclines and sulfonamides in sandy soils after repeated fertilization with liquid manure. Environ Toxicol Chem 24(4):861–868

Harvey CA et al (2014) Extreme vulnerability of smallholder farmers to agricultural risks and climate change in Madagascar. Phil trans Royal Soc London. Ser B, biol Sci 369 (1639):20130089. https://doi.org/10.1098/rstb.2013.0089

Hodgkins SB et al (2014) Changes in peat chemistry associated with permafrost thaw increase greenhouse gas production. Proc Natl Acad Sci U S A 111(16):5819–5824. https://doi.org/10.1073/pnas.1314641111

Hollis A, Ahmed Z (2013) Preserving antibiotics, rationally. New England J Med 369 (26):2474–2476. https://doi.org/10.1056/NEJMp1311479

Hosseini N, Johnston J, Lindenschmidt KE (2017) Impacts of climate change on the water quality of a regulated prairie river. Water 9:199–213. https://doi.org/10.3390/w9030199

Hu HW, Wang JT, Li J, Li JJ, Ma YB, Chen D, He JZ (2016) Field-based evidence for copper contamination induced changes of antibiotic resistance in agricultural soils. Environ Microbiol 18(11):3896–3909. https://doi.org/10.1111/1462-2920.13370

Jechalke S et al (2014) Fate and effects of veterinary antibiotics in soil. Tr Microbiol 22 (9):536–545. https://doi.org/10.1016/j.tim.2014.05.005

Ji X, Shen Q, Liu F, Ma J, Xu G, Wang Y, Wu M (2012) Antibiotic resistance gene abundances associated with antibiotics and heavy metals in animal manures and agricultural soils adjacent to feedlots in Shanghai, China. J Haz Mat 235–6:178–185. https://doi.org/10.1016/j.jhazmat.2012.07.040

John OS (2014) Surveillance of antimicrobial resistance in pigs using *Escherichia coli* faecal isolates. J Microbiol 3(4):78–83

Karcic E, Aljicevic M, Bektas S, Karcic B (2015) Antimicrobial susceptibility/resistance of Streptococcus Pneumoniae. Mater Socio Medica 27:180. https://doi.org/10.5455/msm.2015.27.180-184

Karmakar R et al (2016) Potential effects of climate change on soil properties: a review. Sci Internat 4(2):51–73. https://doi.org/10.17311/sciintl.2016.51.73

Kennedy D (2013) Time to deal with antibiotics. Science 342(6160):777. https://doi.org/10.1126/science.1248056

Khan S, Beattie TK, Knapp CW (2016a) Relationship between antibiotic- and disinfectant-resistance profiles in bacterial harvested from tap water. Chemosphere 152:13–141. https://doi.org/10.1016/j.chemosphere.2016.02.086

Khan S, Knapp CW, Beattie TK (2016b) Antibiotic resistant bacteria found in municipal drinking water. Environ Proc 3(3):541–552. https://doi.org/10.1007/s40710-016-0149-z

Khan S, Beattie TK, Knapp CW (2017) The use of minimum selectable concentrations (MSCs) for determining the selection of antimicrobial resistant bacteria. Ecotoxicology 26(2):283–292. https://doi.org/10.1007/s10646-017-1762-y

Kim C, Lee S-M, Lee C-K (2010) Impacts of climate change on Korean agriculture and its counterstrategies. 한국농촌경제연구원 기본연구보고서: 1–306

Knapp CW, McCluskey SM, Singh BK, Campbell CD, Hudson G, Graham DW (2011) Antibiotic resistance gene abundances correlate with metal and geochemical conditions in archived Scottish soils. PLoS One 6(11):e27300. https://doi.org/10.1371/journal.pone.0027300

Knapp CW, Lima L, Olivares-Rieumont S, Bowen E, Werner D, Graham D (2012) Seasonal variations in antibiotic resistance gene transport in the Almendares river, Havana, Cuba. Front Microbiol 3:396. https://doi.org/10.3389/fmicb.2012.00396

Knapp CW, Callan AC, Aitken B, Shearn R, Koenders A, Hinwood A (2017) Relationship between antibiotic resistance genes and metals in residential soil samples from western Australia. Environ Sci Pollut Res 24:2484. https://doi.org/10.1007/s11356-0160-7997-y

Krishna AK, Govil PK (2007) Soil contamination due to heavy metals from an industrial area of Surat, Gujarat, western India. Environ Monit Assess 124:263. https://doi.org/10.1007/s10661-006-9224-7

Kuchta SL et al (2009) Transport of lincomycin to surface and ground water from manure-amended cropland. J Environ Qual 38(4):1719–1727. https://doi.org/10.2134/jeq2008.0365

Kumar K et al (2005) Antibiotic uptake by plants from soil fertilized with animal manure. J Environ Qual 34(6):2082–2085. https://doi.org/10.2134/jeq2005.0026

Landers TF et al (2012) A review of antibiotic use in food animals: perspective, policy, and potential. Public Health Rep 127(1):4–22. https://doi.org/10.1177/003335491212700103

Le PVV et al (2019) Predicting the direct and indirect impacts of climate change on malaria in coastal Kenya. PLoS One 14(2):1–18. https://doi.org/10.1371/journal.pone.0211258

Lee MH, Lee HJ, Ryu PD (2001) Public health risks: chemical and antibiotic residues. Asian-Australas J Anim Sci 14(3):402–413. https://doi.org/10.5713/ajas.2001.402

Lee J, Lee CS, Hugunin KM, Maute CJ, Dysko RC (2010) Bacteria from drinking water supply and their fate in gastrointestinal tracts of germ-free mice: a phylogenetic comparison study. Water Res 44(17):5050–5058. https://doi.org/10.1016/j.watres.2010.07.027

Lehrer S (2010) Association between malaria incidence and all cancer mortality in fifty U.S. states and the District of Columbia. Anticancer Res 30:1371–1373

Lemieux J-M et al (2008) Simulating the impact of glaciations on continental groundwater flow systems: 1. Relevant processes and model formulation. J Geophys Res 113:F03017. https://doi.org/10.1029/2007JF000928

Levy SB (1997) Antibiotic resistance: an ecological imbalance. Ciba Found Symp 207:1–14

Levy SB (2002) Factors impacting on the problem of antibiotic resistance. J Antimicrob Chemother 4:25–30

Lewis P, Mone MA, Impiglia A (2018) Impacts of climate change on farming systems and livelihoods in the near east and North Africa – with special focus on small-scale family farming. FAO, Quebec, QC

Li B, Yang Y, Ma L, Ju F, Guo F, Tiedje JM, Zhang T (2015) Metagenomic and network analysis reveal wide distribution and co-occurrence of environmental antibiotic resistance genes. J ISME 9:2490–2502. https://doi.org/10.1038/ismej.2015.59

Li D et al (2017) Characterization of Pb^{2+} biosorption by psychrotrophic strain *Pseudomonas* sp. I3 isolated from permafrost soil of Mohe wetland in Northeast China. J Environ Manag 196:8–15. https://doi.org/10.1016/J.JENVMAN.2017.02.076

Lin K, Gan J (2011) Sorption and degradation of wastewater-associated non-steroidal anti-inflammatory drugs and antibiotics in soils. Chemosphere 83(3):240–246. https://doi.org/10.1016/j.chemosphere.2010.12.083

Lozano C et al (2016) *Staphylococcus aureus* in animals and food: methicillin resistance, prevalence and population structure. A review in the African continent. Microorganisms 4(1):12. https://doi.org/10.3390/microorganisms4010012

MacFadden DR, McGough SF, Fisman D, Santillana M, John S, Brownstein JS (2018) Antibiotic resistance increases with local temperature. Nat Clim Chang 8:510–514. https://doi.org/10.1038/s41558-018-0161-6

Manyi-Loh C et al (2018) Antibiotic use in agriculture and its consequential resistance in environmental sources: potential public health implications. Molecules 23(4):795. https://doi.org/10.3390/molecules23040795

Manzetti S, Ghisi R (2014) The environmental release and fate of antibiotics. Mar poll bull 79 (1):7–15. https://doi.org/10.1016/j.marpolbul.2014.01.005

Maron DF, Smith TJS, Nachman KE (2013) Restrictions on antimicrobial use in food animal production: an international regulatory and economic survey. Glob Health 9:48. https://doi.org/10.1186/1744-8603-9-48

Marques AM, Congregad F, Simonpujol DM (1979) Antibiotic and heavy metal resistance of *Pseudomonas aeruginosa* isolates in soils. J Appl Bacteriol 47(2):347–350. https://doi.org/10.1111/j.1365-2672.1979.tb01765.x

Marti R et al (2013) Impact of manure fertilization on the abundance of antibiotic-resistant bacteria and frequency of detection of antibiotic resistance genes in soil and on vegetables at harvest. Appl Environ Microbiol 79(18):5701–5709. https://doi.org/10.1128/AEM.01682-13

Martinez JS (2009) Environmental pollution by antibiotics and by antibiotic resistance determinants. Environ Poll 157(11):2893–2902. https://doi.org/10.1016/j.envpol.2009.05.051

Máthé I, Benedek T, Táncics A, Palatinszky M, Lányi S, Márialgeti K (2012) Diversity, activity, antibiotic and heavy metal resistance of bacteria from petroleum hydrocarbon contaminated soils located in Harghita County (Romania). Intern Biodeter Biodegrad 73:41–49. https://doi.org/10.1016/j.ibiod.2012.05.018

McIntosh JC, Walter LM (2005) Volumetrically significant recharge of Pleistocene glacial meltwaters into epicratonic basins: constraints imposed by solute mass balances. Chem Geol 222(3–4):292–309. https://doi.org/10.1016/J.CHEMGEO.2005.07.010

McIntosh JC, Garven G, Hanor JS (2011) Impacts of Pleistocene glaciation on large-scale groundwater flow and salinity in the Michigan Basin. Geofluids 11(1):18–33. https://doi.org/10.1111/j.1468-8123.2010.00303.x

McLean MA, Angilletta MJ, Williams KS (2005) If you can't stand the heat, stay out of the city: thermal reaction norms of chitinolytic fungi in an urban heat island. J Thermal Biol 30 (5):384–391. https://doi.org/10.1016/j.jtherbio.2005.03.002

Mellon M, Benbrook C, Benbrook KL (2001) Hogging it: estimates of antimicrobial abuse in livestock. Union of Concerned Scientists, UCS Publishing, Cambridge, MA

Mindlin S et al (2016) Resistance of permafrost and modern *Acinetobacter lwoffii* strains to heavy metals and arsenic revealed by genome analysis. BioMed Res Internat 2016:3970831. https://doi.org/10.1155/2016/3970831

Moyane JN et al (2013) Antibiotics usage in food-producing animals in South Africa and impact on human: antibiotic resistance. Afr J Microbiol Res 7(24):2990–2997. https://doi.org/10.5897/AJMR2013.5631

Neuzil CE (2012) Hydromechanical effects of continental glaciation on groundwater systems. Geofluids 12(1):22–37. https://doi.org/10.1111/j.1468-8123.2011.00347.x

Nordor AV, Bellet D, Siwo GH (2018) Cancer – malaria: hidden connections. Open Biol 8 (10):180127

Oelke C, Zhang T (2004) A model study of circum-Arctic soil temperatures. Permafrost Periglacial Proc 15(2):103–121. https://doi.org/10.1002/ppp.485

Oluyege JO, Oluwaniyi TT, Ijasan OC (2015) Composition of antibiotic resistant bacteria from irrigated vegetable farmland. J Microbiol Res 5(5):161–168. https://doi.org/10.5923/j.microbiology.20150505.04

Padol A et al (2015) Occurrence, public health implications and detection of antibacterial drug residues in cow milk. Environ We Int J Sci 7112:22

Pan M, Chu LM (2017) Leaching behavior of veterinary antibiotics in animal manure-applied soils. Sci Total Environ 579:466–473. https://doi.org/10.1016/j.scitotenv.2016.11.072

Park JY, Huwe B (2016) Effect of pH and soil structure on transport of sulfonamide antibiotics in agricultural soils. Environ Poll 213:561–570. https://doi.org/10.1016/j.envpol.2016.01.089

Perry JA, Wright GD (2013) The antibiotic resistance "mobilome": searching for the link between environment and clinic. Front Microbiol 4:138. https://doi.org/10.3389/fmicb.2013.00138

Petrova MA et al (2002) Mercury-resistant bacteria from permafrost sediments and prospects for their use in comparative studies of mercury resistance determinant. Russian J Gen 38 (11):1330–1334. https://doi.org/10.1023/A:1021167615633

Pham Kim D et al (2013) First survey on the use of antibiotics in pig and poultry production in the Red River delta region of Vietnam. Food Public Health 3(5):247–256. https://doi.org/10.5923/j.fph.20130305.03

Pollard AT, Morra MJ (2018) Estrogens: properties, behaviors, and fate in dairy manure-amended soils. Environ Rev 25(4):452–462

Prado T, Pereira WC, Silva DM, Seki LM, Carvalho AP, Asensi MD (2008) Detection of extended-spectrum beta-lactamase-producing *Klebsiella pneumoniae* in effluents and sludge of a hospital sewage treatment plant. Lett Appl Microbiol 46(1):136–141. https://doi.org/10.1111/j.1472-765X.2007.02275.x

Pruden A, Pei R, Storteboom H, Carlson KH (2006) Antibiotic-resistance genes as emerging contaminants: studies in northern Colorado. Environ Sci Technol 40:7445–7450

Prussin AJ, Marr LC (2015) Sources of airborne microorganisms in the built environment. Microbiome 3(1):78. https://doi.org/10.1186/s40168-015-0144-z

Rachid S et al (2000) Effect of subinhibitory antibiotic concentrations on polysaccharide intercellular adhesin expression in biofilm-forming *Staphylococcus epidermidis*. Antimicrob Ag Chemother 44(12):3357–3363. https://doi.org/10.1128/aac.44.12.3357-3363.2000

Reichel R et al (2013) Effects of slurry from sulfadiazine- (SDZ) and difloxacin- (DIF) medicated pigs on the structural diversity of microorganisms in bulk and rhizosphere soil. Soil Biol Biochem 62:82–91. https://doi.org/10.1016/j.soilbio.2013.03.007

Rengel Z (2011) Soil pH, soil health and climate change. In: Singh BP, Cowie AL, Chan KY (eds) Soil health and climate change. Springer, Berlin, pp 69–85. https://doi.org/10.1007/978-3-642-20256-8_4

Reza AS, Jean JS, Lee MK, Liu CC, Bundschuh J, Yang HJ, Lee JF, Lee YC (2010) Implications of organic matter on arsenic mobilization into groundwater: evidence from northwestern (Chapai-Nawabganj), central (Manikganj) and southeastern (Chandpur) Bangladesh. Water Res 44(19):5556–5574

Riesenfeld CS, Goodman RM, Handelsman J (2004) Uncultured soil bacteria are a reservoir of new antibiotic resistance genes. Environ Microbiol 6(9):981–989. https://doi.org/10.1111/j.1462-2920.2004.00664.x

Rochelle PA, Fry JC, Day MJ (1989) Factors affecting conjugal transfer of plasmids encoding mercury resistance from pure cultures and mixed natural suspensions of epilithic bacteria. J Gen Microbiol 135:409–424. https://doi.org/10.1099/00221287-135-2-409

Rodgers K, McLellan I, Peshkur T, Williams R, Tonner R, Hursthouse AS, Knapp CW, Henriquez FL (2018) Can legacy pollution influence antimicrobial resistance in estuarine sediments? The geochemical role!. 96–96. 34th Society for Environmental Geochemistry and Health International Conference on Sustainable Geochemistry, Livingstone, Zambia

Rodgers K, McLellan I, Peshkur T, Williams R, Tonner R, Hursthous AS, Knapp CW, Henriquez FL (2019) Can the legacy of industry pollution influence antimicrobial resistance in estuarine sediments? Environ Chem Lett 17(2):595–607. https://doi.org/10.1007/s10311-018-0791-y

Rodgers K, McLellan I, Peshkur T, Williams R, Tonner R, Knapp CW, Henriquez FL, Hursthouse AS (2020) The legacy of industrial pollution in estuarine sediments: spatial and temporal variability implications for ecosystem stress. Environ Geochem Health 42:1057–1068. https://doi.org/10.1007/s10653-019-00316-4

Rosenzweig C, Casassa G, Karoly JD, Imeson A, Liu C, Menzel A, Rawlins S, Root LT, Seguin B, Tryjanowski P (2007) Assessment of observed changes and responses in natural and managed systems. In: Parry LM, Canziani FO, Palutikof PJ, der Linden JP, Hanson EC (eds) Climate change 2007: impacts, adaptation and vulnerability. Contribution of working group II to the fourth assessment report of the intergovernmental panel on climate change. Cambridge University Press, Cambridge, UK, pp 79–131

Sanderson CE, Fox JT, Dougherty ER, Cameron ADS, Alexander KA (2018) The changing face of water: a dynamic reflection of antibiotic resistance across landscapes. Front Microbiol 9:1–13. https://doi.org/10.3389/fmicb.2018.01894

Schwartz T, Kohnen W, Jansen B, Obst U (2003) Detection of antibiotic-resistant bacteria and their resistance genes in wastewater, surface water, and drinking water biofilms. FEMS Microbiol Ecol 43:325–335. https://doi.org/10.1111/j.1574-6941.2003.tb01073.x

Seiler C, Berendonk TU (2012) Heavy metal driven co-selection of antibiotic resistance in soil and water bodies impacted by agriculture and aquaculture. Front Microbiol 3:399. https://doi.org/10.3389/fmicb.2012.00399

Seneviratne SI et al (2010) Investigating soil moisture-climate interactions in a changing climate: a review. Earth-Sci Rev 99(3–4):125–161. https://doi.org/10.1016/j.earscirev.2010.02.004

Sharma B et al (2015) *Borrelia burgdorferi*, the causative agent of Lyme disease, forms drug-tolerant persister cells. Antimicrob Ag Chemother 59(8):4616–4624. https://doi.org/10.1128/AAC.00864-15

Silbergeld EK, Graham J, Price LB (2008) Industrial food animal production, antimicrobial resistance, and human health. Annu Rev Pub Health 29:151–169. https://doi.org/10.1146/annurev.publhealth.29.020907.090904

Singer AC, Shaw H, Rhodes V, Hart A (2016) Review of antimicrobial resistance in the environment and its relevance to environmental regulators. Front Microbiol 7:1728. https://doi.org/10.3389/fmicb.2016.01728

Smillie CS et al (2011) Ecology drives a global network of gene exchange connecting the human microbiome. Nature 480(7376):241–244. https://doi.org/10.1038/nature10571
Starinsky A, Katz A (2003) The formation of natural cryogenic brines. Geochim Cosmochim Acta 67(8):1475–1484. https://doi.org/10.1016/S0016-7037(02)01295-4
Stepanauskas R, Glenn TC, Jagoe CH, Tuckfield RC, Lindell AH, McArthur JV (2005) Elevated microbial tolerance to metals and antibiotics in metal-contaminated industrial environments. Environ Sci Technol 39:3671–3678
Stepanauskas R, Glenn TC, Jagoe CH, Tuckfield RC, Lindell AH, King CJ, McArthur JV (2006) Coselection for microbial resistance to metals and antibiotics in freshwater microcosms. Environ Microbiol 8:1510–1514
Su JQ, Wei B, Ou-Yang WF, Huang F-Y, Zhao Y, Xu H-J, Zhu Y-G (2015) Antibiotic resistome and its associate with bacterial communities during sewage sludge composting. Environ Sci Technol 49(12):7356–7363. https://doi.org/10.1021/acs.est.5b01012
Tandukar M, Oh S, Tezel U, Konstantinidis KT, Pavlostathis SP (2013) Long-term exposure to benzalkonium chloride disinfectants results in change of microbial community structure and increased antimicrobial resistance. Environ Sci Technol 47(17):9730–9738. https://doi.org/10.1021/es401507k
Thanner S, Drissner D, Walsh F (2016) Antimicrobial resistance in agriculture. MBio 7(2):1–7. https://doi.org/10.1128/mBio.02227-15
Thiele-bruhn S, Peters D (2007) Photodegradation of pharmaceutical antibiotics on slurry and soil surfaces. Landbauforschung Völkenrode 57:13–23
Titus JG (1989) Chapter 16, southeast. In: the potential effects of global climate change on the United States. US Environmental Protection Agency. EPA-230-05-89-050
Tolls J (2001) Sorption of veterinary pharmaceuticals in soils: a review. Environ Sci Technol 35 (17):3397–3406. https://doi.org/10.1021/es0003021
Tonner RL, K Rodgers, T Peshkur, I McLellan, R Williams, A Hursthouse, F Henriquez, C Knapp (2017) Co-selection of antibiotic resistance in Clyde estuary, UK: a result from a legacy of pollution. Society for Applied Microbiology: 3rd Antimicrobial Resistance Conference. SfAM. 23–24 November (UK)
Turetsky MR et al (2000) Organic matter accumulation, peat chemistry, and permafrost melting in peatlands of boreal Alberta. Écosci 3:379–392. https://doi.org/10.1080/11956860.2000.11682608
Van Boeckel TP et al (2015) Global trends in antimicrobial use in food animals. Proc Natl Acad Sci U S A 112(18):5649–5654. https://doi.org/10.1073/pnas.1503141112
van de Steeg J, Tibbo M (2012) Livestock and climate change in the near east region: measures to adapt to and mitigate climate change. FAO, Cairo, Egypt
van Elsas JD, Bailey MJ (2002) The ecology of transfer of mobile genetic elements. FEMS Microbiol Ecol 42(2):187–197. https://doi.org/10.1111/j.1574-6941.2002.tb01008.x
Vardoulakis S, Dimitroulopoulou C, Thornes J et al (2015) Impact of climate change on the domestic indoor environment and associated health risks in the UK. Environ Int 85:299–313. https://doi.org/10.1016/j.envint.2015.09.010
Ventola CL (2015) The antibiotic resistance crisis: causes and threats. PT 40:277–283
Verner D (2012) Adaptation to a changing climate in the Arab countries, Washington DC
Wang Q, Guo M, Yates SR (2006) Degradation kinetics of manure-derived sulfadimethoxine in amended soil. J Ag Food Chem 54(1):157–163. https://doi.org/10.1021/jf052216w
Wang J, Wang J, Zhao Z, Chen J, Lu H, Liu G, Zhou J, Guan X (2017) PAHs accelerate the propagation of antibiotic resistance genes in coastal water microbial community. Environ Pollut 231(1):1145–1152. https://doi.org/10.1016/j.envpol.2017.07.067
West BM, Liggit P, Clemans DL, Francoeur SN (2011) Antibiotic resistance, gene transfer, and water quality patterns observed in waterways near CAFO farms and wastewater treatment facilities. Water Air Soil Poll 217:473–489. https://doi.org/10.1007/s11270-010-0602-y
Whitehead PG, Wade AJ, Butterfield D (2009) Potential impacts of climate change on water quality and ecology in six UK rivers. Hydrol Res 40(2–3):113–122

Willerslev E, Hansen AJ, Poinar HN (2004) Isolation of nucleic acids and cultures from fossil ice and permafrost. Trend Ecol Evol 19(3):141–147. https://doi.org/10.1016/J.TREE.2003.11.010
Witte W (1998) Medical consequences of antibiotic use in agriculture. Science 279(5353):996–997. https://doi.org/10.1126/science.279.5353.996
Wolters B et al (2015) Transferable antibiotic resistance plasmids from biogas plant digestates often belong to the IncP-1ε subgroup. Front Microbiol 5:765. https://doi.org/10.3389/fmicb.2014.00765
Woolhouse M et al (2015) Antimicrobial resistance in humans, livestock and the wider environment. Philos Trans R Soc London Ser B 370(1670):20140083. https://doi.org/10.1098/rstb.2014.0083
World Food Programme (WFP) and Overseas Development Institute (ODI) (2015) Food in an uncertain future – the impacts of climate change on food security and nutrition in the Middle East and North Africa. Cairo
World Health Organisation (2012) The evolving threat of antimicrobial resistance options for action. WHO, Geneva
World Health Organisation (2015) Global action plan on antimicrobial resistance. http://apps.who.int/iris/bitstream/10665/193736/1/9789241509763eng.pdf?ua=1
Wright GD (2007) The antibiotic resistome: the nexus of chemical and genetic diversity. Nature Rev Microbiol 5:175–186. https://doi.org/10.1038/nrmicro1614
Wright GD (2010) Antibiotic resistance in the environment: a link to the clinic. Curr Op Microb 13 (5):589–594. https://doi.org/10.1016/j.mib2010.08.005
Wright MS, Peltier GL, Stepanauskas R, McArthur JV (2006) Bacterial tolerances to metals and antibiotics in metal-contaminated and reference streams. FEMS Microbiol Ecol 58:293–302
Wu X-L et al (2014) Distribution and risk assessment of quinolone antibiotics in the soils from organic vegetable farms of a subtropical city, southern China. Sci Total Environ 487:399–406. https://doi.org/10.1016/j.scitotenv.2014.04.015
Xia XH, Wu Q, Mou XL, Lai YJ (2015) Potential impacts of climate change on the water quality of different water bodies. J Environ Inform 25(2):85–98. https://doi.org/10.3808/jei.201400263
Xiao K-Q et al (2016) Metagenomic profiles of antibiotic resistance genes in paddy soils from South China. FEMS Microbiol Ecol 92(3):fiw023. https://doi.org/10.1093/femsec/fiw023
Yang X, Shang C, Westerhoff P (2007) Factors affecting formation of haloacetonitriles, haloketones, chloropicrin and cyanogens halides during chloramination. Water Res 41:1193–1200
Yang J-F et al (2012) Biological degradation and microbial function effect of norfloxacin in a soil under different conditions. J Environ Sci Health Part B 47(4):288–295. https://doi.org/10.1080/03601234.2012.638886
You Y, Silbergeld EK (2014) Learning from agriculture: understanding low-dose antimicrobials as drivers of resistome expansion. Front Microbiol 5:284. https://doi.org/10.3389/fmicb.2014.00284
Zhang T et al (2001) An amplified signal of climatic change in soil temperatures during the last century at Irkutsk, Russia. Clim Chang 49(1):41–76. https://doi.org/10.1023/A:1010790203146
Zhang T et al (2008) Statistics and characteristics of permafrost and ground-ice distribution in the Northern Hemisphere. Polar Geogr 31(1–2):47–68. https://doi.org/10.1080/10889379909377670
Zhang D-C et al (2013) Isolation and characterization of bacteria from ancient siberian permafrost sediment. Multidisc Digital Pub Inst 2(1):85–106. https://doi.org/10.3390/biology2010085
Zhang A, He X, Meng Y, Guo L, Long M, Yu H, Li B, Fan L, Liu S, Wang H, Zou L (2016) Antibiotic and disinfectant resistance of *Escherichia coli* isolated from retail meats in Sichuan, China. Microb Drug Resist 22(1):80–87. https://doi.org/10.1089/mdr.2015.0061
Zhu Y-G et al (2013) Diverse and abundant antibiotic resistance genes in Chinese swine farms. Proc Natl Acad Sci U S A 110(9):3435–3440. https://doi.org/10.1073/pnas.1222743110

Phyllosphere Microbiome: Plant Defense Strategies

11

Beenish Saleem

Abstract

Aerial parts of plants are colonized by an array of microbes. These microbial species occupy specific niches on the leaf surface driven by localization of nutrients, microbial interactions, and structure of the phylloplane. Earlier studies have demonstrated that the phylloplane microbial populations vary considerably due to different physiological factors, climate change being a significant reason. Inter-generic and interspecific interactions are crucial in determining the colonizers. The physiological activities of the phylloplane colonizers also influence plant growth and protection against diseases. Plant associated microorganisms have been reported to synthesize various compounds that help in alleviating the plant's defense response towards any pathogenic attack. Bacterial–bacterial, bacterial–fungal and fungal–fungal antagonistic interactions have been exploited for crop protection and plant health. Phylloplane microorganisms and their metabolites have been studied for biological control of a plethora of pathogens on a range of crop plants. Phylloplane microorganisms have been reported to change the physiochemical properties of their environment. During the course of colonization, microbes often secrete phenolic compounds having antimicrobial activities that suppress the growth of microflora and act as efficient bio-control agents. Mature leaves are known to secrete more of such compounds as compared to their younger counterparts. The chapter will lead to insights into Phyllosphere microbiology and how it implicates the growth and development of plants, providing protection from diseases, and inducing resistance against phytopathogens. The chapter also throws light on how climate change may affect Phyllosphere microflora which eventually alters the plant's defense response.

B. Saleem (✉)
Amity Institute of Biotechnology, Amity University, Noida, Uttar Pradesh, India

S. A. Lone, A. Malik (eds.), *Microbiomes and the Global Climate Change*,
https://doi.org/10.1007/978-981-33-4508-9_11

Keywords

Phylloplane · Phytopathogens · Plant Defense · Epiphytes · Interactions · Bio-control

11.1 Introduction

The phyllosphere provides large surface area as a sustainable habitat for numerous microorganisms but remains a subject less explored as compared to the rhizosphere. However, in recent years, the role phylloplane colonizers play in plant growth and development, productivity, and raising the standards of plant defense mechanism has gained importance in research fraternity and attracted global attention towards their multifaceted potential in developing eco-friendly agronomic models for plant growth and protection.

A wide variety of microbes, predominantly bacteria, fungi, and yeast are known to colonize various plant surfaces. A number of factors drive the localization and colonization of microbial aggregates on the phylloplane. Trichomes, ridges, and stomatal openings on the leaf surface provide spaces for the microorganisms to aggregate and thrive (Lindow and Leveau 2002). The microbial life on the aerial parts of plants deals with complex system of microbe–microbe interactions, and interactions between the colonizers and host plant (Fig. 11.1). Many of them have been reported to produce phytohormones, promote plant growth and productivity, attenuate plant diseases, induce tolerance against biotic/abiotic stresses, and elicit defense response upon pathogenic attacks (Windham et al. 1986; Vorholt 2012). Phylloplane microorganisms could be very well considered as crucial strategic tools to devise bio-control strategies against a variety of phytopathogens because of the antimicrobial activity of non-pathogenic bacteria and fungi. They are known to facilitate water uptake by reducing leaf surface tension, act as stimulants of plant growth, and provide resistance against abiotic stress like soil salinity, drought, or floods (Lindow and Brandl 2003; Whipps et al. 2008). The enhancement of plant defense system by the action of non-pathogenic bacteria or fungi may act as model for development of sustainable and eco-friendly antimicrobial agents. A number of studies have well elucidated the inhibitory actions of microbes isolated from plant phylloplane. Studies suggest how *Trichoderma* species have been widely considered as leaf-associated microfungi and studied for their efficient antimicrobial properties (Kalita et al. 1996). *Trichoderma viride* was reported to alter the levels and expressions of phenylalanine ammonia lyase, peroxidase, and polyphenol oxidase in rose plants. The enhanced activity of the defense enzymes provided protection against *Diplocarpon rosae* and reduced disease occurrence (Karthikeyan et al. 2007). *Trichoderma harzianum* OTPB3 could develop systemic resistance in *Solanum lycopersicum* against early and late blight disease, inhibiting the in vitro growth of *Alternaria solani* and *Phytophthora infestans* (Chowdappa et al. 2013). Microbial metabolites demonstrate antimicrobial properties and their application could significantly reduce the in vitro and in vivo growth of plant pathogens (Saleem and Paul

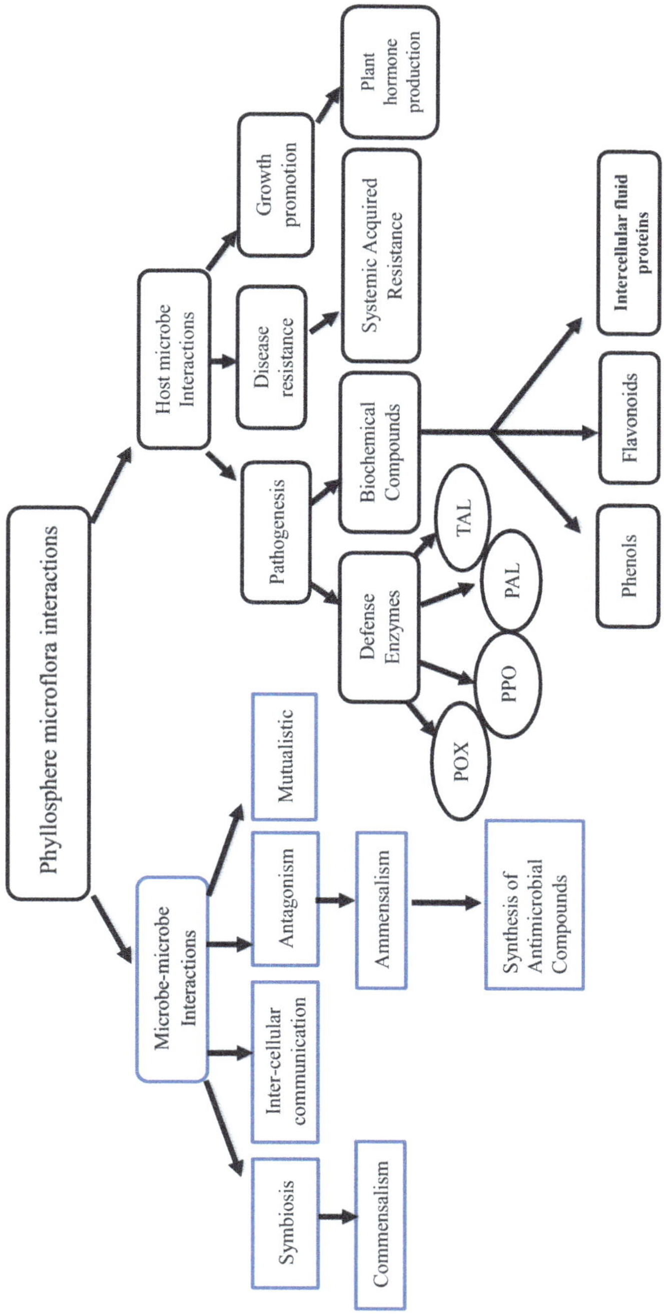

Fig. 11.1 Schematic representation of the inter-microbial and plant–microbial interactions

2015a, b, c). Phylloplane microbes are known to induce a defense response in plants upon pathogen invasion by activating the defense enzymes and phenolic compounds (Pradeep and Jambhale 2002).

The phylloplane colonizers may shift with respect to changes in the environment under which a plant is grown which may include sudden rise or drop in temperature conditions, change in humidity, loss of leaf surface wettability, or ageing and withering of leaves (Nix et al. 2008). Global warming is a growing concern of current times and has been implicated in detrimental aftereffects on the ecosystem. The impact of global warming/climate change may also lead to adverse effects on the plant–microflora relationship (Youssef et al. 2015). Increase in temperature, severity of solar radiations, atmospheric CO_2, and fall in precipitation cause deficiency of nutrients and required moisture on the plant surface which eventually leads to a gradual loss of phyllosphere microbial populations which could be beneficial in the improvisation of plant defense system and protection against severe plant diseases (Tyagi et al. 2014). Insights into underlying problems concerned with microbial loss due to climatic change should help us work on the cause and build a proper management system for better plant growth, quality, and productivity. Principal requirements such as appropriate temperature, soil conditions, relative humidity, photoperiod, etc. should be taken into consideration to avoid loss of beneficial microbial communities from soil and plants. A deeper knowledge into the benefits conferred by phyllosphere microbiota may help us understand the different possibilities and dimensions in which application of leaf-associated microorganisms can be employed. There is a lacuna of information about the survival patterns of human pathogens and non-pathogenic microflora cohabiting plant surfaces under different climatic conditions. Intense studies are required to understand if different phylloplane colonizers respond to the climatic changes differently, and how the shifting pattern of microbial populations varies from one another.

11.2 Phylloplane

The surface of leaf acts as a suitable habitat for a wide array of microorganisms (Hirano and Upper 2000). The term Phyllosphere was proposed by Last (1955) and Ruinen (1956) to specify the leaf surface and its associated microenvironment. It provides the microflora with an ample space to survive and leaches out nutrients which contain sugars and amino acids as major sources of carbon and nitrogen, respectively, needed for their growth and development (Mercier and Lindow 2000). Various plants are the sources of sustenance for fluctuating microbial colonists owing to different physiological conditions (Lindow and Brandl 2003). The abaxial and adaxial leaf surfaces of different plants are known to carry a wide variety of epiphytic communities (Yadav et al. 2011).

11.3 Phylloplane Microbes or Epiphytes

Each part of the plant acts as a host to specific microbial communities. The most abundantly isolated microorganisms from the aerial plant parts belong to the leaf surface. They are usually sourced from air, soil, irrigation water, rain, insects, and are usually specific about the niches they thrive on the plant's surface (Meyer and Leveau 2012).

Phylloplane harbors a significant variety of microflora termed as epiphytes surviving on the surface, whereas the microbes thriving in the internal tissues are called endophytes. Most prominently found microbes are bacteria along with fungi and yeast (Prabakaran et al. 2011). Microflora of the Phyllosphere is specialized at sustaining in the most inevitable conditions like exposure to solar radiations, temperature variations, biotic/abiotic stress, pollution, and lack of humidity (Kishore et al. 2005). The UV radiations act as a crucial basis for selection and survival of phylloplane microbiota and for selecting UV-resistant microbial species (Stockwell et al. 1999). The epiphytes have been found to modify the microenvironment of their host. Aging leads to the wearing of leaf cuticle, alteration in leaf surface wettability is also one of the various physiochemical changes the microbes have to face. *Pseudomonas fluorescens* have been reported to cause increase in the wettability of host leaves for their easy mobility and survival on phylloplane (Knoll and Schreiber 2000). The phyllosphere microbes are also specialized at producing phytohormones thereby promoting plant growth (Vastakaite and Buzaite 2011; Limtong et al. 2014), nitrogen and carbon dioxide fixation (Smith and Goodman 1999), cellulolytic activity (El-Said 2001), detoxification of pollutants, onset of systemic resistance in plants, production of defense enzymes/metabolites, and helping through the plants defense mechanisms against phytopathogens (Harman et al. 2004).

11.4 Inter-Microbial Interactions

The phylloplane microbiome is determined by different genetic and environmental factors, defining the survival of these microbes and shaping them into complex communities, forming microbial ecosystems where they interact and influence each other's growth patterns (Gall 1970). The viability of epiphytic microorganisms decides the establishment of communication and interaction with one another in complex and specialized molecular and physiological mechanisms (Smid and Lacroix 2013). Control of pathogenic growth on the phylloplane is achieved by interactions among epiphytes (saprophytic in nature) and pathogenic microflora, subsequently reducing foliar diseases (Blakeman and Fokkema 1982). It is crucial to ponder upon the importance of these interactions to understand their implications on disease management and crop productivity. Microbial interactions are usually in the form of symbiosis, intra-cellular communications, or interactions upon direct surface contacts which may lead to inhibition of one microbe caused by the other (Gutjahr and Parniske 2013). Metabolic interactions, bio-control properties, secretion of toxins or metabolic compounds lead to the suppression or promotion of growth and help in the sustenance of microbial communities (Wolin et al. 1997). Various microbe–microbe interactions cause inhibition of the epiphytic colonizers due to competition for survival (Mercier and Lindow 2000). Most of the volatile fungal metabolites secreted by various leaf-inhabiting microfungi are known for their mycostatic activity against the growth of several foliar pathogens (Upadhyay 1981). Suppression of phytopathogens by phylloplane residents has managed to gain importance and may, thereby, help in affecting disease occurrence in plants (Patil and Kachapur 2000). Various bacterial–bacterial, bacterial–fungal and fungal–fungal antagonistic interactions have been studied for crop protection and plant health (May et al. 1997; Abdel-Sater 2001). A number of previous investigations have shown various bacterial species inhibiting the growth of certain microfungi and, on the other hand, promoting the growth of others (Schrey et al. 2012).

Pseudomonas species are gram negative, aerobic, polar flagella bearing rods (Srivastava and Shalini 2008). They have been widely acknowledged as significant antifungal agents with substantial investigations undertaken in the area of rhizosphere (Alemu and Alemu 2013). Chitinases are enzymes found in bacteria and other microorganisms that are useful in the biodegradation of polysaccharide "chitin" abundantly found as the main component of the exoskeleton of yeasts and fungi, and thus act as one of the major factors contributing to the bacterial antagonism against microfungi (Hamid et al. 2013). *Bacillus cereus* was found producing two antifungal chitinases against *Botrytis elliptica* causing leaf blight in Lily (Huang et al. 2005). *Enterobacter agglomerans*, a chitinolytic microbe, antagonizes the growth of *Rhizoctonia solani* that causes leaf spot and root rot diseases in tobacco plants (Gonzalez et al. 2011). *Lactobacillus* species are also known to antagonize the growth of *Penicillium*, *Fusarium*, and *Aspergillus* species (Magnusson et al. 2003). *Erwinia herbicola* and *P. syringae* isolated from soybean leaves had antimicrobial activity against *Escherichia coli*, *P. syringae* pv. glycinea, and *Geotrichum candidum*, respectively, thereby suppressing bacterial blight symptoms (Völksch et al. 1996). *P. fluorescens*, *P. aeruginosa*, and *P. asplenii* acted as potent inhibitor of *R. solani* causing sheath blight of Rice (Akter et al. 2014).

11.5 Antimicrobial Activity of Epiphytes

Microorganisms often tend to alter and restructure the colonization patterns on plant parts by antagonizing the growth and affecting the survival of other microbial communities (Beenish 2017). The bacterial–bacterial, bacterial–fungal, and fungal–fungal antagonistic interactions have been extensively investigated for the purpose of better agriculture, crop protection, and improving plant health (Mohamed and Sater 2001). These interactions lead to the inhibition of phytopathogens by phylloplane colonizers which are considered vital for their help in controlling plant diseases (Patil and Kachapur 2000). Earlier studies have reported bacterial species as potent inhibitors of certain microfungi and have also been observed promoting the growth of others (Schrey et al. 2012). A number of bacterial genera like *Pseudomonas* and human pathogens *Serratia* and *Klebsiella* have been found demonstrating antifungal activities against a variety of fungi (Kerr 1999). *Pseudomonas* species are also known to inhibit the activity of phytopathogens like *Botrytis cinerea* and *Alternaria* species (Swadling and Jeffries 1998). Thus, proving to be significant antagonist of prominently found pathogens, they could be used as a potent plant growth promoter and suppressor of plant diseases. *Listeria denitrificans* (E2), *Pseudomonas fluorescens* (C37 and C92), and *Xanthomonas campestris* (D119) retrieved from the phyllosphere of *Lolium perenne* (S24) showed inhibitory activity against *Drechslera dictyoides* (Drechsler) Shoemaker (Austin et al. 1977). *Bacillus* strains isolated from phylloplane were reported of possessing antimicrobial properties and were capable of reducing foliar diseases significantly (Halfeld-Vieira et al. 2008). The phylloplane bacterium *Ochrobactrum anthropi* BMO-111 was found effective in significantly reducing the occurrence of blister blight disease of tea, it antagonized the growth of causal pathogen *Exobasidium vexans* in vitro as well as on the plant (Sowndhararajan et al. 2013). *Aureobasidium pullulans, Cladosporium cladosporioides, Epicoccum purpurascens, Fusarium oxysporum,* and *Myrothecium roridum* caused the inhibition of *Colletotrichum gloeosporioides* whereas *Aspergillus terreus, Cephalosporium roseo-griseum,* and *Penicillium oxalicum* were capable of suppressing the growth of *Puccinia psidii*, a potent phytopathogen with extensive host range (Pandey et al. 1993). *Trichoderma harzianum* and *Trichoderma pseudokoningii* were observed suppressing the growth of *Colletotrichum destructivum* of cowpea (Akinbode and Ikotun 2011). Kawamata et al. (2004) isolated fungal microflora from phylloplane of rice plants and randomly selected them for investigation of antimicrobial activity against rice leaf blast. It was reported that the phylloplane isolates belonged to the species of *Fusarium, Trichoderma, Cladosporium, Penicillium, Pestalotiopsis,* and *Epicoccum* which caused disease suppression upon inoculation of plant with conidial/hyphal suspension of both phylloplane fungus and pathogen simultaneously. *Fusarium sp.* isolates significantly inhibited *Magnaporthe grisea* and formed inhibition zones 3–5 mm wide in vitro against the pathogen. A wide array of microfungal species are considered to be the major reason for foliar diseases. In order to cope up with fungal phytopathogens without affecting the plant's vigor, it is crucial to make use of bio-control agents that may act against the pathogenic microorganisms, act as plant growth promoters, and do not deteriorate plant health. Various studies have provided ample information about the use of bacteria as an effective bio-control agent against different disease causing fungal phytopathogens, not leading to any deleterious effect on plant health (Thomashow and Weller 1996).

A variety of bacterial–fungal interactions have been studied previously. It is crucial to understand the biological mechanisms underlying this phenomenon which range from physical associations to molecular communications and antibiosis followed by toxic compounds and exudates released by the bacteria which may lead to the fungal growth inhibition (Warmink and Elsas 2009). Antibiosis is caused due to the varied metabolic products secreted by the microorganisms with antagonistic properties. Bacterial microflora has long been recognized for the synthesis of certain volatile compounds like HCN, benzaldehyde, acetaldehyde, and iron-binding compounds known as siderophores as well as lactic acid bacteria that produces cyclic dipeptides and phenyllactic acid which are implicated in imparting antimicrobial properties against different fungal microflora (Schnürer and Magnusson 2005; Weisskopf 2013). Saleem and Paul (2015a, b, c) concluded that *Pseudomonas koreensis* and *Sphingobacterium daejeonense* isolated from the phylloplane of *Solanum lycopersicum* significantly reduced the growth of cohabiting fungal pathogens. Various enzymes, bacteriocins, and lipopeptides produced by the *Bacillus* strains render it possible for the bacteria to antagonize the growth of various bacterial, fungal pathogens (Baruzzi et al. 2011). A key role is played by the antimicrobial peptides produced by microorganisms due to non-specific innate immunity response in inhibiting the growth of pathogens (Cruz et al. 2014). It can be concluded from previous studies that the bacterial–fungal contact inhibition also plays a vital role in limiting microbial growth and may lead to physiological changes in the microbes. In many cases, release of deleterious compounds by them may stimulate the diffusion of specific defensive compounds by the microfungal species as a line of protection against the inhibitory action of the antifungal compounds (Klett 2011). Human pathogens have been found colonizing the phylloplane of various plants, due to the high toxicity, *Klebsiella pneumoniae* and *Serratia fonticola* inhabiting the phylloplane of *Solanum lycopersicum,* were found antagonizing the growth of a variety of fungal pathogens like *Cladosporium cladosporioides*, *Cladosporium herbarum*, *Fusarium oxysporum,* etc. on the leaf surface (Beenish 2017).

11.6 Climate Change and Microbial Colonization

Global warming is a much researched topic, yet there are less explored aspects concerning its harmful effects on the agro-ecosystem, ecology of the phyllosphere, and correlation between phyllosphere colonization and plant's disease resistance (Aydogan et al. 2018). The envelope of gases surrounding our earth is getting adversely thick due to which it tends to contain more and more of heat within the atmosphere eventually causing a rise in temperature. Previous studies suggest loss of species among different microbial communities due to global warming (Kumar 2018). Composition of microflora colonizing the phylloplane may show variations among different plant species and is greatly affected by a number of factors including geographical location, season, and various environmental conditions such as temperature, rain, sun exposure, dryness, etc. (Whipps et al. 2008; Ding and Melcher 2016). Fig. 11.2 shows how abiotic changes, including fluctuations in weather conditions, are responsible for modification of the physicochemical conditions and microbial diversity on the phylloplane which may further add complexity to plant defense and agricultural systems. Studies suggest that the plants which are not tolerant to varying temperature conditions especially in soil polluted with heavy metals show difficulty in adapting to the changes and may demonstrate complex

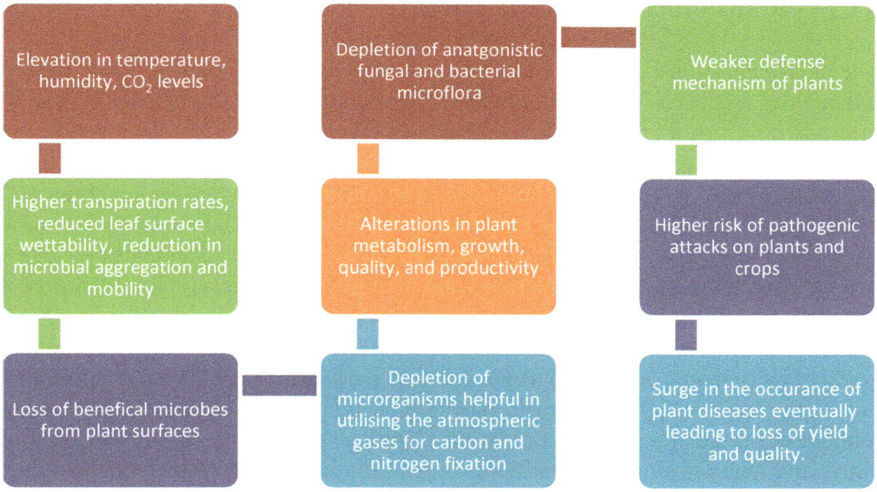

Fig. 11.2 Schematic representation of impact of climate conditions on plants and microflora

response towards the stress factors (Tyagi et al. 2014). Changes in the weather conditions including elevation in temperature, increased transpiration rates, and stomatal conductance whether long term or short term can affect crop quality and interrelationship dynamics between microbes colonizing the crops. It has been reported in a number of studies that any change in climatic conditions under which the plants are grown may disrupt the equilibrium required for feasible microbial sustenance which further leads to a shift of microbial population inhabiting the surface thereby causing alleviation in biodiversity and a rise in plant stress (Lindow and Brandl 2003). Earlier studies have led to the observation that any decrease or increase in the temperature and precipitation usually leads to the depletion of nutrients and altered moisture levels on the leaf surface which may modify the pathobiome by threatening microbial aggregation, abundance, and sustenance (Elad 2009). The irregular wettability on leaf surface usually results in intermittent growth of bacterial and fungal colonists, posing survival threats during dry season (Blakeman and Fokkema 1982). These changes consequently lead to the shifting of varieties of microorganisms from the plants. Some of the beneficial microbes produce antimicrobial compounds and induce systemic acquired resistance against pathogenic microflora by complex plant–microbe interactions upon stress recognition (Brader et al. 2017). Altered phyllosphere microbial populations may in turn strongly affect plant growth and its ability to combat diseases, making the plant more prone to pathogenic invasion (Elad 2009). Loss of plant beneficial microflora may significantly influence the plant's metabolism, growth promotion, and disease suppression.

11.7 Plant–Microbe Interactions

The microbial colonization occurring on the host plants is followed by host–microbe interactions leading to either pathogenesis, resistance to diseases, or growth promotion that develops a cascade of secondary processes by the phylloplane colonizers (Knoll and Schreiber 2000). Microbes surviving on the plants benefit from metabolites and nutrients leached from the surface, utilizing sugars (glucose, sucrose, fructose) as carbon sources (Lindow and Brandl 2003). Frequent production of phytohormones by microbial flora has been vastly investigated and only since few years the role of phyllosphere microbes and their possible effects on the alteration of growth patterns of host has come into perspective (Vastakaite and Buzaite 2011; Limtong et al. 2014). Series of changes in the plant microhabitat and physiology could be an outcome of the colonizer's interactions with the host (Lindow and Brandl 2003). Synthesis of phytohormones has long been considered as a virulence factor. Pathogens on aerial parts of the plant may produce them at the site of infection or stress and quantities of phytohormones often play a crucial role in influencing the pathogenesis (Prasannath 2013). In studies undertaken previously, a rise in the levels of IAA has been noticed at the site and course of infection (Fu and Wang 2011).

Microfungal species associated with the phylloplane have been utilized as bio-control agents to prevent pathogens from infecting the host (Evueh and Ogbebor 2008). Many such fungi have shown to enhance plant growth promotion, crop yield and quality, resistance to stress factors, and the assimilation of nutrients (Windham et al. 1986). These fungi have also been reported to prevent a number of plant diseases such as foliar, root, or fruit diseases, along with invertebrates like nematodes (Shoresh et al. 2010). Microbes synthesize a range of biochemicals that cause localized or systemic resistance thereby accelerating the defense mechanism in host plants. Secretion of antimicrobial compounds on plant surfaces acts as a major line of defense by which plants deter potential pathogens (Lindow and Brandl 2003). These plant–microbe associations can cause significant modifications in the proteome and metabolism of the host (Harman et al. 2004). Foliar bacterial community is significantly affected and altered in terms of population, diversity, and richness upon powdery mildew infection. According to the study by Suda et al. (2009) shifting patterns of bacterial populations showed greater specificity towards plant species. Similarly, considerable changes in leaf-associated fungal and bacterial communities were observed upon powdery mildew infection in oak (*Quercus robur*) due to the causal pathogen *Erysiphe alphitoides*. The investigations further led to the revelation of pathobiotic network of different fungal and bacterial operational taxonomic units connected with each other and directly interacting with *Erysiphe alphitoides*. The study suggested that some of these might have been conferring protection to the oak phyllosphere by inhibiting *E. alphitoides* and thereby reducing the probabilities of powdery mildew occurrence in natural physiological and weather conditions (Jakuschkin et al. 2016). Severity of powdery mildew infection was negatively correlated with the richness and diversity of the fungal microflora present on phyllosphere of pumpkin (*Cucurbita moschata*). Phylloplane that was scarcely colonized by powdery mildew showed greater diversity and richness, whereas a significant decrease was seen in densely colonized leaves (Zhang et al. 2018). Investigations by Wang et al. (2019) suggested that *Protomyces* strain SC29 on the phylloplane of Arabidopsis activates defense responses by initiating yeast-specific MAMPs that can trigger plant immunity and can provide insights into the evolution of fungal virulence, and plant immunity against yeasts. SC29 treated Arabidopsis showed enhancement in immunity against *Botrytis cinerea* infection, associated with activation of MAPK3/6, camalexin, and SA-signalling pathways. *Protomyces* have been evident inhabitants of the phyllosphere as yeasts, but invade their hosts in the hyphal form. The inability of SC29 to cause disease on Arabidopsis, but the ability to persist on the leaf surface suggests a change to survival strategies where phylloplane as a microhabitat plays a key role.

11.8 Elicitation of Plant Defense Response by Fungal Metabolites and Ergosterol

Phylloplane fungi interact with plants in many ways and affect their growth, development, and protection. Fungi growing on the phylloplane can be classified into two groups: residents and casuals. Resident fungi multiply on the surface of healthy leaves without affecting the host. Casual fungi mostly land on the leaf surface due to environmental factors but are unable to grow and survive (Prabakaran et al. 2011). Endophytic fungi are crucial in implicating the functionality of the host plant's endosymbiotic systems which could be nutritional, defense-related, or biotic/abiotic stress toleration (White et al. 2014). Various phylloplane microfungi have been well-known antagonists and are utilized as bio-control agents to prevent phytopathogens infecting the host. As suggested by earlier studies, secretion of antimicrobial biochemicals on aerial plant surfaces is thought to be one of the defensive strategies by which plants combat potential pathogens (Peláez et al. 1998). Phenols, peroxidase, and polyphenol oxidases are defense-related chemicals released by the plants against pathogens. Many of the volatile fungal metabolites secreted by various phylloplane microfungi are known for their antifungal activity against several foliar pathogens (Upadhyay 1981). Thakur and Harsh (2014) worked with ten phylloplane isolates of *Chlorophytum tuberosum* against its potent pathogen *Colletotrichum dematium* and found that volatile secretions of these leaf-inhabiting fungi could effectively reduce the mycelial growth of *C. dematium*. *Trichoderma viride* and *Aspergillus flavus* caused maximum growth inhibition of *Alternaria brassicae* causing leaf spot disease of mustard. The metabolites released by these phylloplane fungi arrested the hyphal growth of the pathogen (Yadav et al. 2011). Volatile and non-volatile compounds released by *Trichoderma* species may significantly inhibit the pathogen's growth by antibiosis, competition, mycoparasitism, or direct contact inhibition (Kuberan et al. 2012; Parizi et al. 2012). Similarly, some *Alternaria* metabolites were also reported to exhibit antimicrobial properties (Lou et al. 2013). These metabolite acted as bio-control agent in controlling fungal disease in grapevine cultivation.

Ergosterol is of fungal origin, a sterol composition in cell membrane, and a common secretion of fungal metabolites (Weete et al. 2010). It acts as an elicitor of defense responses in plants (Felix et al. 1999). As per the study of Kauss and Jeblick (1996) ergosterol in plant cells effectively stimulates hydrogen peroxide (H_2O_2) production in the host. Application of ergosterol has been reported to activate the expression of defense genes in plants against pathogens thereby activating the defense signalling pathways (Lochman and Mikes 2006). It is reported to induce cross-talk between the salicylate and jasmonate signalling pathways, thus increasing the resistance against pathogenic attack. Ergosterol reportedly leads to alteration in the proton fluxes and membrane hyperpolarization mechanism in the motor cells of host plants (Amborabé et al. 2003). *Cladosporium* species have been reported to be amongst the dominant microfungi inhabiting phylloplane of *Solanum lycopersicum* (Saleem and Paul 2016). Granado et al. (1995) treated tomato plants with ergosterol extracted from *Cladosporium fulvum* and found that the plants could perceive ergosterol which elicited cellular alkalinization in them.

11.9 Intercellular Fluid Proteins

These are the proteins found in the apoplast or the intercellular space surrounding plant cells. This is known to be a dynamic environment where variety of metabolic, defensive, and translocation processes occur. Prior investigations have led to the understanding that apoplast is considered to be a primary location where initial resistance responses against pathogenic microbiota take place, e.g. generation of reactive oxygen species by extracellular peroxidases and oxidases, and strengthening of plant cell wall by crosslinking or callose deposition. As a result of the aforementioned processes, an array of proteins and amino acids are found in the leaf apoplastic fluid (O'Leary et al. 2014). Plant–pathogen interactions usually modify the intercellular fluid protein quantities. Rise in the quantities can be observed upon pathogen invasion. A number of new proteins such as P14 were observed in intercellular spaces or apoplast of *Solanum lycopersicum* phylloplane infected with *Cladosporium fulvum*, a potent fungal pathogen (De Wit et al. 1986).

11.10 Phenylalanine Ammonia Lyase (PAL) in Plant Defense

A number of researches have brought to our understanding that induction of resistance via biosynthesis and endogenous distribution of different phenolic compounds, expression of PAL, TAL, POX, PPO, and key enzymes offer a major contribution in the biological control of pathogens and insects, thus, promoting the standards of defense mechanism in plants (Mohamed et al. 2007; Mahmoud et al. 2012).

PAL acts as a key enzyme in the phenyl propanoid pathway. It catalyzes the synthesis of phenols and salicylic acid involved in defense against the pathogens, biotic stress response, and secondary plant metabolism and is associated with posttranslational phosphorylation (Nicholson and Hammerschmidt 1992). Induction of PAL activity in host tissues is followed by pathogen infection and several biotic and abiotic elicitors (Ebel 1986). Concentration of PAL activity may largely vary based on the developmental stage of plant, genotype, environmental factors, pathogen type, and plant–microbe interactions (Dixon and Paiva 1995). PAL activity indicates the degree of host resistance as its rapid accumulation reaches high levels during incompatible interactions when a defense response is induced within the host plant as compared to susceptible responses (compatible interactions) (Bhattacharyya and Warde 1988). Gupta and Kaushik (2002) in their study suggested the enhanced specific PAL activity in infected leaves as compared to healthy leaves of mustard.

11.11 Tyrosine Ammonia Lyase (TAL) in Plant Defense

TAL is a crucial enzyme in the phenyl propanoid pathway and member of the aromatic amino acid lyase family. TAL (transcription activator-like) effectors comprise a novel class of DNA-binding proteins. They are employed by gram negative *Xanthomonas* sp. which is pathogenic to plants. It generally transports various types of effector proteins through a type III secretion system into plant cells where they determine the virulence severity. TALs primarily localize to the nucleus where they associate with the target promoters thereby inducing gene expression in plants (Scholze and Boch 2011). It has high specificity for L-tyrosine, generates 4-coumaric acid as a protein cofactor and antibiotic precursor in microbes (Watts et al. 2006). Earlier studies report changes in the TAL activity in plants due to a number of factors including biotic/abiotic stress and pathogenic attack (Dixon and Paiva 1995). The increase in TAL activity in compost treated tomato plants due to induced systemic resistance was found to be related to biosynthesis of lignin from tyrosine (Fayzalla et al. 2009; Abdel-Fattah and Al-Amri 2012). A significant increase in TAL activity was recorded upon inoculation with plant pathogen *Sclerospora graminicola* which acted as a signal and led to induction of resistance in pearl millet seedlings (Hindumathy 2012).

11.12 Peroxidases (POX)

POXs are hem-containing glycoproteins and are found in microorganisms, plant and animal tissues. They are involved in a variety of plant physiological processes including cell wall modifications, developmental and defense-related mechanisms (Gaspar et al. 1991). Large multigene family encodes the class III plant peroxidases family (POX, EC 1.11.1.7), which further constitutes numerous peroxidase isoenzymes (Hiraga et al. 2001). Endoplasmic reticulum is responsible for targeting the plant peroxidases (class III) towards the outside of the plant cell or to the vacuole (Welinder et al. 2002). Plant peroxidases are synthesized extracellularly and are known to catalyze the generation of reactive oxygen species (ROS) coupled to oxidation of phytohormone indole-3-acetic acid and synthesis of biochemicals associated with plant defense mechanism: salicylic acid, aromatic monoamines, and chitooligosaccharides (Kawano 2003). Studies demonstrate that peroxidases are capable of generating reactive oxygen species during the massive oxidative burst in incompatible interactions which establishes an extremely toxic environment for pathogens to survive, thereby promoting plant protection (Gómez-Vásquez et al. 2004). POX enzymes are considered to play a vital role in defense against pathogenic attack and are also responsible for removal of toxic H_2O_2 from the host cells, thereby deterring oxidative damage and protecting the cells. Increase in POX activity is observed upon foliar infections or inoculations. Rise in POX activity was seen in three cucumber cultivars when the leaves were systemically infected with *Cucumber mosaic virus* (Wood and Barbara 1971). The rise in POX activity due to arbuscular mycorrhizal fungi is reported to be involved in lignification (Lagrimini and Rothstein 1987).

POX isoforms play a crucial part in plant defense mechanism. Previous researches have led to the observation that peroxidase expression levels and its isozyme patterns were significantly affected by biotic/abiotic stress, application of chemicals, and invasion by pathogens in a number of plants (Gasper et al. 1982). Two anionic isozymes of POX were induced in the diseased leaves of Tobacco plants infected with *Tobacco mosaic virus*, and were also later systemically induced in the leaves that remained distant and uninoculated (Lagrimini and Rothstein 1987). Another study demonstrated the presence of four types of POX isoforms in *Pseudomonas syringae* pv. *tomato* infected tomato plants while the control plants showed synthesis of only one type (Bashan et al. 1987). An early localized and systemic increase in peroxidase activity was observed upon *Rhizoctonia* infection in Norway spruce, and the prominent isoforms were found to be basic peroxidases (Nagy et al. 2004).

11.13 Polyphenol Oxidases (PPO)

PPOs have been playing a crucial role in the phenylpropanoid pathway (Kojima and Takeuchi 1989). They are present ubiquitously, which means they can be found in plants, animals, and microorganisms predominantly in bacteria and fungi. They are copper-containing enzymes, nuclear encoded and are known to catalyze the O_2-dependent oxidation of mono and *o*-diphenols to *o*-diquinones. PPOs are highly reactive intermediates and their secondary reactions are responsible for the oxidative browning which is caused during plant senescence, pathogenic invasion, wounding, etc. (Thipyapong et al. 2004). It has been established clearly in different studies that PPOs play a key role in plant defense mechanism since they are largely present at the site of wounding, pathogenic ingression, or insect infestation. They can also be induced due to various abiotic and biotic injuries or signalling molecules (Constabel et al. 1995; Thipyapong and Steffens 1997). PPO activities were significantly increased in SA pre-treated onion plants inoculated with *Stemphylium vesicarium* (Abo-Elyousr et al. 2009). Ibrahim (2012) suggested that the resistance induced by SA against *X. vesicatoria* in tomato led to increased PPO activities. Previous studies provide us information as to how increase or decrease in PPO levels could be responsible in suppressing disease incidences, pathogen invasion, and multiplication in transgenic tomato plants. Significant reduction in bacterial growth was observed in tomato plants with enhanced PPO levels when they were inoculated with bacterial pathogen *Pseudomonas syringae* pv. *tomato*, whereas plants with suppressed PPO activity had higher disease incidence and showed a greater risk of infection (Constabel and Barbehenn 2008).

PPO isoforms are induced in tomatoes after infection with phytopathogens *P. syringae* pv. *tomato* and *Alternaría solani*. The expression and subsequent rise in isozymes levels upon pathogen attack suggest their part in inhibition of diseases (Thipyapong and Steffens 1997). A significant positive correlation was established between the degree of plant resistance and PPO level (Kavitha and Umesha 2008). Raised levels suggest a stronger defense response in host plants. Eight different types of PPO isozymes were induced in tomato plants that were treated with pathogens (Bashan et al. 1987). Similarly, involvement of PPO isozymes was observed in conferring resistance to wheat against *A. triticana* (Tyagi et al. 2000).

11.14 Age Related Resistance (ARR) in Plants

A plethora of factors could be responsible for a plant's ability to protect itself from any pathogen ingress or infection; they may include environmental conditions—temperature, humidity, nature of the infected tissue, or the genotypic correlation between the host species and the pathogen. Apart from the aforementioned reasons the plant's developmental stage also acts as a crucial factor that confers it the ability to fight against pathogenic attacks; however, it is far less frequently taken into consideration. Henceforth, investigating the implications of a plant's development and its age on microbial colonization patterns and disease resistance could be a vital breakthrough in our knowledge of plant–microbe interactions. This development of plant resistance to pathogenic infections with respect to ageing has been recognized by various terms such as "ontogenic resistance," "developmental resistance," "mature seedling resistance," "adult seedling resistance," "age-related resistance," etc. (Whalen 2005). According to Develey-Rivière and Galiana (2007), resistance developed in plants during the course of ageing could prove effective against a variety of pathogens, pathovar, or pathogenic strains, but when race-specific resistance is concerned, the resistance response is mainly dependent upon the functional regulation of genes related to plant resistance. Many of the plant–pathogen interactions are dependent on the developmental stage of the plant as the plant's physiological age acts as a crucial factor in the induction of resistance response against the invading pathogen. Recent researches have made possible to understand how age can cause variations in the microbial colonization and thus may impact a plant's defensive strategies thereof; the younger leaves of *Solanum lycopersicum* are heavily colonized and more defensive against diseases as compared to the ones approaching senescence (Saleem and Paul 2016). This may suggest of a rise in resistance with time as the plant ages, yet an increase in susceptibility, i.e. decrease in resistance could be seen towards senescence, plants already resistant to a pathogen raising their ability to control infection and colonization at a precise growth phase. Various observations have led to the understanding that there is no role of growth stage in development of resistance and a decrease in resistance response of a plant against any pathogen may occur as it matures because of increased susceptibility towards pathogens (Visker et al. 2003). Studies were carried out to determine the impact of antioxidants synthesized by leaves when a pathogenic attack takes place. Changes in quantities were observed as the leaves aged, which may help in mediating a defense response against pathogens (Dat et al. 2000).

Previous studies have also reported the involvement of phylloplane microbes in the biosynthesis of phytohormones which could further play a crucial role in aspects concerned with plant growth promotion, development, reduced susceptibility towards pathogens during plant–microbe interactions, and were found effective in developing ARR in the host plants (Mayda et al. 2000; Mauch-Mani and Mauch 2005; Saleem and Paul 2015a, b, c). The effects of salicylic, jasmonic, and abscisic acids along with ethylene, on the disease resistance capacity of the host plant have been well investigated (Develey-Rivière and Galiana 2007). Eventual development of ARR was also observed in Arabidopsis due to de novo synthesis of salicylic acid as the plants matured (Kus et al. 2002).

11.15 Systemic Acquired Resistance (SAR) in Plants

Previous investigations and extensive analysis of mechanisms related to the induction resistance responses provide essential insights into the wide range of strategies a plant could make use of in order to survive against pathogens and other trigger elements. Survival mechanism to attenuate disease occurrence may include activation of specific genes, synthesis of defense-related proteins, synthesis of antimicrobial compounds, etc. (Develey-Rivière and Galiana 2007). Systemic acquired resistance is a phenomenon which can be mediated by various factors where the plants can be induced by a number of conditions and agents including pathogen ingress, natural/synthetic/biochemicals, wounding, or application of plant extracts (Kessmann et al. 1994; Schweizer et al. 1998; Funnell et al. 2004; Hassan et al. 2009;). The exogenous application of biological or chemical agents on plants can elicit a defense response from the site of pathogenic invasion as well as the parts that were not affected by the pathogen. Non-pathogenic microbes with antimicrobial properties isolated from the phylloplane or any other niche can well be reckoned as effective bio-control agents and could play a significant part in the enhancement of plant defense strategies. A systematic acquired resistance was initiated in *Solanum lycopersicum* plants upon treatment with the bio-control bacteria *Pseudomonas fluorescens* WCS417. Earlier investigations mention that defense response in plants is initiated causing rise in the production and distribution of salicylic acid, both locally and systemically, the biosynthesis of salicylic acid acts as a signal responsible for the onset of SAR (Gaffney et al. 1993; Meuwly et al. 1995). Study of this phenomenon in tobacco (*Nicotiana tabacum*) could determine that effects of SAR persisted for a minimum of 20 days approximately (Ross 1961). Some of the investigations state that establishment of SAR is a salicylic acid dependent process while some studies stress on that, it is mediated by jasmonate-ethylene sensitive pathway. Pieterse et al. (1996) in their study revealed that systemic acquired resistance could be a defense response elicited in the plant when it encounters a pathogen. It was also reported that the resistance response is correlated with the biosynthesis and accumulation of salicylic acid along with distribution and is characterized by the activation of genes that encode pathogenesis-related proteins. Other inducers of systemic resistance include jasmonic acid, its derivative methyl jasmonate and ethylene (Pieterse and Van Loon 1999; Park et al. 2007). In the course of plant–microbe interactions, phylloplane microflora has also been implicated in playing a key role in inducing SAR in plants. Panstruga and Kuhn (2019) reported the efficacy of bacterial strain *Bacillus subtilis* UMAF6639 in initiating a defense response by the activation of jasmonate and salicylic acid dependent pathways. The microorganism originally isolated as a phyllospheric endophyte is responsible for induction of defense responses in melon plants thus offering them protection against cucurbit powdery mildew. The phenomenon generally leads to a notable rise in defense enzyme PAL and POX expressions and activities (Kurth et al. 2014). Investigations carried out by Podile and Prakash (1996) showed that *Bacillus subtilis* AF could significantly inhibit the occurrence of crown rot in *Aspergillus niger* infected soil, this suggests that the bacterium could be effectively considered as a bio-control agent. Similarly, as studies suggest that the exogenous application of biological agents on localized sites may confer resistance to the entire plant, a systemic resistance was initiated for protection against disease causing pathogens including *Pseudomonas syringae* pv. *maculicola* in *Arabidopsis thaliana* by local inoculation of a single leaf with avirulent *P. syringae* pv. *tomato* also found inhabiting the tomato phylloplane carrying the avrRpt2 avirulence gene (Cameron et al. 1994). Inoculums of *Serratia marcescens* strain B2 retrieved from the phylloplane of tomato plants were employed for the treatment of rice plants. The bacterium was effective in significantly inhibiting the in vitro growth of common phytopathogenic fungi including *P. oryzae*, *Botrytis cinerea*, *Rhizoctonia solani* AG-4, and *Fusarium oxysporum* f. sp. *cyclaminis* under greenhouse conditions (Someya et al. 2002).

11.16 Priming and Pathogenesis-Related (PR) Proteins

Priming has long been recognized as a state in which the plant prepares itself for a faster, better, and stronger defense response upon pathogenic invasion or any abiotic stress. This physiological state can be the preliminary step in the establishment of an appropriate defense mechanism and could be induced even by the application of microbial metabolites or synthetic compounds on plants (Conrath et al. 2006). Kauss et al. (1992) brought about the first systematic study of priming in plant cell suspension cultures. Priming is linked with a rise in accumulation or posttranslational modification of cellular signalling proteins in inactive state thereby playing a key role in signal amplification. Study by Van Peer et al. (1991) showed that the first evidence proving priming is an important factor in microorganism-mediated induced systemic resistance (ISR) was provided from experiments with carnation (*Dianthus caryophyllus*), where treatment with *Pseudomonas fluorescens* strain WCS417r mediated a rapid rise in phytoalexin levels upon inoculation with *Fusarium oxysporum* f.sp. *dianthi.* Frequent exposure to stress could initiate or modulate these otherwise dormant signalling proteins, followed by activating the signal amplification further leading to quicker and stronger activation of the defense mechanism and systemic resistance (Conrath et al. 2006). Earlier findings suggest an important role rhizobacteria play in activating the defense responses in *Arabidopsis*, it, however, does not directly affect the defense genes, but induce priming to enhance the expression of jasmonic acid and ethylene-inducible genes upon infection by *Pseudomonas syringae* (Van Wees et al. 1999; Verhagen et al. 2004). β-amino butyric acid (BABA) was reported to prime salicylic acid-inducible *PR-1* expression in *Arabidopsis* (Zimmerli et al. 2001; Ton et al. 2009). Challenging *Arabidopsis* with BABA or avirulent bacteria generally initiates priming against *P. syringae* pv. *tomato* DC3000 correlating with raised levels of SA-dependent gene transcripts of *PR1*, *PR2,* and *PR5* upon infection. The phenomenon indicates alterations in the regulatory mechanisms of defense gene expression (Slaughter et al. 2012). Elevation in the expression and quantities of phenylalanine ammonia lyase (PAL) and other PR proteins was also observed followed by the application of exogenous chemicals that may mimic the effect of pathogen invasion (Kohler et al. 2002). This clearly demonstrates that the plant prepares itself by accelerating the production of defense enzymes and biochemicals upon stimulation by chemicals, metabolites, or phytopathogens.

Proteins triggered in the plants upon pathogen invasion or similar conditions are called pathogenesis-related proteins (PRs). They are induced in the plants as a line of defense against the ingression (Van Loon et al. 1994). Pathogen invasion may refer to pathogen attack by various invaders including fungi, bacteria, viruses, insects, and herbivores. PR proteins may also be induced by the application of chemicals that resemble the effect of pathogenic attack e.g. ethylene, jasmonic acid, salicylic acid, along with that wounding of the tissue can also accumulate PR proteins at the site of injury or infection (Sels et al. 2008). Phytohormones have also been considered as major factors determining the onset of PR proteins. The PR proteins are considered to be exclusive products of SAR genes and their accumulation in the host plant is a sign of pathogen mediated systemic acquired resistance. Phytopathogens upon infecting the host tissue activate the defense-related genes which could be expressed in both infected and non-infected tissues as SAR is developed and helps in conferring protection to the entire plant (Van Loon et al. 2006). The PRs are regulated by the defense regulatory SAR and ISR-mediating hormones such as salicylic acid, jasmonic acid, and ethylene. It, henceforth, suggests that they could play a vital role in alleviating the effects of ingress by pathogens (Van Loon et al. 2006). ROS (reactive oxygen species) led biosynthesis of proteins and their induction by plant cell wall fragments has also gained importance, the proteins are highly toxic to invading pathogen due to their hydrolytic, proteinase-inhibitory, and membrane permeabilizing properties (Edreva 2005). Thus, it could be well perceived by previous studies that the term "pathogenesis-related proteins" was specified to indicate proteins triggered by a range of pathogens as well as stress conditions either stimulated by pathogens or induced by the application of biochemicals/agents that mimic the effect of pathogen infection initiating similar kind of stress (Van Loon et al. 1994).

11.17 Changes Induced in Total Phenols and Flavonoids by Microorganisms

The type, quality, and quantity of biochemical substances present on the phylloplane could be subjected to fluctuations and modifications as a result of the metabolic activity of microorganisms present on the leaf surface (Blakeman and Atkinson 1981; Morris and Rouse 1985). Earlier researches suggest that microorganism invasion on plants bears a direct association with the quantities of simple and polyphenols (Tyagi and Chauhan 1982). A number of studies revealed that the composition of phenolic content in a plant and degree of its resistance to diseases shared a positive correlation. Attenuation in the biosynthesis of phenolic compounds in plants was observed generally after attack by plant pathogens (Farkas and Kiraly 1962). Bhatia et al. (1972) studied the ability of tomato plants to resist infection caused by *A. solani.* It was found to be significantly higher in the plant variety with increased amount of phenolics in the leaf, stem, and roots as compared to the susceptible variety which comprised of lower quantities. Owing to this phenomenon, the investigations carried out by Saleem and Paul (2015a, b, c) on *Solanum lycopersicum* showed a distinct difference in the phenolic and flavonoid content of control plants and plants challenged with microfungi isolated from the phylloplane. The study demonstrated significant ($p \leq 0.05$) changes in flavonoids and phenolic composition of leaves upon inoculation with *Aspergillus candidus, A. niger, A. flavus, Fusarium oxysporum, Rhizoctonia solani, Trichoderma harzianum, Alternaria alternate, A. citrifolia, Cladosporium cladosporioides, C. herbarum, Curvularia lunata,* and *Penicillium expansum.* Phenolics and the related oxidative enzymes have generally been considered as few of the most significant biochemical parameters determining the viability of plants, disease resistance, and differentiating between resistant and susceptible genotypes (Pradeep and Jambhale 2002). Phenolic compounds have attracted a number of studies previously, however, the role of flavonoids in plant defense has not been largely investigated, exceptions being catechins and proanthocyanidins (Feucht and Treutter 1999). Plants may synthesize increased amounts of flavonoids and phenolic compounds possessing effective antioxidant activities that may further lead towards devising various defensive and disease fighting strategies. Phenolic compounds may act as antimicrobial agents due to various factors which may include (a) binding of the quinone nucleus to –SH and NH2 groups in the bacterial cell or (b) disturbance caused in the electron transport chain (Owens Jr 1953).

Flavonoids involved in plant defense are generally "preformed" or "induced." The preformed flavonoids are present in healthy plants having no instances of disease, while induced flavonoids are a consequence of plant–pathogen interaction (Treutter 2006). It is evident from previous studies that they could promote plant protection against pathogens, insects, and herbivores by increasing the toxicity and reducing their nutritive value followed by change in the palatability and decrease in digestibility (Harborne and Williams 2000). Flavonoids are generally considered as vital factors in the enhancement of plant resistance response against pathogenic microbiota, though the area of research has been explored scarcely. Studies suggest that antioxidative properties of flavonoids could be responsible for their antimicrobial properties which can be non-specific and helpful against ROS, which are generated by the host plant followed by pathogenic infection, and are a major governing factor for oxidative and pathological stress (Dai et al. 1996). Flavonoids rush to the site of ingress when a pathogen invades the plant's tissues and a hypersensitive reaction is elicited by them ultimately causing programmed cell death. They could also be directly employed in suppressing the growth of pathogenic microflora by inhibiting the plant cell wall, digesting enzymes of the pathogens by chelation of metals essential for their activity (Treutter 2005). Blount et al. (1992) reported that flavonoids could act as potent antimycotic agents. The antimicrobial activity is largely based on inhibiting spore development, mycelial growth, or hyphae elongation. Studies revealed the presence of proanthocyanidins or small quantities of dihydroquercetin could render resistance to the barley mutants against *Fusarium* sp. (Skadhauge et al. 1997).

11.18 Conclusion

The role which associated microbial communities play in implicating the outcome of host–pathogen interactions is widely acknowledged, but little has been investigated in this regard. Not much information exists on how phyllosphere microbes affect the defensive strategies of the host plant. It is of great importance that leaf associated microbiota is capable of limiting infections, may act as barriers against pathogens, induces defense response, and could significantly contribute to the non-host resistance against non-adapted pathogens. Phylloplane microbes possess agricultural importance in acting as bio-control agents against a variety of pathogenic microorganisms. Intensive investigations may open channels leading to the application of beneficial phylloplane microbes in the development of bio-control for crop protection and plant health.

The chapter provides an insight into the functional correlation between phyllosphere microflora, changes in plant physiology, and implication on the plants defense response. It describes how immensely helpful phyllosphere microbes and their metabolites could act in designing bio-control strategies for combating diseases in agriculturally important plants and crops. The bacterial and fungal isolates could also be further studied individually or as consortium in providing resistance to plants against various biotic and abiotic stresses. It throws light on the molecular dynamics of how the human pathogens could provide resistance against phytopathogens by acting as bio elicitors of defense response in plants. The elevated concentration of defensive metabolites and enhancement of defense enzymes activity prove beneficial to the plant defense dynamics and help in understanding the role of phylloplane microbiota in improvising the disease resistance mechanism of plants and crops. Production of effective bio-control will prove to be environment friendly and cost effective as compared to the conventional approach and will be highly accepted for reducing the usage of chemical alternatives. A number of previous and current researches testify the extent of possibilities phyllosphere microbiome may open for designing models for crop protection. Leaf associated microorganisms may be instrumental in providing protection against diseases and may have broad-spectrum activity against various pathogens. Their antimicrobial and plant growth promoting properties are thus of conspicuous agronomic interest, however no comprehensive information is available and the concept remains less explored.

11.19 Future Prospects

Improvisation and maintenance of conditions under which plants or crops are grown in order to maintain the natural community structure of microbiome on plant surfaces. Intensive research in employment of epiphytes and human pathogens singly or in consortia could prove beneficial in the development of non-toxic and biodegradable antimicrobial agents. Betterment of quality, productivity, and stress tolerance in plants can be achieved. Foliar or root application of microbial metabolites may further help in reducing incidences of disease occurrence.

References

Abdel-Fattah GM, Al-Amri SM (2012) Induced systemic resistance in tomato plants against *Fusarium oxysporum* f.sp. *lycopersici* by different kinds of compost. Afr J Biotechnol 11 (61):12454–12463

Abdel-Sater MA (2001) Antagonistic interactions between fungal pathogen and leaf surface fungi of onion (Allium cepa L.). Pak J Biol Sci 4(7):838–842

Abo-Elyousr KA, Hussein MAM, Allam ADA, Hassan MH (2009) Salicylic acid induced systemic resistance on onion plants against *Stemphylium vesicarium*. Arch Phytopathol Plant Protect 42 (11):1042–1050

Akinbode OA, Ikotun T (2011) Potential of two *Trichoderma* species as antagonistic agents against *Colletotrichum destructivum* of cowpea. Afr J Microbiol Res 5(5):551–554

Akter S, Kadir J, Juraimi AS, Saud HM, Elmahdi S (2014) Isolation and identification of antagonistic bacteria from phylloplane of rice as biocontrol agents for sheath blight. J Environ Biol 35 (6):1095

Alemu F, Alemu T (2013) Antifungal activity of secondary metabolites of *Pseudomonas fluorescens* isolates as a biocontrol agent of chocolate spot disease (*Botrytis fabae*) of faba bean in Ethiopia. Afr J Microbiol Res 7(47):5364–5373

Amborabé BE, Rossard S, Pérault JM, Roblin G (2003) Specific perception of ergosterol by plant cells. C R Biol 326(4):363–370

Austin B, Dickinson CH, Goodfellow M (1977) Antagonistic interactions of phylloplane bacteria with *Drechslera dictyoides* (Drechsler) shoemaker. Can J Microbiol 23(6):710–715

Aydogan EL, Moser G, Müller C, Kämpfer P, Glaeser SP (2018) Long-term warming shifts the composition of bacterial communities in the phyllosphere of *Galium album* in a permanent grassland field-experiment. Front Microbiol 9:144

Baruzzi F, Quintieri L, Morea M, Caputo L (2011) Antimicrobial compounds produced by *Bacillus* spp. and applications in food. Sci Microb Pathog Commun Curr Res Technol Adv 2 (1):1102–1111

Bashan Y, Okon Y, Henis Y (1987) Peroxidase, polyphenoloxidase, and phenols in relation to resistance against *Pseudomonas syringae* pv. *tomato* in tomato plants. Can J Bot 65(2):366–372

Beenish S (2017) Leaf age correlation to phyllosphere microbe microbe plant microbe interactions on *Solanum lycopersicum*. Amity Institute of Biotechnology, Amity University, Noida. http://hdl.handle.net/10603/188984

Bhatia IS, Uppal DS, Bajaj KC (1972) Study of phenolic contents of resistant and susceptible varieties of tomato *Lycopersicum esculantum* in relation to early blight disease. Indian Phytopathol 25:231–235

Bhattacharyya MK, Warde WB (1988) Phenylalanine ammonia lyase activity in soybean hypocotyls and leaves following infection with *Phytophthora megasperma* f. sp. *glycinea*. Can J Bot 66(1):18–23

Blakeman JP, Atkinson P (1981) Antimicrobial substances associated with the aerial surfaces of plants. In: Blakeman JP (ed) Microbial ecology of the phylloplane. Academic Press, London, pp 245–263

Blakeman JP, Fokkema NJ (1982) Potential for biological control of plant diseases on the phylloplane. Annu Rev Phytopathol 20(1):167–190

Blount JW, Dixon RA, Paiva NL (1992) Stress responses in alfalfa (*Medicago sativa* L.) XVI. Antifungal activity of medicarpin and its biosynthetic precursors; implications for the genetic manipulation of stress metabolites. Physiol Mol Plant Pathol 41(5):333–349

Brader G, Compant S, Vescio K, Mitter B, Trognitz F, Ma LJ, Sessitsch A (2017) Ecology and genomic insights into plant-pathogenic and plant-nonpathogenic endophytes. Annu Rev Phytopathol 55:61–83

Cameron RK, Dixon RA, Lamb CJ (1994) Biologically induced systemic acquired resistance in *Arabidopsis thaliana*. Plant J 5(5):715–725

Chowdappa P, Kumar SM, Lakshmi MJ, Upreti KK (2013) Growth stimulation and induction of systemic resistance in tomato against early and late blight by *Bacillus subtilis* OTPB1 or *Trichoderma harzianum* OTPB3. Biol Control 65(1):109–117

Conrath U, Beckers GJM, Flors V, García-Agustín P, Jakab G, Mauch F, Newman MA, Pieterse CMJ, Poinssot B, Pozo MJ, Pugin A, Schaffrath U, Ton J, Wendehenne D, Zimmerli L, Mauch-Mani B (2006) Priming: getting ready for battle. Mol Plant-Microbe Interact 19(10):1062–1071

Constabel CP, Barbehenn R (2008) Defensive roles of polyphenol oxidase in plants. Chapter 12. In: Schaller A (ed) Induced plant resistance to herbivory. Springer Verlag, New York, pp 253–270

Constabel CP, Bergey DR, Ryan CA (1995) Systemin activates synthesis of wound inducible tomato leaf polyphenol oxidase via the octadecanoid defense signaling pathway. Proc Natl Acad Sci U S A 92(2):407–411

Cruz J, Ortiz C, Guzman F, Fernández-Lafuente R, Torres R (2014) Antimicrobial peptides: promising compounds against pathogenic microorganisms. Curr Med Chem 21(20):2299–2321

Dai GH, Nicole M, Andary C, Martinez C, Bresson E, Boher B, Geiger JP (1996) Flavonoids accumulate in cell walls, middle lamellae and callose-rich papillae during an incompatible interaction between *Xanthomonas campestris* pv. *malvacearum* and cotton. Physiol Mol Plant Pathol 49(5):285–306

Dat J, Vandenabeele S, Vranova E, Montagu MV, Inze D, Breusegem FV (2000) Dual action of the active oxygen species during plant stress responses. Cell Mol Life Sci 57(5):779–795

De Wit PJ, Buurlage MB, Hammond KE (1986) The occurrence of host-, pathogen-and interaction-specific proteins in the apoplast of *Cladosporium fulvum* (syn. *Fulvia fulva*) infected tomato leaves. Physiol Mol Plant Pathol 29(2):159–172

Develey-Rivière MP, Galiana E (2007) Resistance to pathogens and host developmental stage: a multifaceted relationship within the plant kingdom. New Phytol 175(3):405–416
Ding T, Melcher U (2016) Influences of plant species, season and location on leaf endophytic bacterial communities of non-cultivated plants. PLoS One 11(3):e0150895
Dixon RA, Paiva NL (1995) Stress-induced phenylpropanoid metabolism. Plant Cell 7 (7):1085–1097
Ebel J (1986) Phytoalexin synthesis: the biochemical analysis of the induction process. Annu Rev Phytopathol 24(1):235–264
Edreva A (2005) Pathogenesis-related proteins: research progress in the last 15 years. Gen Appl Plant Physiol 31(1–2):105–124
Elad Y (2009) Effect of climate change on phyllosphere microflora: plant-microorganism interactions. IOP Conf Ser Earth Environ Sci 6(37):372014
El-Said AH (2001) Phyllosphere and phylloplane fungi of banana cultivated in Upper Egypt and their cellulolytic ability. Mycobiology 29(4):210–217
Evueh GA, Ogbebor NO (2008) Use of phylloplane fungi as biocontrol agent against *Colletotrichum* leaf disease of rubber (*Hevea brasiliensis* Muell. Arg.). Afr J Biotechnol 7 (15):2569–2572
Farkas GL, Kiraly Z (1962) Role of phenolic compounds in the physiology of plant disease resistance. Phytopathology 44(2):105–150
Fayzalla EA, Abdel-Fattah GM, Ibrahim AS, El-Sharkawy HHA (2009) Induction of resistance in tomato plants against *Fusarium oxysporum* f.sp. *lycopersici* by vesicular arbuscular mycorrhizae (VAM). J Agric Sci Mansoura Univ 34:9787–9799
Felix G, Duran JD, Volko S, Boller T (1999) Plants have a sensitive perception system for the most conserved domain of bacterial flagellin. Plant J 18(3):265–276
Feucht W, Treutter D (1999) The role of flavan-3-ols and proanthocyanidins in plant defense. In: Inderjit S, Dakshini KMM, Foy CL (eds) Principles and practices in plant ecology. CRC Press, Boca Raton, pp 307–338
Fu F, Wang Q (2011) Removal of heavy metal ions from wastewaters: a review. J Environ Manag 92(3):407–418
Funnell DL, Lawrence CB, Pedersen JF, Schardl CL (2004) Expression of the tobacco β-1,3-glucanase gene, *PR-2d*, following induction of SAR with *Peronospora tabacina*. Physiol Mol Plant Pathol 65(6):285–296
Gaffney T, Friedrich L, Vernooij B, Negrotto D, Nye G, Uknes S, Ward E, Kessmann H, Ryals J (1993) Requirement of salicylic acid for the induction of systemic acquired resistance. Science 261(5122):754–756
Gall LS (1970) Significance of microbial interactions in control of microbial ecosystems. Biotechnol Bioeng 12(3):333–340
Gaspar T, Penel C, Hagege D, Greppin H (1991) Peroxidase in plant growth, differentiation and developmental processes. Chapter 4. In: Lobarzewski J, Greppin H, Pennel C, Gaspar T (eds) Biochemical, molecular and physiological aspects of plant peroxidases. University M. Curie-Sklodowska, University of Geneva, Lublin (Poland) and Geneva (Switzerland), pp 249–280
Gasper T, Penel C, Thorpe T, Greppin H (1982) Peroxidases 1970–1980. A survey of their biochemical and physiological roles in higher plants. University of Geneva Press, Geneva, p 324
Gómez-Vásquez R, Day R, Buschmann H, Randles S, Beeching JR, Cooper RM (2004) Phenylpropanoids, phenylalanine ammonia lyase and peroxidases in elicitor-challenged cassava (*Manihot esculenta*) suspension cells and leaves. Ann Bot 94(1):87–97
Gonzalez M, Pujol M, Metraux JP, Gonzalez-Garcia VICENTE, Bolton MD, Borras-Hidalgo ORLANDO (2011) Tobacco leaf spot and root rot caused by *Rhizoctonia solani* Kühn. Mol Plant Pathol 12(3):209–216
Granado J, Felix G, Boller T (1995) Perception of fungal sterols in plants (subnanomolar concentrations of ergosterol elicit extracellular alkalinization in tomato cells). Plant Physiol 107(2):485–490

Gupta SK, Kaushik CD (2002) Metabolic changes in mustard leaf and siliqua wall due to the infection of *Alternaria* blight (*Alternaria brassicae*). Crucif Newslett 24:85–86

Gutjahr C, Parniske M (2013) Cell and developmental biology of arbuscular mycorrhiza symbiosis. Annu Rev Cell Dev Biol 29:593–617

Halfeld-Vieira DA, Vieira Júnior JR, Romeiro RDS, Silva HSA, Baracat-Pereira MC (2008) Induction of systemic resistance in tomato by the autochthonous phylloplane resident *Bacillus cereus*. Pesq Agrop Brasileira 41(8):1247–1252

Hamid R, Khan MA, Ahmad M, Ahmad MM, Abdin MZ, Musarrat J, Javed S (2013) Chitinases: an update. J Pharm Bioallied Sci 5(1):21

Harborne JB, Williams CA (2000) Advances in flavonoid research since 1992. Phytochemistry 55 (6):481–504

Harman GE, Howell CR, Viterbo A, Chet I, Lorito M (2004) *Trichoderma* species—opportunistic, avirulent plant symbionts. Nat Rev Microbiol 2(1):43–56

Hassan MAE, Bereika MFF, Abo-Elnaga HIG, Sallam MAA (2009) Direct antimicrobial activity and induction of systemic resistance in potato plants against bacterial wilt disease by plant extracts. Plant Pathol J 25(4):352–360

Hindumathy CK (2012) The defense activator from yeast for rapid induction of resistance in susceptible pearl millet hybrid against downy mildew disease. Int J Agric Sci 4(2):196–201

Hiraga S, Sasaki K, Ito H, Ohashi Y, Matsui H (2001) A large family of class III plant peroxidases. Plant Cell Physiol 42(5):462–468

Hirano S, Upper C (2000) Bacteria in the leaf ecosystem with emphasis on *Pseudomonas syringae*—a pathogen, ice nucleus, and epiphyte. Microbiol Mol Biol Rev 64(3):624–653

Huang CJ, Wang TK, Chung SC, Chen CY (2005) Identification of an antifungal chitinase from a potential biocontrol agent, *Bacillus cereus* 28-9. BMB Rep 38(1):82–88

Ibrahim YE (2012) Activities of antioxidants enzymes in salicylic acid treated tomato against *Xanthomonas vesicatoria*. Afr J Microbiol Res 6(27):5678–5682

Jakuschkin B, Fievet V, Schwaller L, Fort T, Robin C, Vacher C (2016) Deciphering the pathobiome: intra-and interkingdom interactions involving the pathogen *Erysiphe alphitoides*. Microb Ecol 72(4):870–880

Kalita P, Bora LC, Bhagabati KN (1996) Phylloplane microflora of citrus and their role in management of citrus canker. Indian Phytopathol 49(3):234–237

Karthikeyan M, Bhaskaran R, Mathiyazhagan S, Velazhahan R (2007) Influence of phylloplane colonizing biocontrol agents on the black spot of rose caused by *Diplocarpon rosae*. J Plant Interact 2(4):225–231

Kauss H, Jeblick W (1996) Influence of salicylic acid on the induction of competence for H_2O_2 elicitation (comparison of ergosterol with other elicitors). Plant Physiol 111(3):755–763

Kauss H, Theisinger-Hinkel E, Mindermann R, Conrath U (1992) Dichloroisonicotinic and salicylic acid, inducers of systemic acquired resistance, enhance fungal elicitor responses in parsley cells. Plant J 2(5):655–660

Kavitha R, Umesha S (2008) Regulation of defense-related enzymes associated with bacterial spot resistance in tomato. Phytoparasitica 36(2):144–159

Kawamata H, Narisawa K, Hashiba T (2004) Suppression of rice blast by phylloplane fungi isolated from rice plants. J Gen Plant Pathol 70(2):131–138

Kawano T (2003) Roles of the reactive oxygen species-generating peroxidase reactions in plant defense and growth induction. Plant Cell Rep 21(9):829–837

Kerr JR (1999) Bacterial inhibition of fungal growth and pathogenicity. Microb Ecol Health Dis 11 (3):129–142

Kessmann H, Staub T, Hofmann C, Maetzke T, Herzog J, Ward E, Uknes S, Ryals J (1994) Induction of systemic acquired disease resistance in plants by chemicals. Annu Rev Phytopathol 32(1):439–459

Kishore GK, Pande S, Podile AR (2005) Phylloplane bacteria increase seedling emergence, growth and yield of field-grown groundnut (*Arachis hypogaea* L.). Lett Appl Microbiol 40(4):260–268

Klett PF (2011) Bacterial-fungal interactions: hyphens between agricultural, clinical, environmental, and food microbiologists. Microbiol Mol Biol Rev 75(4):583–609

Knoll D, Schreiber L (2000) Plant–microbe interactions: wetting of ivy (*Hedera helix* L.) leaf surfaces in relation to colonization by epiphytic microorganisms. Microb Ecol 40(1):33–42

Kohler A, Schwindling S, Conrath U (2002) Benzothiadiazole-induced priming for potentiated responses to pathogen infection, wounding, and infiltration of water into leaves require the *NPR1/NIM1* gene in *Arabidopsis*. Plant Physiol 128(3):1046–1056

Kojima M, Takeuchi W (1989) Detection and characterization of p-coumaric acid hydroxylase in Mungbean, *Vigna mungo*, seedlings. J Biochem 105(2):265–270

Kuberan T, Vidhyapallavi RS, Balamurugan A, Nepolean P, Jayanthi R, Premkumar R (2012) Isolation and biocontrol potential of phylloplane *Trichoderma* against *Glomerella cingulata* in tea. J Agric Technol 8:1039–1050

Kumar D (2018) Effect of climate change on microbial growth. J Bacteriol Mycol Open Access 6 (2):158

Kurth F, Mailänder S, Bönn M, Feldhahn L, Herrmann S, Große I, Tarkka MT (2014) Streptomyces-induced resistance against oak powdery mildew involves host plant responses in defense, photosynthesis, and secondary metabolism pathways. Mol Plant-Microbe Interact 27 (9):891–900

Kus JV, Zaton K, Sarkar R, Cameron RK (2002) Age-related resistance in Arabidopsis is a developmentally regulated defense response to *Pseudomonas syringae*. Plant Cell 14 (2):479–490

Lagrimini LM, Rothstein S (1987) Tissue specificity of tobacco peroxidase isozymes and their induction by wounding and tobacco mosaic virus infection. Plant Physiol 84(2):438–442

Last FT (1955) Seasonal incidence of *Sporobolomyces* on cereal leaves. Trans Br Mycol Soc 38 (3):221–239

Limtong S, Kaewwichian R, Yongmanitchai W, Kawasaki H (2014) Diversity of culturable yeasts in phylloplane of sugarcane in Thailand and their capability to produce Indole-3-acetic acid. World J Microbiol Biotechnol 30(6):1785–1796

Lindow SE, Brandl MT (2003) Microbiology of the Phyllosphere. Appl Environ Microbiol 69 (4):1875–1883

Lindow SE, Leveau JH (2002) Phyllosphere microbiology. Curr Opin Biotechnol 13(3):238–243

Lochman J, Mikes V (2006) Ergosterol treatment leads to the expression of a specific set of defense-related genes in tobacco. Plant Mol Biol 62(1–2):43–51

Lou J, Fu L, Peng Y, Zhou L (2013) Metabolites from *Alternaria* fungi and their bioactivities. Molecules 18(5):5891–5935

Magnusson J, Ström K, Roos S, Sjögren J, Schnürer J (2003) Broad and complex antifungal activity among environmental isolates of lactic acid bacteria. FEMS Microbiol Lett 219(1):129–135

Mahmoud MA, Al-Sohaibani SA, Al-Othman MR, Abd El-Aziz ARM (2012) Biochemical screening of chocolate spot disease on faba bean caused by *Botrytis fabae*. Afr J Microbiol Res 6 (32):6122–6129

Mauch-Mani B, Mauch F (2005) The role of abscisic acid in plant–pathogen interactions. Curr Opin Plant Biol 8(4):409–414

May RB, Völksch, Kampmann G (1997) Antagonistic activities of epiphytic bacteria from soybean leaves against *Pseudomonas syringae* pv. *glycinea in vitro* and *in planta*. Microb Ecol 34 (2):118–124

Mayda E, Mauch-Mani B, Vera P (2000) Arabidopsis *dth9* mutation identifies a gene involved in regulating disease susceptibility without affecting salicylic acid-dependent responses. Plant Cell 12(11):2119–2128

Mercier J, Lindow SE (2000) Role of leaf surface sugars in colonization of plants by bacterial epiphytes. Appl Environ Microbiol 66(1):369–374

Meuwly P, Molders W, Buchala A, Metraux JP (1995) Local and systemic biosynthesis of salicylic acid in infected cucumber plants. Plant Physiol 109(3):1107–1114

Meyer KM, Leveau JH (2012) Microbiology of the phyllosphere: a playground for testing ecological concepts. Oecologia 168(3):621–629

Mohamed A, Sater A (2001) Antagonistic interactions between fungal pathogen and leaf surface fungi of onion (*Allium cepa*). Pak J Biol Sci 4(7):838–842

Mohamed C, Arbia A, Azza R (2007) Phenolic compounds and their role in biocontrol and resistance of chickpea to fungal pathogenic attacks. Tunis J Plant Prot 2(1):7–21

Morris CE, Rouse DI (1985) Role of nutrients in regulating epiphytic bacterial populations. In: Windels CE, Lindow SE (eds) Biological control of the phylloplane. The American Phytopathological Society, St. Paul, MN, pp 63–82

Nagy NE, Fossdal CG, Dalen LS, Lönneborg A, Heldal I, Johnsen O (2004) Effects of *Rhizoctonia* infection and drought on peroxidase and chitinase activity in Norway spruce (*Picea abies*). Physiol Plant 120(3):465–473

Nicholson RL, Hammerschmidt R (1992) Phenolic compounds and their role in disease resistance. Annu Rev Phytopathol 30(1):369–389

Nix SS, Burpee LL, Jackson KL, Buck JW (2008) Short-term temporal dynamics of yeast abundance on the tall fescue phylloplane. Can J Microbiol 54(4):299–304

O'Leary BM, Rico A, McCraw S, Fones HN, Preston GM (2014) The infiltration-centrifugation technique for extraction of apoplastic fluid from plant leaves using *Phaseolus vulgaris* as an example. JoVE 94:e52113

Owens WA Jr (1953) Age and mental abilities: a longitudinal study. Genet Psychol Monogr 48:3–54

Pandey RR, Arora DK, Dubey RC (1993) Antagonistic interactions between fungal pathogens and phylloplane fungi of guava. Mycopathologia 124(1):31–39

Panstruga R, Kuhn H (2019) Mutual interplay between phytopathogenic powdery mildew fungi and other microorganisms. Mol Plant Pathol 20(4):463–470

Parizi TE, Ansari M, Elaminejad T (2012) Evaluation of the potential of *Trichoderma viride* in the control of fungal pathogens of Roselle (*Hibiscus sabdariffa* L.) inávitro. Microb Pathog 52 (4):201–205

Park SW, Kaimoyo E, Kumar D, Mosher S, Klessig DF (2007) Methyl salicylate is a critical mobile signal for plant systemic acquired resistance. Science 318(5847):113–116

Patil PV, Kachapur MR (2000) Role of phyllosphere microorganisms in the control of sunflower rust caused by *Puccinia helianthi*. Integr Plant Dis Manag Sustain Agric:11–15

Peláez F, Collado J, Arenal F, Basilio A, Cabello A, Matas MD, Hernández P (1998) Endophytic fungi from plants living on gypsum soils as a source of secondary metabolites with antimicrobial activity. Mycol Res 102(6):755–761

Pieterse CMJ, Van Loon LC (1999) Salicylic acid-independent plant defense pathways. Trends Plant Sci 4:52–58

Pieterse CMJ, Van Wees SC, Hoffland E, Van Pelt JA, Van Loon LC (1996) Systemic resistance in Arabidopsis induced by biocontrol bacteria is independent of salicylic acid accumulation and pathogenesis-related gene expression. Plant Cell 8(8):1225–1237

Podile AR, Prakash AP (1996) Lysis and biological control of *Aspergillus niger* by *Bacillus subtilis* AF 1. Can J Microbiol 42(6):533–538

Prabakaran M, Merinal S, Panneerselvam A (2011) Investigation of phylloplane mycoflora from some medicinal plants. Eur J Exp Biol 1(2):219–225

Pradeep T, Jambhale ND (2002) Relationship between phenolics, polyphenol oxidase and peroxidase, and resistance to powdery mildew in *Zizyphus*. Indian Phytopathol 55(2):195–196

Prasannath K (2013) Pathogenicity and virulence factors of phytobacteria. Sch Acad J Biosci 1 (1):24–33

Ross AF (1961) Systemic acquired resistance induced by localized virus infections in plants. Virology 14(3):340–358

Ruinen J (1956) Occurrence of *Beijerinckia* species in the phyllosphere. Nature 177(4501):220–221

Saleem B, Paul PK (2015a) Antifungal activity of bacteria on phylloplane of tomato. J Funct Environ Bot 5(2):116–127

Saleem B, Paul PK (2015b) Phytohormones from phylloplane microbes: a review. Trends Biosci 8 (17):4450–4458
Saleem B, Paul PK (2015c) Phylloplane microfungi altering total phenols and flavonoids in *Solanum lycopersicum* leaves. Trends Biosci 8(22):6125–6133
Saleem B, Paul PK (2016) Microbial colonization of tomato phylloplane is influenced by leaf age. J Funct Environ Bot 6(1):8–15
Schnürer J, Magnusson J (2005) Antifungal lactic acid bacteria as biopreservatives. Trends Food Sci Technol 16(1–3):70–78
Scholze H, Boch J (2011) TAL effectors are remote controls for gene activation. Curr Opin Microbiol 14(1):47–53
Schrey SD, Eric Erkenbrack E, Früh E, Fengler S, Hommel K, Horlacher N, Schulz D, Ecke M, Kulik A, Fiedler HP, Hampp R, Tarkka MT (2012) Production of fungal and bacterial growth modulating secondary metabolites is widespread among mycorrhiza-associated streptomycetes. BMC Microbiol 12(1):164
Schweizer P, Buchala A, Dudler R, Métraux JP (1998) Induced systemic resistance in wounded rice plants. Plant J 14(4):475–481
Sels J, Mathys J, De Coninck B, Cammue B, De Bolle MF (2008) Plant pathogenesis related (PR) proteins: a focus on PR peptides. Plant Physiol Biochem 46(11):941–950
Shoresh M, Harman GE, Mastouri F (2010) Induced systemic resistance and plant responses to fungal biocontrol agents. Annu Rev Phytopathol 48:21–43
Skadhauge B, Thomsen KK, Von Wettstein D (1997) The role of the barley testa layer and its flavonoid content in resistance to *Fusarium* infections. Hereditas 126(2):147–160
Slaughter A, Daniel X, Flors V, Luna E, Hohn B, Mauch-Mani B (2012) Descendants of primed arabidopsis plants exhibit resistance to biotic stress. Plant Physiol 158(2):835–843
Smid EJ, Lacroix C (2013) Microbe–microbe interactions in mixed culture food fermentations. Curr Opin Biotechnol 24(2):148–154
Smith K, Goodman R (1999) Host variation for interactions with beneficial plant-associated microbes. Annu Rev Phytopathol 37(1):473–491
Someya N, Nakajima M, Tadaaki HIBI, Yamaguchi I, Akutsu K (2002) Induced resistance to rice blast by antagonistic bacterium, *Serratia marcescens* strain B2. J Gen Plant Pathol 68 (2):177–182
Sowndhararajan K, Marimuthu S, Manian S (2013) Biocontrol potential of phylloplane bacterium *Ochrobactrum anthropi* BMO-111 against blister blight disease of tea. J Appl Microbiol 114 (1):209–218
Srivastava R, Shalini (2008) Antifungal activity of *Pseudomonas fluorescens* against different plant pathogenic fungi. Internet J Microbiol 7(2):2789–2796
Stockwell VO, McLaughlin RJ, Henkels MD, Loper JE, Sugar D, Roberts RG (1999) Epiphytic colonization of pear stigmas and hypanthia by bacteria during primary bloom. Phytopathology 89(12):1162–1168
Suda W, Nagasaki A, Shishido M (2009) Powdery mildew-infection changes bacterial community composition in the phyllosphere. Microbes Environ 24(3):217–223
Swadling IR, Jeffries P (1998) Antagonistic properties of two bacterial biocontrol agents of Grey Mould disease. Biocontrol Sci Tech 8(3):439–448
Thakur S, Harsh NSK (2014) Efficacy of volatile metabolites of phylloplane fungi of *Rauwolfia serpentina* against *Alternaria alternata*. Curr Res Environ Appl Mycol 4(2):152–156
Thipyapong P, Steffens JC (1997) Tomato polyphenol oxidase: differential response of the polyphenol oxidase F promoter to injuries and wound signals. Plant Physiol 115(2):409–418
Thipyapong P, Hunt MD, Steffens JC (2004) Antisense downregulation of polyphenol oxidase results in enhanced disease susceptibility. Planta 220(1):105–117
Thomashow LS, Weller DM (1996) Current concepts in the use of introduced bacteria for biological disease control: mechanisms and antifungal metabolites. In: Plant-microbe interactions. Springer, Boston, MA, pp 187–235

Ton J, Ent SVD, Hulten MV, Pozo M, Oosten VV, Van Loon LC, Mauch-Mani B, Turlings TCJ, Pieterse CMJ (2009) Priming as a mechanism behind induced resistance against pathogens, insects and abiotic stress. Induc Resist Plants Against Insects Dis 44:3–13

Treutter D (2005) Significance of flavonoids in plant resistance and enhancement of their biosynthesis. Plant Biol 7(6):581–591

Treutter D (2006) Significance of flavonoids in plant resistance: a review. Environ Chem Lett 4 (3):147–157

Tyagi VK, Chauhan SK (1982) The effect of leaf exudates on the spore germination of phylloplane mycoflora of chilli (*Capsicum annuum L.)* cultivars. Plant Soil 65(2):249–256

Tyagi M, Kayastha AM, Sinha B (2000) The role of peroxidase and polyphenol oxidase isoenzymes in wheat resistance to *Alternaria triticina*. Biol Plant 43(4):559–562

Tyagi S, Singh R, Javeria S (2014) Effect of climate change on plant-microbe interaction: an overview. Eur J Mol Biotechnol 3:149–156

Upadhyay RK (1981) Effect of volatile phylloplane fungal metabolites on the growth of a foliar pathogen. Experientia 37(7):707–708

Van Loon LC, Pierpoint WS, Boller TH, Conejero V (1994) Recommendations for naming plant pathogenesis-related proteins. Plant Mol Biol Rep 12(3):245–264

Van Loon LC, Rep M, Pieterse CMJ (2006) Significance of inducible defense-related proteins in infected plants. Annu Rev Phytopathol 44:135–162

Van Peer R, Niemann GJ, Schippers B (1991) Induced resistance and phytoalexin accumulation in biological control of *Fusarium* wilt of carnation by *Pseudomonas* sp. strain WCS417r. Phytopathology 81(7):728–734

Van Wees SCM, Luijendijk M, Smoorenburg I, Van Loon LC, Pieterse CMJ (1999) Rhizobacteria-mediated induced systemic resistance (ISR) in *Arabidopsis* is not associated with a direct effect on expression of known defense-related genes but stimulates the expression of the jasmonate-inducible gene *Atvsp* upon challenge. Plant Mol Biol 41(4):537–549

Vastakaite V, Buzaite O (2011) Plant growth promoting properties of culturable phylloepiphytic pine bacteria. Biologija 57(1):33–36

Verhagen BWM, Glazebrook J, Zhu T, Chang HS, Van Loon LC, Pieterse CMJ (2004) The transcriptome of rhizobacteria-induced systemic resistance in *Arabidopsis*. Mol Plant-Microbe Interact 17(8):895–908

Visker MHPW, Keizer LCP, Budding DJ, Van Loon LC, Colon LT, Struik PC (2003) Leaf position prevails over plant age and leaf age in reflecting resistance to late blight in potato. Phytopathology 93(6):666–674

Völksch B, Nüske J, May R (1996) Characterization of two epiphytic bacteria from soybean leaves with antagonistic activities against *Pseudomonas syringae* pv. *glycinea*. J Basic Microbiol 36 (6):453–462

Vorholt JA (2012) Microbial life in the phyllosphere. Nat Rev Microbiol 10(12):828–840

Wang K, Sipilä T, Rajaraman S, Safronov O, Laine P, Auzane A, Salojärvi J (2019) A novel phyllosphere resident Protomyces species that interacts with the Arabidopsis immune system. bioRxiv:594028. https://doi.org/10.1101/594028

Warmink JA, Elsas JD (2009) Migratory response of soil bacteria to *Lyophyllum* sp. strain karsten in soil microcosms. Appl Environ Microbiol 75(9):2820–2830

Watts KT, Mijts BN, Lee PC, Manning AJ, Schmidt-Dannert C (2006) Discovery of a substrate selectivity switch in tyrosine ammonia-lyase, a member of the aromatic amino acid lyase family. Chem Biol 13(12):1317–1326

Weete JD, Abril M, Blackwell M (2010) Phylogenetic distribution of fungal sterols. PLoS One 5(5): e10899

Weisskopf L (2013) The potential of bacterial volatiles for crop protection against phytopathogenic fungi. In: Microbial pathogens and strategies for combating them: science, technology and education, vol 2. Formatex Research Center, Badajoz, pp 1352–1363

Welinder KG, Justesen AF, Kjaersgård IV, Jensen RB, Rasmussen SK, Jespersen HM, Duroux L (2002) Structural diversity and transcription of class III peroxidases from *Arabidopsis thaliana*. Eur J Biochem 269(24):6063–6081

Whalen MC (2005) Host defense in a developmental context. Mol Plant Pathol 6(3):347–360
Whipps J, Hand P, Pink D, Bending GD (2008) Phyllosphere microbiology with special reference to diversity and plant genotype. J Appl Microbiol 105(6):1744–1755
White JF, Torres MS, Johnson H, Irizarry I, Tadych M (2014) A functional view of plant microbiomes: endosymbiotic systems that enhance plant growth and survival. In: Advances in endophytic research. Springer, New Delhi, pp 425–439
Windham MT, Elad Y, Baker R (1986) A mechanism for increased plant growth induced by *Trichoderma* spp. Phytopathology 76(5):518–521
Wolin MJ, Miller TL, Stewart CS (1997) Microbe-microbe interactions. In: The rumen microbial ecosystem. Springer, Dordrecht, pp 467–491
Wood KR, Barbara DJ (1971) Virus multiplication and peroxidase activity in leaves of cucumber (*Cucumis sativus* L.) cultivars systemically infected with W strain of *Cucumber mosaic virus*. Physiol Plant Pathol 1(1):73–81
Yadav RKP, Karamanoli K, Vokou D (2011) Bacterial populations on the phyllosphere of Mediterranean plants: influence of leaf age and leaf surface. Front Agric China 5(1):60–63
Youssef NH, Couger MB, McCully AL, Criado AEG, Elshahed MS (2015) Assessing the global phylum level diversity within the bacterial domain: a review. J Adv Res 6(3):269–282
Zhang Z, Luo L, Tan X, Kong X, Yang J, Wang D, Zhang D, Jin D, Liu Y (2018) Pumpkin powdery mildew disease severity influences the fungal diversity of the phyllosphere. Peer J 6:e4559
Zimmerli L, Métraux JP, Mauch-Mani B (2001) ß-Aminobutyric acid-induced protection of *Arabidopsis* against the necrotrophic fungus *Botrytis cinerea*. Plant Physiol 126(2):517–523

Part III

Climate Change and Agriculture

Understanding Methanogens, Methanotrophs, and Methane Emission in Rice Ecosystem

12

Sandeep K. Malyan, Smita S. Kumar, Ajeet Singh, Om Kumar, Dipak Kumar Gupta, Ajar Nath Yadav, Ram Kishor Fagodiya, Shakeel A. Khan, and Amit Kumar

Abstract

Rising concentration of methane (CH_4), nitrous oxide, carbon dioxide, and chlorofluorocarbons in the atmosphere result in global warming. These greenhouse gases (GHGs) trap the infrared radiations remitted from the Earth. The global mean temperature is rising more rapidly than ever due to presence of higher concentration of GHGs in the atmosphere. Anthropogenic activities such as fossil fuel burning, biomass combustion, industrialization, modern agricultural system, etc., are the key factors responsible for rising GHGs concentration. After carbon dioxide, CH_4 is the major GHG contributing to the global warming. CH_4

S. K. Malyan
Research Management and Outreach Division, National Institute of Hydrology, Roorkee, Uttarakhand, India

S. S. Kumar
Department of Environmental Sciences, J.C. Bose University of Science and Technology, YMCA, Faridabad, Haryana, India

A. Singh
Division of Biochemistry, ICAR-Indian Agricultural Research Institute, New Delhi, India

O. Kumar · S. A. Khan
Centre for Environment Science and Climate Resilient Agriculture, ICAR-Indian Agricultural Research Institute, New Delhi, India

D. K. Gupta
ICAR-Indian Agricultural Research Institute, Gauria Karma Barhi, Jharkhand, India

A. N. Yadav
Akal College of Agriculture, Eternal University, Sirmour, Himachal Pradesh, India

R. K. Fagodiya
ICAR-Central Soil Salinity Research Institute, Karnal, India

A. Kumar (✉)
Central Muga Eri Research and Training Institute, Central Silk Board, Jorhat, Assam, India

S. A. Lone, A. Malik (eds.), *Microbiomes and the Global Climate Change*,
https://doi.org/10.1007/978-981-33-4508-9_12

is produced by methanogens by complex processes known as methanogenesis. Methanogens are strictly anaerobic bacteria and they can persist in extreme environmental conditions. Rice is the stable food for more than 50% of global population. Rice is generally cultivated in subtropical regions and it is reported that continuous flooded environment is better for higher production. Flooding condition of the rice creates favorable environment for methanogenic bacteria. Under anaerobic environment, methanogens consume soil organic matter as carbon source and emit CH_4 gas to atmosphere. CH_4 emission from rice soil is the net balances of two processes: production by methanogens and oxidation by methanotrophs. Methanotrophs are obligate aerobic bacteria which consume CH_4 as the source of carbon and help oxidation of CH_4 to carbon dioxide. In rice ecosystem, population of methanogenic and methanotrophic bacteria depends upon several biotic and abiotic factors which are discussed in this chapter.

Keywords

Rice · Methane · Methanogens · Methanotrophs · Mitigation · Global warming

12.1 Introduction

Methane (CH_4) is the simplest hydrocarbon greenhouse gas (GHG) produced by methanogens (archaea) under anaerobic environment (Liu et al. 2019a; Malyan et al. 2019b). CH_4 is colorless, odorless, non-toxic gas with tetrahedral structure. The density of CH_4 and air at standard temperature and pressure is 0.714 and 1.225 kg/m^3, respectively, and it is much lighter than the air. Methane, carbon dioxide, and nitrous oxides are the major greenhouse which emitted from agriculture soil including rice and play major role in global warming (2019a; IPCC 2014). Global atmospheric concentration of CH_4 was below 800 ppb before industrialization and its rose above 1850 ppb in 2019 (Fig. 12.1). Higher anthropogenic emission of CH_4 play significant role in rising atmospheric concentration. Rice field, biomass burning, fuel production industry, landfill, waste treatment, and livestock are the major anthropogenic sources of the CH_4 gas (Kumari et al. 2020; Kumar et al. 2020a; Mukherjee et al. 2018; Kumar et al. 2016a, b; Malyan et al. 2016a, b, c; Pathak et al. 2016) (Fig. 12.2). Saunois et al. (2016) reported that the CH_4 emission form rice cultivation emits about 115–243 Tg CH_4 year^{-1} and it is the leading contributor to anthropogenic emissions (Table 12.1).

Rice is the grain of the grass species *Oryza glaberrima* (African rice) and *Oryza sativa* (Asian rice). Rice is the stable food for more than half of the world population (Pramanik and Kim 2017) and its global production was 769.7 million tonnes in 2017 (FAO 2019). China, India, and Indonesia were the three biggest rice producing countries in the world and they account for 61% of the total production (Fig. 12.3). All the three greenhouse gases i.e. CH_4, CO_2 and N_2O emits from the rice field (Kumar et al. 2017; Malyan et al. 2019a; Kumar et al. 2020b). Standing water throughout the cropping period after root establishment is considered favorable

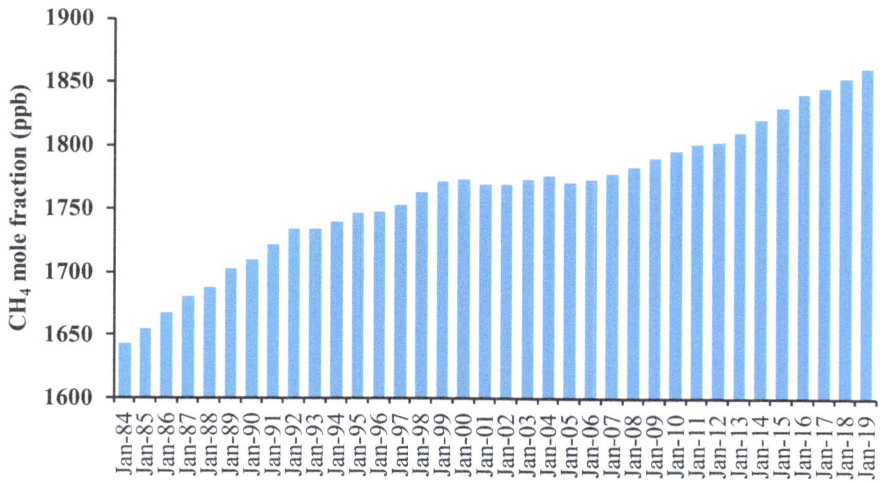

Fig. 12.1 Trend of CH_4 concentration in atmosphere (Source: Ed Dlugokencky, NOAA/ESRL online link https://www.esrl.noaa.gov/gmd/ccgg/trends_ch4/)

Fig. 12.2 Major sources of anthropogenic methane emission to atmosphere

condition for better rice production (Malyan 2017; Kumar and Malyan 2016; Malyan et al. 2016a; Gupta et al. 2015; Sethunathan et al. 2002). The standing water create anaerobic environment in the soil which stimulate the population of CH_4 producing bacteria (methanogens) (Mishra et al. 1997; Smartt

Table 12.1 Major anthropogenic contributor for global atmospheric methane emissions

S. No	Source	Anthropogenic emissions(in Tg (CH_4)/ year
01	Rice cultivation	115–243
02	Landfill	77–133
03	Ruminants	
04	Waste treatment	
05	Energy	
06	Biomass burning	15–53
07	Othe	

Source: (Saunois et al. 2016)

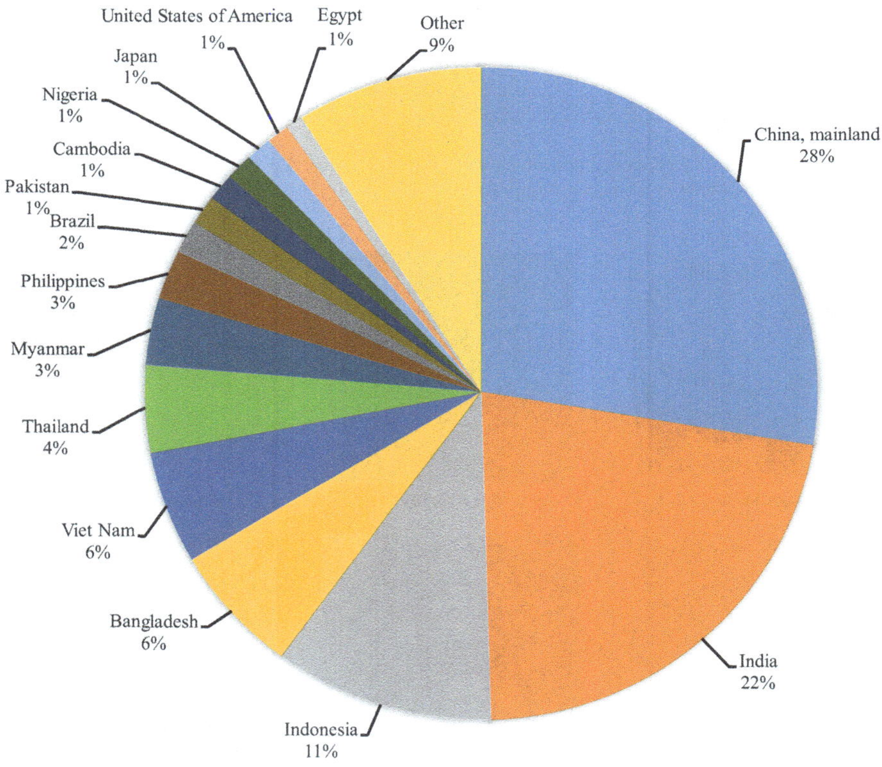

Fig. 12.3 Global rice production (2017) (Source: FAO 2019)

et al. 2018; Malyan et al. 2016c; Ke et al. 2014; Conrad 2007). Neue (1993) quoted that CH_4 emission form rice field was first reported by Harrison and Aiyer in the year 1913. The emission of CH_4 is actually the net balance of two processes, the production by methanogens and its oxidation by methanotrophs (Malyan et al. 2016c). CH_4 emission from rice field is effected by several biotic (methanogens, methanotrophs, other microorganisms' population) and abiotic (soil pH, temperature, texture, organic carbon matter, water content, etc.) factors (Mona et al. 2021; Malyan et al. 2019b; Gupta et al. 2016a; Hussain et al. 2015; Bhatia et al. 2013; Liesack et al. 2000). In the present chapter, we focused on the microbial diversity, mechanism, and influencing factors for CH_4 emission from rice soils.

12.2 Methanogens and Methane Production in Rice Field

The process of CH_4 production by methanogens is known as methanogenesis, it takes place under anaerobic conditions (Serrano-Silva et al. 2014; Conrad 2007; Le Mer and Roger 2001). In the biogenic formation of CH_4 in rice ecosystem, the soil organic matter is consumed by methanogens as carbon source and they release CH_4 gas as by-product (Gupta et al. 2016b; Dubey et al. 2014; Le Mer and Roger 2001). Plants litter, roots, weed biomass, dead microorganisms, animal waste, and organic fertilizers are the main source of soil organic matter (Kimura et al. 2004). The soil organic matter is converted into acetate by three process: hydrolysis, acidogenesis, and acetogenesis (Dubey 2005). Methanogens prefer to consume acetate as substrate for the production of CH_4. Generally, methanogens are mesophilic and the optimum temperature for the CH_4 production is 25 °C (Dunfield et al. 1993). Some of the genera of methanogens are also found in extreme environmental conditions such as geothermal sediments, hot springs, and hypersaline sediments (Nazaries et al. 2013; Dubey 2005). The rate of CH_4 production in soil by methanogens depends upon several factors such as environmental conditions, substrate availability, and the presence of other competing microbial community (Serrano-Silva et al. 2014; Dubey 2005; Roy and Conrad 1999). Methanogens are divided into five categorized based on substrate utilization (Table 12.2). About 80% of the methanogens prefer to utilized acetate as the C source, while 10–30% of methanogens prefer formate and hydrogen/carbon dioxide as the C source (Conrad 2007).

Methane producing bacteria use NH_4^+ as N source, however, there are few methane producing microbes which are having N-fixing gene (nif): *Methanobacteriales, Methanococcales, and Methanomicrobiales* (Serrano-Silva et al. 2014; Dubey 2005). Methane producing Achaea needs unique coenzymes such as coenzyme (F_{420}), coenzyme M (CoM), coenzyme B (CoB), ferredoxin (Fd), methanofuran (MFR), and tetrahydromethanopterin (H_4MPT), to complete methanogenesis (Nazaries et al. 2013).

Table 12.2 Types of methanogens on the basis substrates consumption

Substrates	Product formed	Trophic group
$4H_2 + CO_2$	$CH_4 + 2H_2O$	Hydrogenotrophs
4HCOOH	$CH_4 + 3CO_2 + 2H_2O$	Formatotrophs
Acetate	$CH_4 + CO_2$	Actetotrophs
$4CH_3OH$	$3CH_4 + CO_2 + 2H_2O$	Methylotrophs
$CH_3CHOHCH_3 + CO_2$	CH_4+ $4CH_3CHOHCH_3 + 2H_2O$	Alcoholotrophs

Source: Malyan et al. (2016c)

12.3 Methanotrophs and Methane Oxidation in Rice Soil

Methanotrophs are gram-negative aerobic bacteria which play significant role in controlling atmospheric CH_4 levels. Methanotrophs oxidize CH_4 to CO_2 via methanemonooxygenase (MMO) enzyme (Dubey 2005; Hanson and Hanson 1996). Rice cultivation is a main anthropogenic source of CH_4 emission at global level. The population of methanotrophs in rice is diverse and it depends upon the prevailing biotic and abiotic factors at that particular time (Conrad 2007; Liesack et al. 2000). The commonly reported methanotrophs from rice fields include *Methylobcater, Methylomonas, Methylocystis, Methylococcus, Methylosinus, Methylocaldum, Methylocystis, Methylomicrobium*, etc. (Chen and Murrell 2010; Mancinelli 1995; Trotsenko and Khmelenina 2005). All the identified methanotrophs are broadly categorized in two major groups based on their assimilating compounds (Fig. 12.4): type I and type II (Fazli et al. 2013). The methanotrophic bacteria which are phylogenetically identified as γ-proteobacteria assimilate one carbon compounds via the ribulose monophosphate cycle are known as type I methanotrophs (Fazli et al. 2013). The methantrophic bacteria which are phylogenetically identified as α-proteobacteria assimilate C1 intermediates via serine pathways are known as type II methanotrophs (Fazli et al. 2013). In rice field, the population of methanotrophs in rhizosphere is much higher than the bulk soil (Dubey 2005; Gilbert and Frenzel 1995). The concentration of oxygen and CH_4 affect the niches of type I and type II methanotrophs in rice. The aerenchyma assists the transportation of oxygen from atmosphere to rhizospheric zone. The more the arenchyma size more oxygen sinks towards rhizospheric zone which deplete the anaerobic condition and enhance the aerobic environment in rice-rhizosphere. In aerobic environment the dominance of methanotrophs type I was observed

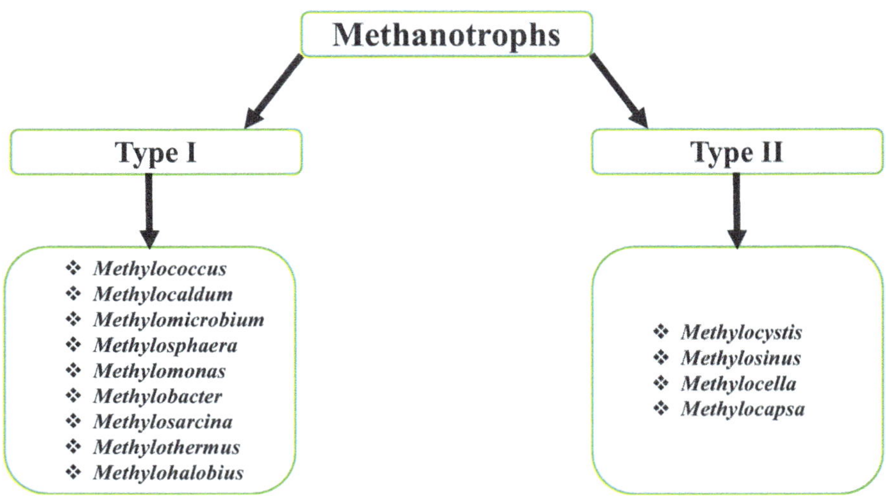

Fig. 12.4 Major groups of methanotrophs (Source: Conrad 2007)

(Bhattacharyya et al. 2019). Methanotrophs type I are active in the environment with low CH_4 and high oxygen concentration (Mayumi et al. 2010). The population of methanotroph type II organisms is found in anaerobic environment such as bulk soil in rice ecosystem. Therefore, in continuous flooded bulk soil, methanotrophs type II organisms are more active while in intermitted rice ecosystem methanotrophs type I organisms are more active.

Until recently, it was believed that methanotrophs cannot consume carbon-carbon bond compounds such as acetate as the sole source of energy and therefore methanotrophs were consider as obligate methylotrophic (Conrad 2007). However, Dedysh et al. (2005) reported that *Methylocella* spp. of methanotrophs use acetate as carbon and energy source and they prefer acetate compound over CH_4. Hence, the presence of available carbon source also effects the population of methanotrophs in rice ecosystem. The population of aerobic methanotrophs is measured generally by most probable number (MPN) counts method (Singh and Pandey 2013). Generally, MPN count per gram soil ranges from 10^4 to 10^7 (Conrad 2007) but the aerobic methanotrophs population in the rhizosphere zone is much higher than this order. In the flooded environment methanotrophic bacteria became inactive due to lack to oxygen.

12.4 Methane Oxidation in Rice Ecosystem

Methane oxidation in rice ecosystem is a biological process carried out by aerobic and anaerobic methanotrophs. CH_4/methanol serves as an energy source for the growth of methanotrophs (Hanson and Hanson 1996; Semrau et al. 2010). Aerobic methanotrophs are obligate (use only CH_4 as C and energy source) and facultative (utilize multi-carbon substrates) in nature (Dedysh and Dunfield 2011) having mesophilic characteristics (optimum growth range 20 °C–40 °C) and neutrophilic characteristics (optimum growth range pH 6–8) (Whittenbury et al. 1970). Oxygen availability plays important role for aerobic methanotrophy (Bodegom et al. 2001). Aerobic methanotrophs can be low-affinity methanotrophs (oxidize high CH_4 concentrations(>100 ppm)) and high-affinity methanotrophs (oxidize low CH_4 concentrations (1.8 ppm) (Malyan et al. 2016c; Alam and Jia 2012).

The aerobic methane oxidation is completed through the sequential conversion of methane into carbon dioxide utilizing different enzymes. Firstly, CH_4 is converted into methyl aldehyde (CH_3CHO) by methane monooxygenase enzymes. The CH_3CHO is subsequently converted into formaldehyde by methanol dehydrogenase and formaldehyde is further converted into formate by the activity of formaldehyde dehydrogenase and this formate finally converts into carbon dioxide through dehydrogenase activity. Methane monooxygenase is of two types, i.e. particulate or membrane-bound form (pMMO) and soluble cytoplasmic form (sMMO) (Semrau et al. 2010) and serves as the process initiation catalysts and rate limiting enzyme. MMOs is multi-substrate enzyme that can oxidize propylene (PP) (Inubushi et al. 2002) and is inhibited by acetylene (C_2H_2) (Hanson and Hanson 1996). MMO can also be inhibited by $NH4^+$ due to competitive nature of ammonia with methane

(Dunfield and Knowles 1995) as MMO is also able to convert ammonia into nitrite through hydroxylamine intermediate which are further toxic to methanotrophs and lead to inhibition of methane oxidation ability in rice soil (Eller and Frenzel 2000).

The anaerobic methane oxidation is completed by Achaea, sulfate-reducing bacteria and starts by physical association of anaerobic methanotrophs (Serrano-Silva et al. 2014; Chowdhary and Dick 2013; Nazaries et al. 2013). The sulfate reducing bacteria oxidize CH_4 to CO_2 and sulfate acts as an electron acceptor. Nitrite is also one of the electron acceptors in rice flooded soil as sulfate (Beal et al. 2009) and Fe and Mn in marine environment for anaerobic methane oxidation (Malyan et al. 2016c).

Nitrogen fertilization through synthetic fertilizers and organic fertilizers increases the $NH^{+}{}_4$ and $NO^{-}{}_3$ concentration in rice soils which inhibits the CH_4 oxidation. This inhibition promotes the chance of carbon reduction in form of CH_4 emission (Fagodiya et al. 2017). The ammonium-based fertilization stimulates growth and activity of methanotrophs in the rice-rhizosphere (Bodelier et al. 2000a, b). Nitrate inhibit only in very high concentrations due to osmotic effects (Bodelier and Laanbroek 2004a). Krüger and Frenzel (2003) reported that with decrease in mineral nitrogen in rice field, methane oxidation decreased up to zero. Bodelier and Laanbroek (2004b) concluded that mineral nitrogen can function as limiting factors for growth of methanotrophic bacteria and ultimately regulated methane oxidation.

12.5 Factors Affecting Methane Emission in Rice Ecosystems

12.5.1 Soil Temperature

For the best activity of mesophilic microorganism usually temperature range of 20–35 °C is considered as ideal. The microbial community of CH_4 oxidation bacteria is highly sensitive to temperature change (He et al. 2012). He et al. (2012) conducted sediment incubation study with three temperatures (4 °C, 10 °C, and 21 °C) to identify methanotrophs activity and CH_4 oxidation. The CH_4 oxidation at 21 °C was highest (37.4μmol g^{-1} day^{-1}) in the uppermost sediment layer (0–1 cm) (He et al. 2012). The rate of CH_4 oxidation increased with increasing temperature from 5 °C to 20 °C were also reported by Whalen et al. (1990). But some of the contrasting findings were also reported by Bender and Conrad (1995), they observed the maximum CH_4 oxidation at 0 °C.

12.5.2 Soil Organic Matter

Soil organic matter affect the cumulative microbial community and the functions of the ecosystem (Tveit et al. 2013). Several findings reported that addition of organic matter such as straw, farm yard manure, residues of other crops, etc., increases the CH_4 emissions (Gupta et al. 2016b; Hussain et al. 2015; Bhatia et al. 2013a). In fact, there are few studies which reported that the addition of organic matter in the form of biochar help in CH_4 mitigation in rice (Purakayastha et al. 2019; Panwar et al. 2019; Xiao et al. 2018; Pandey et al. 2014). In one recent study, Wu et al. (2019b) reported that biochar application increased the CH_4 oxidation potential of methanotrophic bacteria. Nitrate and ammonium promoted the methanotrophs type II and type I respectively, which enhanced the CH_4 oxidation potential for prolonged duration on biochar application (Wu et al. 2019b; Feng et al. 2012).

12.5.3 Soil Texture

Soil texture directly influences the population of microbes. Soil porosity changes with the soil texture and its effects the CH_4 oxidation process (Shukla et al. 2013). Soil with greater porosity (such as sandy) is the favorable environment for CH_4 oxidation by methanotrophs type I.

12.5.4 Application of Fertilizers

Fertilizer type, method, and dose are the critical component affecting for microbial community dynamics in rice soil. Fertilizer applied to soil is not completely utilized by crop (Ranjan and Yadav 2019). The activity of methanotrophs is affected by the type of fertilizer applied. The period of ammonium ion in the soil affect the CH_4 oxidation. In short term, ammonium ion has no effect on CH_4 oxidation potential of methanotrophs. Ammonium ions in long term act as inhibitors of CH_4 oxidation potential of methanotrophs in rice soil (Shukla et al. 2013; Banger et al. 2012). Role of ammonium ions is different in different CH_4 environmental conditions. Un-saturated soil such as in upland rice have low soil CH_4 and under such environment increased in ammonium ions deceased the rate of CH_4 oxidations (Shukla et al. 2013; Hütsch et al. 1994; Steudler et al. 1989). The role of ammonium ions in rice soils was not conclusive as different studies concluded positive as well as negative control (Shukla et al. 2013; Bodelier et al. 2000a, b) especially in case of flooded and non-flooded rice soils (Table 12.3).

12.6 Mitigation of Methane Emission from Rice Ecosystem

Methane emission from rice soil can be mitigated by modifying water content, fertilizer application, tillage practice, by selecting suitable rice cultivars, organic matter management, etc. (Malyan et al. 2020; Samoy-Pascual et al. 2019; Setyanto et al. 2018; Tariq et al. 2017). Some of the significant tools of CH_4 mitigation from rice are discussed below:

Table 12.3 Factors affecting methane emission from rice field

S. No	Factors	References
01	Soil temperature	(Centeno et al. 2017; Gaihre et al. 2016; He et al. 2012; Schütz et al. 1990)
02	Soil organic matter	(Dhanuja et al. 2019; Wu et al. 2019a; Bhattacharyya and Barman 2018)
03	Soil texture	(Singh et al. 2018)
04	Rice cultivar	(Malyan et al. 2019b; Zheng et al. 2018)
05	Fertilizers	(Kong et al. 2019; Liu et al. 2019b; Sun et al. 2019; Buragohain et al. 2018; Hussain et al. 2015)
06	Method of transplanting	(Li et al. 2019; Wang et al. 2018a; Simmonds et al. 2015;)
07	Water management	(Fertitta-Roberts et al. 2019; Jiang et al. 2019)

12.6.1 Irrigation Management

Methane is produced by methanogenic bacteria in anaerobic environment when soil redox potential is less than −150 mV (Khosa et al. 2011). Water is one of the important natural resource, which needs immediate attention to enhance the water use efficiency under changing environment (Pathak et al. 2014). At the same time water management also reduces the methane emission from the rice soil. Irrigation significantly affects the field pore space and soil oxygen concentration. Water management practices such as intermittent drainage, midseason drainage, alternate wetting and drying, controlled irrigation, etc. have been documented globally for reducing CH_4 emissions as compared to continuous flooding (Table 12.4). Intermitted drainage enhanced the oxygen diffusion in the soil and therefore rose the soil redox potential which result in CH_4 emissions reduction (Haque et al. 2017). Kim et al. (2014) and Kudo et al. (2014) stated that intermitted drainage reduced cumulative CH_4 emission by 43.52% and 47%, respectively, as compared to continuous flooding (Table 12.4). Haque et al. (2017) also reported 54–58% reduction in CH_4 emissions as compared to continuous flooding (Table 12.4). Alternate wetting and drying (AWD) is novel water and environment saving technology. AWD reduced CH_4 emissions (Table 12.4) with decreasing the economical yield. In AWD, irrigation water is given after the standing water disappearance in the crop. The number of days between irrigation and non-flooding varied from 1 to 10 days in AWD and it depended upon soil type, environmental conditions, and crop growth. Samoy-Pascual et al. (2019) observed that AWD can minimize cumulative CH_4 emissions by 20–73% as compared to traditional flooding system in Philippines (Table 12.4). Oo et al. (2018) conducted field experiment at Tamil Nadu Rice Research Institute, India, and reported that AWD can reduce 24–41% CH_4 emission as compared to traditional irrigation method (Table 12.4). In recent studies, 26% and 49% CH_4 mitigation by AWD were also demonstrated by Tran et al. (2018) and Chidthaisong et al. (2018), respectively (Table 12.4). Tariq et al. (2017) stated that early drainage is also effective irrigation methodology for CH_4 reduction as compared to continuous flooding (Table 12.4).

Table 12.4 Methane mitigation through several water management practices as compared to continuous flooding in rice

References	Practice name	Mitigation (%)	Remarks
Samoy-Pascual et al. (2019)	Alternate wetting & drying (AWD)	20–73	Experiment was carried out at Philippine Rice research institute, Philippines
Wu et al. (2019a)	Midseason drying-flooding	52.26	Study was conducted at Taoyuan, China
Setyanto et al. (2018)	AWD	35	Study was carried out in Central Java, Indonesia
Oo et al. (2018)	AWD	24–41	Study was conducted at Tamil Nadu Rice research institute, India
Tran et al. (2018)	AWD	26	Study was conducted at Huong Tra district, Central Vietnam
Chidthaisong et al. (2018)	AWD	49	Experiment was carried out in Prachin Buri, Thailand
Haque et al. (2017)	Intermitted drainage	54–58	Experiment was conducted in Gyeongsang National University, Jinju, South Korea
Tariq et al. (2017)	Early drainage	89–92	Experiment was carried out at University of Copenhagen, Denmark
Peyron et al. (2016)	Dry seeding with delay flooding	59	Experiment was carried out at Italian Rice research Center, Castello d'Agogna, Italy
Kudo et al. (2014)	Intermittent drainage	47	Study was conducted at Kanagawa, Japan
Kim et al. (2014)	Intermittent drainage	43.52	Study was conducted at Suwon, Korea
Li et al. (2014)	Midseason aeration	12–27	Study was conducted at Jurong city of China
Ma et al. (2013)	Midseason drainage	37–51	Study was conducted at Jiangxi Province, China

12.6.2 Rice Cultivar

Selection of appropriate cultivar is most important for achieving the goal of sustainable agricultural. There is variation in the rate of CH_4 production, oxidation, and transportation among different rice cultivar (Aulakh et al. 2000; Hussain et al. 2015). Generally cumulative CH_4 emissions of long duration cultivar are higher than the short or medium duration rice cultivar (Malyan et al. 2019b). Soil redox potential affects the activity of both CH_4 production by methanogens and oxidation by methanotrophic bacteria. Size of aerenchyma varied among the rice cultivar and it affects the rate of methane emission and uptake through methanogens and methanotrophs under prevailing anaerobic/aerobic environment in rice soil (Nishiuchi et al. 2012).

12.6.3 Methane Mitigation Through Azolla

Azolla is nitrogen fixation aquatic fern and its importance for CH_4 emission reduction is explored in several studies globally (Table 12.5). Biologically nitrogen fixation, control of weed, and source of organic matter are other significant role of *Azolla* in rice ecosystem (Singh et al. 2016). Bharati et al. (2000) and Malyan et al. (2019b) reported that *Azolla* application in rice with optimum dose of nitrogen fertilizer reduced CH_4 emissions in the range of 15–42% (Table 12.5). *Azolla* is photosynthetic fern and it liberates oxygen in flooded water and it results in higher dissolved oxygen concentration in flooded water. The higher dissolved oxygen suppress the activity of methanogens bacteria and enhances the CH_4 oxidation simultaneously which result in CH_4 emissions reduction. Yang et al. (2019); Xu et al. (2017), and Liu et al. (2017) demonstrated the CH_4reduction by *Azolla* in China (Table 12.5).

12.6.4 Other Interventions for Methane Mitigation in Rice

Industrial by-products such as fly ash, phosphogypsum, steel slag have been reported to reducing the methane emission from rice fields (Kumar et al. 2020a). Steel slag is having high content of iron oxides and free (Fe^{3+}) form. Fe^{3+} contributes in mitigation of methane from rice soil as it competes with H_2/acetate for electrons during the degradation of soil organic matter (Wang et al. 2018b; Alpana et al. 2017; Wang et al. 2015). Silica present in steel slag supports aerenchyma enlargement and make more diffusion of oxygen towards the rhizosphere and hence reducing the anaerobic condition, further reducing methanogenesis and enhancing rhizospheric methane oxidation (Ali et al. 2012b, 2015). The phosphogypsum increases the sulfur-reducing bacteria in paddy soil as sulfate is the important constituent in it. These sulfur-reducing bacteria will compete with methanogens for organic matters, thus reducing the methane emission (Sun et al. 2018; Ozuolmez et al. 2015; Ali et al. 2007). The intermittent-irrigation, alternate drying and wetting, direct seeded rice, single aeration, etc. are the important agronomic practices to manage the methanogens and methanotrophs population for mitigating methane emission (Kumar et al. 2016a; Bhatia et al. 2013).

Table 12.5 Methane mitigation by *Azolla* as compared to full dose of chemical nitrogen fertilizers

Country (references)	Mitigation (%) as compared to dose chemically N dose
India (Bharati et al. 2000; Malyan et al. 2019b)	15–42
China (Liu et al. 2017; Xu et al. 2017; Yang et al. 2019)	11–33
Japan (Kimani et al. 2018)	34
Bangladesh (Ali et al. 2012a)	11–13

12.7 Conclusion

Methane is one of the major greenhouse gases emitted from the rice fields and play a significant role in global warming. Methanogens are basically responsible for methane production, while methanotrophs are responsible for methane oxidation. Enhancing methane oxidations or reducing productions are the two important aspects to curtail the methane emission from the rice soils. The population of methanogens is highly influenced by various factors such as organic matter, population of the substrate competitor microbes, pH, temperature, plant rhizosphere environment, rice cultivar, irrigation, etc. Thus, these managements are highly important to control the methane emission from rice fields. Besides this, applications of some industrial by-products are also found suitable for methane mitigation from rice fields.

References

Alam SM, Jia Z (2012) Inhibition of methane oxidation by nitrogenous fertilizers in a paddy soil. Front Microbiol 3:246. https://doi.org/10.3389/fmicb.2012.00246

Ali M, Farouque M, Haque M, Ul Kabir A (2012a) Influence of soil amendments on mitigating methane emissions and sustaining Rice productivity in Paddy soil ecosystems of Bangladesh. J Environ Sci Nat Resour 5:179–185. https://doi.org/10.3329/jesnr.v5i1.11574

Ali M, Farouque M, Haque M, ul Kabir A (2012b) Influence of soil amendments on mitigating methane emissions and sustaining rice productivity in paddy soil ecosystems of Bangladesh. J. Environ. Sci. & Nat Resour 5:179–185

Ali MA, Lee C-H, Kim P-JJKJoEA (2007) Effect of Phospho-gypsum on reduction of methane emission from rice paddy soil. Korean J Environ Agric 26:131–140

Ali MA, Kim P, Inubushi K (2015) Mitigating yield-scaled greenhouse gas emissions through combined application of soil amendments: a comparative study between temperate and subtropical rice paddy soils. Sci Total Environ 529:140–148

Alpana S, Vishwakarma P, Adhya T, Inubushi K, Dubey S (2017) Molecular ecological perspective of methanogenic archaeal community in rice agroecosystem. Sci Total Environ 596:136–146

Aulakh MS, Bodenbender J, Wassmann R, Rennenberg H (2000) Methane transport capacity of rice plants. II. Variations among different rice cultivars and relationship with morphological characteristics. Nutr Cycl Agroecosyst 58:367–375. https://doi.org/10.1023/A:1009839929441

Banger K, Tian H, Lu C (2012) Do nitrogen fertilizers stimulate or inhibit methane emissions from rice fields? Glob Chang Biol 18:3259–3267. https://doi.org/10.1111/j.1365-2486.2012.02762.x

Bender M, Conrad R (1995) Effect of CH4 concentrations and soil conditions on the induction of CH4 oxidation activity. Soil Biol Biochem 27:1517–1527. https://doi.org/10.1016/0038-0717(95)00104-M

Bharati K, Mohanty SR, Singh DP, Rao VR, Adhya TK (2000) Influence of incorporation or dual cropping of Azolla on methane emission from a flooded alluvial soil planted to rice in eastern India. Agric Ecosyst Environ 79:73–83. https://doi.org/10.1016/S0167-8809(99)00148-6

Bhatia A, Kumar A, Das TK, Singh J, Jain N, Pathak H (2013) Methane and nitrous oxide emissionsfrom soils under direct seeded rice. Int J Agric Stat Sci 9(2):729–736

Bhatia A, Kumar A, Kumar V, Jain N (2013a) Low carbon option for sustainable agriculture. Indian Farming 63(2):18–22

Bhattacharyya P, Barman D (2018) Crop residue management and greenhouse gases emissions in tropical Rice lands, in: soil management and climate change: effects on organic carbon. Nitrogen Dyn Greenhouse Gas Emissions. https://doi.org/10.1016/B978-0-12-812128-3.00021-5

Bhattacharyya P, Dash PK, Swain CK, Padhy SR, Roy KS, Neogi S, Berliner J, Adak T, Pokhare SS, Baig MJ, Mohapatra T (2019) Mechanism of plant mediated methane emission in tropical lowland rice. Sci Total Environ 651:84–92. https://doi.org/10.1016/j.scitotenv.2018.09.141
Bodegom P, Stams F, Mollema L, Boeke S, Leffelaar P (2001) Methane oxidation and the competition for oxygen in the rice rhizosphere. Appl Environ Microbiol 67:3586–3597
Bodelier PLE, Laanbroek HJ (2004a) Nitrogen as a regulatory factor of methane oxidation in soils and sediments. FEMS Microbiol Ecol 47:265–277
Bodelier PLE, Hahn AP, Arth IR, Frenzel P (2000a) Effects of ammonium-based fertilisation on microbial processes involved in methane emission from soils planted with rice. Biogeochemistry 51:225–257
Bodelier PLE, Roslev P, Henckel T, Frenzel P (2000b) Stimulation by ammonium-based fertilizers of methane oxidation in soil around rice roots. Nature 403:421–424. https://doi.org/10.1038/35000193
Bodelier PLE, Laanbroek HJ (2004b) Nitrogen as a regulatory factor of methane oxidation in soils and sediments. FEMS Microbiol Ecol 47(3):265–277
Buragohain S, Sarma B, Nath DJ, Gogoi N, Meena RS, Lal R (2018) Effect of 10 years of biofertiliser use on soil quality and rice yield on an Inceptisol in Assam, India. Soil Res. https://doi.org/10.1071/SR17001
Centeno CAR, Alberto MCR, Wassmann R, Sander BO (2017) Assessing diel variation of CH4 flux from rice paddies through temperature patterns. Atmos Environ 167:23–39. https://doi.org/10.1016/j.atmosenv.2017.08.007
Chen Y, Murrell JC (2010) Ecology of aerobic methanotrophs and their role in methane cycling. In: Handbook of hydrocarbon and lipid microbiology. https://doi.org/10.1007/978-3-540-77587-4_229
Chidthaisong A, Cha-un N, Rossopa B, Buddaboon C, Kunuthai C, Sriphirom P, Towprayoon S, Tokida T, Padre AT, Minamikawa K (2018) Evaluating the effects of alternate wetting and drying (AWD) on methane and nitrous oxide emissions from a paddy field in Thailand. Soil Sci Plant Nutr. https://doi.org/10.1080/00380768.2017.1399044
Chowdhary TR, Dick RP (2013) Ecology of aerobic methanotrophs in controlling methane fluxes from wetlands. Appl Soil Ecol 65:8–22
Conrad R (2007) Microbial ecology of methanogens and methanotrophs. Adv Agron. https://doi.org/10.1016/S0065-2113(07)96005-8
Dedysh SN, Dunfield PF (2011) Chapter three — facultative and obligate methanotrophs: how to identify and differentiate them. Methods Enzymol 495:31–44
Dedysh SN, Knief C, Dunfield PF (2005) Methylocella species are facultatively methanotrophic. J Bacteriol. https://doi.org/10.1128/JB.187.13.4665-4670.2005
Dhanuja C, Saxena DK, Abbasi T, Abbasi SA (2019) Effect of application of vermicompost on methane emission and grain yield of China Ponni paddy crop. Paddy Water Environ. https://doi.org/10.1007/s10333-018-00687-x
Dubey SK (2005) Microbial ecology of methane emission in rice agroecosystem: a review. Appl Ecol Environ Res. https://doi.org/10.15666/aeer/0302_001027
Dubey SK, Singh A, Watanabe T, Asakawa S, Singla A, Arai H, Inubushi K (2014) Methane production potential and methanogenic archaeal community structure in tropical irrigated Indian paddy soils. Biol Fertil Soils 50:369–379. https://doi.org/10.1007/s00374-013-0858-7
Dunfield P, knowles R, Dumont R, Moore TR (1993) Methane production and consumption in temperate and subarctic peat soils: response to temperature and pH. Soil Biol Biochem. https://doi.org/10.1016/0038-0717(93)90130-4
Dunfield PF, Knowles R (1995) Kinetics of inhibition of methane oxidation by nitrate, nitrite, and ammonium in a humisol. Appl Environ Microbiol 61:3129–3135
Eller G, Frenzel P (2000) Changes in activity and community structure of methane-oxidizing bacteria over the growth period of rice Appl. Environ Microbiol 67:2395–2403
Fagodiya RK, Pathak H, Kumar A, Bhatia A, Jain N (2017) Global temperature change potential of nitrogen use in agriculture: a 50-year assessment. Sci Rep 7:44928

FAO (2019) Neumega (IL-11) label, FDA label. https://doi.org/10.1192/bjp.112.483.211-a
Fazli P, Man HC, Md Shah UK, Idris A (2013) Characteristics of methanogens and methanotrophs in rice fields: a review. Asia-Pacific J Mol Biol Biotechnol 21(1):3–17
Feng Y, Xu Y, Yu Y, Xie Z, Lin X (2012) Mechanisms of biochar decreasing methane emission from Chinese paddy soils. Soil Biol Biochem 46:80–88. https://doi.org/10.1016/j.soilbio.2011.11.016
Fertitta-Roberts C, Oikawa PY, Darrel Jenerette G (2019) Evaluating the GHG mitigation-potential of alternate wetting and drying in rice through life cycle assessment. Sci Total Environ 653:1343–1353. https://doi.org/10.1016/j.scitotenv.2018.10.327
Gaihre YK, Wassmann R, Villegas-pangga G, Sanabria J, Aquino E, Sta Cruz PC, Paningbatan EP (2016) Effects of increased temperatures and rice straw incorporation on methane and nitrous oxide emissions in a greenhouse experiment with rice. Eur J Soil Sci 67:868–880. https://doi.org/10.1111/ejss.12389
Gilbert B, Frenzel P (1995) Methanotrophic bacteria in the rhizosphere of rice microcosms and their effect on porewater methane concentration and methane emission. Biol Fertil Soils. https://doi.org/10.1007/BF00336586
Gupta DK, Bhatia A, Kumar A, Chakrabarti B, Jain N, Pathak H (2015) Global warming potential of rice (*Oryza sativa*)-wheat (*Triticum aestivum*) cropping system of the Indo-Gangetic Plains. Indian J Agric Sci 85(6):807–816
Gupta DK, Bhatia A, Das TK, Singh P, Kumar A, Jain N, Pathak H (2016a) Economic analysis of different greenhouse gas mitigation technologies in rice–wheat cropping system of the Indo-Gangetic plains. Curr Sci 110(5):867–873
Gupta DK, Bhatia A, Kumar A, Das TK, Jain N, Tomer R, Malyan SK, Fagodiya RK, Dubey R, Pathak H (2016b) Mitigation of greenhouse gas emission from rice-wheat system of the Indo-Gangetic plains: through tillage, irrigation and fertilizer management. Agric Ecosyst Environ 230:1–9. https://doi.org/10.1016/j.agee.2016.05.023
Hanson RS, Hanson TE (1996) Methanotrophic bacteria. Microbiological 60:439–471
Haque MM, Biswas JC, Kim SY, Kim PJ (2017) Intermittent drainage in paddy soil: ecosystem carbon budget and global warming potential. Paddy Water Environ. https://doi.org/10.1007/s10333-016-0558-7
He R, Wooller MJ, Pohlman JW, Quensen J, Tiedje JM, Leigh MB (2012) Shifts in identity and activity of methanotrophs in Arctic Lake sediments in response to temperature changes. Appl Environ Microbiol 78:4715–4723. https://doi.org/10.1128/AEM.00853-12
Hussain S, Peng S, Fahad S, Khaliq A, Huang J, Cui K, Nie L (2015) Rice management interventions to mitigate greenhouse gas emissions: a review. Environ Sci Pollut Res 22:3342–3360. https://doi.org/10.1007/s11356-014-3760-4
Hütsch BW, Webster CP, Powlson DS (1994) Methane oxidation in soil as affected by land use, soil pH and N fertilization. Soil Biol Biochem 26:1613–1622. https://doi.org/10.1016/0038-0717(94)90313-1
Inubushi K, Sugii H, Watanabe I et al (2002) Nutr Cycl Agroecosyst 64:71. https://doi.org/10.1023/A:1021136431176
Jiang Y, Carrijo D, Huang S, Chen J, Balaine N, Zhang W, van Groenigen KJ, Linquist B (2019) Water management to mitigate the global warming potential of rice systems: a global meta-analysis. F Crop Res 234:47–54. https://doi.org/10.1016/j.fcr.2019.02.010
Ke X, Lu Y, Conrad R (2014) Different behaviour of methanogenic archaea and Thaumarchaeota in rice field microcosms. FEMS Microbiol Ecol 87:18–29. https://doi.org/10.1111/1574-6941.12188
Khosa MK, Sidhu BS, Benbi DK (2011) Methane emission from rice fields in relation to management of irrigation water. J Environ Biol 32(2):169–172
Kim GY, Gutierrez J, Jeong HC, Lee JS, Haque MDM, Kim PJ (2014) Effect of intermittent drainage on methane and nitrous oxide emissions under different fertilization in a temperate paddy soil during rice cultivation. J Korean Soc Appl Biol Chem 57:229–236. https://doi.org/10.1007/s13765-013-4298-8

Kimani SM, Cheng W, Kanno T, Nguyen-Sy T, Abe R, Oo AZ, Tawaraya K, Sudo S (2018) Azolla cover significantly decreased CH4 but not N2O emissions from flooding rice paddy to atmosphere. Soil Sci Plant Nutr 64:68–76. https://doi.org/10.1080/00380768.2017.1399775

Kimura M, Murase J, Lu Y (2004) Carbon cycling in rice field ecosystems in the context of input, decomposition and translocation of organic materials and the fates of their end products (CO2 and CH4). Soil Biol Biochem. https://doi.org/10.1016/j.soilbio.2004.03.006

Kong D, Li S, Jin Y, Wu S, Chen J, Hu T, Wang H, Liu S, Zou J (2019) Linking methane emissions to methanogenic and methanotrophic communities under different fertilization strategies in rice paddies. Geoderma 347:233–243. https://doi.org/10.1016/j.geoderma.2019.04.008

Krüger M, Frenzel P (2003) Effects of N-fertilization on CH4 oxidation and production, and consequences for CH4 emissions from microcosms and rice fields. Glob Change Biol 9:773–784

Kudo Y, Noborio K, Shimoozono N, Kurihara R (2014) The effective water management practice for mitigating greenhouse gas emissions and maintaining rice yield in Central Japan. Agric Ecosyst Environ 186:77–85. https://doi.org/10.1016/j.agee.2014.01.015

Kumar SS, Malyan SK (2016) Nitrification inhibitors: a perspective tool to mitigate greenhouse gas emission from rice soils. Curr World Environ 11:423–428. https://doi.org/10.12944/cwe.11.2.10

Kumar A, Tomer R, Bhatia A, Jain N, Pathak H (2016a) Greenhouse gas mitigation in Indian agriculture. In: Pathak H, Chakrabarti B (eds) Climate change and agriculture technologies for enhancing resilience. ICAR-IARI, New Delhi, pp 137–149

Kumar SS, Malyan SK, Kumar A, Bishnoi NR (2016b) Optimization of Fenton's oxidation by Box-Behnken design of response surface methodology for landfill leachate. J Mater Environ Sci 7:4456–4466

Kumar A, Bhatia A, Fagodiya RK, Malyan SK, Meena B (2017) Eddy covariance flux tower: a promising technique for greenhouse gases measurement. Adv Plants Agric Res 7:337–340. https://doi.org/10.15406/apar.2017.07.00263

Kumar SS, Kumar A, Singh S, Malyan SK, Baram S, Sharma J, Singh R, Pugazhendhi A (2020a) Industrial wastes: fly ash, steel slag and phosphogypsum-potential candidates to mitigate greenhouse gas emissions from paddy fields. Chemosphere 241:124824. https://doi.org/10.1016/j.chemosphere.2019.124824

Kumar A, Medhi K, Fagodiya RK, Subrahmanyam G, Mondal R, Raja P, Malyan SK, Gupta DK, Gupta CK, Pathak H (2020b) Molecular and ecological perspectives of nitrous oxide producing microbial communities in agro-ecosystems. Rev in Environ Sci Bio/Technol 19(4):717–750

Kumari S, Fagodiya RK, Hiloidhari M, Dahiya RP, Kumar A (2020) Methane production and estimation from livestock husbandry: a mechanistic understanding and emerging mitigation options. Sci Tot Envir 136135. https://doi.org/10.1016/j.scitotenv.2019.136135

Le Mer J, Roger P (2001) Production, oxidation, emission and consumption of methane by soils: a review. Eur J Soil Biol 37:25–50. https://doi.org/10.1016/S1164-5563(01)01067-6

Li H, Guo HQ, Helbig M, Dai SQ, Zhang MS, Zhao M, Peng CH, Xiao XM, Zhao B (2019) Does direct-seeded rice decrease ecosystem-scale methane emissions?—a case study from a rice paddy in Southeast China. Agric For Meteorol 272–273:118–127. https://doi.org/10.1016/j.agrformet.2019.04.005

Li X, Ma J, Yao Y, Liang S, Zhang G, Xu H, Yagi K (2014) Methane and nitrous oxide emissions from irrigated lowland rice paddies after wheat straw application and midseason aeration. Nutr Cycl Agroecosyst 100:65–76. https://doi.org/10.1007/s10705-014-9627-8

Liesack W, Schnell S, Revsbech NP (2000) Microbiology of flooded rice paddies. FEMS Microbiol Rev 24:625–645. https://doi.org/10.1016/S0168-6445(00)00050-4

Liu J, Xu H, Jiang Y, Zhang K, Hu Y, Zeng Z (2017) Methane emissions and microbial communities as influenced by dual cropping of Azolla along with early Rice. Sci Rep 7:1–14. https://doi.org/10.1038/srep40635

Liu J, Zang H, Xu H, Zhang K, Jiang Y, Hu Y, Zeng Z (2019a) Methane emission and soil microbial communities in early rice paddy as influenced by urea-N fertilization. Plant Soil. https://doi.org/10.1007/s11104-019-04091-0

Liu Y, Tang H, Muhammad A, Huang G (2019b) Emission mechanism and reduction countermeasures of agricultural greenhouse gases – a review. Greenh Gases Sci Technol. https://doi.org/10.1002/ghg.1848

Ma J, Ji Y, Zhang G, Xu H, Yagi K (2013) Timing of midseason aeration to reduce CH4 and N2O emissions from double rice cultivation in China. Soil Sci Plant Nutr 59:35–45. https://doi.org/10.1080/00380768.2012.730477

Malyan SK (2017) Reducing methane emission from rice soil through microbial interventions. ICAR-Indian Agricultural Research Institute, New Delhi-110012, India

Malyan SK, Kumar A, Kumar J, Kumar SS (2016a) Water management tool in rice to combat two major environmental issues: global warming and water scarcity. In: Kumar S, Beg MA (eds) Environmental concerns of 21st century: Indian and global context. Book Age Publications, India, pp 45–58

Malyan SK, Bhatia A, Kumar O, Tomer R (2016b) Impact of nitrogen fertilizers on methane emissions from flooded Rice. Curr World Environ 11:846–850. https://doi.org/10.12944/cwe.11.3.20

Malyan SK, Bhatia A, Kumar A, Gupta DK, Singh R, Kumar SS, Tomer R, Kumar O, Jain N (2016c) Methane production, oxidation and mitigation: a mechanistic understanding and comprehensive evaluation of influencing factors. Sci Total Environ 572:874–896. https://doi.org/10.1016/j.scitotenv.2016.07.182

Malyan SK, Kumar A, Baram S, Kumar J, Singh S, Kumar SS, Yadav AN (2019a) Role of fungi in climate change abatement through carbon sequestration. In: Yadav A, Singh S, Mishra S, Gupta A (eds) Recent advancement in white biotechnology through fungi. Fungal biology. Springer, Cham. https://doi.org/10.1007/978-3-030-25506-0_11

Malyan SK, Bhatia A, Kumar SS, Fagodiya RK, Pugazhendhi A, Duc PA (2019b) Mitigation of greenhouse gas intensity by supplementing with Azolla and moderating the dose of nitrogen fertilizer. Biocatal Agric Biotechnol 20:101266. https://doi.org/10.1016/j.bcab.2019.101266

Malyan SK, Singh S, Bachheti A, Chahar M, Sah MK, Narender KA, Yadav AN, Kumar SS (2020) Cyanobacteria: a perspective paradigm for agriculture and environment. In: Yadav AN et al (eds) Trends of microbial biotechnology for sustainable agricultural and biomedicine systems: diversity and functional perspectives. Elsevier. https://doi.org/10.1016/B978-0-12-820526-6.00014-2

Mancinelli RL (1995) The regulation of methane oxidation in soil. Annu Rev Microbiol. https://doi.org/10.1146/annurev.micro.49.1.581

Mayumi D, Yoshimoto T, Uchiyama H, Nomura N, Nakajima-Kambe T (2010) Seasonal change in methanotrophic diversity and populations in a rice field soil assessed by DNA-stable isotope probing and quantitative real-time PCR. Microbes Environ. https://doi.org/10.1264/jsme2.ME10120

Mishra S, Rath AK, Adhya TK, Rao VR, Sethunathan N (1997) Effect of continuous and alternate water regimes on methane efflux from rice under greenhouse conditions. Biol Fertil Soils 24:399–405. https://doi.org/10.1007/s003740050264

Mona S, Malyan SK, Saini N, Deepak B, Pugazhendhi A, Kumar SS (2021) Towards sustainable agriculture with carbon sequestration, and greenhouse gas mitigation using algal biochar. Chemosphere 275:129856. https://doi.org/10.1016/j.chemosphere.2021.129856

Mukherjee J, Mridha N, Mondal S, Chakraborty D, Kumar A (2018) Identifying suitable soilhealth indicators under variable climate scenarios: a ready reckoner for soil management. In: Bal S, Mukherjee J, Choudhury B, Dhawan A (eds) Advances in crop environment interaction. Springer, Singapore

Nazaries L, Murrell JC, Millard P, Baggs L, Singh BK (2013) Methane, microbes and models: fundamental understanding of the soil methane cycle for future predictions. Environ Microbiol 15:2395–2417. https://doi.org/10.1111/1462-2920.12149

Neue H-U (1993) Methane emission from rice fields. Bioscience 43:466–474. https://doi.org/10.2307/1311906

Nishiuchi S, Yamauchi T, Takahashi H, Kotula L, Nakazono M (2012) Mechanisms for coping with submergence andwaterlogging in rice. Rice. https://doi.org/10.1186/1939-8433-5-2

Oo AZ, Sudo S, Inubushi K, Mano M, Yamamoto A, Ono K, Osawa T, Hayashida S, Patra PK, Terao Y, Elayakumar P, Vanitha K, Umamageswari C, Jothimani P, Ravi V (2018) Methane and nitrous oxide emissions from conventional and modified rice cultivation systems in South India. Agric Ecosyst Environ. https://doi.org/10.1016/j.agee.2017.10.014

Ozuolmez D, Na H, Lever MA, Kjeldsen KU, Jørgensen BB, Plugge CM (2015) Methanogenic archaea and sulfate reducing bacteria co-cultured on acetate: team work or coexistence? Front Microbiol 6:492

Pandey A, Mai VT, Vu DQ, Bui TPL, Mai TLA, Jensen LS, de Neergaard A (2014) Organic matter and water management strategies to reduce methane and nitrous oxide emissions from rice paddies in Vietnam. Agric Ecosyst Environ 196:137–146. https://doi.org/10.1016/j.agee.2014.06.010

Panwar NL, Pawar A, Salvi BL (2019) Comprehensive review on production and utilization of biochar. SN Appl Sci 1. https://doi.org/10.1007/s42452-019-0172-6

Pathak H, Jain N, Bhatia A, Kumar A, Chatterjee D (2016) Improved nitrogen management: a key to climate change adaptation and mitigation. Ind J Fert 12:151–162

Peyron M, Bertora C, Pelissetti S, Said-Pullicino D, Celi L, Miniotti E, Romani M, Sacco D (2016) Greenhouse gas emissions as affected by different water management practices in temperate rice paddies. Agric Ecosyst Environ. https://doi.org/10.1016/j.agee.2016.07.021

Pramanik P, Kim PJ (2017) Contrasting effects of EDTA applications on the fluxes of methane and nitrous oxide emissions from straw-treated rice paddy soils. J Sci Food Agric 97:278–283. https://doi.org/10.1002/jsfa.7727

Purakayastha TJ, Bera T, Bhaduri D, Sarkar B, Mandal S, Wade P, Kumari S, Biswas S, Menon M, Pathak H, Tsang DCW (2019) A review on biochar modulated soil condition improvements and nutrient dynamics concerning crop yields: pathways to climate change mitigation and global food security. Chemosphere:345–365. https://doi.org/10.1016/j.chemosphere.2019.03.170

Ranjan R, Yadav R (2019) Targeting nitrogen use efficiency for sustained production of cereal crops. J Plant Nutr 42:1086–1113. https://doi.org/10.1080/01904167.2019.1589497

Roy R, Conrad R (1999) Effect of methanogenic precursors (acetate, hydrogen, propionate) on the suppression of methane production by nitrate in anoxic rice field soil. FEMS Microbiol Ecol. https://doi.org/10.1016/S0168-6496(98)00092-0

Samoy-Pascual K, Sibayan EB, Grospe FS, Remocal AT, T-Padre A, Tokida T, Minamikawa K (2019) Is alternate wetting and drying irrigation technique enough to reduce methane emission from a tropical rice paddy? Soil Sci. Plant Nutr 65:203–207. https://doi.org/10.1080/00380768.2019.1579615

Saunois M, Bousquet P, Poulter B, Peregon A, Ciais P, Canadell JG, Dlugokencky EJ, Etiope G, Bastviken D, Houweling S, Janssens-Maenhout G, Tubiello FN, Castaldi S, Jackson RB, Alexe M, Arora VK, Beerling DJ, Bergamaschi P, Blake DR, Brailsford G, Brovkin V, Bruhwiler L, Crevoisier C, Crill P, Covey K, Curry C, Frankenberg C, Gedney N, Höglund-Isaksson L, Ishizawa M, Ito A, Joos F, Kim HS, Kleinen T, Krummel P, Lamarque JF, Langenfelds R, Locatelli R, Machida T, Maksyutov S, McDonald KC, Marshall J, Melton JR, Morino I, Naik V, O'Doherty S, Parmentier FJW, Patra PK, Peng C, Peng S, Peters GP, Pison I, Prigent C, Prinn R, Ramonet M, Riley WJ, Saito M, Santini M, Schroeder R, Simpson IJ, Spahni R, Steele P, Takizawa A, Thornton BF, Tian H, Tohjima Y, Viovy N, Voulgarakis A, Van Weele M, Van Der Werf GR, Weiss R, Wiedinmyer C, Wilton DJ, Wiltshire A, Worthy D, Wunch D, Xu X, Yoshida Y, Zhang B, Zhang Z, Zhu Q (2016) The global methane budget 2000-2012. Earth Syst Sci Data 8:697–751. https://doi.org/10.5194/essd-8-697-2016

Schütz H, Seiler W, Conrad R (1990) Influence of soil temperature on methane emission from rice paddy fields. Biogeochemistry 11:77–95. https://doi.org/10.1007/BF00002060

Semrau J, Dispirito AA, Yoon S (2010) Methanotrophs and copper. FEMS Microbiol Rev 34:496–531

Serrano-Silva N, Sarria-Guzmán Y, Dendooven L, Luna-Guido M (2014) Methanogenesis and methanotrophy in soil: a review. Pedosphere 24:291–307. https://doi.org/10.1016/S1002-0160(14)60016-3

Setyanto P, Pramono A, Adriany TA, Susilawati HL, Tokida T, Padre AT, Minamikawa K (2018) Alternate wetting and drying reduces methane emission from a rice paddy in Central Java, Indonesia without yield loss. Soil Sci Plant Nutr. https://doi.org/10.1080/00380768.2017.1409600

Shukla PN, Pandey KD, Mishra VK (2013) Environmental determinants of soil methane oxidation and methanotrophs. Crit Rev Environ Sci Technol 43:1945–2011. https://doi.org/10.1080/10643389.2012.672053

Simmonds MB, Anders M, Adviento-Borbe MA, van Kessel C, McClung A, Linquist BA (2015) Seasonal methane and nitrous oxide emissions of several rice cultivars in direct-seeded systems. J Environ Qual. https://doi.org/10.2134/jeq2014.07.0286

Singh JS, Pandey VC (2013) Fly ash application in nutrient poor agriculture soils: impact on methanotrophs population dynamics and paddy yields. Ecotoxicol Environ Saf. https://doi.org/10.1016/j.ecoenv.2012.11.011

Singh JS, Kumar A, Rai AN, Singh DP (2016) Cyanobacteria: a precious bio-resource in agriculture, ecosystem, and environmental sustainability. Front Microbiol. https://doi.org/10.3389/fmicb.2016.00529

Singh NK, Patel DB, Khalekar GD (2018) Methanogenesis and Methane Emission in Rice/Paddy Fields. pp 135–170. https://doi.org/10.1007/978-3-319-99076-7_5

Smartt AD, Brye KR, Norman RJ (2018) Methane emissions among hybrid Rice cultivars in the mid Southern United States. Ann Adv Agric Sci:2. https://doi.org/10.22606/as.2018.21001

Steudler PA, Bowden RD, Melillo JM, Aber JD (1989) Influence of nitrogen fertilization on methane uptake in temperate forest soils. Nature 341:314–316. https://doi.org/10.1038/341314a0

Sun L, Ma Y, Liu Y, Li J, Deng J, Rao X, Zhang Y (2019) The combined effects of nitrogen fertilizer, humic acid, and gypsum on yield-scaled greenhouse gas emissions from a coastal saline rice field. Environ Sci Pollut Res 26:19502–19511. https://doi.org/10.1007/s11356-019-05363-z

Sun W, Xiao E, Pu Z, Krumins V, Dong Y, Li B, Hu M (2018) Paddy soil microbial communities driven by environment-and microbe-microbe interactions: a case study of elevation-resolved microbial communities in a rice terrace. Sci Total Environ 612:884–893

Tariq A, Jensen LS, de Tourdonnet S, Sander BO, de Neergaard A (2017) Early drainage mitigates methane and nitrous oxide emissions from organically amended paddy soils. Geoderma. https://doi.org/10.1016/j.geoderma.2016.08.022

Tran DH, Hoang TN, Tokida T, Tirol-Padre A, Minamikawa K (2018) Impacts of alternate wetting and drying on greenhouse gas emission from paddy field in Central Vietnam. Soil Sci Plant Nutr. https://doi.org/10.1080/00380768.2017.1409601

Trotsenko YA, Khmelenina VN (2005) Aerobic methanotrophic bacteria of cold ecosystems. FEMS Microbiol Ecol. https://doi.org/10.1016/j.femsec.2005.02.010

Tveit A, Schwacke R, Svenning MM, Urich T (2013) Organic carbon transformations in high-Arctic peat soils: key functions and microorganisms. ISME J 7:299–311. https://doi.org/10.1038/ismej.2012.99

Wang W, Sardans J, Lai DY, Wang C, Zeng C, Tong C, Liang Y, Peñuelas J (2015) Effects of steel slag application on greenhouse gas emissions and crop yield over multiple growing seasons in a subtropical paddy field in China. Field Crops Res 171:146–156

Wang Z, Gu D, Beebout SS, Zhang H, Liu L, Yang J, Zhang J (2018a) Effect of irrigation regime on grain yield, water productivity, and methane emissions in dry direct-seeded rice grown in raised beds with wheat straw incorporation. Crop J 6:495–508. https://doi.org/10.1016/j.cj.2018.05.004

Wang C, Wang W, Sardans J, Singla A, Zeng C, Lai DYF, Peñuelas J (2018b) Effects of steel slag and biochar amendments on CO_2, CH_4, and N_2O flux, and rice productivity in a subtropical Chinese paddy field. Environ Geochem Health 41:1419–1431 (2019). https://doi.org/10.1007/s10653-018-0224-7

Whalen SC, Reeburgh WS, Sandbeck KA (1990) Rapid methane oxidation in a landfill cover soil. Appl Environ Microbiol 56:3405–3411

Whittenbury R, Phillips KC, Wilkinson JF (1970) Enrichment, isolation and some properties of methaneutilizing bacteria. J Gen Microbiol 61:205–218

Wu X, Wang W, Xie K, Yin C, Hou H, Xie X (2019a) Combined effects of straw and water management on CH 4 emissions from rice fields. J Environ Manag 231:1257–1262. https://doi.org/10.1016/j.jenvman.2018.11.011

Wu Z, Song Y, Shen H, Jiang X, Li B, Xiong Z (2019b) Biochar can mitigate methane emissions by improving methanotrophs for prolonged period in fertilized paddy soils. Environ Pollut 253:1038–1046. https://doi.org/10.1016/j.envpol.2019.07.073

Xiao Y, Yang S, Xu J, Ding J, Sun X, Jiang Z (2018) Effect of biochar amendment on methane emissions from paddy field under water-saving irrigation. Sustainability 10. https://doi.org/10.3390/su10051371

Xu H, Zhu B, Liu J, Li D, Yang Y, Zhang K, Jiang Y, Hu Y, Zeng Z (2017) Azolla planting reduces methane emission and nitrogen fertilizer application in double rice cropping system in southern China. Agron Sustain Dev 37. https://doi.org/10.1007/s13593-017-0440-z

Yang YD, Xu HS, Li DY, Liu JN, Nie JW, Zeng ZH (2019) Methane emissions responding to Azolla inoculation combined with midseason aeration and N fertilization in a double-rice cropping system. Environ Sci Pollut Res. https://doi.org/10.1007/s11356-019-05342-4

Zheng H, Fu Z, Zhong J, Long W (2018) Low methane emission in rice cultivars with high radial oxygen loss. Plant Soil 431:119–128. https://doi.org/10.1007/s11104-018-3747-x

Soil Microflora and its Role in Diminution of Global Climate Change

13

Nilanjana Majumder and Koushik Biswas

Abstract

From the last two decades, the issue of climate change and its potential impacts on air, water and soil health has consistently been reminding us of the biggest scientific and political challenges of twenty-first century at global level. There are a number of strategies initiated to reduce the adverse effect of climate change on biodiversity, interaction between hosts and pathogens, population dynamics of microbiomes, aquatic ecosystem and several abiotic factors, namely temperature, rainfall, moisture status of air and soil, etc. Majority of these strategies are emphasized upon the technologies based on chemical and biological conversions. Microbial biodiversity of soil and its potential exploitation towards the mitigation of global climate change is a matter of utmost importance that needs to be explored. The proper utilization of beneficial microorganisms present in soil could fix the major dilemma related to the uncertainty of agricultural production happened due to erratic environment and natural calamities. The maximum exploration of soil microbial community could be employed in various arenas of sustainable agriculture such as suppression of soil-borne plant pathogens, enhancer in plant growth and development, biofertilizer cum symbiotic nitrogen fixer, inducer of systemic acquired resistance, alleviation of biotic and abiotic stresses in plants, remediation of heavy metals from contaminated soils, soil resilience, degradation of polyethylene component from soil. Considering the importance of soil microflora and its role in nature, there is a need to promote the investigations on these aspects with possible modifications which can improve the ecology and environment for better life in the world.

N. Majumder · K. Biswas (✉)
Department of Biotechnology, Visva-Bharati (A Central University) Santiniketan, Bolpur, West Bengal, India

S. A. Lone, A. Malik (eds.), *Microbiomes and the Global Climate Change*,
https://doi.org/10.1007/978-981-33-4508-9_13

Keywords

Climate · Soil microbe · Global warming · Mitigation

13.1 Introduction

In modern world, change in climate and environment due to human activities is a well-discussed topic when it comes to the point of research about extinction in plant and animal species and change in biodiversity. Extinction of species, communities and habitats are considered as a subject of detailed research, thus well understood and documented compared to microorganisms. The role of microorganisms are usually less discussed in the context of climate change, as well as the influence of climate change on the general mechanisms of microbial activity remains hardly analysed (Cavicchioli et al. 2019). Despite being invisible to the naked eye, the microbial abundance (total $\sim10^{30}$ bacteria and archaea) and diversity perform their role in maintaining a healthy global ecosystem and account for the life support system of the biosphere. The terrestrial ecosystems undergo a principal role in such climate-feedbacks as they both release and absorb greenhouse gases, precisely carbon dioxide, methane and nitrous oxide, while acting as a significant carbon sink by storing in soil and living plants (Bardgett et al. 2008). The net effect of climate change on total amount of ecosystem carbon budgets is dependent on the equilibrium between photosynthesis (autotrophic respiration) and respiration (heterotrophic soil microbial respiration). Thus, the immense diversity of soil microbes constitutes a significant role in emission of greenhouse gases and in elevation of environmental warming due to their biological processes involved in plant–microbe interaction on carbon-nitrogen cycling. All these observations confirm the need of including the study and research on terrestrial microorganisms especially soil microbes, for the microbial mitigation of soil to maintain the agricultural sustainability, their interactions with autotrophic and heterotrophic living organisms, as well as their biological mechanisms related to climate changes (Cavicchioli et al. 2019).

13.2 Terrestrial Microbiome

Terrestrial plants account for a significantly large biomass and also act as the source of roughly half of the net global primary production (Cavicchioli et al. 2019). Different microbial communities reside in different terrestrial ecosystems depending on their distribution over space and time, different land use types and disturbances introduced by human (Singh et al. 2010). The plant–microbe interaction with mycorrhizal fungi and nitrogen fixing bacteria is well understood in the terrestrial ecosystem whereas the involvement of heterotrophic microbial communities in climate change is still somehow unclear (Singh et al. 2010). The soil ecosystem is a significant reservoir of dynamic carbon and for this reason is an efficient determinant of pattern in climatic changes. Soil carbon comprises two-third of the total Carbon content of terrestrial ecosystem, i.e. more than the total combined amount of Carbon in atmosphere and vegetation. The microbial exercises have been one of the basic determinants of convergences of ozone harming substances, for example, Carbon dioxide (CO_2), Nitrous oxide (N_2O) and Methane (CH_4) in environment. Microorganisms have been major players in the cycling of ozone depleting substances and after their development to earthly condition also since their advancement in the sea more than 3.5 billion years ago. Microorganisms are probably going to get influenced by the climatic changes just as they can control the atmosphere through their association with the motions among soil and environment in ozone harming substances (Dutta et al. 2016). As per the IPCC environmental change report (IPCC 2007), in 2005, the normal worldwide convergence of CO_2 in the air was almost 380 ppm which outperformed the past most extreme focus for the past 6.5 million years by around 80 ppm. The Special Report on Global Warming of 1.5 °C (SR15) published by IPCC on the year 2018, depicted that limiting the warming close to 1.5 °C would require to decrease the net emissions by around 45% by 2030.As the worldwide increment in ozone depleting substance outflow has got one of the most fundamental logical and political difficulties, the better comprehension of the natural components occurring between land, sea and environment liable for worldwide carbon trades has got one of the greatest imperative.

Both immediate and aberrant impacts have been applied by environmental change on the mechanisms of soil microorganisms that discharge ozone depleting substances to the air and award their commitment on a worldwide temperature alteration. The immediate impacts grasp the effect on soil organisms and creation of ozone harming substance bringing about rise in temperature, changing precipitation cycles and other outrageous climatic occasions. Atmosphere driven changes may result in plant profitability just as decent variety in vegetation that may adjust soil conditions, particularly the stock of carbon to soil and the action of microbial networks occupied with arrival of carbon (Singh et al. 2010). Soil microorganisms largely affect plant efficiency which could be characterized in two primary recognizable systems as immediate consequences for plants by means of root-related living beings that structure mutualistic or pathogenic associations with plants, and backhanded impacts by means of the activity of free-living organisms that change the paces of supplement supply (Van der Heijden et al. 2008).

13.3 Changes in Climate Affect Soil Microflora

Numerous studies have shown that shifts in species interactions according to climate change are likely to alter biodiversity and the function of terrestrial ecosystems, but very few studies have focused on soil communities (Singh et al. 2010). Soil microbes interact with one another as well as with plants in a myriad of ways that shape and maintain ecosystem properties which can shape landscape patterns of plant and animal abundance, diversity and composition.

13.3.1 Temperature and Thermal Adaptation on Soil Microbes

The relative plenitude and capacity of soil networks get adjusted due to climatic changes as soil community individuals differ in their physiology, growth rates and temperature lability. Worldwide climatic changes, for example, warming is directly altering the microbial soil respiration rates, and the temperature sensitive procedures interceded by the soil organisms. The job of elevated temperature in microbial metabolism has got critical consideration in some on-going investigations (Bradford 2013; Frey et al. 2013; Hagerty et al. 2014). The inherent temperature sensitivity of microbial action bringing about little changes in microbial network creation is characterized as the factor by which microbial metabolic action raises with a 10 °C elevation in temperature. In any case, utilizing just the factor to clarify temperature affectability in models may lead to poor forecasts as this relationship covers a lot of considerable interactions that have influence on the temperature sensitivity of microbial mechanisms, for example, decomposition. Further, while decomposition of soil organic matter, soil respiration and growth of microbial biomass are up-regulated with temperature, these reactions to experimental warming are frequently brief in field studies. The brief impacts of warming on soil microbial networks have been recognized to happen, as labile soil carbon substrates are depleted by expanded microbial action in soil and as a result of adjusting the microbial networks either adapt and move in synthesis or limit their biomass in light of the modified condition (Van der Heijden et al. 2008; Bradford 2013). As far as long term effect of increasing high temperature on soil microbiome is concerned, the combustion of fossil fuel could be proven as the most threatening ailment to the living world. The burning of fossil fuel derived from the degradation of biomass of microbes, plants and animals over millions of years is now considered as the prime factor of rising in atmospheric CO_2 concentration and liberation of greenhouse gases. As a result, the increasing rate of global warming phenomenon is not only altering the physical and chemical properties of soil but also the complex network association among various microorganisms and with plants and animals. These changes are clearly experienced by the behavioural response of microbiomes to the surrounding environments (e.g. rising rate of microbial respiration and acidification of soil micro-climate) (Cavicchioli et al. 2019). Experimental evidences of induced warming showed an alteration in microbial community abundance of Gram-positive and Gram-negative bacteria (Van der Heijden et al. 2008). Often short term laboratory experiment and long term field studies showed contradictory results on thermal compensation prompting in debate on thermal acclimation (Van der Heijden et al. 2008). Laboratory experiments demonstrate on more than 15 years old mid-latitude forest showed soil microbial respiration thermally adapting to seasonal variation in temperature supported by the contribution of substrate depletion and reduced soil biomass (Bradford et al. 2008) (Fig. 13.1).

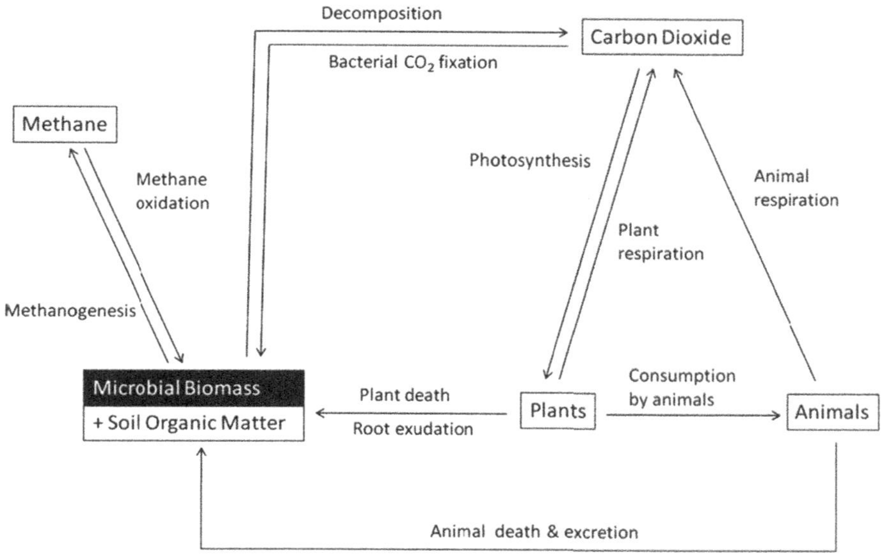

Fig. 13.1 The major processes mediated by soil microorganism of terrestrial carbon cycle (Adapted from Gougoulias et al. 2014)

13.3.2 Precipitation

As indicated by some on-going experiments, it has been discovered that warming had a differential impact on both bacterial and fungal community structures under every precipitation system. In the meantime, the warming effect on the diversity and assorted variety of bacterial and fungal communities fluctuated among precipitation treatments. These discoveries show that the impact of warming on the development and structure of a soil microbial community might be controlled by controlling the precipitation and the adjustments in the microbial community structure, which were influenced by treatment-incited changes in soil moisture. Certain microbial groups may have special adaptive characteristics to desiccation, for example, thickened cell walls, coming about in greater resistance to variances in soil moisture. As the soil is anaerobic and that moisture controls nutrient diffusion, soil moisture could be a basic factor influencing microbial community structure by means of connecting moisture based mechanistic processes.

The alteration of diversity and structure manipulated by precipitation or total soil moisture is much more prominent in fungal communities (Singh et al. 2010). As the fungal communities show more interaction with plants due to their symbiotic behaviour, the effect of changes in soil properties due to alteration in precipitation tends to manipulate the fungal communities which might emerge as a reason for a combined 34% soil diversity. An acidic soil condition under increased precipitation may act as environmental stress for fungal communities which may consequently alter soil fungal diversity (Li et al. 2018). The major alterations in bacterial communities were observed in specific dominant phyla (e.g., *Proteobacteria and Acidobacteria*), occupying unique niches. Some climate change studies suggested that climate change drivers could change the structure and composition of soil microbial communities by selecting species that can thrive in the new environments (Zhang et al. 2016). As actual climate change can be more gradual than the changes observed in short term experiments, these studied results might not reflect the long term changes in the soil microbial communities and for that, further monitoring is required to confirm whether the detected changes in pattern will persist under more gradual climate changes over a longer time scale.

13.3.3 Elevation of Carbon Dioxide

Arrival of progressively powerful ozone harming substances, for example, Methane and Nitrous oxide by soil organisms is supported by expanded barometrical degree of CO_2 (Pathak et al. 2012). Truth be told, higher CO_2 levels increment efflux of methane as well as down direct the take-up of methane up to 30% by soil microorganisms (Dutta et al. 2016). Nonetheless, raised degrees of carbon dioxide prompt significant modifications in the microbial networks of tree leaves which are an intense wellspring of supplements for the little phytophagous creatures, may have results on the natural way of life all things considered microorganisms. Also, increment in microbial respiration happens due to up-regulated plant efficiency that happens because of raised carbon dioxide, particularly under nutrient rich conditions (Curtis and Wang 1998). This expands the carbon transition to roots, their symbionts and heterotrophic microorganisms through effectively degradable sugars and amino acids as root exudates (Zak et al. 1993). The outcomes of quickened carbon motion from roots to soil and carbon exchange for microbial networks are complicated to anticipate, due to their generous variety with factors like plant character, soil ripeness, soil nourishment web associations and a scope of other biological system properties (Bardgett et al. 2005). In any case, some potential results in connection to the soil microorganisms and carbon exchange incorporate the up-regulation of soil carbon as simplified natural carbon by respiration, because of stimulated microbial action and their abundance, which consequently gives more carbon substrate to soil microbial networks. A key component by which carbon cycle criticism is impacted by microorganisms and their digestion for atmosphere driven adjustments in vegetation synthesis is through changes in the quality and amount of natural issue going into the soil as plant litter. Slow-developing plants, for example, evergreen bushes, produce low quality litter which is poor in supplements and wealthy in headstrong mixes, for example, lignin and phenolic acids, decay gradually because of weakening of microbial movement; though, quickly developing plants, for example, graminoids and N-fixers, produce generally top notch supplement rich litter which breaks down quickly because of expanded degree of microbial action. The view of the significance for carbon trade of such connection between change in atmosphere, vegetation and related soil microbial working is poorly understood and therefore requires further research (Bardgett et al. 2008).

13.3.4 Resistance Development in Several Harmful Plant Pathogens

The predominant view is that natural changes including rising air CO_2 concentration, warming, and altered precipitation designs will build the frequency of plant invasions and worsen the negative impacts of these factors on the environment. As talked about before (Wolfe and Klironomos 2005), migration of plant species into novel territories may adjust the capacity of native soil food webs by (1) changing the quality, amount and timing of litter deposition and rhizodeposition; (2) causing direct release of novel antimicrobial substances; (3) by essentially modifying food relations by introducing alternate methods of nutrient acquiring, for example, nitrogen fixation; or (4) by modifying soil structure or physical properties because of novel or prevailing rooting habits.

Dispersal of organisms into new environments could bring about change in the abundance and elimination of taxonomic groups in soil. These impacts could affect environment diversity and functionality and could make both wild and maintained plant systems increasingly vulnerable to invasive pathogens or may decline subterranean challenge, permitting pathogenic soil organisms to thrive. In any case, the functional redundancy of soil taxonomic groups common to many soil communities may buffer against critical shifts in the productivity of wild and maintained plant systems. This implies that huge changes to various groups of soil organisms may eject each other out, but still these soil community movements will have no net impact on ecosystem productivity (Bradford et al. 2002).

Notwithstanding indirect effects on soil life forms caused by invasive plant species, ecological changes may also legitimately elevate the potential for distribution of soil organisms (Pariaud et al. 2009). The oomycete *Phytophthora cinnamomi*, an invasive root pathogen endemic to New Guinea-Sulawesi, for example, has as of now invaded in at least 76 countries, that too as the dominant part of worldwide biodiversity hotspots. *P. cinnamomi* has caused massive elimination of Quercus spp. also, Castanea in the USA and Eucalyptus spp. in Australia over critical geographic regions (Dunstan et al. 2010). Models show that expansion in soil temperature related with climate change is probably going to expand the decent variety of *P. cinnamomi* in both Europe (Bergot et al. 2004) and Australia. Warming of soil is likewise likely to cause comparable directional dispersal of plant parasitic nematodes, worms and numerous other soil life forms.

13.4 Effects of Soil Microflora on Climate Change

13.4.1 Carbon Dioxide Emission

The measure of environmental carbon discharged because of microbial disintegration of soil, which is exceptionally delicate to climatic patterns, is 7.5–multiple times in contrast with what is yearly radiated because of anthropogenic capacities overall (Crowther et al. 2015). Additionally, microorganisms in soil contribute 5.5×10^{12} kg (55 billion tons) of carbon dioxide yearly through the deterioration of plant organic matter, which is itself far more than what is discharged by people. Soil biotic collaborations are along these lines considered huge on the grounds that they are critical in intervening soil microbial relationship to environmental change (Weiman 2015). Plants discharge ~50% of fixed carbon into the soil, which is experiment in woods accessible for microbial development. Notwithstanding microorganisms utilizing exudates as a vitality source, exudates can free natural mixes from minerals that are utilized for microbial respiration, consequently expanding carbon discharge disturbing natural incorporation of minerals (Cavicchioli et al. 2019). Anthropogenic environmental changes may support a good uprising in carbon discharge which can be anticipated by transient variety happened in deterioration of natural processes and generation of CO_2, resulting in prompt changes in microbial network and carbon use. This may further bring about decreased microbial biomass and thus diminished free carbon for microbial processes as shown in a 26-year soil warming experiment in woods (Weiman 2015). Overhauled information on reactions in microbial network and practical systems is required for better understanding over-the-ground plant-biomass reactions to CO_2 levels (Fig. 13.2).

The polar permafrost, known as the greatest earthly repository of carbon mixes (Zimmerman and Labonte 2015), should be particularly referenced with regard to environmental change. Regardless of possessing just 9% of the landmass of the earth, the permafrost contains about 25–50% of worldwide soil natural carbon. Around 1.7×10^{15} kg of carbon remain caught in permafrost, though the environment contains roughly 8.50×10^{14} kg (Atkin et al. 2015). Be that as it may, a

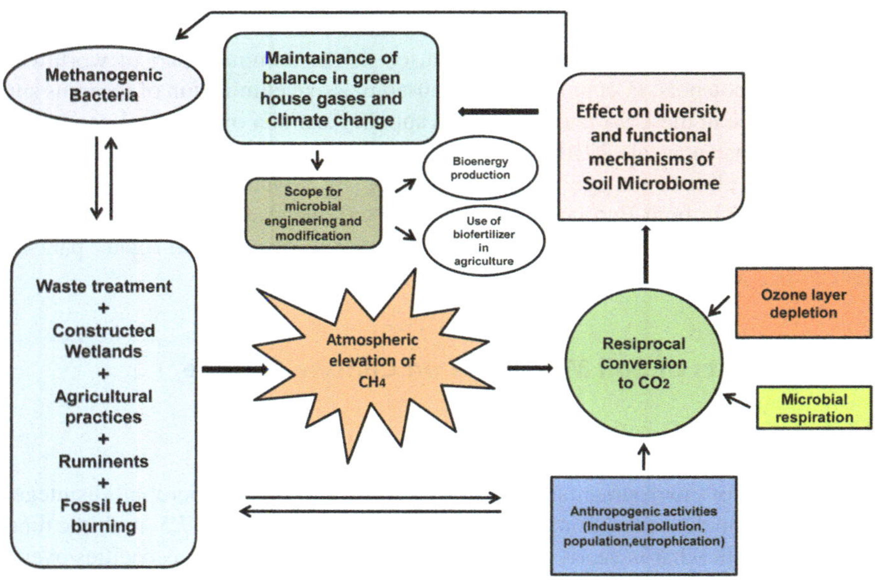

Fig. 13.2 Generation of Carbon dioxide and Methane and their impact on soil microbiome

worldwide temperature alteration starts the softening of the idle permafrost layers and subsequently, the microbial network which stayed lethargic for a considerable length of time inside these layers is activated, and their biotic exercises, for example, respiration, methanogenesis and multiplication are provoked (Zimmerman and Labonte 2015). Microbes that are available in the defrosting soil layer along these lines produce a few chemicals and proteins associated with disaggregation of enormous carbon atoms, for example, plant cellulose into monosaccharides that can fill in as wellspring of vitality for the microorganisms. Every step of these procedures realizes the emanation of ozone depleting substances (e.g., CO_2 and CH_4) by change of soil carbon (Zimmerman and Labonte 2015).

13.4.2 Methane Emission

Worldwide discharge of CH_4 is all the more straightforwardly related to microorganisms when contrasted with CO_2 (Singh et al. 2010) (Fig. 13.2). Microorganisms are liable for about 85% of worldwide methane discharge determined by rise of temperature (Steffen et al. 2015). In this unique circumstance, an uncommon notice must be given to springy marsh soils, as enormous number of microbial qualities and proteins are liable for arrival of methane which can be exemplified by the catalyst methyl-coenzyme M reductase, engaged with change of CO_2 into methane. Microorganisms are equipped for disposing of air methane even at low focuses and expend about 60% of methane delivered internationally and happen in both land and sea (Steffen et al. 2015). In such manner, the microorganisms of local prairie fields merit a notice, which assume the significant job of settling carbon transition. Modification in examples of precipitation emerging because of environmental change influences these organisms. This thus has the potential for influencing soil carbon store, along these lines invigorating significant outcomes on the carbon balance in such environments (Zimmerman and Labonte 2015). Dissolving and decrease of permafrost permit microbial decay of recently solidified carbon, radiating CO_2 and CH_4 (Cavicchioli et al. 2019). Seaside permafrost scraped area can prompt the preparation of huge amounts of carbon to the sea, consequently bringing about conceivably more prominent CO_2 discharges happening through expanded microbial remineralization causing a positive input circle that goes about as a trigger of environmental change. Anaerobic CH_4 creation by methanogens and CO_2 generation by a scope of microorganisms through expanding water-soaked soils are advanced by dissolving of permafrost (Dutta et al. 2016). An investigation of 7 years' laboratory analysis shows more prominent measure of CO_2 and methane emanation by methanogens engaged with permafrost defrosting in anoxic condition than oxic condition foreseeing climatic change before the finish of century (Knoblauch et al. 2018).

13.4.3 Role of Ruminants, Earthworms and Herbivores

The measure of natural issue that is cycled back to soil is influenced by herbivores (invertebrates and mammals). Consequently microbial biomass and movement additionally get influenced, for example, grasshoppers bring about expanded microbial cell movement by lessening plant biomass and plant nitrogen request. Environmental change may crumble herbivory, winding up in generally speaking adjustments to worldwide carbon and nitrogen cycles. Detritivores (for example, Earthworms) can impact the arrival of ozone depleting substance by implication influencing plants by expanding soil fruitfulness and decent variety of soil microorganisms. Earthworms guarantee alteration of soils through bolstering, tunnelling and affidavit of waste materials. The anaerobic condition of the night crawler gut sustains microorganisms that experience denitrification and produce nitrous oxide (Cavicchioli et al. 2019).

Worms expand soil fertility, despite the fact that the consolidated impacts of raised temperature and lightened precipitation on detritivore bolstering and microbial respiration may decrease emanations, their quality can bring about net ozone harming substance outflows. Ruminant creatures are considered as the biggest single wellspring of anthropogenic CH_4 discharges, as 19–48 times more noteworthy emanation of carbon dioxide is started from ruminant meat generation contrasted with plant-based high-protein nourishments. Ventures for non-ruminant meat creation (for example, pigs, poultry and fish) may discharge 3–10 times more methane than high-protein plant nourishments (Cavicchioli et al. 2019).

13.5 Agriculture

As 40% of terrestrial land has been used as agricultural fields, elevation of such lands may increase substantial changes in Carbon, Nitrogen and Phosphorus cycles, leading to significant changes through loss in biodiversity of microorganisms. Use of both plant associated and animal associated microorganisms for increasing sustainability in agriculture and to alleviate the effects of climate change on agriculture has become a burning topic of research and for this, the better understanding of effect of climate change on microbes is required.

13.5.1 Methane Emission from Different Agricultural Activities

Methane is the second most significant anthropogenic greenhouse gas simply after CO_2 regarding complete atmosphere controlling and microbial use, accurately known as methanotrophy, which is the largest terrestrial sink. Hence, to more readily foresee methane emissions, it will be fundamental to comprehend the reaction of CH_4 flux to atmosphere change. Natural emanations (~250 million tons of CH_4 every year) are ruled by microbial methanogenesis, a process that is completed by a group of anaerobic archaea in wetlands, sea, rumens and termite guts. Be that as it may, these common sources are enlarged by emissions from human activities (unequivocally rice cultivation, landfill, petroleum derivative extraction and live-stock cultivation) (about 320 million tons of CH_4 every year), which aside from certain emanations from non-renewable energy source extractions are additionally determined by microorganisms. Methanotrophic microorganisms fill in as a critical buffer to sublimate the gigantic measures of methane delivered in a portion of these conditions. The alleged 'low-affinity' methanotrophs, which are active just at a CH_4 concentration of under 40 ppm; belonging to class Gamma-proteobacteria, can regularly devour a huge extent of CH_4 released in soil before discharged into the environment. For CH_4 as of now in the atmosphere, the 'high affinity' methanotrophic microorganisms, belonging to the class Alpha-proteobacteria, (active at a concentration of under 12 ppm) may likewise act as a net CH_4 sink and expel roughly 30 million tons CH_4 from the climate every year. Rice is considered as a prime food source for half of the overall population and rice paddies are considered as a wellspring of around 20% of cultivating CH_4 surges, paying little heed to covering just generally 10% of arable land. Anthropogenic ecological change has been foreseen to two-fold CH_4 outflows from rice fields before the end of the century.

One truth that must be referenced with respect to this territory is that farming is considered as a gigantic wellspring of the strong ozone harming substance Nitrous oxide (N_2O) outflow, discharged by oxidation and decrease of nitrogen interceded by organisms. The compound N_2O reductase in rhizobacteria living in vegetable root knobs and other soil microorganisms may likewise change over N_2O to Nitrogen. The rate at which microbial nitrogen cycle happen and Nitrogen is discharged (deterioration, mineralization, nitrification, denitrification and fixation) can be activated by environmental change (Cavicchioli et al. 2019).

13.5.2 Fossil Fuel Combustion and Use of Fertilizers

Emissions of black carbon and organic carbon from cooking and heating with biofuels and from diesel transportation, and emission of sulphates and nitrates from fossil fuel combustion, can increase seasonal mean AODs (aerosol optic density) to as high as 0.4 (larger during volcanic periods), tends to decrease 10–15% of incident solar radiation at the surface. In addition to this, elevated utilization of fertilizers in fields together increases availability of environmental nitrogen, leading to a severe threat on ecosystem sustainability (Cavicchioli et al. 2019; Nisbet et al. 2019).

13.5.3 Eutrophication

The combined effects of climatic change and use of fertilizers in agricultural fields might imply unpredictable potential effects on microbial competitiveness (Cavicchioli et al. 2019). In a study in deep lake of Zurich revealed that reducing the use of phosphorus rich fertilizers increases non-nitrogen fixing cyanobacterial growth, elevation of annual warming might also result in persistence of toxic cyanobacteria in lake due to absence of predation and reduction in annual mixing (Posch et al. 2012). Human activities that have direct effect on microorganisms include greenhouse gas emissions (CO_2, methane and N_2O), pollution (precisely eutrophication), use of land in agriculture and growth in population which positively manipulates climate change, pollution, various agricultural practices and the multiplication of pathogenic microbes. In small lakes, located in climate sensitive regions, eutrophication causes elevated emission of methane, raises the need of considering lake microorganisms on greenhouse gas emission (Cavicchioli et al. 2019).

13.6 Microbial Mitigation of Climate Change

The ecological community of all types of microorganisms present in soil microflora, termed as 'soil microbiome', has a great level of changes in their population, diversity and functionality due to the continuous effect of climate change. So, the microbial shifting in various directions due to the alternation of climate is still under the subject of research for most scientific communities. Due to higher potentiality level of evolution, they can stably modify their ecological response as the climate deviates from its native state. For example, the consistent elevation of temperature leads to the reduction of respiration rate in soil microbes (Bradford et al. 2008) or the changes in precipitation pattern now make the microbial community much more tolerant to adverse condition than earlier time (Wallenstein and Hall 2012). These kind of incidents now make us to think that the maximum buffering capacity of terrestrial living communities with the continuous change could be utilized in practical only by monitoring and exploiting microbial processes undergoing on soil microbiome under soil. Therefore, it is necessary to discuss and open a third eye of knowledge that could explore a number of approaches dealing with microbial modelling for minimizing the effect of global climate change in sustainable manner. There are number of approaches concerned with the engineering of microbial communities through bioremediation with a great potential under complex environment. Here, we are highlighting some physical literature of most promisable and convenient techniques to suggest a way out that scientists can imply to.

13.6.1 Management of Soil-Borne Plant Pathogens

Soil microbial diversity is a mixture of heterogenous population of fungi, bacteria, viruses, algae, nematodes, protozoans and actinomycetes representing a great variability in a single microflora. Since last 50 years of science and research discovery, several types of beneficial effects of soil microbes have been studied and implemented in agriculture and plant health management (Subba Rao and Gaur 2000). As per as management of plant pathogen is concerned, some specific suppression of mechanisms among microbes and plants are well established and that includes (1) Plant-microbe (Biocontrol agent) interaction; (2) Microbe (pathogen)-

microbe (Biocontrol agent) interaction; (3) Plant metabolites (Biocontrol agent)—microbe (pathogen) interaction; (4) Induced systemic resistance. Release of biocontrol products in efficient way is dependent not only on its types or structural configuration, but also on switching on or off of the complex interactions of microbes and plants as mentioned (Sangeetha et al. 2012).

13.6.1.1 Exploitation of Pseudomonas, Bacillus and Other Rhizobacterial Species

Majority of the Gram-negative *Pseudomonads*, namely *P. fluorescens, P. syringae, P. aeruginosa, P. putida, P. aureofaciens, P. chlororaphis, P. brassicacearum, P. varvalensis* are well known for their biocontrol activity against a wide range of plant pathogens. The typical colonization level touching 10^5–10^6 CFU (colony forming unit)/g value at root region is sufficient enough to shield the plant against *Pythium spp.* attack. Though the laboratory based *Pseudomonas* culture growth reached to root up to 10^7–10^8 CFU/g, but no longer preserve the colonization rate with equal potential (Landa et al. 2003).

Another report confirmed that the combinatorial application of *P. fluorescens* (Pfbv22) and *Bacillus subtilis* (Babv57) was highly effective to suppress the nematode population in banana plant significantly. In that case, the active antibiotic compound (2,4-DAPG) is the prime agent to suppress the pathogenic invasion after its introduction where suppression is almost lost when *Pseudomonas spp.* is eliminated. In similar way, the microbial formulations of *Pseudomonas* and *Bacillus spp.* were proved to be highly challenging against various *Phytophthora, Fusarium* and *Pythium spp.* in some vegetables and flowering plants (Sangeetha et al. 2012). In other hand, the Gram-positive rod shaped bacteria, *Bacillus sp.*is highly efficient for its antagonistic activities against several fungal pathogens as their endospores are inherently tolerant to desiccation and heat. Besides their defensive actions (such as parasitism, antibiosis or induced systemic resistance), some species of these Rhizobacteria are highly capable of promoting plant growth by releasing specific volatiles (Ryu et al. 2003) and even acquisition of photosynthetic energy by *B. subtilis* (GB03) in *Arabidopsis thaliana* (Zhang et al. 2008).

13.6.1.2 Exploitation of Rhizobacteria Against Other Biotic Stresses

The prime hormone related to biotic and abiotic stress is ethylene, mainly derived from the precursor of 1-amino cyclopropane-carboxylic acid (ACC) and that also act as a plant growth inhibitor. But another key enzyme, ACC-deaminase (1-amino cyclopropane-1-carboxylate deaminase) is reported for its blocking mechanism on ACC in stressed plant and this noble enzyme is interestingly higher in several PGPR strains of fluorescent Pseudomonads (Wang et al. 2000; Mayak et al. 2004). In addition to the disease suppression by narrowing down the deleterious enzyme activity, some rhizobacteria known as endo-mycorrhizae/vesicular arbuscular mycorrhizal fungi (VAM) are also suppressing the pathogen attack by strengthening structural defense mechanism. VAM fungi basically develop a complex network of fungal arbuscules or vesicles (hyphae) and thus prevent pathogenic entry through root in the host plant. Beside this, VAM fungi also restricts the pathogen growth by different ways including by enhancement of nutrient supply to plant and boosting host immunity, development of root lignification and sometimes altering the chemical composition of defense responsive enzymes, namely chitinase and isoflavonoids (Garcia-Garrido and Ocampo 1989).

13.6.1.3 Induction of Systemic Resistance

It has been reported that several strains of root colonizing microbes are potentially acting as an elicitor to induce hypersensitive response for host plant defense (Haas and Defago 2005). The development of this type of resistance in host plant due to the interaction between plant roots and some rhizobacteria is known as induced systemic resistance (ISR). It is very promising that a number of chemical elicitors are synthesized in PGPR strains after immediate induction triggered by pathogens and they are mostly salicylic acid, siderophores, cyclic lipopeptides, lipopolysaccharides, 2,4-diacetyl phloroglucinol (Phl) and other volatile substances (Ongena et al. 2004).

13.6.2 Microbial Exploitation for Sustainable Agriculture and Plant Growth Enhancement

As a part of natural ecosystem, microbial diversity is solely depending on the internal network between microbial growth and development with several biotic and abiotic factors influencing the microbiota. In other way, microbial diversity has a great impact on the quality and quantity of various plant derived metabolites. For instance, the synthesis principle of different flavon precursors in cocoa beans and tea leaves are influenced by development of microflora in and out of the reaction centre. In similar way, some rhizobacteria (PGPR) are tremendously helpful in stable supply of major nutrients to plant roots by means of asymbiotic, symbiotic or associative fixing mechanisms. Some of the important areas of implications of microbial diversity in the development of agriculture and plant metabolism can be drawn for literature exploration.

13.6.2.1 Development of Biofertilizer or Nitrogen Fixer

A spectrum of plant growth promoting rhizobacteria are responsible for fixing atmospheric and terrestrial nutrient component by various mechanisms. Some notable microbial members are Cyanobacteria, Actinomycetes, free-living diazotrophic bacteria (*Azospirillum and Azotobacter*) and phosphate solubilizing bacteria (*Microbacterium, Pseudomonas*) (Peoples et al. 1995). Microbial exploitation for enhancement of biological nitrogen fixing (BNF) bacteria (*Rhizobium and Bradyrhizobium*) from nodules of legume crops is currently of great concern. As the abundance of these microbial communities in the rhizosphere not only increase the availability of nitrogen and phosphorus by expanding the root surface area but also suppress the harmful effects of incoming plant pathogenic microbes, these beneficial functions, majority of rhizobacteria are also reported to synthesize auxin series phytohormones and indole acetic acid inside their cell, which is considered to have an implicit role in promoting plant growth (Wani et al. 2007).

13.6.2.2 Phytohormone Synthesis

A number of scientific literatures reported that several phytohormones are (Auxin, GA, Cytokinin, Ethylene) produced in higher rate in plant when their root rhizosphere is associated with PGPRs. For instance, some bacteria like *Comamonas acidovorans, Alcaligenes piechaudii* and *Agrobacterium sp.* are associated with increased production of Indole-3-acetic acid (IAA) in lettuce (Salisbury 1994), where *Aeromonas veronii* and *Enterobacter cloacae* are related to enhance IAA production in *Oryza sativa* (Mehnaz et al. 2001). On other hand, the synthesis of natural cytokinin is induced when *P. fluorescens* and *Rhizobium leguminosarum* are associated with the growth factor in soybean (de Salamone et al. 1996) and in rape or lettuce (Noel et al. 1996). As per as the increased production of gibberellin is concerned, *Penicillium corylophilum, Penicillium corylophilum, Penicillium funiculosum, Penicillium cyclopium, Aspergillus flavus and Aspergillus niger* are mostly reported PGPRs found in plant rhizosphere (Hasan 2002). Besides the magical production of phytohormones by these rhizobacteria, some of them are capable of synthesizing ACC-deaminase, a component that actively breaks down the precursor of ethylene biosynthetic pathway, called ACC. The most discussed rhizobacteria of such category are *Variovorax paradoxus, Pseudomonas sp., Enterobacter cloacae, Bacillus pumilus.* The secretion level of Ethylene, the main villain of root lengthening and overall plant growth could be reduced by naturally culturing these PGPR groups (Saleh and Glick 2001). As per as economic and ecological point of view, the beneficial application of a major groups of Rhizobacteria and PGPRs can be considered to replace the excessive use of agrochemicals.

13.6.3 Bioremediation

The demand of pure drinking water is increasing day by day worldwide. But the mode of ground water contamination with the help of spreading pollution in soil, namely lands filter, mining and agrochemicals making the drinking water hardly reachable in an adequate quantity to a majority of world population. But the activity of microflora available in soil is quite promising in degrading harmful pollutants and decreasing the impact of poor sanitation, For instance, exploration of *Enterobacter cloacae* by any effective technique of bioremediation could drastically reduce the contamination of selenium in agricultural and industrial waste water (Schlanger 2017).

Bioremediation by using microbial communities deals with biological cleaning by recycling the waste materials into non-toxic or useful forms, with the help of metabolic activities of microbes rather than other potent organisms. On the other way, microbial bioremediation is mostly involved with single or multiple microbial events like eradication, detoxification, degradation, immobilization of different chemicals or biological wastes or physical hazardous materials which are continuously generated or deposited in environment by the activity of human or other organisms.

13.6.3.1 Microbial Bioremediation: Overview and Types

In this context, we will discuss about some most promising approaches or strategies, that are coming into the light of scientific perception and some of the notable approaches are bioaugmentation, bio-stimulation, bio-attenuation, biopiling and bioventing. In a brief, the potential approaches of microbial bioremediation need to be addressed before entering into the area of possibilities of bioremediation in relation to soil microbes.

As per as bio-stimulation is concerned, a specific concentrated nutrient medium or a combination of media is supplied to the site of soil or ground water area where beneficial microorganisms are in action and it is performed to stimulate their growth and development which escalate their bioremediation activity. In addition to this medium, several other factors like oxygen, temperature, pH and a little amount of pollutant suitable for optimum microbial activity are provided to the microbiome to accelerate their enzyme or biomass production (Adams et al. 2015). Bio-attenuation is the natural eradication process of pollutant concentration of surroundings by means of biological (biodegradation), physical (dispersion, dilution) or chemical (ion exchange) forces. Some of the tiny bugs or microbes have the ability to uptake harmful chemicals or pollutants from soil or ground water and metabolize them into less toxic or harmless gasses or solvents. Another term, bioaugmentation is also popular in relation to bioremediation that is one form of biodegradation. Basically it is the addition of a natural population of microbes to the natural microbial population exist in soil or ground water contaminated with organic or inorganic pollutants and thereby enhance their degradation capacity to reduce pollution rate (Malik and Ahmed 2012). This strategy is much more convenient and realistic way of constructing bioengineering microbes in order to completely destroy the complex pollutant. The genetically engineered microorganisms can be customized through DNA manipulation and are highly efficient than microorganisms of natural species to breakdown certain compounds or pollutants with a faster rate.

13.6.3.2 Bioremediation of Heavy Metals

The heavy metals directed contamination in soil and waterbodies is getting special attraction because persistent accumulation and impact on human body are quite dangerous than the efficacy of other available contaminants in ecosystem. Among them, cadmium (Cd), arsenic (As), nickel (Ni), cobalt (Co), caesium (Cs), strontium (Sr), chromium (Cr), mercury (Hg), lead (Pb), silver (Ag), zinc (Zn), copper (Cu), vanadium (V), tin (Sn) and thallium (Tl) are considered as the most potent toxic heavy metals dangerous to ecosystem (Wang and Chen 2006). The deleterious impact of heavy metals is quite high because its ionic form have a high electrostatic affinity with cellular components and causing severe damages in the form of destabilization in the molecular structure of vital biomolecules (RNA, DNA and enzymes), which results mutagenesis, cancer or genetic disorders at extreme stage (Abatenh et al. 2017). Several studies reported that a number of microorganisms have an inherent capacity to reduce or neutralize the efficiency of most of the potentially toxic heavy metals by activating certain phenomena such as bioaccumulation, bioaugmentation, biosorption, biodegradation, combating resistance to heavy metals, etc. (Table 13.1). Therefore in conjugation of phytoremediation, microbial efficiency to block the negative effect of heavy metals could be enhanced by adopting various feasible genetic engineering approaches in future scientific studies.

13.6.3.3 Other Scopes and Possibilities: Synthetic Biology Approach

Besides several genetic engineering approaches, the recent advancement of synthetic biology to reprograming microorganisms in order to divert their genetic potential towards the remediation of pollutants is considered as a steady strategy at global level. To adopt such strategies, it is very important to know the key signals generated and circulated through a complex network established by several genes in microbial metabolism. To precisely and accurately control the candidate genes or gene cluster during their switching on and off moment, understanding the entire map of regulatory network involved with a number of genes, enzymes and proteins in microbial biosystem is extremely necessary. To customize such synthetic regulatory networks, major challenges appeared; of them designing of several well characterize regulatory parts sensitive to genetic and environmental factors, constructing of circuits require large DNA fragments with many cellular regions and finally debugging of circuit to evaluate the cellular response when exposed to abiotic stress under subject are of great importance. By building and reprogramming such internal gene circuits, it is quite possible to deploy a specific bacterial species or community (e.g. *Pseudomonas*) naturally associated with plants having phytoremediation potential (e.g. legumes) and accelerate the bioremediation process in much faster rate from waste contaminated areas. This type of bio-programming approach could be proven as a most indispensable leading bioremediation technique in the areas of plastic degradation, reduction of use of industrial chemicals, biofuel production, heavy metal detoxification and even minimizing the effect of agrochemicals used in agriculture (McAdams and Shapiro 1995).

13.7 Conclusion

The existence of soil microbial diversity is the most powerful and secret weapon to explore its potentiality for combating continuously elevating and most threating global climate change. As most of the techniques related to bioremediation driven by soil microbial communities are not so old, but it is very essential to imply this technology up to a maximum level by flourishing our knowledge on microbial reactions and possible modifications to restore adverse climate change. The increasing rate of climate change in terms of precipitation, irregular happening and shifting of seasons, global warming, raising of sea water level are directly or indirectly influencing the frequency of disease development and plant health. The severe effect of changing climate can damage our soil to extreme level and that can further depress

Table 13.1 Major soil microorganisms used in bioremediation

Organisms	S. no.	Genus/species	Application in bioremediation
Bacteria	1.	*Bacillus subtilis* strain NAP1, NAP2, NAP4 (Abatenh et al. 2017)	Used to degrade contaminants which are taken as food by the bacteria in oil-based paints to negligible level in environment. Optimum levels of nutrient and other chemical substances are supplied for fast activation of microbial enzymes
		Bacillus spp. ETL-2012, *B. pumilus*HKG212 (Abatenh et al. 2017)	Used for textile dye, Sulphonated di-azo dye and RNB dye degradation
		Bacillus cereus, B. Coagulans, B. Brevis (Abatenh et al. 2017)	Diesel, crude oil degradation from ocean and terrestrial environment
	2.	*Pseudomonas aeruginosa, P. fluorescens, P. putida, P. cepacia* (Abatenh et al. 2017)	Monocyclic aromatic hydrocarbons (e.g. benzene, xylene) in diesel, crude oil and oil-based product; textile dye, heavy metal (Fe^{2+}, Cu^{2+}, Pb^{2+}, Mn^{2+}) degradation
	3.	*Staphylococcus aureus, Klebsiella oxytoca* (Abatenh et al. 2017)	Contaminants in vat dyes and textile effluents degradation
	4.	*Enterobacter cloacae* (Banerjee et al. 2015)	Heavy metal (Pb^{2+}, Cd^{2+}, Ni^{2+}) in soil and waste water and agricultural waste degradation
	5.	*Citrobacter sp.* (Abatenh et al. 2017)	Diesel, crude oil degradation
	6.	*Cupriavidus metallidurans* (Perpetuo et al. 2011)	Heavy metal mainly cadmium degradation
	7.	*Acinetobacter sp., Arthrobacter sp., Photobacterium sp., Flavobacterium sp., Achromobacter sp.* (Abatenh et al. 2017; Perpetuo et al. 2011)	Heavy metals and pesticides (Chlorpyrifos and methyl parathion), polycyclic aromatic hydrocarbons (e.g.-toluene) degradation
Fungi	8.	*Penicillium chrysogenum, P. funiculosum, Penicillium ochrochloron* (Abatenh et al. 2017)	MAHs and PAHs, industrial dyes degradation
	9.	*Aspergillus niger, A. Fumigates* (Abatenh et al. 2017; Perpetuo et al. 2011)	Hydrocarbon in industrial waste degradation
	10.	*Cunninghamella elegans* (Abatenh et al. 2017)	Heavy metals degradation
	11.	*Fusarium sp.* (Abatenh et al. 2017)	Oil degradation
Yeast	12.	*Saccharomyces cerevisiae* (Abatenh et al. 2017; Perpetuo et al. 2011)	Heavy metals degradation (Lead, Mercury and Nickel)
	13.	*Candida sp.* (Abatenh et al. 2017; Perpetuo et al. 2011)	Hydrocarbons in industrial waste

(continued)

Table 13.1 (continued)

Organisms	S. no.	Genus/species	Application in bioremediation
Algae	14.	*Chlorella vulgaris* (Zeraatkar et al. 2016)	Uranium, nickel degradation
	15.	*Chlamydomonas reinhardtii* (Zeraatkar et al. 2016; Chekroun et al. 2014)	Cadmium, chromium; herbicide (Prometryne, Fluroxypyr) degradation
	16.	*Scenedesmus sp.* (Zeraatkar et al. 2016; Chekroun et al. 2014)	Chromium, organic pollutants in oil degradation
	17.	*Cladophora sp.* (Zeraatkar et al. 2016)	Cu^{2+} degradation

our ability to grow food and our good health, wildlife ecosystem and our planet's sustainability. Taking into consideration of above serious matter, microbial bioremediation tailoring with the use of modern genetic engineering and synthetic biology tools adopted with soil microbiota could only can tackle environmental pollution and replenish the green planet with clean air, water and soil. So, the better understanding on the microbial diversity and its role in plant disease suppression could be the most promising weapon to defeat the adverse effect of global climate change in near future. Molecular microbiology has disseminated its implication to uncover the qualitative and quantitative accounts of soil microbiota. Therefore, disseminating the highest level of applications of soil microbiota by studying its qualitative and quantitative accounts at molecular level would be taken with highest priority for restoring the native state of global climatic change.

Author Contribution NM carried out the literature survey on interaction between soil microorganisms and climate change, their impact on environment and wrote that section of the manuscript. KB wrote the section explaining microbial mitigation on climate change and upgraded the manuscript with his valuable suggestions and immense support.

Acknowledgement The authors acknowledge Department of Biotechnology, Visva-Bharati for providing an excellent environment during manuscript preparation.

References

Abatenh E, Gizaw B, Tsegaye Z et al (2017) Application of microorganisms in bioremediation review. J Environ Microbiol 1(1):02–09

Adams GO, Fufeyin PT, Okoro SE et al (2015) Bioremediation, biostimulation and bioaugmention: a review. Int J Environ Bioremed Biodegrad 3(1):28–39

Atkin OK et al (2015) Global variability in leaf respiration in relation to climate, plant functional types and leaf traits. New Phytol 206:614–636

Banerjee G, Pandey S, Ray AK et al (2015) Bioremediation of heavy metals by a novel bacterial strain enterobacter cloacae and its antioxidant enzyme activity, flocculant production, and protein expression in presence of lead, cadmium, and nickel. Water Air Soil Pollut 226:91

Bardgett RD, Bowman WD, Kaufmann R et al (2005) A temporal approach to linking aboveground and belowground ecology. Trends Ecol Evol 20:634–641
Bardgett RD, Freeman C, Ostle NJ (2008) Microbial contributions to climate change through carbon cycle feedbacks. ISME J 2(8):805–814
Bergot M, Cloppet E, Pérarnaud V et al (2004) Simulation of potential range expansion of oak disease caused by *Phytophthora cinnamomi* under climate change. Glob Chang Biol 10 (9):1539–1552
Bradford MA (2013) Thermal adaptation of decomposer communities in warming soils. Front Microbiol 4:333
Bradford MA, Jones TH, Bardgett RD et al (2002) Impacts of soil faunal community composition on model grassland ecosystems. Science 298(5593):615–618
Bradford MA, Davies CA, Frey SD et al (2008) Thermal adaptation of soil microbial respiration to elevated temperature. Ecol Lett 11:1316–1327
Cavicchioli R, Ripple WJ, Timmis KN et al (2019) Scientists' warning to humanity: microorganisms and climate change. Nat Rev Microbiol 17(9):569–586
Chekroun KB, Sánchez E, Baghour M (2014) The role of algae in bioremediation of organic pollutants Int. Res J Public Environ Health 1(2):19–32
Crowther TW, Thomas SM, Maynard DS et al (2015) Biotic interactions mediate soil microbial feedbacks to climate change. PNAS 112(22):7033–7038
Curtis PS, Wang X (1998) A meta-analysis of elevated CO2 effects on woody plant mass, form, and physiology. Oecologia 113:299–313
de Salamone IEG, Dobereiner J, Urquiaga S et al (1996) Biological nitrogen fixation in Azospirillum strain-maize genotype associations as evaluated by the 15N isotope dilution technique. Biol Fertil Soils 23:249–256
Dunstan WA, Rudman T, Shearer BL (2010) Containment and spot eradication of a highly destructive, invasive plant pathogen (*Phytophthora cinnamomi*) in natural ecosystems. Biol Invasions 12(4):913–924
Dutta H, Dutta A et al (2016) The microbial aspect of climate change. Energ Ecol Environ 1 (4):209–232
Frey SD, Lee J, Mellilo JM et al (2013) The temperature response of soil microbial efficiency and its feedback to climate. Nat Clim Change Lett 3:395–398
Garcia-Garrido JM, Ocampo JA (1989) Effect of VA mycorrhizal infection of tomato on damage caused by *Pseudomonas syringae*. Soil Biol Biochem 21:165–167
Gougoulias C, Clark JM, Shaw LJ (2014) The role of soil microbes in the global carbon cycle: tracking the below-ground microbial processing of plant-derived carbon for manipulating carbon dynamics in agricultural systems. J Sci Food Agric 94:2362–2371
Haas D, Defago G (2005) Biological control of soil-borne pathogens by fluorescent pseudomonads. Nat Rev Microbiol 3:307–319
Hagerty SB, van Groenigen KJ, Allison SD et al (2014) Accelerated microbial turnover but constant growth efficiency with warming in soil. Nat Clim Chang 4:903–906
Hasan HAH (2002) Gibberellin and auxin-indole production by plant root-fungi and their biosynthesis under salinity-calcium interaction. Rostlinna Vyroba 48(3):101–106
IPCC (2007) Climate change 2007: the physical science basis
Knoblauch C, Beer C, Liebner S et al (2018) Methane production as key to the greenhouse gas budget of thawing permafrost. Nat Clim Change 8:309–312
Landa BB, Mavrodi DM, Thomoshow LS et al (2003) Interactions between strains of 2,4-diacetylphloroglucinol-producing Pseudomonas fluorescens in the rhizosphere of wheat. Phytopathology 93:982–994
Li G, Kim S, Han SH et al (2018) Precipitation affects soil microbial and extracellular enzymatic responses to warming. Soil Biol Biochem 120:212–221
Malik ZA, Ahmed S (2012) Degradation of petroleum hydrocarbons by oil field isolated bacterial consortium. Afr J Biotechnol 11(3):650–658

Mayak S, Tirosh T, Glick BR (2004) Plant growth promoting bacteria that confer resistance in tomato and pepper to salt stress. Plant Physiol Biochem 167:650–656
McAdams HH, Shapiro L (1995) Circuit simulation of genetic networks. Science 269 (5224):650–656
Mehnaz S, Mirza MS, Haurat J et al (2001) Isolation and 16S rRNA sequence analysis of the beneficial bacteria from the rhizosphere of rice. Can J Microbiol 472:110–117
Nisbet EG et al (2019) Very strong atmospheric methane growth in the four years 2014–2017: implications for the Paris agreement. Global Biogeochem Cycles 33:318–342
Noel TC, Sheng C, Yost CK et al (1996) *Rhizobium leguminosarum* as a plant growth promoting rhizobacterium: direct growth promotion of canola and lettuce. Can J Microbiol 42:279–283
Ongena M, Duby F, Rossignol F et al (2004) Stimulation of the lipoxygenase pathway is associated with systemic resistance induced in bean by a nonpathogenic *Pseudomonas* strain. Mol Plant-Microbe Interact 17:1009–1018
Pariaud B, Ravigné V, Halkett F et al (2009) Aggressiveness and its role in the adaptation of plant pathogens. Plant Pathol 58(3):409–424
Pathak H, Aggarwal PK, Singh SD (2012) Climate change impact, adaptation and mitigation in agriculture: methodology for assessment and applications. IARI, New Delhi
Peoples MB, Herridge DF, Ladha JK (1995) Biological nitrogen fixation: an efficient source of nitrogen for sustainable agricultural production. Plant Soil 174:3–28
Perpetuo EA, Souza CB, Nascimento CAO (2011) Engineering Bacteria for bioremediation. In: Carpi A (ed) Progress in molecular and environmental bioengineering - from analysis and modeling to technology applications. IntechOpen, London
Posch T, Köster O, Salcher MM et al (2012) Harmful filamentous cyanobacteria favoured by reduced water turnover with lake warming. Nat Clim Change Lett 2(11):809–813
Ryu CM, Farag MA, Hu CH et al (2003) Bacterial volatiles promote growth of Arabidopsis. Proc Natl Acad Sci U S A 100:4927–4932
Saleh SS, Glick BR (2001) Involvement of gacS and rpoS in enhancement of the plant growth promoting capabilities of *Enterobacter cloacae* CAL2 and UW4. Can J Microbiol 47:698–705
Salisbury FB (1994) The role of plant hormones. In: Wilkinson RE (ed) Plant-environment interactions. Marcel Dekke, Nova Iorque, pp 39–81
Sangeetha G, Sundaramoorthy S, Kurucheve V (2012) Exploitation of soil microbial diversity in suppressing soil borne plant pathogens and enhancing crop productivity. In: Sinha A et al (eds) Modern trends in microbial biodiversity of natural ecosystem. Biotech Books publisher, New Delhi, pp 23–43
Schlanger Z (2017) Exposure to the bacteria in soil can be good for mental health, and could treat depression and prevent PTSD — Quartz. Soil has a microbiome, and it may double as an antidepressant. Quartz, New York, pp 1–9
Singh BK, Bardgett RD, Smith P et al (2010) Microorganisms and climate change: terrestrial feedbacks and mitigation options. Nat Rev Microbiol 8(11):779–790
Steffen W, Hughes L, Pearce A (2015) Climate change 2015: growing risks, critical choices. https://www.climatecouncil.org.au/climate-change-2015-growing-risks-critical-choices
Subba Rao NS, Gaur DY (2000) Microbial diversity-management and exploitation for sustainable agriculture. In: Yadav JSP et al (eds) Extended summaries: international conference on managing natural resources for sustainable agricultural production in the 2st century. IARI, New Delhi, pp 48–58
Van der Heijden MG, Bardgett RD, van Straalen NM (2008) The unseen majority: soil microbes as drivers of plant diversity and productivity in terrestrial ecosystems. Ecol Lett 11(3):296–310
Wallenstein MD, Hall EK (2012) A trait-based framework for predicting when and where microbial adaptation to climate change will affect ecosystem functioning. Biogeochemistry 109:35–47
Wang J, Chen C (2006) Biosorption of heavy metal by *Saccharomyces cerevisiae*. Biotechnol Adv 24(427):451
Wang C, Knill E, Glick BR et al (2000) Effect of transferring 1-aminocyclopropane-1-carboxylic acid (ACC) deaminase genes into Pseudomonas fluorescens strain CHA0 and its gacA

derivative CHA96 on their growth promoting and disease-suppressive capacities. Can J Microbiol 46:898–907

Wani PA, Khan MS, Zaidi A (2007) Effect of metal tolerant plant growth promoting *Bradyrhizobium sp.* (Vigna) on growth, symbiosis, seed yield and metal uptake by greengram plants. Chemosphere 70:36–45

Weiman S (2015) Microbes help to drive global carbon cycling and climate change. Microbe Mag 10(6):233–238

Wolfe BE, Klironomos JN (2005) Breaking new ground: soil communities and exotic plant invasion. Bioscience 55(6):477–487

Zak DR, Pregitzer KS, Curtis PS et al (1993) Elevated atmospheric CO2 and feedback between carbon and nitrogen cycles. Plant Soil 151:105–111

Zeraatkar AK, Ahmadzadeh H, Talebi AF et al (2016) Potential use of algae for heavy metal bioremediation, a critical review. J Environ Manag 181:817–831

Zhang H, Xie X, Kim MS et al (2008) Soil bacteria augment Arabidopsis photosynthesis by decreasing glucose sensingand abscisic acid levels in plants. Plant J 56:264–273

Zhang Y, You Q, Chen C et al (2016) Impacts of climate change on streamflows under RCP scenarios: a case study in Xin River Basin, China. Atmos Res 178–179:521–534

Zimmerman BL, Labonte B (2015) Climate change and the microbial methane banquet. Clim Alert 27(1):1–6

Role of Microorganisms in Plant Adaptation Towards Climate Change for Sustainable Agriculture

14

Mohammad Tarique Zeyad, Mahendra Vikram Singh Rajawat, Murugan Kumar, Abdul Malik, Mohammad Anas, Waquar Akhter Ansari, Bansh Narayan Singh, Devendra Singh, and Anil Kumar Saxena

Abstract

Plant beneficial microorganisms (PBMs) have a tendency to colonize soil and various parts of plant (especially in root) for augmenting the nutrients in soil as well as secretion of other biomolecules. In either ways PBMs help to improve plant productivity and immunity for increasing tolerance capability or adaptation towards diverse climatic conditions. Although earlier reports have exhibited significant contribution of PBMs for the development of induced systemic resistance against abiotic stresses like low/high temperature, salinity, moisture deficit and acidic conditions. PBMs play crucial role in nutrients transportation from soil to plant that leads to the reduction in application of chemical fertilizers and low accumulation of toxic elements in agricultural lands. Reducing or minimizing the use of chemical fertilizers may also help to decrease the occurrence of contamination by fertilizers and maintains ion balance in soil, as a result soil health will be improved. In addition, exopolysaccharide secretion and biofilm formation by PBMs alter the physico-chemical properties of rhizospheric soil that also impact higher plant response to abiotic stress, such as moisture deficit, metal toxicity,

Mohammad Tarique Zeyad and Mahendra Vikram Singh Rajawat contributed equally with all other contributors.

M. T. Zeyad
National Bureau of Agriculturally Important Microorganisms, Kushmaur Mau, Uttar Pradesh, India

Department of Agricultural Microbiology, Faculty of Agricultural Sciences, Aligarh Muslim University, Aligarh, Uttar Pradesh, India

M. V. S. Rajawat · M. Kumar (✉) · W. A. Ansari · B. N. Singh · D. Singh · A. K. Saxena
National Bureau of Agriculturally Important Microorganisms, Kushmaur Mau, Uttar Pradesh, India

A. Malik · M. Anas
Department of Agricultural Microbiology, Faculty of Agricultural Sciences, Aligarh Muslim University, Aligarh, Uttar Pradesh, India

S. A. Lone, A. Malik (eds.), *Microbiomes and the Global Climate Change*,
https://doi.org/10.1007/978-981-33-4508-9_14

chilling injury, saline, and low/high temperature. PBMs mediated adaptation or tolerance in plants towards different climatic conditions might be accompanied through different mechanisms like induction of cold/heat shock proteins and osmoprotectants. Application of these microorganisms might be effective for alleviation of climate change in various crops, thus developing an emerging approach towards sustainable agriculture. These microorganisms might also be utilized as model to decipher stress tolerance mitigation and responding processes that can be established for crop plants to cope with the stress caused by climate change.

Keywords

PBMs · Climate change · Biofilm formation · Exopolysaccharide production · Heat shock proteins

14.1 Introduction

Agriculture is one of the worst affected areas due to changing climatic scenario. In many developing countries, agriculture plays a key role in the economic development and also provides self-employment (Emami et al. 2018; Damianos et al. 2018). According to an estimate, it was found that developing countries such as India, Pakistan, Bangladesh, Iran, Afghanistan, Sri Lanka, Nepal and Bhutan are losing crops, equivalent to at least US$10 billion annually due to land degradation (Amellal and Bartoli 1998).

In tropical regions, different types of biotic and abiotic stresses affect agriculture. The enhancement in abiotic and biotic stresses plays a significant role in stagnation or decline in crop productivity. There are many reports that suggest reduction in the yields of paddy and wheat crops due to increasing water stress, salt stress and enhanced air temperature in many regions of south Asia. Several researchers have reported that Himalayan mountain glaciers lost their mass over the last few decades and are set to experience a rise in temperature by up to 2.5 °C by 2050 (Bolch et al. 2012; Gardner et al. 2013; Zhao et al. 2016). The necessity of irrigation is estimated to rise by 10% in arid and semi-arid regions with every 1 °C increase in temperature. In addition to high temperature, many others crucial natural disaster like floods, heat waves, droughts, cold waves, extreme rainfall, cyclones and elevated CO_2 are responsible for serious economic losses that are to be observed following global warming. All of these factors are responsible for causing serious negative impact on agricultural practices and water resources.

In India drought is main abiotic stress that affects about two-third farming area of arid and semi-arid ecosystem, soil alkalinity/salinity, low pH, high temperature, and toxicity of metals are the other factors. Evaporation and transpiration are very high while rainfall is sporadic in the arid condition, limiting the productive agriculture in those areas (Qureshi et al. 2008). It is a key challenge for the development of cost effective, resourceful and easily adjustable methods for the management of abiotic stress. Comprehensive research is under way to develop new approaches to abiotic stress worldwide.

There is a great deal of research in development of new approaches to address abiotic stresses through developing heat and drought-tolerant species, shifting the crop calendar, resource management practices, etc. (Venkateswarlu and Singh 2015; Mickelbart et al. 2015). Many recent research exhibit that microorganisms can also play beneficial role in the management of crops with abiotic stress. The beneficial role of microbes such as disease control, management of nutrients and growth promotion is well studied. In rhizosphere or endo-rhizosphere, the advantageous microorganisms may colonize and enhance plant growth through numerous different direct and indirect mechanisms (Verma et al. 2016; Singh et al. 2019; Ansari et al. 2020; Meena et al. 2017). Microorganisms are becoming increasingly important in handling biotic and abiotic stresses.

This chapter describes the role of microbes in amelioration of different abiotic stresses like drought, heat, salt, water-logging, etc. caused under changing climatic scenario.

14.2 Adaptation to Stress by Microbes

Different kinds of stress factors cause significant effect on the growth and metabolism of microbes. The selection and application of microbes in stress environment therefore needs intensive research and technology development. Consequently, research focus should be on the occurrence and functional diversity of microbes which are agriculturally important in different types of extreme environmental conditions. Occurrence of *Rhizobium*, *Bradyrhizobium*, *Azospirillum*, *Azotobacter*, *Bacillus*, *Paenibacillus, Pseudomonas, Burkholderia*, *Gallionella, Microbacterium, Achromobacter,* etc. has been reported from different extreme environmental conditions such as acid soils, alkaline soils, saline soils, contaminated soils, desert ecosystem and highly eroded hill slopes of India (Grover et al. 2011). Several genes are responsible in the adaptation of microbes to stress conditions that are complex and multifaceted regulatory processes. In certain microorganisms, which are existing under extreme conditions such as hypersaline soils, and thermal springs, the optimum cellular metabolic process occurs at very high temperature and salinity, respectively, (Li et al. 2017) whereas other microbes adopt various mechanisms to combat the environmental stress. Under osmotic stress condition, most of microorganisms secrete osmoprotectants (K^+, glutamate, glycine, proline, proline betaine, ectoine, trehalose, etc.) in order to modulate their osmolarity in cytoplasm (Talibart et al. 1994; Jawahar et al. 2019). Bacteria contain large amount of highly anionic polysaccharides (membrane derived oligosaccharides) in the periplasmic space region which are too large for the diffusion of proteins through porins and perhaps help in maintaining turgor pressure of periplasm of those microorganisms. Vriezen et al. (Vriezen et al. 2006) explored many physiological factors in

Rhizobium for stability during heat stress and revealed the survival of *Sinorhizobium meliloti* USDA 1021 in high desiccation environmental conditions, when bacterial cells were dried in stationary stage by increasing the drying temperature and availability of phosphate and chloride.

Production of exopolysaccharides (EPS) under stress conditions by bacteria is also one of the key ways to mitigate moisture stress and variations in water potential. EPS production by *Pseudomonas* sp. under moisture stress resulted in increased retention of water and regulation of the diffusion in microbial environment for carbon sources (Sandhya et al. 2009). EPS possess an exclusive water retention and cementing characteristics that play a crucial role in the development and maintenance of soil aggregates. Additionally, EPS also possess regulation of nutrients as well as flow of water across the roots of plant via biofilm formation (Noirot-Gros et al. 2018).

The challenge is triggered due to high temperature in the tropical and subtropical regions for crop production and microbial colonization. Due to high temperature, aggregation and denaturation of cellular proteins occur. Microorganism synthesizes heat shock proteins (HSPs) which are a specific group of polypeptides produced in response to rapid increase in temperature. The heat shock proteins consist of molecular chaperons (*Dna*K, *Dna*J, *Gro*ES, *Gro*EL, *Clp*B, *Clp*A, *Clp*X), small heat shock proteins (sHSP), and proteases. Molecular chaperons facilitate the correct folding of proteins and proteases are essential for the degradation of irreversibly damaged proteins (Mogk et al. 2019). Induction of HSPs occurs when thermotolerant bacteria are exposed to high temperature (Ali et al. 2009). These chaperones play important role in the survival of cell during heat shock for consequent recovery. Correspondingly, bacteria respond in cold temperature by induction of cryoprotective proteins (Govrin et al. 2019).

Microorganisms could play a key role in stress management once their distinct characteristics of tolerance to extreme limits have been understood, their prevalence and genetic diversity have been understood, and various methods for studying their distribution in agricultural production have been developed. These microbes also provide impressive systems for recognizing the strategies of stress tolerance that can be consequently engineered into crops.

14.3 Alleviation of Abiotic Stress in Plants by Rhizospheric Microbes

Rhizospheric and endophytic microorganisms enhance the plant growth and can also confer some degree of tolerance to crop plants in various types of abiotic stresses like high temperature, drought, salinity, metal toxicity, chemical toxicity, and chilling. Different bacterial genera such as *Burkholderia, Enterobacter, Methylobacterium, Achromobacter*, *Paenibacillus*, *Pseudomonas, Azospirillum*, *Microbacterium*, *Rhizobium*, *Variovorax*, *Pantoea, Bacillus,* etc. have been reported to contribute towards growth of the host plants under different abiotic stress conditions (Table 14.1). Application of these microorganisms could be beneficial in agriculture

Table 14.1 Microorganisms conferring abiotic stress tolerance to crop plants

Organisms	Crop	Type of stress	Mechanism	References
Pantoea agglomerans	*Triticum aestivum*	Drought	Rhizosphere soil aggregation through EPS	(Cherif-Silini et al. 2019)
Pseudomonas sp.	*Eleusine coracana*	Drought	Enhance ACC deaminase activity due to presence of *acdS* gene was amplified by PCR	(Chandra et al. 2018)
Rhizobium sp.	*Medicago sativa*	Drought	Over production of IAA production	(Defez et al. 2017)
Bacillus thuringiensi	*Triticum*	Drought	Production of volatile organic compound	(Timmusk et al. 2014)
Rhizobium sp.	*Helianthus*	Drought	Soil aggregation through EPS	(Naseem et al. 2018)
Azospirillum lipoferum	*Triticum*	Drought	Improved growth, grains yield, relative water content	(Agami et al. 2017)
Streptomyces laurenti and *Penicillium* sp.	*Sorghum bicolor*	Drought	Solubilization of phosphorus, Zinc and other PGP activities	(Kour et al. 2020)
AM Fungi	*Sorghum*	Drought	Improved uptake of nutrient	(Symanczik et al. 2018)
Paraphaeosphaeria quadriseptata	*Arabidopsis*	Drought	Induction of HSP	(McLellan et al. 2007)
Pseudomonas sp.	*Pisum sativum*	Drought	Decreased ethylene production	(Arshad et al. 2008)
Pseudomonas mendocina and *Glomus intraradices*	*Lactuca sativa*	Drought	Improved antioxidant status	(Kohler et al. 2008)
Pseudomonas putida P45	*Helianthus*	Drought	Improved soil aggregation due to EPS production	(Sandhya et al. 2009)
Bacillus megaterium and *Glomus* sp.	*Trifolium*	Drought	IAA and proline production	(Marulanda et al. 2007)
Bacillus subtilis	*Arabidopsis thaliana* and *Brassica campestris*	Drought and salinity	Attributes PGP activities	(Woo et al. 2020)

(continued)

Table 14.1 (continued)

Organisms	Crop	Type of stress	Mechanism	References
P. polymyxa and *Rhizobium tropici*	*Phaseolus vulgaris*	Drought	Change in hormone balance and stomatal conductance	(Figueiredo et al. 2008)
Trichoderma parareesei	*Brassica napus*	Salinity and drought	Expression of genes related to abscisic acid (ABA) under drought stress, and ethylene (ET) under salt stress	(Poveda 2020)
Bacillus megaterium, Arthrobacter	*Solanum lycopersicum*	Salinity	Synthesis of indole acetic acid	(Fan et al. 2016)
B. Amylolequifaciens, B. insolitus Microbacterium sp., *P. syringae*	*Triticum*	Salinity	Restricted Na^+ influx	(Ashraf et al. 2004)
Klebsiella sp.	*Triticum*	Salinity	Increase plant biomass and chlorophyll content	(Singh et al. 2015)
Pseudomonas fluorescens	*Arachis hypogaea*	Salinity	Synthesis of ACC deaminase	(Saravanakumar and Samiyappan 2007)
Burkholderia phytofirmans PsJN	*Vitis vinifera*	Low temperature	Synthesis of ACC deaminase	(Ait Barka et al. 2006)
P. Putida	*Brassica napus*	Low temperature	Synthesis of ACC deaminase	(Chang et al. 2007)
Pseudomonas sp. AMK-P6	*Sorghum bicolor*	Heat	Induction of heat shock proteins and improved plant biochemical status	(Ali et al. 2009)
Pantoea agglomerans		Metal	Rhizosphere soil aggregation through EPS	(Mohite et al. 2018)
PGPR	*Cicer arietinum*	Metal	Sequestration of metal ions	(Abbaszadeh-Dahaji et al. 2016)
Alcaligenes faecalis, Bacillus cereus	*Sorghum bicolor*	Ni & cd	Enhance the anti-oxidative enzymes	(El-Meihy et al. 2019)
Bacillus cereus, *Providencia rettgeri,* and *Myroides odoratimimus*	*Sorghum bicolor*	Chromium and heat	Enhance plant growth, antioxidant status (SOD, CAT, APX) and decreased proline and malondialdehyde contents	(Bruno et al. 2020)

Paenibacillus polymyxa	*Zea mays*	pH	Phosphate solubilization	(Mohd Din et al. 2020)
Pseudomonas putida, Enterobacter cloacae, P. putida	*Solanum lycopersicum*	Flooding	Synthesis of ACC deaminase	(Grichko and Glick 2001)
AM fungi & *Bradyrhizobium*	*Pterocarpus Officinalis*	Flooding	Development of adv. Roots, aerenchyma and hypertrophied lenticels	(Fougnies et al. 2008)
Rhizospheric microbes	*Medicago sativa, Festuca arundinacea, Salix miyabeana*	Persistent organic pollutants (POPs)	PGP activities and degradation of POP	(NJB et al. 2020)

thus opening an innovative and emerging approach to alleviate abiotic stresses. The presence of the bacteria in the plant increased the acclimation to cooling temperatures resulting in lower cell damage, higher photosynthetic activity, and accumulation of cold-stress-related metabolites such as starch, proline, and phenolic compounds. (Ait Barka et al. 2006; Miliute et al. 2015).

Different mechanisms involved in tolerance of plants stress through microbes have been proposed (Fig. 14.1). Secretion of indole acetic acid, gibberellins, and many other factors through plant growth promoting (PGP) rhizobacteria resulted in the improvement in root length, number of root tips, and surface area of root, leading to improved nutrient uptake thus significant improvement of plant under stress conditions (Egamberdieva and Kucharova 2009). Some PGP strains produce antioxidant enzymes and cytokinin that help in the degradation of reactive oxygen species (ROS) and accumulation of abscisic acid (ABA). Activities of antioxidant compounds are corelated to tolerance in response to oxidative stress (Anjum et al. 2017). Inoculation of bacterium *Azospirillum brasilense* Sp245 in wheat (*Triticum aestivum*) under drought condition resulted in increase in water status, yields of grains, and quality of minerals (Ca, K, and Mg) at the time of harvesting. Bakhshandeh et al. (Bakhshandeh et al. 2020) observed enhancement of seed germination, seedling growth, and uptake of potassium in soybean (*Glycine max* (L.) Merr.) by four PGP microorganisms (PGPM) (*Bacillus megaterium*, *B. cereus*, *Trichoderma longibrachiatum,* and *T. simmonsii*) under salinity stress. The use of these PGPMs can be recommended for sustainable agriculture either because of their ability to promote seed germination or by increased seedling growth in optimal and stressful conditions.

The Responsive to Dehydration (RD) and Early Responsive to Dehydration (ERD) are those genes which express in response to drought stress. The protein encoded by these genes has broad structural and functional variability with a specific class of protein acting as connector of stress response mechanisms. The functional characterization of ERD genes (ERD 1 to 16) has been established, they display varied and heterogenous biochemical functions and are present in various sub-cellular compartments. ERD1 codes for chloroplast ATP dependent protease, ERD2 and ERD4 encode for cytosolic HSP70 and membrane proteins, respectively. However, the functions of ERD3 and ERD7 are not known, mitochondrial proline dehydrogenase and carbohydrate transporter are encoded by ERD5 and ERD6, respectively, ERD8 generates HSP81 and ERD16 codes for ubiquitin extension protein 1, UBQ1. ERD9, ERD11, and ERD13 belong to glutathione-S-transferase family, and ERD10 and ERD14 are the members of the protein late glutathione-S-transferase family. At last, ERD15 is a hydrophilic protein that possesses a PAM2 domain and interferes with polyA-binding proteins (PABP).

PGPR strains produce 1-Aminocyclopropane-1-carboxylate (ACC) deaminase that confer the Induced Systemic Tolerance (IST) against drought and salt stresses in plants (Mayak et al. 2004). Many characteristics of the plant life cycle are controlled by ethylene levels and ethylene biosynthesis is subject to tight regulation, transcriptional and post-transcriptional environmental control factors, including abiotic and biotic stress (Shrivastava and Kumar 2015). S-adenosylmethionine is

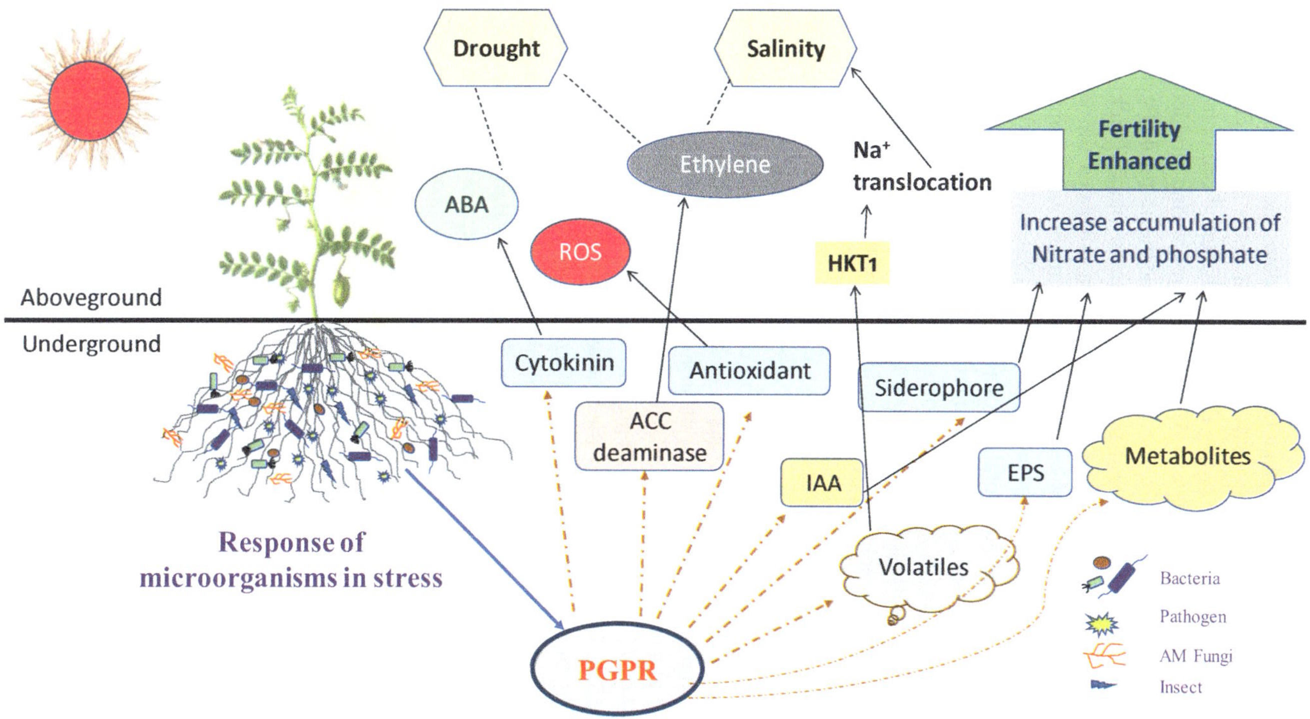

Fig. 14.1 Schematic illustration of the different plant–microbe interactions under abiotic stress

converted into 1-aminocyclopropane-1-carboxylate (ACC) during the biosynthetic process of ethylene. ACC is an instant precursor of 1-aminocyclopropane-1-carboxylate synthase (ACS). Ethylene, plant hormone regulates the process of homeostasis and causes the reduction of root and shoot growth under stress conditions. Plant ACC is separated and decomposed by the action of bacterial cell to provide nitrogen and energy source in the presence of ACC deaminase enzyme produced by bacteria. Moreover, by the removal of ACC, bacteria overcome the harmful effect of ethylene, ameliorate stress of plant, and enhance the growth of plant (Glick et al. 2007). Inoculation of ACC deaminase producing bacteria enhances the length of roots which might be beneficial in absorbing relatively more water content from deep soil under drought stress environmental conditions (Salem et al. 2018).

Physico-chemical and structural characteristic of soil change in the rhizosphere region due to complex and diverse interactions among soil, roots, water, and microorganism (Jia et al. 2018). Exopolysaccharides secreted by microorganisms can adhere with the particles of soil to produce micro- and macro-aggregates. Fungal hyphae and plant roots are fit in the pores in between micro-aggregates and thereby stabilizing macro-aggregates.

Bacteria producing exopolysaccharides inoculated into plants reveal enhanced resistance to stress of water due to improved structure of soil (Kumar and Verma 2018). Exopolysaccharide reduces the saline condition in the soil by binding with the cations including Na^+ and thus decreased its availability to plants. Chen et al. (Chen et al. 2007) studied the accumulation of proline in relation with the drought and salt tolerance in plants. The genes *pro*BA derived from *Bacillus subtilis* inserted into *Arabidopsis thaliana* displayed high production of free proline as a result increased the tolerance to osmotic stress in the transgenic crops. Enhanced production of proline along with reduced electrolyte leakage, selective uptake of potassium (K^+) ion, and relative content of water resulted in the tolerance of salt in *Zea mays* co-inoculated with the *Pseudomonas* and *Rhizobium* (Bano and Fatima 2009). Proline helps plant to protect membranes and proteins against the adverse effect of temperature and high concentration of inorganic ions. It also acts as hydroxyl radical scavenger and a protein-compatible hydrotope (Iqbal et al. 2016). Proline accumulates buffer cellular redox potential under environmental conditions (Priya et al. 2019).

Rhizobium metabolism is critical for signaling of plant growth, yield, adaptation to abiotic stress and its management had a significant agronomical impact on the leguminous crops (Schillaci et al. 2019). Figueiredo et al. (Figueiredo et al. 2008) described the growth of plant, content of nitrogen, and nodule formation of *Glycine max* under drought stress because of co-inoculation of rhizobia *Rhizobium tropici* and *P. polymyxa*. *Phaseolus vulgaris* (common bean) crops treated with *Rhizobium etli* over-expressing trehalose-6-phosphate synthase gene had a greater number of nodules with enhanced nitrogenase activity and high biomass in comparison to plants treated with wild type *R. etli*. The plants treated with wild type *R. etli* plants died after 3 weeks while as transgenic plants treated with *R. etli* over-expressing trehalose-6-phosphate synthase gene completely recovered after drought stress. A large number (7200) of expressed sequence tags (ESTS) were analyzed by DNA microarray from the plant nodules treated with bacterial strain over-expressing trehalose-6-phosphate synthase gene display upregulation of genes associated in stress tolerance, indicating a signaling process for trehalose. Many volatile organic compounds (VOCs) released from *Bacillus* (Asari et al. 2016) are bacterial determinants associated in induced systemic tolerance (IST). The volatiles produced by PGPR down-regulate the expression of hkt1 (High Affinity K^+ Transporter 1) in root tissues while in shoots up-regulates, orchestrating lower Na^+ amounts and under salt stress, it causes recirculation of Na^+ in entire plant (Zhang et al. 2008). In *A. thaliana,* colonization of root by *P. chlororaphis* O6 induces tolerance in the plants against biotic and abiotic stresses by generating volatile metabolites, 2R, 3R-butanediol. Salicylic acid (SA), ethylene, and jasmonic acid signaling pathways are required with *Arabidopsis* mutant lines that stimulated tolerance to drought.

The rate of photosynthesis, plant water relations, flowering, and fruits is affected due to high temperature in temperate and tropical crops. Similarly, in many plant species cold temperature is significant factor limiting the efficiency and geographical distribution that includes various agricultural crops. *P. putida* NBRI0987 isolated from drought affected chickpea rhizosphere was responsible for excess production of stress sigma (S) (*rpo*S) while cultivated under high temperature stress at 40 °C in comparison to 30 °C (Srivastava et al. 2008). In the sorghum seedlings, *Pseudomonas* sp. Strain AMK-P6 increases tolerance to high temperature due to the production of proteins with high molecular weight in leaves that improve production of biomass and biochemical status with regard to amino acid, chlorophyll contents, proline, and sugar (Ali et al. 2009). *Burkholderia phytofirmans* PsJN PGP bacterium is capable of epiphytic and endophytic colonization of grapevine tissues. Plantlets with bacterial colonization exhibited significantly greater amounts of phenolic compounds, proline, and starch. PGP characteristics like HCN production, indole acetic acid production, phosphorous solubilization, and production of siderophore at 15 °C and 4 °C was reported in low temperature tolerant *Serratia marcescens* strain SRM and *Pantoea dispersa* strain 1A (Grover et al. 2011). Plant biomass and uptake of nutrients in wheat seedlings in wheat plants significantly increased due to seed inoculation with bacterial strains in cold stress conditions (Selvakumar et al. 2008; Majeed et al. 2018). Cold shock proteins (CSP) have molecular weight approximately 7–10 kDa in bacteria and exhibit binding activity with nucleic acid, sufficient to their role as RNA chaperones. The chaperone appearances of CSP might be significant for inducing growth in stress conditions and throughout ages of high level of metabolic activity. A number of abiotic stresses like cold, heat and water scarcity affecting yields under field conditions in transgenic *Arabidopsis*, maize and rice were shown to increase tolerance due to expression of bacterial CSPs (*Csp* A and *Csp* B) (Castiglioni et al. 2008; Yu et al. 2017). These studies exhibited that the protein family has the capability to carry broad stress tolerance which also increases the yield of grains under both natural and managed environmental conditions.

Rhizobacteria, colonizing sites, showing common stress condition are probably more adaptive and may employ better plant growth promoters under adverse environmental situations (Sattiraju et al. 2019). *Paenibacillus polymyxa* is reported to display more efficient growth in relatively severe and deprived conditions (Timmusk et al. 2014). Two drought and salinity tolerant species of genus *Trichoderma*, (*Trichoderma parareesei* and *T. harzianum*) were able to enhance the hormonal pathways of abscisic acid under drought and ethylene under salt stress (Poveda 2020). Isolation and characterization of native microbes from stress affected soils (i.e. screening based on stress tolerance as well as PGP characters) might be helpful in quick selection of effective bacterial strains that could be applied to develop bioinoculants for improved yield crops in stressful environment.

14.4 Symbiotic Fungi for Reduction of Abiotic Stress

Symbiotic association of arbuscular mycorrhizal (AM) with plants in abiotic stress alleviation is often utilized for improving drought stress and plant resistance to deficiency of water by changing physiology and genes expression in plants (Ruiz-Lozano and Azcón 2000; Bencherif et al. 2019). Many reports showed that AM-induced improvement in drought tolerance, including both increased dehydration and desiccation tolerance (Blomstedt et al. 2018). Inoculation of AM fungi decreases the malondialdehyde content of leaf as well as soluble proteins and increases catalase (CAT), peroxidase (POD), superoxide dismutase (SOD) activities, ensuing enhanced osmotic and drought stress of citrus plants grafted with mycorrhiza (Wu and Xia 2006; Liu et al. 2019). *Glomus versiforme* inoculated in citrus plants, enhanced adjustment of osmotic stability in plant under condition of drought stress by increased level of Ma^{2+}, K^{+}, Ca^{2+}, and non-structural carbohydrates resulting in increase of drought tolerance (Wu and Xia 2006).

The function of abscisic acid (ABA) in AM-mediated plant stress response has been significantly examined (Aroca et al. 2008). In non-AM plant shoots, content of ABA increased considerably on addition of exogenous ABA and associated with stress marker genes. On the other hand, inclusion of exogenous ABA reduced the ABA content in shoots of AM crops and did not show additional increase of expression. *Pseudomonas mendocina* (PGPR) and *Glomus intraradices* or *G. mosseae* inoculated in lettuce increased antioxidant catalase in drought stress environment and showing that it should be employed as inoculants to alleviative oxidative damage (Kohler et al. 2008). A gene (Gi14–3-3) was identified from *Glomus intraradices* that encodes a protein in in-vitro condition and responsible for drought stress (Porcel et al. 2006). These proteins are important to control both the signaling mechanism and effector proteins that were reported in the defense of host plants from drought stress. Glutathione and ascorbate played a crucial role in the mitigation and physiological functions in the plants under low water availability. Such compounds are accumulated in lavender plants at very low level and colonized with indigenous drought-tolerant fungal strains (*Glomus intraradices* and *Glomus* sp.) revealed tolerance in plants against drought stress (Marulanda et al. 2007).

The association of mycorrhiza with lavender plants resulted in improved levels of root biomass, water, N and K. The enhancement in salt tolerance by AM has been commonly related with AM-induced improvements in the acquisition of phosphorus and growth of plant. The AM symbiosis has improved plant resistance to salinity stress, even at higher quality than the drought stress. Involving AM *Glomus* sp., which showed increased phosphate and Na^+ levels in shoots in comparison with non-inoculated controls, was confirmed by increased growth in saline soil conditions (Giri and Mukerji 2004). Similarly, in a parallel finding, AM fungi substantially declined the level and concentration of antioxidant defense enzymes such as catalase (CAT), superoxide dismutase (SOD), glutathione reductase (GR), ascorbate peroxidase (APX) as well as non-enzymatic group of antioxidants accumulated and extracted in pyto-organs of medicinal plant *Sesbania sesban* raised in soils treated with varying levels of salinity (Abd Allah et al. 2015). In sorghum, colonization of *Gigaspora margarita* promotes stomatal conductivity when exposed to drought stress in saline soil and also improves survival under double stress. Acetaldehyde accumulation, a harmful intermediate in ethanol production, in roots has been reported as a major factor contributing to flood damage in susceptible species (Osundina 1997). Colonization of AM by *Glomus intraradices* greatly contributed in flood resistance of *Pterocarpus officinalis* seedlings by increasing plant growth and uptake of phosphorus in leaf. Flooding affects nodules both on submerged portion of stem as well as on adventitious roots (Fougnies et al. 2007). Mycorrhizal *Casuarina equisetifolia* plants could be better adapted to floods compared to uninoculated plants. This may be due to a higher diffusion of oxygen and ethanol removal from increased development of random roots, hypertrophies, lenticels, and aerenchymatous tissue at the root zone and the submerged portion of the stem (Rutto et al. 2002; Toro et al. 2018). *Trifolium* plants with AM fungi displayed improved tolerance against flooding, as evidenced by osmotic adaptation and proline improvement in the tissues of plant (Neto et al. 2006). Therefore, it has been reported that arbuscular mycorrhizal fungi (AMF), endophytic symbiotic dark septate fungi (DSF), which grows in plants under stress conditions like those of alpine and dry grassland habitats, enhanced its resistance to heat and drought and promote nutrient uptake (Porras-Alfaro et al. 2008). An endophytic fungus, *Piriformospora indica* has been shown to tolerate drought in *A. thaliana* by expressing a range of stress-related genes in leaf of plant. This led to greater up-regulation of the levels of phospholipase D, calcineurin B-like proteins (CBL 1) and histone acetyltransferase (HAT) in *P. indica* (Sherameti et al. 2008). *Paraphaeosphaeria quadriseptata* fungus isolated from rhizosphere increased tolerance in plants towards heat stress in *A. thaliana* by inducing HSP101 and HSP70, the preserved systems of the response to stress (McLellan et al. 2007).

14.5 Dual Symbiotic Systems for Alleviating of Abiotic Stress

There has also been reports of significant function of the virus in attaining tolerance to plants towards abiotic stress conditions. *Cuvularia* sp. belongs to endophytic fungi isolated from *Dichanthelium lanuginosum* and cultivated in geothermal soil. They could develop a thermal tolerance (a constant soil temperature of 50 °C for 3 days and an intermittent soil temperature of 65 °C for 10 days (Redman et al. 2002)). The fungus as well as plant could not be able to survive at temperatures above 38 °C. The potential of *Curvularia protuberata* to possess heat tolerance on the host plant was associated with the involvement of virus known as *Curvularia thermotolerance* virus (CatahTV). The wild fungal isolated from the fungus infected with the virus showed tolerance to intermittent soil temperatures of up to 65 °C for 2 weeks (10 h heat per day), while inoculation of plant without virus infected and non-symbiotic plant became chlorotic. The efficacy of *C. protuberata* isolated from a monocot giving the properties of heat tolerance to tomatoes (a dicot) was examined and the same outcomes as those produced by *D. lanuginosum* were examined (Marquez et al. 2007). Thermal sensitivity could be provided by various possible pathways. The fungal endophytes release melanin in the cell wall of host pant, which can release heat along with hyphae and/or composite of oxygen radicals produced during stress of heat (Davidson et al. 1996). Additionally, endophyte can also act as a "biological trigger" that allows symbiotic plants to stimulate stress response faster and more vigorously than non-symbiotic plants (Redman et al. 1999).

14.6 Conclusion

Microorganisms associated with plants can play an essential role in the development of resistance to abiotic stresses. The occurrence of these microbes including endophytic and rhizospheric bacteria as well as symbiotic fungi regulated the stresses through the process of different mechanisms such as increasing osmotic response and triggering of new genes in the plants. Genetic engineering and plant breeding are important methods for the development of stress tolerant crops but it is a time-consuming process. On the other hand, the utilization of microbes to improve stress in plants could be more suitable, cost effective, and ecofriendly choice that could be easily available in a short time period. Future research work is required in this emerging field to evaluate and utilize the potential organisms in field conditions.

References

Abbaszadeh-Dahaji P, Omidvari M, Ghorbanpour M (2016) Increasing phytoremediation efficiency of heavy metal-contaminated soil using PGPR for sustainable agriculture. In: Plant-microbe interaction: an approach to sustainable agriculture. Springer Singapore, Singapore, pp 187–204. http://link.springer.com/10.1007/978-981-10-2854-0_9

Abd Allah EF, Hashem A, Alqarawi AA, Bahkali AH, Alwhibi MS (2015) Enhancing growth performance and systemic acquired resistance of medicinal plant Sesbania sesban (L.) Merr using arbuscular mycorrhizal fungi under salt stress. Saudi J Biol Sci 22(3):274–283. https://linkinghub.elsevier.com/retrieve/pii/S1319562X15000625

Agami RA, Ghramh HA, Hashem M (2017) Seed inoculation with Azospirillum lipoferum alleviates the adverse effects of drought stress on wheat plants. J Appl Bot Food Qual 90:165–173

Ait Barka E, Nowak J, Clement C (2006) Enhancement of chilling resistance of inoculated grapevine plantlets with a plant growth-promoting rhizobacterium, burkholderia phytofirmans strain PsJN. Appl Environ Microbiol 72(11):7246–7252. http://aem.asm.org/cgi/doi/10.1128/AEM.01047-06

Ali SZ, Sandhya V, Grover M, Kishore N, Rao LV, Venkateswarlu B (2009) Pseudomonas sp. strain AKM-P6 enhances tolerance of sorghum seedlings to elevated temperatures. Biol Fertil Soils 46(1):45–55. http://link.springer.com/10.1007/s00374-009-0404-9

Amellal B, Bartoli H (1998) Colonization of wheat roots by an exopolysaccharide-producing pantoea agglomerans strain and its effect on rhizosphere soil aggregation. Appl Environ Microbiol 64(10):3740–3747. http://www.ncbi.nlm.nih.gov/pubmed/9758793

Anjum SA, Ashraf U, Tanveer M, Khan I, Hussain S, Shahzad B et al (2017) Drought induced changes in growth, osmolyte accumulation and antioxidant metabolism of three maize hybrids. Front Plant Sci 08:69. http://journal.frontiersin.org/article/10.3389/fpls.2017.00069/full

Ansari WA, Krishna R, Zeyad MT, Singh S, Yadav A. (2020) Endophytic actinomycetes-mediated modulation of defense and systemic resistance confers host plant fitness under biotic stress conditions. In: Singh R, Manchanda G, Maurya I, Wei Y (eds) Microbial versatility in varied environments. Springer, Singapore. https://doi.org/10.1007/978-981-15-3028-9_10

Aroca R, Vernieri P, Ruiz-Lozano JM (2008) Mycorrhizal and non-mycorrhizal Lactuca sativa plants exhibit contrasting responses to exogenous ABA during drought stress and recovery. J Exp Bot 59(8):2029–2041. https://academic.oup.com/jxb/article-lookup/doi/10.1093/jxb/ern057

Arshad M, Shaharoona B, Mahmood T (2008) Inoculation with Pseudomonas spp. containing ACC-deaminase partially eliminates the effects of drought stress on growth, yield, and ripening of pea (Pisum sativum L.). Pedosphere 18(5):611–620. https://linkinghub.elsevier.com/retrieve/pii/S1002016008600557

Asari S, Matzén S, Petersen MA, Bejai S, Meijer J (2016) Multiple effects of Bacillus amyloliquefaciens volatile compounds: plant growth promotion and growth inhibition of phytopathogens. FEMS Microbiol Ecol 92(6):fiw070. https://academic.oup.com/femsec/article-lookup/doi/10.1093/femsec/fiw070

Ashraf M, Hasnain S, Berge O, Mahmood T (2004) Inoculating wheat seedlings with exopolysaccharide-producing bacteria restricts sodium uptake and stimulates plant growth under salt stress. Biol Fertil Soils 40(3):157–162. http://link.springer.com/10.1007/s00374-004-0766-y

Bakhshandeh E, Gholamhosseini M, Yaghoubian Y, Pirdashti H (2020) Plant growth promoting microorganisms can improve germination, seedling growth and potassium uptake of soybean under drought and salt stress. Plant Growth Regul 90(1):123–136. http://link.springer.com/10.1007/s10725-019-00556-5

Bano A, Fatima M (2009) Salt tolerance in Zea mays (L). following inoculation with rhizobium and pseudomonas. Biol Fertil Soils 45(4):405–413. http://link.springer.com/10.1007/s00374-008-0344-9

Bencherif K, Dalpé Y, Lounès Hadj-Sahraoui A (2019) Arbuscular mycorrhizal fungi alleviate soil salinity stress in arid and semiarid areas. In: Giri B, Varma A (eds) Microorganisms in saline environments: strategies and functions. Soil biology, vol 56. Springer, Cham, pp 375–400. http://link.springer.com/10.1007/978-3-030-18975-4_16

Blomstedt C, Griffiths C, Gaff D, Hamill J, Neale A (2018) Plant desiccation tolerance and its regulation in the foliage of resurrection "flowering-plant" species. Agronomy 8(8):146. http://www.mdpi.com/2073-4395/8/8/146

Bolch T, Kulkarni A, Kaab A, Huggel C, Paul F, Cogley JG et al (2012) The state and fate of Himalayan glaciers. Science 336(6079):310–314. http://www.sciencemag.org/cgi/doi/10.1126/science.1215828

Bruno LB, Karthik C, Ma Y, Kadirvelu K, Freitas H, Rajkumar M (2020) Amelioration of chromium and heat stresses in Sorghum bicolor by Cr6+ reducing-thermotolerant plant growth promoting bacteria. Chemosphere 244:125521. https://linkinghub.elsevier.com/retrieve/pii/S0045653519327614

Castiglioni P, Warner D, Bensen RJ, Anstrom DC, Harrison J, Stoecker M et al (2008) Bacterial RNA chaperones confer abiotic stress tolerance in plants and improved grain yield in maize under water-limited conditions. Plant Physiol 147(2):446–455. http://www.plantphysiol.org/lookup/doi/10.1104/pp.108.118828

Chandra D, Srivastava R, Glick BR, Sharma AK (2018) Drought-tolerant Pseudomonas spp. improve the growth performance of finger millet (Eleusine coracana (L.) Gaertn.) under non-stressed and drought-stressed conditions. Pedosphere 28(2):227–240. https://linkinghub.elsevier.com/retrieve/pii/S100201601860013X

Chang W-S, van de Mortel M, Nielsen L, Nino de Guzman G, Li X, Halverson LJ (2007) Alginate production by Pseudomonas putida creates a hydrated microenvironment and contributes to biofilm architecture and stress tolerance under water-limiting conditions. J Bacteriol 189 (22):8290–8299. http://jb.asm.org/cgi/doi/10.1128/JB.00727-07

Chen M, Wei H, Cao J, Liu R, Wang Y, Zheng C (2007) Expression of bacillus subtilis proBA genes and reduction of feedback inhibition of proline synthesis increases proline production and confers osmotolerance in transgenic arabidopsis. BMB Rep 40(3):396–403. http://koreascience.or.kr/journal/view.jsp?kj=E1MBB7&py=2007&vnc=v40n3&sp=396

Cherif-Silini H, Thissera B, Bouket AC, Saadaoui N, Silini A, Eshelli M et al (2019) Durum wheat stress tolerance induced by endophyte pantoea agglomerans with genes contributing to plant functions and secondary metabolite arsenal. Int J Mol Sci 20(16):3989. https://www.mdpi.com/1422-0067/20/16/3989

Damianos D, Dimara E, Hassapoyannes K, Skuras D (2018) Greek agriculture in a changing international environment. Routledge, London. https://www.taylorfrancis.com/books/9780429844560

Davidson JF, Whyte B, Bissinger PH, Schiestl RH (1996) Oxidative stress is involved in heat-induced cell death in Saccharomyces cerevisiae. Proc Natl Acad Sci 93(10):5116–5121. http://www.pnas.org/cgi/doi/10.1073/pnas.93.10.5116

Defez R, Andreozzi A, Dickinson M, Charlton A, Tadini L, Pesaresi P et al (2017) Improved drought stress response in alfalfa plants nodulated by an IAA over-producing rhizobium strain. Front Microbiol 8:2466. http://journal.frontiersin.org/article/10.3389/fmicb.2017.02466/full

Egamberdieva D, Kucharova Z (2009) Selection for root colonising bacteria stimulating wheat growth in saline soils. Biol Fertil Soils 45(6):563–571. http://link.springer.com/10.1007/s00374-009-0366-y

El-Meihy RM, Abou-Aly HE, Youssef AM, Tewfike TA, El-Alkshar EA (2019) Efficiency of heavy metals-tolerant plant growth promoting bacteria for alleviating heavy metals toxicity on sorghum. Environ Exp Bot 162:295–301. https://linkinghub.elsevier.com/retrieve/pii/S0098847219302497

Emami M, Almassi M, Bakhoda H, Kalantari I (2018) Agricultural mechanization, a key to food security in developing countries: strategy formulating for Iran. Agric Food Secur 7(1):24. https://agricultureandfoodsecurity.biomedcentral.com/articles/10.1186/s40066-018-0176-2

Fan P, Chen D, He Y, Zhou Q, Tian Y, Gao L (2016) Alleviating salt stress in tomato seedlings using Arthrobacter and Bacillus megaterium isolated from the rhizosphere of wild plants grown on saline–alkaline lands. Int J Phytoremediation 18(11):1113–1121. http://www.tandfonline.com/doi/full/10.1080/15226514.2016.1183583

Figueiredo MVB, Burity HA, Martínez CR, Chanway CP (2008) Alleviation of drought stress in the common bean (Phaseolus vulgaris L.) by co-inoculation with Paenibacillus polymyxa and Rhizobium tropici. Appl Soil Ecol 40(1):182–188. https://linkinghub.elsevier.com/retrieve/pii/S0929139308000802

Fougnies L, Renciot S, Muller F, Plenchette C, Prin Y, de Faria SM et al (2007) Arbuscular mycorrhizal colonization and nodulation improve flooding tolerance in Pterocarpus officinalis Jacq. seedlings. Mycorrhiza 17(3):159–166. http://link.springer.com/10.1007/s00572-006-0085-2

Fougnies L, Renciot S, Muller F, Plenchette C, Prin Y, de Faria SM et al (2008) Arbuscular mycorrhizal colonization and nodulation improve flooding tolerance in Pterocarpus officinalis jacq. Seedlings. In: Dakora FD, Chimphango SBM, Valentine AJ, Elmerich C, Newton WE (eds) Biological nitrogen fixation: towards poverty alleviation through sustainable agriculture.

Current plant science and biotechnology in agriculture, vol 42. Springer, Dordrecht, pp 133–134. http://link.springer.com/10.1007/978-1-4020-8252-8_49
Gardner AS, Moholdt G, Cogley JG, Wouters B, Arendt AA, Wahr J et al (2013) A reconciled estimate of glacier contributions to sea level rise: 2003 to 2009. Science 340(6134):852–857. http://www.sciencemag.org/cgi/doi/10.1126/science.1234532
Giri B, Mukerji KG (2004) Mycorrhizal inoculant alleviates salt stress in Sesbania aegyptiaca and Sesbania grandiflora under field conditions: evidence for reduced sodium and improved magnesium uptake. Mycorrhiza 14(5):307–312. http://link.springer.com/10.1007/s00572-003-0274-1
Glick BR, Todorovic B, Czarny J, Cheng Z, Duan J, McConkey B (2007) Promotion of plant growth by bacterial ACC deaminase. CRC Crit Rev Plant Sci 26(5–6):227–242. http://www.tandfonline.com/doi/abs/10.1080/07352680701572966
Govrin R, Obstbaum T, Sivan U (2019) Common source of cryoprotection and osmoprotection by osmolytes. J Am Chem Soc 141(34):13311–13314. https://pubs.acs.org/doi/10.1021/jacs.9b06727
Grichko VP, Glick BR (2001) Amelioration of flooding stress by ACC deaminase-containing plant growth-promoting bacteria. Plant Physiol Biochem 39(1):11–17. https://linkinghub.elsevier.com/retrieve/pii/S0981942800012122
Grover M, Ali SZ, Sandhya V, Rasul A, Venkateswarlu B (2011) Role of microorganisms in adaptation of agriculture crops to abiotic stresses. World J Microbiol Biotechnol 27 (5):1231–1240. http://link.springer.com/10.1007/s11274-010-0572-7
Iqbal MJ, Maqsood Y, Abdin ZU, Manzoor A, Hassan M, Jamil A (2016) SSR markers associated with proline in drought tolerant wheat germplasm. Appl Biochem Biotechnol 178 (5):1042–1052. http://link.springer.com/10.1007/s12010-015-1927-1
Jawahar G, Rajasheker G, Maheshwari P, Punita DL, Jalaja N, Kumari PH et al (2019) Osmolyte diversity, distribution, and their biosynthetic pathways. In: Plant signaling molecules. Elsevier, Amsterdam, pp 449–458. https://linkinghub.elsevier.com/retrieve/pii/B9780128164518000289
Jia T, Guo T, Cao M, Chai B (2018) Effects of heavy metals on phyllosphere and rhizosphere microbial community of Bothriochloa ischaemum. Appl Sci 8(9):1419. http://www.mdpi.com/2076-3417/8/9/1419
Kohler J, Hernández JA, Caravaca F, Roldán A (2008) Plant-growth-promoting rhizobacteria and arbuscular mycorrhizal fungi modify alleviation biochemical mechanisms in water-stressed plants. Funct Plant Biol 35(2):141. http://www.publish.csiro.au/?paper=FP07218
Kour D, Rana KL, Kaur T, Sheikh I, Yadav AN, Kumar V et al (2020) Microbe-mediated alleviation of drought stress and acquisition of phosphorus in great millet (Sorghum bicolour L.) by drought-adaptive and phosphorus-solubilizing microbes. Biocatal Agric Biotechnol 23:101501. https://linkinghub.elsevier.com/retrieve/pii/S1878818119318559
Kumar A, Verma JP (2018) Does plant—microbe interaction confer stress tolerance in plants: a review? Microbiol Res 207:41–52. https://linkinghub.elsevier.com/retrieve/pii/S0944501317307954
Li Y, Xu L, Gong H, Ding B, Dong M, Li Y (2017) A microbial exopolysaccharide produced by sphingomonas species for enhanced heavy oil recovery at high temperature and high salinity. Energy Fuel 31(4):3960–3969. https://pubs.acs.org/doi/10.1021/acs.energyfuels.6b02923
Liu Q, Ren T, Zhang Y, Li X, Cong R, White PJ et al (2019) Yield loss of oilseed rape (Brassica napus L.) under nitrogen deficiency is associated with under-regulation of plant population density. Eur J Agron 103:80–89. https://linkinghub.elsevier.com/retrieve/pii/S1161030118307275
Majeed A, Muhammad Z, Ahmad H (2018) Plant growth promoting bacteria: role in soil improvement, abiotic and biotic stress management of crops. Plant Cell Rep 37(12):1599–1609. http://link.springer.com/10.1007/s00299-018-2341-2
Marquez LM, Redman RS, Rodriguez RJ, Roossinck MJ (2007) A virus in a fungus in a plant: three-way symbiosis required for thermal tolerance. Science 315(5811):513–515. http://www.sciencemag.org/cgi/doi/10.1126/science.1136237

Marulanda A, Porcel R, Barea JM, Azcón R (2007) Drought tolerance and antioxidant activities in lavender plants colonized by native drought-tolerant or drought-sensitive glomus species. Microb Ecol 54(3):543–552. http://link.springer.com/10.1007/s00248-007-9237-y

Mayak S, Tirosh T, Glick BR (2004) Plant growth-promoting bacteria confer resistance in tomato plants to salt stress. Plant Physiol Biochem 42(6):565–572. https://linkinghub.elsevier.com/retrieve/pii/S0981942804000762

McLellan CA, Turbyville TJ, Wijeratne EMK, Kerschen A, Vierling E, Queitsch C et al (2007) A rhizosphere fungus enhances arabidopsis thermotolerance through production of an HSP90 inhibitor. Plant Physiol 145(1):174–182. http://www.plantphysiol.org/lookup/doi/10.1104/pp.107.101808

Meena VS, Meena SK, Verma JP, Kumar A, Aeron A, Mishra PK et al (2017) Plant beneficial rhizospheric microorganism (PBRM) strategies to improve nutrients use efficiency: a review. Ecol Eng 107:8–32. https://linkinghub.elsevier.com/retrieve/pii/S0925857417303865

Mickelbart MV, Hasegawa PM, Bailey-Serres J (2015) Genetic mechanisms of abiotic stress tolerance that translate to crop yield stability. Nat Rev Genet 16(4):237–251. http://www.nature.com/articles/nrg3901

Miliute I, Buzaite O, Baniulis D, Stanys V (2015) Bacterial endophytes in agricultural crops and their role in stress tolerance: a review. Zemdirbyste-Agriculture 102(4):465–478. http://www.zemdirbyste-agriculture.lt/1024_str60/

Mogk A, Ruger-Herreros C, Bukau B (2019) Cellular functions and mechanisms of action of small heat shock proteins. Annu Rev Microbiol 73(1):89–110. https://www.annualreviews.org/doi/10.1146/annurev-micro-020518-115515

Mohd Din ARJ, Rosli MA, Mohamad Azam Z, Othman NZ, Sarmidi MR (2020) Paenibacillus polymyxa role involved in phosphate solubilization and growth promotion of Zea mays under abiotic stress condition. Proc Natl Acad Sci India Sect B Biol Sci 90:63–71. http://link.springer.com/10.1007/s40011-019-01081-1

Mohite BV, Koli SH, Patil SV (2018) Heavy metal stress and its consequences on exopolysaccharide (EPS)-producing Pantoea agglomerans. Appl Biochem Biotechnol 186 (1):199–216. http://link.springer.com/10.1007/s12010-018-2727-1

Naseem H, Ahsan M, Shahid MA, Khan N (2018) Exopolysaccharides producing rhizobacteria and their role in plant growth and drought tolerance. J Basic Microbiol 58(12):1009–1022. http://doi.wiley.com/10.1002/jobm.201800309

Neto D, Carvalho LM, Cruz C, Martins-Loução MA (2006) How do mycorrhizas affect C and N relationships in flooded Aster tripolium plants? Plant Soil 279(1–2):51–63. http://link.springer.com/10.1007/s11104-005-6333-y

NJB B, Gonzalez E, Desjardins D, Labrecque M, Pitre FE (2020) Co-cropping with three phytoremediation crops influences rhizosphere microbiome community in contaminated soil. Sci Total Environ 711:135067. https://linkinghub.elsevier.com/retrieve/pii/S0048969719350594

Noirot-Gros M-F, Shinde S, Larsen PE, Zerbs S, Korajczyk PJ, Kemner KM et al (2018) Dynamics of aspen roots colonization by pseudomonads reveals strain-specific and mycorrhizal-specific patterns of biofilm formation. Front Microbiol 9:853. http://journal.frontiersin.org/article/10.3389/fmicb.2018.00853/full

Osundina MA (1997) Nodulation and growth of mycorrhizal Casuarina equisetifolia J.R. and G. First in response to flooding. Biol Fertil Soils 26(2):95–99. http://link.springer.com/10.1007/s003740050349

Porcel R, Aroca R, Cano C, Bago A, Ruiz-Lozano JM (2006) Identification of a gene from the arbuscular mycorrhizal fungus glomus intraradices encoding for a 14-3-3 protein that is up-regulated by drought stress during the AM symbiosis. Microb Ecol 52(3):575–582. http://link.springer.com/10.1007/s00248-006-9015-2

Porras-Alfaro A, Herrera J, Sinsabaugh RL, Odenbach KJ, Lowrey T, Natvig DO (2008) Novel root fungal consortium associated with a dominant desert grass. Appl Environ Microbiol 74 (9):2805–2813. http://aem.asm.org/cgi/doi/10.1128/AEM.02769-07

Poveda J (2020) Trichoderma parareesei favors the tolerance of rapeseed (Brassica napus L.) to salinity and drought due to a chorismate mutase. Agronomy 10(1):118. https://www.mdpi.com/2073-4395/10/1/118

Priya M, Sharma L, Singh I, Bains TS, Siddique KHM, Bindumadhava H et al (2019) Securing reproductive function in mungbean grown under high temperature environment with exogenous application of proline. Plant Physiol Biochem 140:136–150. https://linkinghub.elsevier.com/retrieve/pii/S0981942819301913

Qureshi AS, McCornick PG, Qadir M, Aslam Z (2008) Managing salinity and waterlogging in the Indus Basin of Pakistan. Agric Water Manag 95(1):1–10. https://linkinghub.elsevier.com/retrieve/pii/S0378377407002429

Redman RS, Freeman S, Clifton DR, Morrel J, Brown G, Rodriguez RJ (1999) Biochemical analysis of plant protection afforded by a nonpathogenic endophytic mutant of colletotrichum magna. Plant Physiol 119(2):795–804. http://www.plantphysiol.org/lookup/doi/10.1104/pp.119.2.795

Redman RS, Sheehan KB, Stout RG, Rodriguez RJ, Henson JM (2002) Thermotolerance generated by plant/fungal symbiosis. Science 298(5598):1581. http://www.ncbi.nlm.nih.gov/pubmed/12446900

Ruiz-Lozano JM, Azcón R (2000) Symbiotic efficiency and infectivity of an autochthonous arbuscular mycorrhizal Glomus sp. from saline soils and Glomus deserticola under salinity. Mycorrhiza 10(3):137–143. http://link.springer.com/10.1007/s005720000075

Rutto KL, Mizutani F, Kadoya K (2002) Effect of root-zone flooding on mycorrhizal and non-mycorrhizal peach (Prunus persica Batsch) seedlings. Sci Hortic (Amsterdam) 94 (3–4):285–295. https://linkinghub.elsevier.com/retrieve/pii/S0304423802000080

Salem G, Stromberger ME, Byrne PF, Manter DK, El-Feki W, Weir TL (2018) Genotype-specific response of winter wheat (Triticum aestivum L.) to irrigation and inoculation with ACC deaminase bacteria. Rhizosphere 8:1–7. https://linkinghub.elsevier.com/retrieve/pii/S2452219818300405

Sandhya V, SK. Z. A, Grover M, Reddy G, Venkateswarlu B (2009) Alleviation of drought stress effects in sunflower seedlings by the exopolysaccharides producing Pseudomonas putida strain GAP-P45. Biol Fertil Soils 46(1):17–26. http://link.springer.com/10.1007/s00374-009-0401-z

Saravanakumar D, Samiyappan R (2007) ACC deaminase from Pseudomonas fluorescens mediated saline resistance in groundnut (Arachis hypogea) plants. J Appl Microbiol 102(5):1283–1292. http://doi.wiley.com/10.1111/j.1365-2672.2006.03179.x

Sattiraju KS, Kotiyal S, Arora A, Maheshwari M (2019) Plant growth-promoting microbes: contribution to stress management in plant hosts. In: Environmental biotechnology: for sustainable future. Springer Singapore, Singapore, pp 199–236. http://link.springer.com/10.1007/978-981-10-7284-0_8

Schillaci M, Gupta S, Walker R, Roessner U (2019) The role of plant growth-promoting bacteria in the growth of cereals under abiotic stresses. In: Root biology - growth, physiology, and functions. IntechOpen, London. https://www.intechopen.com/books/root-biology-growth-physiology-and-functions/the-role-of-plant-growth-promoting-bacteria-in-the-growth-of-cereals-under-abiotic-stresses

Selvakumar G, Kundu S, Joshi P, Nazim S, Gupta AD, Mishra PK et al (2008) Characterization of a cold-tolerant plant growth-promoting bacterium Pantoea dispersa 1A isolated from a sub-alpine soil in the North Western Indian Himalayas. World J Microbiol Biotechnol 24(7):955–960. http://link.springer.com/10.1007/s11274-007-9558-5

Sherameti I, Tripathi S, Varma A, Oelmüller R (2008) The root-colonizing endophyte pirifomospora indica confers drought tolerance in arabidopsis by stimulating the expression of drought stress–related genes in leaves. Mol Plant-Microbe Interact 21(6):799–807. http://apsjournals.apsnet.org/doi/10.1094/MPMI-21-6-0799

Shrivastava P, Kumar R (2015) Soil salinity: a serious environmental issue and plant growth promoting bacteria as one of the tools for its alleviation. Saudi J Biol Sci 22(2):123–131. https://linkinghub.elsevier.com/retrieve/pii/S1319562X14001715

Singh RP, Jha P, Jha PN (2015) The plant-growth-promoting bacterium Klebsiella sp. SBP-8 confers induced systemic tolerance in wheat (Triticum aestivum) under salt stress. J Plant Physiol 184:57–67. https://linkinghub.elsevier.com/retrieve/pii/S0176161715001662

Singh BN et al (2019) Importance and utilization of plant-beneficial rhizobacteria in agriculture. In: Singh D., Prabha R. (eds) Microbial interventions in agriculture and environment. Springer, Singapore. https://doi.org/10.1007/978-981-32-9084-6_8

Srivastava S, Yadav A, Seem K, Mishra S, Chaudhary V, Nautiyal CS (2008) Effect of high temperature on pseudomonas putida NBRI0987 biofilm formation and expression of stress sigma factor RpoS. Curr Microbiol 56(5):453–457. http://link.springer.com/10.1007/s00284-008-9105-0

Symanczik S, Lehmann MF, Wiemken A, Boller T, Courty P-E (2018) Effects of two contrasted arbuscular mycorrhizal fungal isolates on nutrient uptake by Sorghum bicolor under drought. Mycorrhiza 28(8):779–785. http://link.springer.com/10.1007/s00572-018-0853-9

Talibart R, Jebbar M, Gouesbet G, Himdi-Kabbab S, Wróblewski H, Blanco C et al (1994) Osmoadaptation in rhizobia: ectoine-induced salt tolerance. J Bacteriol 176(17):5210–5217. http://jb.asm.org/lookup/doi/10.1128/jb.176.17.5210-5217.1994

Timmusk S, Abd El-Daim IA, Copolovici L, Tanilas T, Kännaste A, Behers L et al (2014) Drought-tolerance of wheat improved by rhizosphere bacteria from harsh environments: enhanced biomass production and reduced emissions of stress volatiles. PLoS One 9(5):e96086. https://dx.plos.org/10.1371/journal.pone.0096086

Toro G, Pinto M, Pimentel P (2018) Root respiratory components of Prunus spp. rootstocks under low oxygen: regulation of growth, maintenance, and ion uptake respiration. Sci Hortic (Amsterdam) 239:259–268. https://linkinghub.elsevier.com/retrieve/pii/S030442381830373X

Venkateswarlu B, Singh AK (2015) Climate change adaptation and mitigation strategies in rainfed agriculture. In: Climate change modelling, planning and policy for agriculture. Springer India, New Delhi, pp 1–11. http://link.springer.com/10.1007/978-81-322-2157-9_1

Verma P, Yadav AN, Khannam KS, Kumar S, Saxena AK, Suman A (2016) Molecular diversity and multifarious plant growth promoting attributes of Bacilli associated with wheat (Triticum aestivum L.) rhizosphere from six diverse agro-ecological zones of India. J Basic Microbiol 56(1):44–58. http://doi.wiley.com/10.1002/jobm.201500459

Vriezen JAC, Bruijn FJ, Nusslein K (2006) Desiccation responses and survival of Sinorhizobium meliloti USDA 1021 in relation to growth phase, temperature, chloride and sulfate availability. Lett Appl Microbiol 42(2):172–178. http://doi.wiley.com/10.1111/j.1472-765X.2005.01808.x

Woo O-G, Kim H, Kim J-S, Keum HL, Lee K-C, Sul WJ et al (2020) Bacillus subtilis strain GOT9 confers enhanced tolerance to drought and salt stresses in Arabidopsis thaliana and Brassica campestris. Plant Physiol Biochem 148:359–367. https://linkinghub.elsevier.com/retrieve/pii/S0981942820300401

Wu Q-S, Xia R-X (2006) Arbuscular mycorrhizal fungi influence growth, osmotic adjustment and photosynthesis of citrus under well-watered and water stress conditions. J Plant Physiol 163(4):417–425. https://linkinghub.elsevier.com/retrieve/pii/S0176161705001665

Yu T-F, Xu Z-S, Guo J-K, Wang Y-X, Abernathy B, Fu J-D et al (2017) Improved drought tolerance in wheat plants overexpressing a synthetic bacterial cold shock protein gene SeCspA. Sci Rep 7(1):44050. http://www.nature.com/articles/srep44050

Zhang H, Kim M-S, Sun Y, Dowd SE, Shi H, Paré PW (2008) Soil bacteria confer plant salt tolerance by tissue-specific regulation of the sodium transporter HKT1. Mol Plant-Microbe Interact 21(6):737–744. http://apsjournals.apsnet.org/doi/10.1094/MPMI-21-6-0737

Zhao L, Ding R, Moore JC (2016) The High Mountain Asia glacier contribution to sea-level rise from 2000 to 2050. Ann Glaciol 57(71):223–231. https://www.cambridge.org/core/product/identifier/S0260305500000239/type/journal_article

Novel Approaches for Genome Editing to Develop Climate Smart Crops

15

Reena Yadav, Radhakrishnan Thankappan, and Abhay Kumar

Abstract

We face the hypercritical trial of generating adequate food for the rising human populace living in an altering and unsteady climate. As climate change remains to perform a leading role in agriculture and food surety, there exists necessity for advancement in development of climate smart, robust crop plants providing high micronutrient and high yield constancy. Climate smart crops emphasize on resilient ones which are prime to food and revenue fortification under progressive climate alteration and unpredictability. Plant transformation has certified important insights into plant biology and improved industrial agriculture, but transformation and regeneration remain arduous for most crops. Genome editing provides innovative opportunities for improving crop production but counts on regeneration and genetic transformation as process bottlenecks. In this chapter we converse high-throughput genome editing through application of other biologically dynamic molecules into plant cells, DNA transfer and gene expression without integration, non-Agrobacterium measures to deliver DNA into plant cells, synthetic approaches along with "Modular" Agrobacterium strains. Synthetic biology must be implemented for de novo strategy of transformation methods, as targeted genome editing utilizes artificial nucleases with capacity to escalate plant improvement by delivering the mechanisms to customize genomes promptly in a precise and anticipated manner. The clustered regularly interspaced short palindromic repeats (CRISPR)/CRISPR-associated protein 9 (Cas9) system, a lately established means for the introduction of site-specific double stranded DNA breaks along with two well-established genome editing stages: zinc finger nucleases (ZFNs) and transcription activator-like effector nucleases (TALENs)

R. Yadav (✉) · R. Thankappan
ICAR-Directorate of Groundnut Research, Junagadh, Gujarat, India

A. Kumar
ICAR-National Research Centre on Litchi, Muzaffarpur, Bihar, India

S. A. Lone, A. Malik (eds.), *Microbiomes and the Global Climate Change*,
https://doi.org/10.1007/978-981-33-4508-9_15

are described with possible applications in development of climate smart crops. Genome editing is a potential game changer in crop enhancement when climate resilience is the need of the hour. This chapter shall summarize the ongoing trends and future prospects of genome editing as a promising technique in developing climate smart crops.

Keywords

Genome editing · ZFN · TALEN · CRISPR/Cas9 · Base editing · CRISPR/Cpf1 · Climate smart crops

15.1 Introduction

The Intergovernmental Panel on Climate Change reported that if the increment rate of global heating continues, global temperatures would begin to increase by 1.5 °C between 2030 and 2052 (Intergovernmental Panel on Climate Change 2018). Approximations by FAO suggest that agricultural yield has to surge by the year 2050 to an extent of 60% in order to content the likely food and feed demands by then. Agriculture sector should essentially hence renovate itself so as to provide for a mounting worldwide populace and offer the stability for economic progress and poverty declination. Climate change will create more challenges to this concern due to antagonistic influences on agriculture by altered weather circumstances and consequent biophysical fallouts (Van Meijl et al. 2018). Alterations in climate change comprise intensification of temperature, variations in rainfall, hence instabilities in water accessibility, sunlight duration and evaporation, rise in sea level and salinity, agitations in ecosystems which ultimately lead to enhanced frequency and strength of "extreme events" with extensive influences on agriculture and forestry (Tripathi et al. 2016). These aspects impact cereal production not merely over heat and water stress but also linked with frost, water logging, wind pace, diseases, and pest dynamics (Porter et al. 2014). The degree of these influences will not only subject to the strength and periodicity of the fluctuations but even depend on their merger, which are further indefinite and will be decided by local circumstances. In changing climate situation without adaptation, productivity of wheat, maize, and rice is expected to decline in tropical in addition to temperate areas (Challinor et al. 2014). Wheat production has already shown a decline in most areas and representative data show that production will decrease by 6% for every 1 °C increase in the overall mean temperature (Asseng et al. 2014; Porter et al. 2014). The United Nations Food and Agriculture Organization (FAO) approximates that about 900 million inhabitants on earth are facing unceasing food shortage. Food security is significantly impacted due to rapid growth in global population during this ongoing global climate variation.

At a global scale, there is an immediate requirement to contemplate the correlations between food safety, agriculture, and climate variation. There is increasing certainty that present agricultural procedures are not admissible, since they incline to dissipate valued resources and damage the environment. Hybridization and selection methods are mainly involved in crop improvement by traditional breeding. On the contrary, due to extensive breeding sequence of crops and the expected arbitrary outcomes of conventional breeding, these measures are not enough to accommodate necessity for crops with assortment of attributes to deal with mounting food scarcity we may deal within upcoming years. These factors have modified traditional plant science research viewing platform and demand-based plant breeding advancement to further emphasize strategies and procedures that enable plants to perform well on restraining capital. These considerations have changed the viewing platform of conventional plant science research and progression of demand-based plant breeding to additionally emphasis on strategies and procedures which let plants to be thriving well on restraining resources. Traditional breeding methods are mainly crop produce centered instead of endurance during miscellaneous stress (Barnabás et al. 2008), whereas the need of the hour is to recognize, make up, and establish germplasm which can endure undue weather conditions with produce consistency in "good" as well as "bad" years (Chapman et al. 2012; Fischer and Edmeades 2010; Keating et al. 2010). Therefore, genetic variability is being extensively exploited by researchers to generate significant yield of tolerant cultivars. Novel methods involving transgenics, genomics, and phenomics are occupying a progressively imperative part in enhancing cereal crop yields (Tester and Langridge 2010) to develop next-generation crop plants which use nutrients and water efficiently, and provide sustainable crops over a varied extent of environmental circumstances; they are well defined as climate smart crops.

In the recent years, distinguished technological advancements in plant sciences have revealed the potentials to meet the challenges ahead of food production and agriculture (Bailey-Serres et al. 2019). At present, a series of approaches are accessible for developing sustainable and resilient crop production countering climate change comprising high-throughput single nucleotide polymorphism (SNP) genotyping, next-generation sequencing, genomic selection, and trait mapping. These sophisticated procedures are crucial for a comprehensive understanding of the trait variations and the transformative innovation needed to improve the exertion of plant breeding. The "pyramid" scheme for incorporating appropriate alleles and gene combinations for the functional and molecular construction that governs crop yield traits confronting numerous environmental stresses (heat, drought, salinity, etc.). This proposes different gene combinations to be applied to advance crop production under optimum and strained circumstances and significance of the precise determination of vital structural and regulatory candidate genes expressing desirable traits for applications via genome editing tools (Zafar et al. 2020). Targeted genome editing methods, for instance, Clustered Regularly Interspaced Short Palindromic Repeats (CRISPR)/CRISPR-associated protein (Cas), Zinc Finger Nuclease (ZFNs), and Transcription Activator-Like Effector Nucleases (TALENs) along with base editing and forthcoming novel technologies like DNA-free genome editing and CRISPR/Cpf1 system have the outlook of boosting the possibility for crop enhancement. The use of genome editing in crop biology by addressing specific areas delivers an effective and feasible approach for the development of enhanced crop

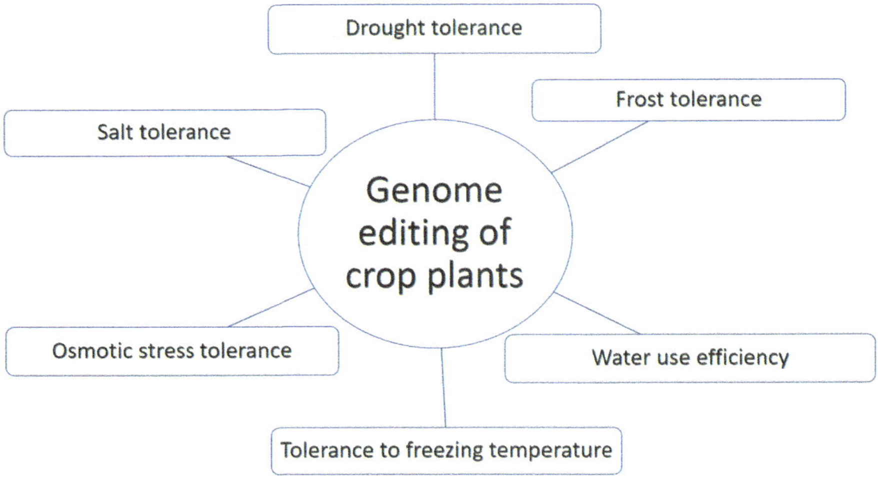

Fig. 15.1 Key areas/traits to be addressed by genome editing to develop climate smart plants/crops

varieties with superior crop longevity and thus improved resistance to climate stress, leading to the generation of climate smart crops (Fig. 15.1). In this chapter, we encapsulate the in progress site-specific genome-editing techniques which can be utilized for sensitive crops and converse their present and upcoming purposes for development of climate smart crops.

15.2 Basic Genome-Editing Techniques

Earliest genome-editing experiments were executed by Rouet and Smih in the year 1994 with exceptional meganuclease, I-SecI which explained that deliberate DSBs by meganucleases can be utilized to attain confined mutagenesis in addition to the inclusion of homologous donor sequence at aimed genomic location. In view of these facts, the area of genome editing has been shaped forward enormously via programmable nucleases. We will emphasize on practical characteristics of different genome-editing techniques and comprehend information with respect to growth prospects for enhanced abiotic stress resistant variants in plants (Fig. 15.2).

15.2.1 Zinc Finger Nucleases (ZFNs)

Zinc finger nucleases (ZFNs) are synthesized proteins utilized for gene targeting. They comprise a DNA-cleaving endonuclease component bound with zinc finger domains designed to attach to a precise DNA sequence. They are mainly customized to make insertions or deletions at cleaved positions in the genomes of organisms. Zinc finger nucleases are the earliest target-precise protein reagents which

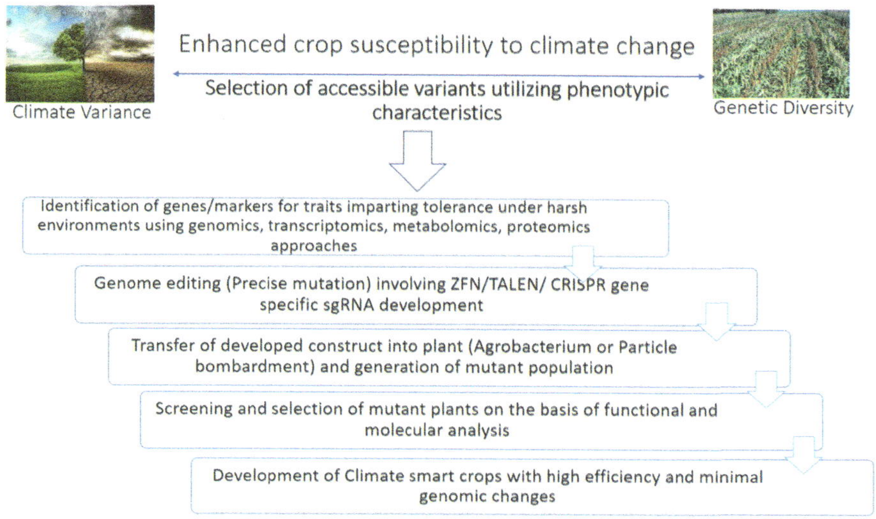

Fig. 15.2 Flowchart demonstration of steps involved in generating climate smart crops using genome-editing technology

metamorphosed the area of genome editing. The combination of ZNFs and FokI was much earlier recognized as a restriction enzyme, though its purpose for genome editing began in 2002 to bring on mutations and targeted genomic break in Drosophila (Bibikova et al. 2002). These are artificial restriction enzymes designed to connect engineered zinc finger DNA-binding domains to a DNA-cleavage domain in order to recognize and target unique DNA sequences within complex genomes (Fig. 15.3). These components may be used by taking advantage of the endogenous DNA repair mechanism to directly alter the genomes of complex organisms. ZFNs have a property of binding to target site of the DNA by precisely recognizing three base pairs (Carroll 2011). A certain zinc finger would, subsequently, be lacking in accuracy to link a distinctive genomic target, nevertheless, distinctive constructs consist of three to four fingers altering a ZFN target site of 18–24 bp (9–12 bp for each half). This stretch is adequate to target specific sites even for bulky genomes of higher species. In principle, the target site may be of any dimension but the context-specific assemblage of ZFN elements confines the size in reality (Teimourian and Abdollahzadeh 2015). A spacer of 5–7 bp can be created by attaching two DNA-binding ZFNs with FokI monomer which aids in performance of FokI for dimerization and produce DSBs. (Voytas and Gao 2014). The method principally introduces targeted DSBs utilizing ZFNs which encourage DNA repair phenomena, constituting error-sensitive non-homologous end joining (NHEJ) and homology-directed repair (HDR) in cells (Yeh et al. 2019). So far ZFNs have been utilized for improving plants such as Arabidopsis (Lloyd et al. 2005; Zhang et al. 2010; De Pater et al. 2009), tobacco (Petolino et al. 2010; Weinthal et al. 2010), maize (Shukla et al.

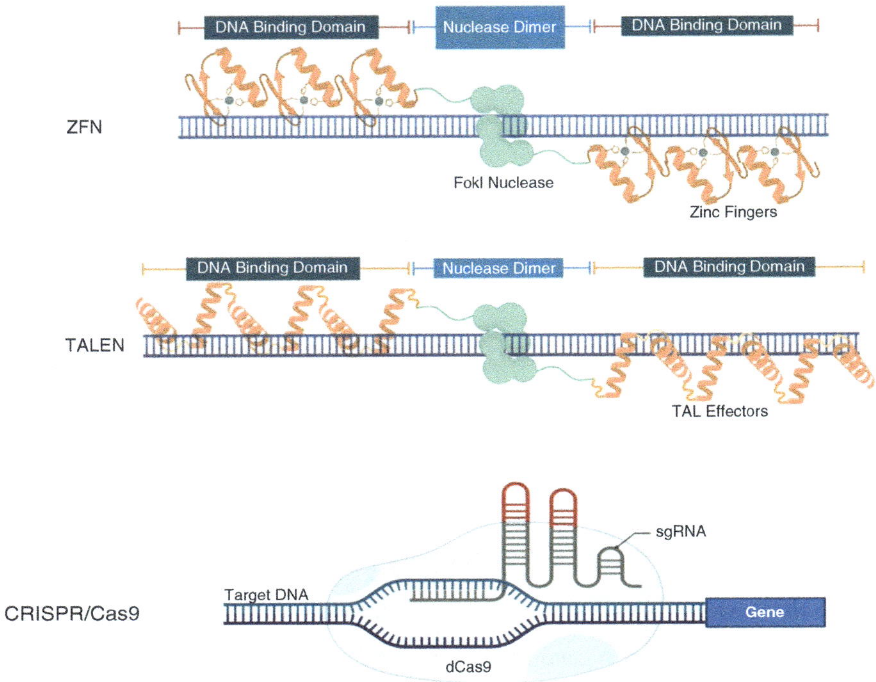

Fig. 15.3 Schematic representation of basic genome-editing techniques: ZFN, TALENs, and CRISPR/Cas9

2009), oil crops such as rapeseed and soybean, floral crop like petunia, rice, and fruit varieties like fig and apple (Zhang et al. 2018).

ZFN-mediated transgene incorporation was applied for trait stacking in maize, which leads to incorporation of numerous beneficial traits collectively to generate better possibilities for crop enhancement (Osakabe and Osakabe 2015). Another utility of ZFNs is found in detecting safe locations for gene introduction in crops like rice, and the specific position should act as reliable loci to undergo additional trait stacking and gene insertion. In the area of refining climate change resistance, ZFNs have made diminutive influence by editing host plant genes involved as they are complicated to be engineered and trying to be multiplexed. A custom-designed zinc finger nuclease accompanied with a heat-shock promoter has been applied to mutate an AP2/ERF family transcription factor gene, ABA-INSENSITIVE 4, responsible for abiotic stress responses in Arabidopsis (Osakabe et al. 2010). However, the configuration of ZFNs remains to be complex and technically challenging procedure with often low efficacy due to off-target cleavages. Gene editing specificity can be enhanced by attenuating DNA-cleavage kinetics by selectively reducing enzyme activity at low-affinity off-target sites which may bring about a decrease in off-target indels of up to 3000-fold (Miller et al. 2019) and by diversifying the structure of zinc finger nucleases utilizing new linkers and architectures enable dense targeting of selected genomic regions with ZFNs leading to greater extent of activity and specificity (Paschon et al. 2019).

15.2.2 Transcription Activator-Like Effector Nucleases—TALENs

The next type of site-directed mutagenesis, transcription activator-like effector nucleases (TALENs), initially identified in a bacterium named Xanthomonas, a plant pathogen (Boch and Bonas 2010). They perform on an equivalent principle as ZFNs, fusions of transcriptional activator-like effector (TALE) repeats and the FokI restriction enzyme target one nucleotide rather than three at the target site, making TALENs exceedingly precise with flexibility in target designing while the number of probable target locations increases comparative to those that can be targeted by ZFNs (Joung and Sander 2013). As soon as TAL effectors enter the host cells, they pass into the nucleus and activate transcription while binding to sequences specific in the host gene promoters. Tandem amino acid repeats of 34 residues in length facilitate DNA identification by TAL effectors. Due to simpler operation, the genes improved by TALENs have been effectively used in extensive variety of plants including wheat, rice, Arabidopsis along with various horticultural crops, for instance, tomato and potato (Xiong et al. 2015). The first reported TALEN-mediated modified crop was rice, which involved disruption of the gene OsSWEET14 (bacterial blight susceptibility gene), subsequent mutant rice plants were found to be bacterial blight resistant (Li et al. 2012). Equivalently, TALENs have been applied to improve wheat by knocking out three TaMLO homeologs so as to generate wheat plants resistant to powdery mildew (Gruner et al. 2020). Knocking out of GL2 gene in maize developed mutants with a possibility to be surface manured due to glossy phenotype and low epicuticular wax in the leaves (Andorf et al. 2019). In sugarcane, TALEN system is utilized to knockout a conserved section in the Caffeic Acid O-methyl-transferase (COMT) as target and successfully co-edit 107 out of 109 copies of COMT genes. It is a major accomplishment via multiplexed genome editing in plants and did not show significant difference to wild-type plants about good agronomic performances while in field trials. Altered cell wall composition improved the saccharification behavior drastically (up to 44%) of sugarcane while producing biofuel (Metje-Sprink et al. 2020). Fatty acid desaturase (FAD) genes are disrupted using TALENs to improve the nutritional values of soybean with greater oleic acid and less linoleic acid amounts enhancing the heat stability and shelf life of soybean oil (Haun et al. 2014).

TALE-based DNA-binding components have shown added efficiency than zinc fingers due to their higher specificity and lesser off-target activity, also they are less cytotoxic along with extensive targeting array (Mussolino et al. 2011). The other significant advantage of TALENs over ZFNs and CRISPR/Cas is that it can effortlessly target any small DNA fragments, for instance, DNA sites encoding microRNA and enhancers that may be missing in ZFNs or CRISPR/Cas targetable sites. Though, few of the limitations for the extensive application of these nucleases are complications in protein synthesis, design, and validation along with challenges of construction of TALE repeats. Overall, TALENs is a crucial discovery in genome editing and has the ability to produce climate adaptive crop plants through the incorporation of possible foreign genes, as mentioned in Table 15.2, and to resolve negative aspects of current plant biotechnology.

15.2.3 CRISPR/Cas9 System

CRISPR/Cas9 is known as the genome-editing tool of the third generation, particularly the type II CRISPR/SpCas9 arrangement which is attained from *Streptococcus pyogenes.* CRISPR/Cas technique is categorized as simple, efficient, potent to target multiple genes and cost effective as compared with ZFNs and TALENs (Khan 2019). Preparing DNA-binding motif in specification with target sequence necessitates profound information of biochemistry and protein engineering which can be simplified by means of small RNA sequences in place of specified DNA sequence. CRISPR was revealed as immune system for prokaryotes which defends cells through specifically attacking and terminating extraneous DNA which could be plasmids or viruses (Brouns et al. 2008; Marraffini and Sontheimer 2008). In year 2013, CRISPR/*Cas9* was initially employed to edit plant genes and since then it is prevalent gene editing tool. Efficient CRISPR-Cas position involves an arrangement of matching repeats incorporated with intruder DNA-targeting spacers which interpret crRNA mechanisms and an operon of cas genes translating Cas protein subunits (Simon et al. 2018). Through studying CRISPR spacers within natural atmospheres, viruses may be connected to their archaeal or bacterial hosts. Three stages of adaptive immunity are: (a) addition of a spacer sequence in the CRISPR arrangement in the form of a small sequence of the invading DNA, (b) transcription of precursor crRNA (pre-crRNA) which matures to produce distinct crRNAs, individually consist of a recurring segment and an intruder directing spacer segment, and (c) Sites corresponding to the crRNA spacer sequence of foreign nucleic acid are cleaved by crRNA-directed Cas proteins. The CRISPR-associated (Cas) gene structure categorizes CRISPR systems, although researchers continue to refine the subset, category, and type classification framework. Two major classes of CRISPR systems exist, individually with distinct forms of CRISPR. Class 1 includes CRISPR structures Type I and Type III, usually exist in Archaea. Types II, IV, V, and VI are all part of Class 2 CRISPR arrangements. The most commonly utilized method is the *Streptococcus pyogenes* type II CRISPR-Cas9 system, due to its simple NGG PAM sequence rule, S. pyogenes Cas9 (spCas9) is used in different applications (Adli 2018). To date, the CRISPR/Cas9 technique has been applied in more than 20 crop species for various yield enhancement, biotic and abiotic stress control traits (Jaganathan et al. 2018). Predominantly null alleles or gene knockouts were attained through trivial insertions and deletions producing frame-shift mutations or by early introduction of stop codons. A seminal research by (Shan et al. 2014) reported that three rice genes, mitogen-activated protein kinase (OsMPK2), betaine aldehyde dehydrogenase2 (OsBADH2), and phytoene desaturase (OsPDS) were altered via

CRISPR/Cas9. In proceeding years numerous studies reported modified genomes in other significant crops such as wheat (*Triticum aestivum*), barley (*Hordeum vulgare*), soybean (*Glycine max*), maize (*Zea mays*) tomato (*Solanum lycopersicum*), and potato (*S. tuberosum*). This technique has also been applied in maize for knocking out Zmzb7 gene to get albino plants with the sgRNA projected to emphasize on the eighth exon of Zmzb7, whereas for expression maize U3 promoter was employed. Later, 31% mutation efficacy was observed in T0 lines developed through agrobacterium mediated transformation of maize embryos (Feng et al. 2015). CRISPR/Cas9 has been utilized to selectively mutate and generate heritability in modified rice lines by specifically targeting OsPMS3, OsMSH1, OsDERF1, OsEPSPS, and OsMYB5 genes (Yang et al. 2016). CRISPR/Cas9 technology has been applied to improve nutritional profiles of crops by targeting FAD2 to enhance oleic acid quantity whereas declining polyunsaturated fatty acids in Camelina sativa, an evolving oil seed plant (Jiang et al. 2017). DEP1, GS3, Gn1a, and IPA1 genes of rice cultivar Zhonghua11 were modified for raising mutants with improved grain quantity and bigger grain size with dense erect panicles (Li et al. 2016). CRISPR/Cas9 applied to knock out the maize waxy gene Wx1 codes the granule bound to the NDP-glucose-starch glucosyltransferase (GBSS), which is responsible for the development of amylose. Seed starch of wild-type (WT) comprises ~25% amylose and ~75% amylopectin while Wx1 altered seed starch is ~100% amylopectin. Thus, lack of amylose formed a high amylopectin maize or waxy corn with enhanced digestibility and higher applicability in bio-industries involved in producing stabilizer and thickener for food commerce along with paper and adhesive making (Gao et al. 2020). Various efforts have been made to apply CRISPR/Cas9 in order to develop biotic stress resistant varieties. The CRISPR/Cas9 produced Taedr1 wheat plants by modifying EDR1 homeologs in wheat concurrently and resulting mutated plants developed resistance to *Blumeria graminis* sp. *tritici (Bgt)*, powdery mildew fungi and not causing mildew-induced cell death (Yin and Qiu 2019). *Magnaporthe oryzae* causes rice blast, one of the most destructive diseases impacting rice production globally. Targeted knockout of OsERF922 by means of CRISPR/Cas9 produced rice Oserf922 null mutants which exhibited greater resistance to rice blast and not disturbing other key agronomic qualities (Wang et al. 2016). *Xanthomonas citri* subsp. *citri (Xcc)* causes citrus canker and CsLOB1 is a susceptibility gene for citrus canker. CRISPR/Cas9-mediated alteration of single allele effector binding elements of the CsLOB1 gene in Duncan grapefruit relieved the signs to some degree while alteration of both CsLOB1 promoter alleles resulted in higher tolerance to citrus canker in Wanjincheng orange (Sun et al. 2019). In a current study, the mutated plants exhibited a significantly reduced vulnerability to citrus canker in Wanjincheng orange by knocking out CsWRKY22, a marker gene for pathogen-induced immunity that uses CRISPR/Cas9 (Wang et al. 2019a). Moreover, CRISPR/Cas12a (Cpf1), alternative improved CRISPR/Cas method, was introduced to alter the Duncan grapefruit gene CsPDS with greater effectiveness and fewer off-target outcomes (Jia et al. 2019).

Precise genome editing of host aspects has been effectively implemented using CRISPR/Cas9 technologies to establish recessive genetic resistance to many viral diseases in plants. CRISPR/Cas9 genome editing for rice tungro spherical virus resistance in plants has been documented recently by knocking out the eIF4 G gene from rice (Macovei et al. 2018). Similarly, eIF4E1 was mutated in tomato cultivar Micro-Tom using CRISPR/Cas9 technology and homozygous lines to add allelic variations in the eIF4E gene and demonstrated resistance to PepMoV and exhibited normal plant growth and development after virus challenge (Yoon et al. 2020). Polyphenol oxidase (PPO) enzyme is accountable for browning in numerous vegetables and fruits. By precise editing of a PPO gene through CRISPR/Cas9 system transport of Ribonucleoprotein complexes, browning of Potato tubers declined significantly (González et al. 2020). An effective CRISPR/Cas9 based gene knockout system was established in watermelon and was remarkably applied to produce herbicide resistance in germplasm (Tian et al. 2018). In addition, CRISPR/Cas9-mediated gene integration and substitution approaches have also shaped various abiotic stress-resilient characteristics in major crops (Table 15.1). Analysis on the OsPYL abscisic acid receptor gene family showed that CRISPR/Cas9 editing pyl1/4/6 triple knockout in rice had enhanced grain produce, high temperature resistance, and decreased preharvest germination related to wild-type (Miao et al. 2018). CRISPR/Cas9 genome editing has been introduced in wheat protoplasts for two genes associated with abiotic stress, i.e. wheat dehydration responsive element linking protein 2 (TaDREB2) and wheat ethylene responsive factor 3 (TaERF3) (Kim et al. 2018). One of the major drawbacks of CRISPR/Cas9 systems is that they are vulnerable to off-target incidents contributing to the risk of Cas9 cutting several positions in the genome accidentally, together with the intended target sequence (Zhang et al. 2015). Several methods are being considered to decline CRISPR/Cas9 off-target action, counting inducible ways by restricting the accessibility duration and Cas9 concentration along with diverse sgRNA designs (Zhang et al. 2015). An approach includes creating a chimeric blend of one FokI catalytic domain and the catalytically inactive Cas9 protein (dCas9). The dCas9 serves as a targeting component to take the adjacent FokI domain to form homodimers with the exact spacer sequence that generates DSBs and enhances the precision of the cleavage, since it involves 40 bp of exclusive sequence and an explicit gap flanked by the two monomers, thereby preventing off-target events (Sedeek et al. 2019). Numerous algorithmic Cas9 target selection tools are developed to minimize off-target effects, for instance, http://zifit.partners.org, www.e-crisp.org. and http://tools.genome-engineering.org. The advancement of other genome extensive methods may also consider the thermodynamic properties of the sgRNA to intensify the accuracy (Doudna and Charpentier 2014). CRISPR-Cas9 system is also compliant to multiplexing leading to mutations into multiple genes concurrently. CRISPR-Cas9 system has unlocked an extensive array of prospects in elementary and pragmatic research in plant biotechnology due to its high competence, effortless targeting, and possibilities of multiplexed editing.

Table 15.1 Application of genome editing tools in different plant species to improve climate change related stress resistance

Target trait	Plant species	Targeted sequence(s)	Results	Method	Method
Drought tolerance	*Arabidopsis thaliana*	MIR169a	Improved drought tolerance	CRISPR/Cas9	Zhao et al. (2016)
	Oryza sativa	OsDERF1, OsPMS3, OsEPSPS, OsMSH1, OsMYB5	Improved drought tolerance	CRISPR/Cas9	Zhang et al. (2014)
	Tomato	SlMAPK3	Improved drought tolerance		Wang et al. (2017)
	Zea mays	ARGOS8	Improved grain yield under field drought stress conditions	CRISPR/Cas9	Shi et al. (2017)
	Oryza sativa	Os SAPK2	Improved drought tolerance	CRISPR/Cas9	Lou et al. (2017)
	Oryza sativa	OsNAC14	Improved drought tolerance	CRISPR/Cas9	Shim et al. (2018)
	Tomato	SlNPR1	Drought resistance	CRISPR/Cas9	Li et al. (2019)
	Arabidopsis thaliana	AREB1	ABA signaling-mediated drought tolerance	CRISPR/dCas9 (epigenome editing)	Roca-Paixão et al. (2019)
Salt tolerance	*Oryza sativa*	OsRAV2	Salt stress tolerance	CRISPR/Cas9	Duan et al. (2016)
	Oryza sativa	OsERF71	Drought and salinity tolerance		Bi and Yang (2017)
	Oryza sativa	OsRR22	Improvement in salinity tolerance	CRISPR/Cas9	Zhang et al. (2019)
Thermotolerance	*Zea mays*	TMS5	Thermosensitive male-sterile	CRISPR/Cas9	Li et al. (2017)
Cold tolerance	*Oryza sativa*	TIFY1a and TIFY1b	May confer cold adaptation	CRISPR/Cas9	Huang et al. (2017)
	Oryza sativa	OsANN3	Confer cold tolerance	CRISPR/Cas9	Shen et al. (2017)

Table 15.2 Potential target genes for genome editing in some important crops

Crop plant	Target Trait	Gene	Reference
Rice	Drought tolerance	osa-miR162, osa-miR164	Fang et al. (2014), Tian et al. (2015)
Rice	Cold tolerance	CCR4-associated factor 1B	Chou et al. (2014)
Maize	Heat tolerance	TapAPX	Padaria et al. (2014)
Chickpea	Drought tolerance	ERECTA-like kinase	Varshney et al. (2014)
Rice	Drought tolerance	OsbZIP46CA1	Tang et al. (2012)
Arabidopsis thaliana	Salinity tolerance	AtNHX1	Tabata et al. (2000)
	Heat and oxidative stress tolerance	RPN1a	Wang et al. (2009)
African rice (*Oryza glaberrima)*	Thermotolerance	OgTT1	Li et al. (2015)
Triticum aestivum	Salinity tolerance	*TaPUB1*	Wang et al. (2019b)
Banana	Cold and salt tolerance	MaSWEET-4d, MaSWEET-1b, and MaSWEET-4b	Miao et al. (2017a)
Banana	Cold and salt tolerance	MaAPS1 and MaAPL3	Miao et al. (2017b)
Cassava	Salt, osmosis, cold, drought tolerance	MeKUPs	Ou et al. (2018)
Cassava	Drought tolerance	MeMAPKKK	Ye et al. (2017)
Cotton	Drought tolerance	GhPIN1–3 and GhPIN2	He et al. (2017)
Cotton	Drought tolerance	GhRDL1	Dass et al. (2017)
Papaya	Drought, heat, and cold tolerance	CpDreb2	Arroyo-Herrera et al. (2016)
Papaya	Heat and cold tolerance	CpRap2.4a and CpRap2.4b	Figueroa-Yañez et al. (2016)
Sugarcane	Drought and chilling tolerance	ScNsLTP	Chen et al. (2017)
Rice	Flood tolerance	Sub1A, SK1, and SK2	Fukao et al. (2006), Hattori et al. (2009)
Wheat and barley	Low temperature tolerance	VRN1 at the FR1 locus and CBFs at the FR2 locus	Dhillon et al. (2010), Stockinger et al. (2007), Knox et al. (2010), Francia et al. (2007)

15.2.4 Editing by Nucleobase Modification (Base Editors)

Base editing is an advanced technique which aids in specific nucleotide replacements without interruption in a gene or demanding a donor sequence. Base editors are a union of catalytically inactive CRISPR–Cas9 (Cas9 nickase, Cas9 variants or dCas9) along with an adenosine or cytosine deaminase component with a capacity to alter a specific base into alternative one. These base editors can create single-base switch or modification by not generating double stranded breaks in the DNA, thus restraining the incidence of indels. A base alteration can produce best trait alterations in crop plants aiding enhancement of agriculturally significant crops (Eid et al. 2018). Base editors are a combination of proteins associated with the DNA directing component and another catalytic component which causes deamination of an adenine or cytosine in the genome (Mishra et al. 2020). The DNA-targeting component can be any one of Cas9 nickase or catalytically inactive Cas9 endonuclease (dCas9) directed by a sgRNA fragment. The dCas9 comprises Asp^{10}Ala and His^{840}Ala transformations which disable nuclease property of the enzyme while the DNA attachment capability is retained. Attachment of dCas9-sgRNA to the designated DNA forms an "R-loop" structure where a trivial length of DNA unwinds which is a stretch of roughly 5–8 nucleotides forming the catalytic or editing space for dCas9 bound deaminase to alter the cytosines. Two forms of DNA base editors are: cytosine base editors (CBEs) and adenine base editors (ABEs).

To accomplish the alteration of uridine from cytidine, a cytidine deaminase was attached via a linker to the N-terminus of a partially active form of Cas9, in which one of the nuclease components is deactivated by the mutation D10A to produce a nickase (nCas9). To achieve uridine modification from cytidine, a cytidine deaminase was attached by a linker to the N-terminus of a partially active version of Cas9, in which one of the nuclease components is deactivated to create a nickase (nCas9) by the mutation D10A. Nickase and cytidine deaminase mixture activates DNA mismatch repair mechanism, ensuing a guided conversion of a base pair of C: G into T: A. In addition to base substitutions, indels may also occur occasionally because of the DNA nicks made on the unedited strand by nCas9 (Gaudelli et al. 2017).

Three rice genes (OsCDC48, OsNRT1.1B, and OsSPL14), one wheat gene (TaLOX2), and one maize gene (ZmCENH3) are mutated by transfection through protoplasts of plasmids with unique codon-dependent plant base editor, PBE and distinct gRNAs. Efficient gene editing with a rate of up to 43.48% for these genes was observed (Zong et al. 2017). A combination of CRISPR-Cas9 and activation-mediated cytidine deaminase (Target-AID) mediated multiple herbicide-resistance point mutations in rice and created marker-free tomato plants with homozygous inheritance DNA substitutions, suggesting the viability of base editing for crop production (Shimatani et al. 2017). Opportunities of base editing are also explored in plants by placing the VQR-Cas9 nickase in place of Cas9 nickase of BE3 (CBE-P1, plant version) and the base editor CBE-P3 induced editing at the extended targets site PMS3 (Hua et al. 2018). A key limitation of base editors resulting in poor specificity and editing efficacy is that any Cytosine which is existing in the activity window of roughly 4–5 nucleotides (or up to 9 nt) can possibly be altered by cytosine deaminase base editors. Consequently, trials are on to produce base editors showing high-precision with small catalytic range which can exactly edit one cytidine residue with more precision and proficiency (Tan et al. 2019). Off-targets are observed when extra cytosines are edited proximal to the target base which has been significantly lowered in cells of human beings via high-fidelity base editor (HF-BE3), generated by mutating base editor BE3 (Rees et al. 2017). Prime requirement for base editing is the existence of a specific PAM sequence (NGG PAM for SpCas9) and target nucleobase within a minor base-editing range to be effective (Gaudelli et al. 2017). To improve the PAM compatibility and enhance the scope of base editing, novel base editors are developed, ABE and CBE via Cas9 alternatives which identify PAMs except the NGG domain (Endo et al. 2019; Wang et al. 2019c). However, it is a comparatively novel gene editing method with high competence to improve crops specifically and hasten the improvement process during the imminent scenario of climate change. It also suggests potentials for breeding crops with climate resilience which can boost comprehensive food security and diversify approaches to be discovered in upcoming time to improve genetic diversity among crop plants.

15.3 Novel Technical Breakthroughs

Continuous technological novelties are being addressed to the genome-editing tools in order to overcome off-target effects, efficiency issues of homologous recombination, limiting protospacer adjacent motif (PAM) arrangements with further complications.

15.3.1 DNA-Free Genome-Editing System

Genome editing in conventional way follows the transfer and insertion or deletion or substitutions in host plant genome via the DNA cassettes including editing mechanisms. Editing complexes may integrate haphazardly, and hence may create unintended genetic variations. Genome-editing tools upon extended expression rise off-target responses in plants due to increased nucleases in edited plants. Furthermore, the presence of external DNA into plant genomes increases monitoring apprehensions as in case of Genetically Modified organisms (Jones 2015). So, DNA-free genome editing arose as a revolutionary technology in the year 2015 with lowered probabilities of unwanted off-target alterations to accomplish existing and upcoming agriculture requirements to evade intensifying concerns by gene technology regulations. DNA-free genome editing is done by means of TALENs and CRISPR/Cas9 while components get transiently expressed by mRNAs encrypting for TALENs or Cas nuclease and gRNAs or through RNPs direct transfer. Pre-assembled complexes which are active during delivery are used in instance of RNPs, while the transcribed editors and the complex have to assemble in case of RNAs. DNA-free genome editing can be carried out via particle bombardment as well as protoplast mediated transformation (Ran et al. 2017; Toda et al. 2019). Effective DNA-free genome editing in crops was initially achieved by pre-assembled complexes of isolated Cas9 proteins and gRNA upon delivery into protoplasts of *Arabidopsis thaliana*, *Nicotiana tabacum*, rice, and lettuce by specific mutation at efficacy of up to 46% in regenerated plants (Woo et al. 2015). CRISPR/Cas9 RNA and CRISPR/Cas9 RNPs via particle bombardment have been carried into wheat embryos and produced genome edited plants (Zhang et al. 2016; Liang et al. 2018). CRISPR-Cas9 RNPs edited genes specifying granule bound starch synthase (GBSS, EC 2.4.1.242.) in potato protoplasts, mutated at a rate of 9% and generated transgene-free plants (Andersson et al. 2018). Development of a strain constitutively making zeaxanthin and exhibiting enhanced photosynthetic production in *Chlamydomonas reinhardtii* is reported on application of a single-phase transformation by the DNA-free CRISPR-Cas9 method via Cas9 and guide RNAs leading to successive *CpFTSY* and *ZEP* two gene knockouts (Baek et al. 2016). This arrangement is particularly beneficial for vegetative propagated plants or plants with extensive generation process as backcrossing is time intense or impossible for plants such as potato, grapevine, and apple (Andersson et al. 2018). Simultaneously, DNA-free genome-editing system has to overcome certain disadvantages due to feasibility issues in most of species, largely owing to inapt *in vitro* practices. Besides, lower efficiency in relation to standard methods and a selection procedure only possible by sequencing, resulting in higher costs for the technique, additional optimizations may lead to better *in vitro* protocols and lowering of costs.

15.3.2 CRISPR/Cpf1 System

Cpf1, an endonuclease of the type V CRISPR classification, demonstrates excessive possibilities when compared with TALEN, ZFN, and CRISPR/Cas9. Cpf1 can re-engineer at the same location upon editing and may perform like a twofold nuclease with endo-ribonuclease property to develop crRNA and endo-deoxyribonuclease property for cutting at target site and generating double stranded breaks (DSBs) (Alok et al. 2020). Since a single crRNA array has potential to target various loci in the genome along with enabling deletions/insertions of genes, nucleobase editing and locus tagging in monocots along with dicot plants with rarer off-target reactions, it provides Cpf1 ability to multiplex. Size of Cpf1 endonuclease is too minor relative to Cas9, thus smaller CRISPR RNA (crRNA) is needed for accurate performance. A single RNA guides Cpf1 effector nuclease protein which attaches upstream of the protospacer adjacent motif (PAM) and cleaves the DNA on the proximal end of the PAM, creating 5 base pair sticky ends instead of blunt ends, which boosts the features of Cas9 to a great level (Zetsche et al. 2015). Trans-activating crRNA is not required while dealing with Cpf1-linked CRISPR into matured crRNAs by the CRISPR/Cpf1 arrangement. The transport of Cpf1 proteins complex with synthetically prepared or *in vitro* transcribed target-precise crRNAs to tobacco and soybean protoplasts generated mutations successfully with AOC gene in tobacco and FAD2 paralogue gene in soybean (Kim et al. 2017). A successful study demonstrated site genome editing with cotton endogenous gene *Cloroplastos alterados* (GhCLA) in allotetraploid cotton delivering 87% efficacy and no off-targets upon applying CRISPR/Cpf1 arrangement (Li et al. 2019). Reported studies back the conclusion that CRISPR/Cpf1 mechanism is extremely precise and effective for genome editing in plants providing alternative for CRISPR/Cas9 system in generating climate responsive crops.

15.4 Outcomes of Genome Editing in Understanding Climate Stress Tolerance

The climate fluctuating conditions, counting elevated temperature, drought stress, and salinity stress are posing utmost risk to worldwide food safety. These impromptu and dynamically changing conditions do not seem to be checked by existing techniques due to inefficiency in constructing required alleles in a functioning mode in a rapid manner. On contrary, the emergence of genome-editing technology with its capability of highly precise genetic alteration has canceled these downsides. Variants of genome editing and CRISPR/Cas system have by now established abiotic stress tolerance by specific modification of significant sensitive and tolerant genes in various plant species (Table 15.1). Climate change related stresses are multifaceted traits, which are dealt by multiple genes and there is a considerable communication amid mechanisms of numerous metabolic, regulatory, and signaling phenomenon along with cellular and molecular aspects leading to abiotic stress management (Waqas et al. 2019). Owing to complexities, the advent of innovative techniques, for example, multiplex editing, DNA-free editing involving RNPs and base editing deliver a wider source of genome-editing techniques for operative moderation against drought as well as additional abiotic stress conditions. Due to the multifaceted effects of drought conditions, the practice of genome modification has been lately applied towards drought tolerance. In plants, overexpression of some transcription factors and genes associated to drought signaling simplify collection of metabolites along with signaling molecules to improve drought resilience (Fang and Xiong 2015). Certain sensitive (S) genes increase drought like circumstances in crops by lowered antioxidant activity, hormonal disproportion, and ROS production. For instance, *Oryza sativa* drought induced SINA protein 1 (OsDIS1), drought and salt tolerant protein 1 (OsDST) and stress related ring finger protein 1 (OsSRFP1) regulate drought stress in adverse direction while upon silencing these genes enhanced antioxidant activity and reduced H_2O_2 concentration resulting in drought tolerance (Joshi et al. 2020). Consequently, drought tolerance could be achieved naturally by targeting adversely expressing genes or sensitive (S) genes via genome-editing techniques. ABA performs as a prime aspect of response against drought in crops by regulating the associated genes expression and avoid water loss by controlling stomatal opening or closing (Osakabe et al. 2014). AREB1 regulates large number of genes prior to ABA-signaling pathway and turn out to be key indicator of ABA production, osmotic regulation, and antioxidant signaling

(Barbosa et al. 2013). It is also described that the ARGOS family genes regulate ethylene signaling negatively and bring upon improved drought resilience and yield in dry environments (Shi et al. 2017). Therefore, AREB1 and ARGOS family genes act as functioning targets for enhancing drought tolerance in crop plants. Various transcriptional factors and genes are linked with heat stress response or tolerance have been confirmed by means of virus induced gene silencing (VIGS). For instance, TRV-VIGS dependent CabZIP63 gene silencing reduced tolerance to high temperature in case of pepper, indicating that CabZIP63 regulates positively for heat tolerance (Singh et al. 2019). Salinity stress is another significant climate factor which is a major risk to agriculture globally and the prerequisite to advance cultivars is credentials of characteristics that reinforce salinity tolerance (Pareek et al. 2020). Knowledge of stress resistance phenomena involved in salinity tolerance in various halophytic organisms, mechanisms for osmo-sensing in plants, roles of receptor-like kinases, phospholipase C, aquaporins, mechanizable calcium channels, and membrane-bound histidine kinases as osmo-sensors in stress discernment may aid as dominant regulators in various stress responses and act as key components for the advancement of stress-resilient plants. These participating phenomena provide required perceptions so as to advance insights into defining and understanding sustainable intensification through genome editing towards climate smart agriculture globally.

15.5 Conclusions and Future Implications

Genome-editing system can significantly enable the learning of various genomic mechanisms and engineering tolerance against climate change related adversities among diverse crop species. The contents of this chapter will provide aid to recognize the current scientific developments, information breaches, complications with technical variations which are essential for the effective exploitation of various tools involved in genome editing for crop enhancement in this changing climate scenario. We have defined the basic principles, recent applications, and possible improvements in basic genome-editing techniques for the enhancement of crops, and presented the comparatively novel nucleobase editing and CRISPR/Cpf1 methods, with excessive possibilities in improving crop farming. Provided the accessibility of a range of genome-editing techniques conferring diverse claims, it becomes significant to select the most appropriate method for a specific species and motive. After suitable genome-editing tool decision, the precise sequences to be targeted are designed and established in appropriate vectors, and then suitable carrier (DNA, RNA, or RNPs) for transfer is designated. As the designated cells receive the genetic carrier, the specific sequences will undergo alteration and regeneration from edited cells will eventually produce tolerant plants. It is observed that certain plants do not endure protoplast-based transformations readily and tissue-culture based regeneration is trying or restricted to a limited number of model plants. In such instances, the usage of immature embryos or pollens which may be induced to propagate *in vitro* will be advantageous as they do not necessitate regeneration.

Genome editing being a highly progressive molecular biology procedure with ability to generate precisely aimed alterations in an organism. Among various studies the robustness and versatility of CRISPR-Cas9 as well as CRISPR/Cpf1 systems when related to ZFN and TALENs has been established in numerous biological frameworks. The technical developments of CRISPR-Cas9 are being executed for generating knockouts and knock-ins along with activation or repression of genes. Due to its high efficacy in applied system, various progressions have been realized in a brief duration since its finding, which comprise DNA-free arrangements (RNPs), manifold Cas9 alternatives, precise base editing, gene multiplexing, gene stacking and ways to rise the occurrence of HDR. The advancements of innovative regulatory components from natural prevailing mechanisms like genes, promoters, cis-regulatory components, sRNAs, and epigenetic alterations have a wide range of potential to enable the designing of signaling, regulatory or metabolic pathways to develop climate smart plants. On the other hand, developments in the CRISPR/Cas associated bioinformatics devices have also raised applications of this significant genome-editing procedure. Altogether, the swift pace of expansion and evolving applicability of CRISPR-Cas9 mechanisms ascertain their enormous influence in deciphering gene monitoring systems through abiotic stress response for adaptation and crop enhancement plans to generate stress resistant crops during this changing climate scenario. Nevertheless, the task of informing the community adequately concerning CRISPR/Cas facilitated technique for crop enhancement cannot be overlooked, since it is vital to take laboratory study to the public, with the aim of delivering sustainable climate smart agriculture. Critical regulations evidently demarcate a borderline amidst the gene-edited plants with external DNA and without any external DNA, making the later subject to be evaded from regulations, and thus, use of these plants can be uncomplicated. One more view offers that the policymakers should focus on the definite features of the ultimate product instead of on the methodology to recognize the probable hazards in any case. Altogether, one must not neglect the responsibility for appropriate valuation and supervision of dangers, directives of the gene drive problem, and crucial safety measures associated with this technology in the area of replication and human genetics. The climate change exigencies demand for excessive innovation and flexibility in development systems of crop resilience. The ongoing progress in the development of genome-editing tools and the new breakthroughs assure genome editing to perform a crucial part in racing up crop improvement and in fulfilling the increasing global requisite for food.

References

Adli M (2018) The CRISPR tool kit for genome editing and beyond. Nat Commun 9:1911

Alok A, Sandhya D, Jogam P, Rodrigues V, Bhati KK, Sharma H, Kumar J (2020) The rise of the CRISPR/Cpf1 system for efficient genome editing in plants. Front Plant Sci 11:264

Andersson M, Turesson H, Olsson N, Fält AS, Ohlsson P, Gonzalez MN, Samuelsson M, Hofvander P (2018) Genome editing in potato via CRISPR-Cas9 ribonucleoprotein delivery. Physiol Plant 164:378–384

Andorf C, Beavis WD, Hufford M, Smith S, Suza WP, Wang K, Woodhouse M, Yu J, Lübberstedt T (2019) Technological advances in maize breeding: past, present and future. Theor Appl Genet 132:817–849

Arroyo-Herrera A, Figueroa-Yáñez L, Castaño E, Santamaría J, Pereira-Santana A, Espadas-Alcocer J, Sánchez-Teyer F, Sánchez-Teyer F, Alcaraz L, López-Gómez R, Sánchez-Calderón L, Rodríguez-Zapata L (2016) A novel Dreb2-type gene from Carica papaya confers tolerance under abiotic stress. Plant Cell Tissue Organ Cult 125:119–133

Asseng S, Ewert F, Martre P, Rotter RP, Lobell DB, Cammarano D, Kimball BA, Ottman MJ, Wall GW, White JW, Reynolds MP, Alderman PD, Prasad PVV, Aggarwal PK, Anothai J, Basso B, Biernath C, Challinor AJ, DeSanctis G, Doltra J, Fereres E, Garcia-Vila M, Gayler S, Hoogenboom G, Hunt LA, Izaurralde RC, Jabloun M, Jones CD, Kersebaum KC, Koehler AK, Müller C, Naresh Kumar S, Nendel C, O'Leary G, Olesen JE, Palosuo T, Priesack E, Eyshi Rezaei E, Ruane AC, Semenov MA, Shcherbak I, Stöckle C, Stratonovitch P, Streck T, Supit I, Tao F, Thorburn PJ, Waha K, Wang E, Wallach D, Wolf J, Zhao Z, Zhu Y (2014) Rising temperatures reduce global wheat production. Nat Clim Chang 5:143–147

Baek K, Kim DH, Jeong J, Sim SJ, Melis A, Kim JS, Jin ES, Bae S (2016) DNA-free two-gene knockout in *Chlamydomonas reinhardtii* via CRISPR-Cas9 ribonucleoproteins. Sci Rep 6:30620

Bailey-Serres J, Parker JE, Ainsworth EA, Oldroyd GED, Schroeder JI (2019) Genetic strategies for improving crop yields. Nature 575:109–118

Barbosa EGG, Barbosa EGG, Leite JP, Marin SRR, Marinho JP, de Fátima Corrêa Carvalho J, Fuganti-Pagliarini R, Farias JRB, Neumaier N, Marcelino-Guimarães FC, de Oliveira MCN, Yamaguchi-Shinozaki K, Nakashima K, Maruyama K, Kanamori N, Fujita Y, Yoshida T, Nepomuceno AL (2013) Overexpression of the ABA-dependent AREB1 transcription factor from *Arabidopsis thaliana* improves soybean tolerance to water deficit. Plant Mol Biol Rep 31:719–730

Barnabás B, Jäger K, Fehér A (2008) The effect of drought and heat stress on reproductive processes in cereals. Plant Cell Environ 31:11–38

Bi H, Yang B (2017) Gene editing with TALEN and CRISPR/Cas in rice. Prog Mol Biol Transl 149:81–98

Bibikova M, Golic M, Golic KG, Carroll D (2002) Targeted chromosomal cleavage and mutagenesis in Drosophila using zinc-finger nucleases. Genetics 161:1169–1175

Boch J, Bonas U (2010) Xanthomonas AvrBs3 family-type III effectors: discovery and function. Annu Rev Phytopathol 48:419–436

Brouns SJJ, Jore MM, Lundgren M, Westra ER, Slijkhuis RJH, Snijders APL, Dickman MJ, Makarova KS, Koonin EV, Oost JV (2008) Small CRISPR RNAs guide antiviral defense in prokaryotes. Science 321:960–964

Carroll D (2011) Genome engineering with zinc-finger nucleases. Genetics 188(4):773–782

Challinor A, Watson J, Lobell D, Howden SM, Smith DR, Chhetri N (2014) A meta-analysis of crop yield under climate change and adaptation. Nat Clim Chang 4:287–291

Chapman SC, Chakraborty S, Dreccer MF, Howden SM (2012) Plant adaptation to climate change—opportunities and priorities in breeding. Crop Pasture Sci 63:251–268

Chen Y, Ma J, Zhang X et al (2017) A novel non-specific lipid transfer protein gene from sugarcane (NsLTPs), obviously responded to abiotic stresses and signaling molecules of SA and MeJA. Sugar Tech 19:17–25

Chou WL, Huang LF, Fang JC, Yeh CH, Hong CY, Wu SJ, Lu CA (2014) Divergence of the expression and subcellular localization of CCR4-associated factor 1 (CAF1) deadenylase proteins in *Oryza sativa*. Plant Mol Biol 85:443–458

Dass A, Abdin MZ, Reddy VS, Leelavathi S (2017) Isolation and characterization of the dehydration stress-inducible GhRDL1 promoter from the cultivated upland cotton (Gossypium hirsutum). J Plant Biochem Biotechnol 26:113–119

De Pater S, Neuteboom LW, Pinas JE, Hooykaas PJJ, Van Der Zaal BJ (2009) ZFN-induced mutagenesis and gene-targeting in Arabidopsis through *Agrobacterium*-mediated floral dip transformation. Plant Biotechnol J 7:821–835

Dhillon T, Pearce SP, Stockinger EJ, Distelfeld A, Li C, Knox AK, Vashegyi I, Vágújfalvi A, Galiba G, Dubcovsky J (2010) Regulation of freezing tolerance and flowering in temperate cereals: the VRN-1 connection. Plant Physiol 153:1846–1858

Doudna JA, Charpentier E (2014) The new frontier of genome engineering with CRISPR-Cas9. Science 346:1258096

Duan YB, Li J, Qin RY, Xu RF, Li H, Yang YC, Ma H, Li L, Wei PC, Yang JB (2016) Identification of a regulatory element responsible for salt induction of rice OSRAV2 through ex situ and in situ promoter analysis. Plant Mol Biol 90:49–62

Eid A, Alshareef S, Mahfouz MM (2018) CRISPR base editors: genome editing without double-stranded breaks. Biochem J 475:1955–1964

Endo M, Mikami M, Endo A, Kaya H, Itoh T, Nishimasu H, Nureki O, Toki S (2019) Genome editing in plants by engineered CRISPR–Cas9 recognizing NG PAM. Nat Plants 5:14–17

Fang Y, Xiong L (2015) General mechanisms of drought response and their application in drought resistance improvement in plants. Cell Mol Life Sci 72:673–689

Fang Y, Xie K, Xiong L (2014) Conserved *miR164*-targeted NAC genes negatively regulate drought resistance in rice. J Exp Bot 65:2119–2135

Feng C, Yuan J, Wang R, Liu Y, Birchler JA, Han F (2015) Efficient targeted genome modification in maize using CRISPR/Cas9 system. J Genet Genomics 43(1):37–43

Figueroa-Yañez L, Pereira-Santana A, Arroyo-Herrera A, Rodriguez-Corona U, Sanchez-Teyer F, Espadas-Alcocer J, Espadas-Gil F, Barredo-Pool F, Castaño E, Rodriguez-Zapata LC (2016) RAP2.4a is transported through the phloem to regulate cold and heat tolerance in papaya tree (Carica papaya cv. Maradol): implications for protection against abiotic stress. PLoS One 11 (10):e0165030

Fischer RA, Edmeades GO (2010) Breeding and cereal yield progress. Crop Sci 50:85–98

Francia E, Barabaschi D, Tondelli A, Laidò G, Rizza F, Stanca AM, Busconi M, Fogher C, Stockinger EJ, Pecchioni N (2007) Fine mapping of a HvCBF gene cluster at the frost resistance locus Fr-H2 in barley. Theor Appl Genet 115:1083–1091

Fukao T, Xu K, Ronald PC, Bailey-Serres J (2006) A variable cluster of ethylene response factor–like genes regulates metabolic and developmental acclimation responses to submergence in rice. Plant Cell 18(8):2021–2034

Gao H, Gadlage MJ, Lafitte HR, Lenderts B, Yang M, Schroder M, Farrell J, Snopek K, Peterson D, Feigenbutz L, Jones S, St Clair G, Rahe M, Sanyour-Doyel N, Peng C, Wang L, Young JK, Beatty M, Dahlke B, Hazebroek J, Greene TW, Cigan AM, Chilcoat ND, Meeley RB (2020) Superior field performance of waxy corn engineered using CRISPR–Cas9. Nat Biotechnol 38:579–581

Gaudelli NM, Komor AC, Rees HA, Packer MS, Badran AH, Bryson DI, Liu DR (2017) Programmable base editing of AT to GC in genomic DNA without DNA cleavage. Nature 551:464–471

González MN, Massa GA, Andersson M, Turesson H, Olsson N, Fält A-S, Storani L, Décima Oneto CA, Hofvander P, Feingold SE (2020) Reduced enzymatic browning in potato tubers by specific editing of a polyphenol oxidase gene via ribonucleoprotein complexes delivery of the CRISPR/Cas9 system. Front Plant Sci 10:1649

Gruner K, Esser T, Acevedo-Garcia J, Freh M, Habig M, Strugala R, Stukenbrock E, Schaffrath U, Panstruga R (2020) Evidence for allele-specific levels of enhanced susceptibility of wheat *mlo* mutants to the hemibiotrophic fungal pathogen *Magnaporthe oryzae* pv. *Triticum*. Genes 11:517

Hattori Y, Nagai K, Furukawa S, Song X-J, Kawano R, Sakakibara H, Wu J, Matsumoto T, Yoshimura A, Kitano H, Matsuoka M, Mori H, Ashikari M (2009) The ethylene response factors SNORKEL1 and SNORKEL2 allow rice to adapt to deep water. Nature 460 (7258):1026–1030

Haun W, Coffman A, Clasen BM, Demorest ZL, Lowy A, Ray E, Retterath A, Stoddard T, Juillerat A, Cedrone F, Mathis L, Voytas DF, Zhang F (2014) Improved soybean oil quality by targeted mutagenesis of the fatty acid desaturase 2 gene family. Plant Biotechnol J 12:934–940

He P, Zhao P, Wang L, Zhang Y, Wang X, Xiao H, Yu J, Xiao G (2017) The PIN gene family in cotton (Gossypium hirsutum): genome-wide identification and gene expression analyses during root development and abiotic stress responses. BMC Genomics 18:507

Hua K, Tao X, Zhu JK (2018) Expanding the base editing scope in rice by using Cas9 variants. Plant Biotechnol J 17:499–504

Huang XZ, Zeng XF, Li JR, Zhao DG (2017) Construction and analysis of *tifyla* and *tify1b* mutants in rice (*Oryza sativa*) based on CRISPR/Cas9 technology. J Agric Biotechnol 25:1003–1012

Intergovernmental Panel on Climate Change (2018) Summary for policymakers. In: Masson-Delmotte V, Zhai P, Pörtner HO et al (eds) Global warming of 1.5°C. World Meteorological Organization, Geneva, Switzerland, p 32

Jaganathan D, Ramasamy K, Sellamuthu G, Jayabalan S, Venkataraman G (2018) CRISPR for crop improvement: an update review. Front Plant Sci 9:985

Jia H, Orbović V, Wang N (2019) CRISPR-LbCas12a-mediated modification of citrus. Plant Biotechnol J 17:499–504

Jiang WZ, Henry IM, Lynagh PG, Comai L, Cahoon EB, Weeks DP (2017) Significant enhancement of fatty acid composition in seeds of the allohexaploid, Camelina sativa, using CRISPR/Cas9 gene editing. Plant Biotechnol J 15:648–657

Jones HD (2015) Future of breeding by genome editing is in the hands of regulators. GM Crops Food 6(4):223–232

Joshi RK, Bharat SS, Mishra R (2020) Engineering drought tolerance in plants through CRISPR/Cas genome editing. 3 Biotech 10:400

Joung JK, Sander JD (2013) TALENs: a widely applicable technology for targeted genome editing. Nat Rev Mol Cell Biol 14(1):49–55

Keating BA, Carberry PS, Bindraban PS, Asseng S, Meinke H, Dixon J (2010) Eco-efficient agriculture: concepts, challenges, and opportunities. Crop Sci 50:109–119

Khan SH (2019) Genome-Editing Technologies: concept, pros, and cons of various genome-editing techniques and bioethical concerns for clinical application. Mol Ther Nucl Acids 16:326–334

Kim D, Alptekin B, Budak H (2018) CRISPR/Cas9 genome editing in wheat. Funct Integr Genomics 18:31–41

Kim H, Kim S, Ryu J, Kang BC, Kim JS, Kim SG (2017) CRISPR/Cpf1-mediated DNA-free plant genome editing. Nat Commun 8:14406

Knox AK, Dhillon T, Cheng H, Tondelli A, Pecchioni N, Stockinger EJ (2010) CBF gene copy number variation at frost resistance-2 is associated with levels of freezing tolerance in temperate-climate cereals. Theor Appl Genet 121:21–35

Li B, Rui H, Li Y, Wang Q, Alariqi M, Qin L, Sun L, Ding X, Wang F, Zou J, Wang Y, Yuan D, Zhang X, Jin S (2019) Robust CRISPR/Cpf1 (Cas12a)-mediated genome editing in allotetraploid cotton (Gossypium hirsutum). Plant Biotechnol J 17(10):1862–1864

Li J, Zhang H, Si X, Tian Y, Chen K, Liu J, Chen H, Gao C (2017) Generation of thermosensitive male-sterile maize by targeted knockout of the ZmTMS5 gene. J Genet Genomics 44:465–468

Li M, Li X, Zhou Z, Wu P, Fang M, Pan X, Lin Q, Luo W, Wu G, Li H (2016) Reassessment of the four yield-related genes Gn1a, DEP1, GS3, and IPA1 in rice using a CRISPR/Cas9 system. Front Plant Sci 7:377

Li T, Liu B, Spalding M, Weeks DP, Yang B (2012) High-efficiency TALEN-based gene editing produces disease-resistant rice. Nat Biotechnol 30:390–392

Li XM, Chao DY, Wu Y, Huang X, Chen K, Cui LG, Su L, Ye WW, Chen H, Chen HC, Dong NQ, Guo T, Shi M, Feng Q, Zhang P, Han B, Shan JX, Gao JP, Lin HX (2015) Natural alleles of a proteasome α2 subunit gene contribute to thermotolerance and adaptation of African rice. Nat Genet 47:827–833

Liang Z, Chen K, Yan Y, Zhang Y, Gao C (2018) Genotyping genome-edited mutations in plants using CRISPR ribonucleoprotein complexes. Plant Biotechnol J 16:2053–2062

Lloyd A, Plaisier CL, Carroll D, Drews GN (2005) Targeted mutagenesis using zinc-finger nucleases in Arabidopsis. Proc Natl Acad Sci USA 102:2232–2237

Lou D, Wang H, Liang G, Yu D (2017) OsSAPK2 confers abscisic acid sensitivity and tolerance to drought stress in rice. Front Plant Sci 8:993

Macovei A, Sevilla NR, Cantos C, Jonson GB, Slamet-Loedin I, Čermák T, Voytas DF, Choi IR, Chadha-Mohanty P (2018) Novel alleles of rice eIF4G generated by CRISPR/Cas9-targeted mutagenesis confer resistance to Rice tungro spherical virus. Plant Biotechnol J 16:1918–1927

Marraffini LA, Sontheimer EJ (2008) CRISPR interference limits horizontal gene transfer in staphylococci by targeting DNA. Science 322:1843–1845

Metje-Sprink J, Sprink T, Hartung F (2020) Genome-edited plants in the field. Curr Opin Biotechnol 61:1–6

Miao H, Sun P, Liu Q, Miao Y, Liu J, Zhang K, Hu W, Zhang J, Wang J, Wang Z, Jia C, Xu B, Jin Z (2017a) Genome-wide analyses of SWEET family proteins reveal involvement in fruit development and abiotic/biotic stress responses in banana. Sci Rep 7:3536

Miao H, Sun P, Liu Q, Liu J, Xu B, Jin Z (2017b) The AGPase family proteins in banana: genome-wide identification, phylogeny, and expression analyses reveal their involvement in the development, ripening, and abiotic/biotic stress responses. Int J Mol Sci 18(8):1581

Miao C, Xiao L, Hua K, Zou C, Zhao Y, Bressan RA, Zhu JK (2018) Mutations in a subfamily of abscisic acid receptor genes promote rice growth and productivity. Proc Natl Acad Sci USA 115:6058–6063

Miller JC, Patil DP, Xia DF, Paine CB, Fauser F, Richards HW, Shivak DA, Bendaña YR, Hinkley SJ, Scarlott NA, Lam SC, Reik A, Zhou Y, Paschon DE, Li P, Wangzor T, Lee G, Zhang L, Rebar EJ (2019) Enhancing gene editing specificity by attenuating DNA cleavage kinetics. Nat Biotechnol 37:945–952

Mishra R, Joshi RK, Zhao K (2020) Base editing in crops: current advances, limitations and future implications. Plant Biotechnol J 18:20–31

Mussolino C, Morbitzer R, Lutge F, Dannemann N, Lahaye T, Cathomen T (2011) A novel TALE nuclease scaffold enables high genome editing activity in combination with low toxicity. Nucleic Acids Res 39:9283–9293

Osakabe K, Osakabe Y, Toki S (2010) Site-directed mutagenesis in Arabidopsis using custom-designed zinc finger nucleases. Proc Natl Acad Sci USA 107:12034–12039

Osakabe Y, Osakabe K (2015) Genome editing with engineered nucleases in plants. Plant Cell Physiol 56(3):389–400

Osakabe Y, Osakabe K, Shinozaki K, Tran LS (2014) Response of plants to water stress. Front Plant Sci 5:86

Ou W, Mao X, Huang C, Tie W, Yan Y, Ding Z, Wu C, Xia Z, Wang W, Zhou S, Li K, Hu W (2018) Genome-wide identification and expression analysis of the KUP family under abiotic stress in cassava (*Manihot esculenta* Crantz). Front Physiol 9:17

Padaria JC, Vishwakarma H, Biswas K, Jasrotia RS, Singh GP (2014) Molecular cloning and in silico characterization of high temperature stress responsive *pAPX* gene isolated from heat tolerant Indian wheat cv. Raj 3765. BMC Res Notes 7:713

Pareek A, Dhankher OP, Foyer CH (2020) Mitigating the impact of climate change on plant productivity and ecosystem sustainability. J Exp Bot 71(2):451–456

Paschon DE, Lussier S, Wangzor T et al (2019) Diversifying the structure of zinc finger nucleases for high-precision genome editing. Nat Commun 10:1133

Petolino JF, Worden A, Curlee K, Connell J, Strange Moynahan TL, Larsen C, Russell S (2010) Zinc finger nuclease-mediated transgene deletion. Plant Mol Biol 73:617–628

Porter JR, Xie L, Challinor AJ, Cochrane K, Howden SM, Iqbal MM, Lobell DB, Travasso MI (2014) Food security and food production systems. In: Field CB, Barros VR, Dokken DJ, Mach KJ, Mastrandrea MD, Bilir TE, Chatterjee M, Evi KL, Estrada YO, Genova RC, Girma B, Kissel ES, Levy AN, MacCracken S, Mastrandrea PR, White LL (eds) Climate change 2014: impacts, adaptation, and vulnerability. Part A: global and sectoral aspects. Contribution of working group II to the fifth assessment report of the intergovernmental panel on climate change. Cambridge University Press, Cambridge, pp 485–533

Ran Y, Liang Z, Gao C (2017) Current and future editing reagent delivery systems for plant genome editing. Sci China Life Sci 60:490–505

Rees H, Komor A, Yeh W, Lopes JC, Warman M, Edge ASB, Liu DR (2017) Improving the DNA specificity and applicability of base editing through protein engineering and protein delivery. Nat Commun 8:15790

Roca-Paixão JF, Gillet FX, Ribeiro TP, Bournaud C, Lourenço-Tessutti IT, Noriega DD, Melo BP, de Almeida-Engler J, Grossi-de-Sa MF (2019) Improved drought stress tolerance in Arabidopsis by CRISPR/dCas9 fusion with a histone acetyl transferase. Sci Rep 9:8080

Rouet P, Smih F, Jasin M (1994) Introduction of double-strand breaks into the genome of mouse cells by expression of a rare-cutting endonuclease. Mol Cell Biol 14:8096–8106

Sedeek KEM, Mahas A, Mahfouz M (2019) Plant genome engineering for targeted improvement of crop traits. Front Plant Sci 10:114

Shan Q, Wang Y, Li J, Gao C (2014) Genome editing in rice and wheat using the CRISPR/Cas system. Nat Protoc 9:2395

Shen C, Que Z, Xia Y, Tang N, Li D, He R, Cao M (2017) Knock out of the annexin gene *OsAnn3* via CRISPR/Cas9-mediated genome editing decreased cold tolerance in rice. J of Plant Bio 60:539–547

Shi J, Huirong G, Hongyu W, Renee LH, Archibald RL, Yang M, Hakimi SM, Mo H, Habben JE (2017) ARGOS8 variants generated by CRISPR-Cas9 improve maize grain yield under field drought stress conditions. Plant Biotechnol J 15:207–216

Shim JS, Oh N, Chung PJ, Kim YS, Choi YD, Kim JK (2018) Overexpression of OsNAC14 improves drought tolerance in rice. Front Plant Sci 9:310

Shimatani Z, Kashojiya S, Takayama M, Terada R, Arazoe T, Ishii H, Teramura H, Yamamoto T, Komatsu H, Miura K, Ezura H, Nishida K, Ariizumi T, Kondo A (2017) Targeted base editing in rice and tomato using a CRISPR-Cas9 cytidine deaminase fusion. Nat Biotechnol 35:441–443

Shukla VK, Doyon Y, Miller JC, DeKelver RC, Moehle EA, Worden SE, Mitchell JC, Arnold NL, Gopalan S, Meng X, Choi VM, Rock JM, Wu YY, Katibah GE, Zhifang G, McCaskill D, Simpson MA, Blakeslee B, Greenwalt SA, Butler HJ, Hinkley SJ, Zhang L, Rebar EJ, Gregory PD, Urnov FD (2009) Precise genome modification in the crop species *Zea mays* using zincfinger nucleases. Nature 459:437–441

Singh B, Salaria N, Thakur K, Kukreja S, Gautam S, Goutam U (2019) Functional genomic approaches to improve crop plant heat stress tolerance. F1000Research 8:721

Simon JE, Rodriguez AS, Santiago VN (2018) CRISPR-Cas9: a precise approach to genome engineering. Ther Innov Regul Sci 52(6):701–707

Stockinger EJ, Skinner JS, Gardner KG, Francia E, Pecchioni N (2007) Expression levels of barley Cbf genes at the frost resistance-H2 locus are dependent upon alleles at Fr-H1 and Fr-H2. Plant J 51:308–321

Sun L, Nasrullah KF, Nie Z, Wang P, Xu J (2019) Citrus genetic engineering for disease resistance: past, present and future. Int J Mol Sci 20:5256

Tabata S, Kaneko T, Nakamura Y et al (2000) Sequence and analysis of chromosome 5 of the plant Arabidopsis thaliana. Nature 408:823–826

Tester M, Langridge P (2010) Breeding technologies to increase crop production in a changing world. Science 327:818–822

Tan J, Zhang F, Karcher D, Bock R (2019) Engineering of high-precision base editors for site-specific single nucleotide replacement. Nat Commun 10:439

Tang N, Zhang H, Li X, Xiao J, Xiong L (2012) Constitutive activation of transcription factor OsbZIP46 improves drought tolerance in rice. Plant Physiol 158:1755–1768

Teimourian S, Abdollahzadeh R (2015) Technology developments in biological tools for targeted genome surgery. Biotechnol Lett 37:29–39

Tian C, Zuo Z, Qiu JL (2015) Identification and characterization of ABA-responsive microRNAs in rice. J Genet Genom 42:393–402

Tian SW, Jiang LJ, Cui XX, Zhang J, Guo SG, Li MY, Zhang HY, Ren Y, Gong GY, Zong M, Liu F, Chen QJ, Xu Y (2018) Engineering herbicide-resistant watermelon variety through CRISPR/Cas9- mediated base-editing. Plant Cell Rep 37:1353–1356

Toda E, Koiso N, Takebayashi A, Ichikawa M, Kiba T, Osakabe K, Osakabe Y, Sakakibara H, Kato N, Okamoto T (2019) An efficient DNA- and selectable-marker-free genome-editing system using zygotes in rice. Nat Plants 5:363–368

Tripathi A, Tripathi DK, Chauhan DK, Kumar N, Singh GS (2016) Paradigms of climate change impacts on some major food sources of the world: a review on current knowledge and future prospects. Agr Ecosyst Environ 216:356–373

Van Meijl H, Havlik P, Lotze-Campen H, Stehfest E, Witzke P, Domínguez IP, Humpenöder F (2018) Comparing impacts of climate change and mitigation on global agriculture by 2050. Environ Res Lett 13:064021

Varshney RK, Mir RR, Bhatia S, Thudi M, Hu Y, Azam S, Zhang Y, Jaganathan D, You FM, Gao J, Riera-Lizarazu O, Luo MC (2014) Integrated physical, genetic and genome map of chickpea (*Cicer arietinum* L.). Funct Integr Genomics 14:59–73

Voytas DF, Gao C (2014) Precision genome engineering and agriculture: opportunities and regulatory challenges. PLoS Biol 12(6):e1001877

Wang F, Wang C, Liu P, Lei C, Hao W, Gao Y, Liu YG, Zhao K (2016) Enhanced rice blast resistance by CRISPR/Cas9-targeted mutagenesis of the ERF transcription factor gene OsERF922. PLoS One 11:e0154027

Wang J, Meng X, Hu X, Sun T, Li J, Wang K, Yu H (2019c) xCas9 expands the scope of genome editing with reduced efficiency in rice. Plant Biotechnol J 17:709–711

Wang L, Chen L, Li R, Zhao R, Yang M, Sheng J, Lin S (2017) Reduced drought tolerance by CRISPR/Cas9-mediated SlMAPK3 mutagenesis in tomato plants. J Agric Food Chem 65:8674–8682

Wang L, Chen S, Peng A, Xie Z, He Y, Zou X (2019a) CRISPR/Cas9-mediated editing of CsWRKY22 reduces susceptibility to *Xanthomonas citri* subsp. citri in Wanjincheng orange (*Citrus sinensis* (L.) Osbeck). Plant Biotechnol Rep 13:501–510

Wang S, Kurepa J, Smalle JA (2009) The Arabidopsis 26S proteasome subunit RPN1a is required for optimal plant growth and stress responses. Plant Cell Physiol 50:1721–1725

Wang W, Wang W, Wu Y, Li Q, Zhang G, Shi R, Yang J, Wang W (2019b) The involvement of wheat (*Triticum aestivum* L.) U-box E3 ubiquitin ligase TaPUB1 in salt stress tolerance. J Integ Plant Biol 62:631–651

Waqas MA, Kaya C, Riaz A, Farooq M, Nawaz I, Wilkes A, Li Y (2019) Potential mechanisms of abiotic stress tolerance in crop plants induced by thiourea. Front Plant Sci 10:1336

Weinthal D, Tovkach A, Zeevi V, Tzfira T (2010) Genome editing in plant cells by zinc finger nucleases. Trends Plant Sci 15(6):308–321

Woo JW, Kim J, Kwon SI, Corvalan C, Cho SW, Kim H, Kim SG, Kim ST, Choe S, Kim JS (2015) DNA-free genome editing in plants with preassembled CRISPR-Cas9 ribonucleoproteins. Nat Biotechnol 33:1162–1164

Xiong JS, Ding J, Li Y (2015) Genome-editing technologies and their potential application in horticultural crop breeding. Hortic Res 2:15019

Yang QQ, Zhang CQ, Chan ML, Zhao DS, Chen JZ, Wang Q, Li QF, Yu HX, Gu MH, Sun SSM, Liu QQ (2016) Biofortification of rice with the essential amino acid lysine: Molecular characterization, nutritional evaluation, and field performance. J Exp Bot 67:4285–4296

Ye J, Yang H, Shi H, Wei Y, Tie W, Ding Z, Yan Y, Luo Y, Xia Z, Wang W, Peng M, Li K, Zhang H, Hu W (2017) The MAPKKK gene family in cassava: genome-wide identification and expression analysis against drought stress. Sci Rep 7:14939

Yeh CD, Richardson CD, Corn JE (2019) Advances in genome editing through control of DNA repair pathways. Nat Cell Biol 21:1468–1478

Yin K, Qiu JL (2019) Genome editing for plant disease resistance: applications and perspectives. Phil Trans R Soc B 374:20180322

Yoon YJ, Venkatesh J, Lee JH, Kim J, Lee HE, Kim DS, Kang BC (2020) Genome editing of eIF4E1 in tomato confers resistance to pepper mottle virus. Front Plant Sci 11:1098

Zafar SA, Zaidi SS, Gaba Y, Singla-Pareek SL, Dhankher OP, Li X, Mansoor S, Pareek A (2020) Engineering abiotic stress tolerance via CRISPR/ Cas-mediated genome editing. J Exp Bot 71 (2):470–479

Zetsche B, Gootenberg JS, Abudayyeh OO, Slaymaker IM, Makarova KS, Essletzbichler P, Volz SE, Joung J, Oost J, Regev A, Koonin EV, Zhang F (2015) Cpf1 is a single RNA-guided endonuclease of a class 2 CRISPR-Cas system. Cell 163(3):759–771

Zhang A, Liu Y, Wang F, Li T, Chen Z, Kong D, Bi J, Zhang F, Luo X, Wang J, Tang J, Yu X, Liu G, Luo L (2019) Enhanced rice salinity tolerance via CRISPR/Cas9-targeted mutagenesis of the OsRR22 gene. Mol Breed 39:47

Zhang F, Maeder ML, Unger-Wallace E, Hoshaw JP, Reyon D, Christian M, Li X, Pierick CJ, Dobbs D, Peterson T, Joung JK, Voytas DF (2010) High frequency targeted mutagenesis in *Arabidopsis thaliana* using zinc finger nucleases. Proc Natl Acad Sci USA 107:12028–12033

Zhang H, Zhang J, Wei P, Zhang B, Gou F, Feng Z, Mao Y, Yang L, Zhang H, Xu N, Zhu JK (2014) The CRISPR/Cas9 system produces specific and homozygous targeted gene editing in rice in one generation. Plant Biotechnol J 12:797–807

Zhang XH, Tee LY, Wang XG, Huang QS, Yang SH (2015) Off-target effects in CRISPR/Cas9-mediated genome engineering. Mol Ther Nucleic Acids 4:e264

Zhang Y, Liang Z, Zong Y, Wang Y, Liu J, Chen K, Qiu JL, Gao C (2016) Efficient and transgene-free genome editing in wheat through transient expression of CRISPR/Cas9 DNA or RNA. Nat Commun 7:12617

Zhang Y, Massel K, Godwin ID, Gao C (2018) Applications and potential of genome editing in crop improvement. Genome Biol 19:210

Zhao Y, Zhang C, Liu W, Gao W, Liu C, Song G, Li WX, Mao L, Chen B, Xu Y, Li X, Xie C (2016) An alternative strategy for targeted gene replacement in plants using a dual-Sgrna/Cas9 design. Sci Rep 6:23890

Zong Y, Wang Y, Li C, Zhang R, Chen K, Ran Y, Qiu JL, Wang D, Gao C (2017) Precise base editing in rice, wheat and maize with a Cas9-cytidine deaminase fusion. Nat Biotechnol 35:438–440

Part IV

Climate Change and the Environmental Microbiology

Role of Soil Microbial Flora in Remediation of Hydrocarbon Stressed Soils

16

Khalida Khatoon, Mohammad Anas, Zarreena Siddiqui, and Abdul Malik

Abstract

Change in climate has evolved as a global threat to the environment, apparently affecting all ecosystems on the earth. The indiscriminate release of xenobiotics into the environment causes grave and usually irreparable impairment of ecosystems. Hydrocarbons are a group of compounds basically consisting of carbon and hydrogen and are major components of pesticides, oils and natural gas. These substances are known to contribute to the ozone depletion, greenhouse effect and global warming. Their role has also been implicated in increased occurrence of cancer and respiratory disorders, reduced photosynthetic ability in plant systems, extensive damage to the aquatic ecosystems in the form of oil spills, etc. Soil is known to be a natural and preferred sink of contamination among all the environmental compartments. Polycyclic aromatic hydrocarbons (PAHs) enter into the soil through various sources, viz. partial combustion of oil, wood, coal and petroleum, atmospheric wet and dry deposition, petrochemical industries, industrial wastewater, discharge from vehicles, equipment cleaning and maintenance, and through random spills over the surface of soil or into it. In contaminated soils, microorganisms and plants have to face major challenges for their survival and growth. The pollutants residing in soil bring about selection pressure on the native microorganisms. As a result, microbial flora gets inclined towards a specific organism type that evolves a method of degrading and making use of the pollutants such as PAHs. Thus, microorganisms serve as major players in restraining those stresses and recycling elements by mineralizing or fractioning pollutants, even in environment having inadequate availability of major electron donors and acceptors or having nutrient deficiency. This chapter summarizes the

K. Khatoon (✉) · M. Anas · Z. Siddiqui · A. Malik
Department of Agricultural Microbiology, Faculty of Agricultural Sciences, Aligarh Muslim University, Aligarh, Uttar Pradesh, India

S. A. Lone, A. Malik (eds.), *Microbiomes and the Global Climate Change*,
https://doi.org/10.1007/978-981-33-4508-9_16

important role played by microorganisms for the detoxification processes in the soil environment as healthy soil can ally against climate change.

Keywords

Xenobiotics · Polycyclic aromatic hydrocarbons · Bioremediation · Soil restoration

16.1 Introduction

The ecological imbalances and climate change critically impact the condition of living organisms, which is due to quantum increase in pollution resulting from over-utilization of natural resources, excessive usage of harmful chemicals, their high demand and transportation. Several environmental pollutants generated from the anthropogenic activities have been recognized to be toxic for living organisms once discharged into the ecosystem, they have high tendency to remain persistent for longer period of time in soil than in air and water. The soil serves to be a sink or a filter wherein pollutants are collected very fast but decline slowly (Lombi and Hamon 2005).

Polycyclic aromatic hydrocarbons (PAHs) represent one of the main environmental contaminants (Okoh 2006) which, in soil, may be adsorbed to the surface of organic elements and minerals, stabilize within the fissures and pores of soil and may form a continuous layer above the soil surface or may be present in mobile form (Trofimov and Rozanova 2003). Contamination of soil with petroleum oil disturbs certain soil factors, such as moisture content, pH, temperature, minerals and organic matter content. Due to the hydrophobicity of PAHs, polluted soils are characterized by low moisture content, low water holding capacity and electrical conductivity as compared to pristine soils (Nwazue 2011).

Soil can be considered as the preferred natural sink for the contamination among all the environmental compartments and has been focused for looking at ways to mitigate climate change for being a huge repository for carbon as it contains the carbon compounds of dead and decaying animals and plants, etc. The decontamination of polluted soil has become an issue of social and political interest due to its importance in maintenance of human health and protection of environment as well as its economical relevance (Lombi and Hamon 2005). Carbon is necessary for soil microorganisms to be able to flourish, having a total number $\sim 10^{29}$ bacterial cells in terrestrial environments (McMahon and Parnell 2014). They are key drivers of several processes, viz. inorganic nutrient cycling, organic matter decomposition, pollutant degradation and diseases (Swift et al. 1979; Bardgett et al. 2005) and are involved in various symbiotic associations, tightly linked to above ground communities through trophic interactions, plays important role in biogeochemical cycling and plant–soil interactions (Nielsen et al. 2015). Soil microbes regulate the organic carbon content, stored in soil and release it back to the atmosphere, directly influence the stored carbon in plants and soils by regulating macronutrients

(Nitrogen and Phosphorous) which ultimately regulate the soil productivity. On one hand, microbial population majorly contribute in carbon sequestration and on the other hand, they contribute substantially in greenhouse gas emission such as CO_2, CH_4 and N_2O through heterotrophic respiration, methanogenesis and denitrification, respectively. However, microorganisms also provide important opportunities in remediation of problems caused due to human beings in order to get better agricultural outcomes, biofuel production and treatment of pollutants.

Autochthonous microorganisms are proficient in PAHs removal from contaminated soils. Nevertheless, soil conditions, e.g. nutrient deficiency can retard microbial activity and so mineralization of the contaminants. By managing environmental components like dissolved O_2 and water content, pH, temperature and availability of nutrients, bioremediation potential can be improved. Knowledge of contaminated system can be the best advisor for selection of appropriate bioremediation strategy in order to remove the hydrocarbons from the polluted environment efficiently.

16.2 Anthropogenic Pollution

Our environment is extremely complex system divided in abiotic and biotic components, between them exchange of matter and energy constantly takes place. These exchange processes should be balanced. However, quantum increase in pollution in each sphere of earth has led to ecological imbalances and climate change which in turn affects the living condition of microorganisms in a grave way and combating it is becoming a daunting task (Moschella et al. 2005). Its effect will be aggravated by ever increasing human population, like increasing demand for food, global transport, industrialization, etc. which will virtually have impact on every ecosystem on the earth for many generations. Organic contaminants, heavy metals and metalloids are the two primary groups of pollutants in soil, adversely affecting the physiological systems in every animal species tested. Several new classes of chemicals such as halogenated organic compounds, alkyl phenols, polycyclic aromatic hydrocarbons and phthalates are produced for domestic, industrial as well as agricultural use and are predestined to be released into the environment (Colborn et al. 1993; Rhind 2009). The increase in mutagenic compounds due to the anthropogenic activities becomes a serious problem recently. These chemicals can go through several processes in the environment based on their physicochemical properties, e.g. volatile compounds may pollute the air, hydrophilic substances dissolve in water and hydrophobic compounds buildup in soil and sediments (Zenker et al. 2014).

16.3 Environmental Contaminants and their Fate

Environmental contamination is linked with the society since the beginning of industrialization (Blasco and Picó 2009). The source of environmental contamination includes uncontrolled human activities such as urbanization, industrial set-ups, agricultural practices, automobiles, accidental spillages and illegal dumping (Wasi et al. 2013). Soil pollution includes the presence of undesirable chemicals in soil that are either misplaced or present at higher than the normal concentrations. Such chemicals may be discharged into the soil by seepage from a landfill, infiltration of contaminated water into the soil, rupture of underground storage tanks, mining and industrial activities or by sewer and waste mismanagement (Fig. 16.1). The release of such chemicals results in the accumulation of persistent toxic compounds, radioactive substances, salts and chemicals and disease causing agents, adversely affecting plant growth and animal health. Petroleum hydrocarbons, heavy metals, volatile organic compounds (VOCs) and solvents are the most common chemicals which causes soil pollution (Fig. 16.2). Agricultural inputs, viz. fertilizers, herbicides, pesticides as well as antibiotics present in animal manure also potentially pollute the soil system and may accumulate to such an extent that hampers the soil functions. They may also pollute the ground and surface water and ultimately drinking water supplies as well as aquatic ecosystems.

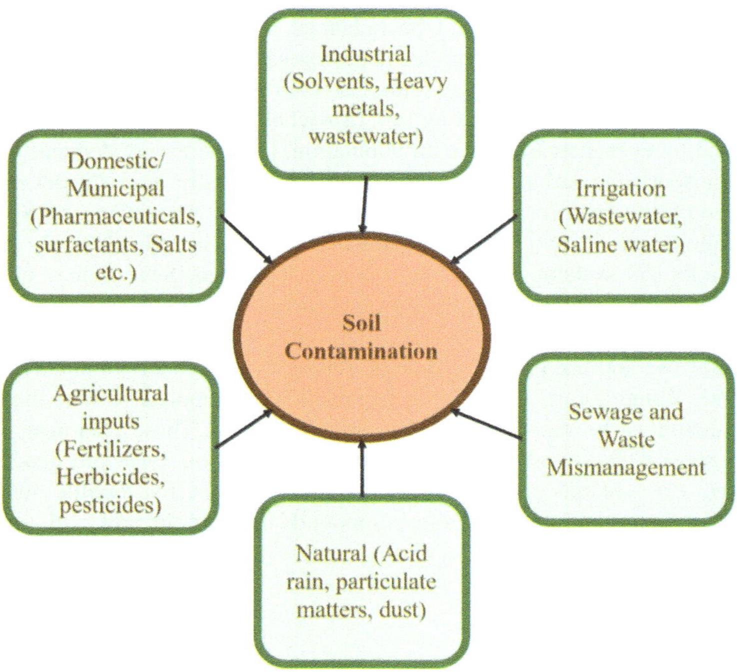

Fig. 16.1 Sources of soil contamination. Source: Yaron et al. (2012)

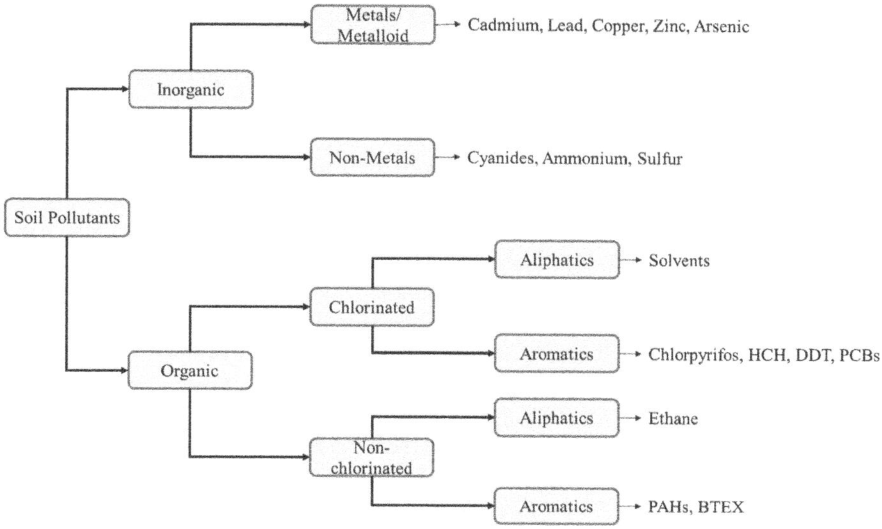

Fig. 16.2 Some examples of pollutants found in soil. Source: Modified from Swartjes (2011)

16.4 Effects of Xenobiotics on Soil Quality

Some of the main ecological functions of soil which are of vital importance are filtration, buffering and transformation of inorganic contaminants such as heavy metals and metalloids (e.g. arsenic, mercury, lead, cadmium, chromium) and organic contaminating agents like PAHs (e.g. benzo[a]pyrene), persistent organic pollutants (e.g. polychlorinated biphenyl) and emerging pollutants (per- and poly-fluoroalkyl substances, polybrominated biphenyls), etc. which bring forward unique issues and challenges to the quality of the environment (ATSDR 2017). These essential services ensure better quality of groundwater and safer food production (Blum 2005). Status of the World's Soil resources Report (FAO and ITPS 2015) identified nutrient imbalances and soil acidification, as the two major issues produced by certain pollutants. Moreover, salinization, another major issue affecting many soils, particularly the soils close to certain industrial activities such as chlor-alkali, textiles, glass, rubber production, pharmaceuticals, oil and gas drilling, detergent and soap manufacture, pigment manufacture, metal processing and ceramic manufacture as well as animal hide processing and leather tanning, makes soil unfit for agricultural use which often becomes useless and barren (Saha et al. 2017). Many beneficial soil organisms such as earthworms can die in soil due to the modification of soil structure as a result of soil pollution. Food crops grown in contaminated soil usually produce lower yields than non-contaminated soil which in turn can be more harmful because of the reduced number of plants on soil will cause more erosion and ultimately polluting the land that might have not been tainted before by spreading the contaminants.

Soil moisture, temperature, mineral and organic matter content and microbial activities are some soil properties and processes which control the toxicological facets of pollutants and are vulnerable to climate change linked with global warming such as increased rainfall, enhanced dry periods, soil erosion and elevation in sea level. The properties of pollutants (shape, size, solubility, molecular structure, acid-base nature of the molecule and charge distribution) themselves are also very important (Gevao et al. 2000).

Pollutants undergo biochemical, microbiological and physicochemical processes due to which they are either retained in the soil or are degraded after entering the soil. Available fractions of such pollutants in soil are usually toxic to soil organisms and humans. Withal, the pollutants will change the makeup of soil and the native microbial population. Microorganisms decompose and recycle the organic contaminants present in the soil and also help plants in absorption of essential nutrients (Naidu et al. 2008; Poggio et al. 2009; Khan et al. 2018).

16.5 Role of Healthy Soil in Climate Change

Soil is considered as a major carbon store, worthy of protection as well as development wherever possible and have a significant role in the distribution and fate of persistent organic pollutants in the environment considering the fact that they have a great retention potential and may serve as re-emission sources for the atmosphere (Harner et al. 2001; Barra et al. 2005). In fact, soil has carbon more than the combined carbon content of the atmosphere and earth's vegetation that is why even a minute imbalance in soil carbon could result in major consequences for atmosphere and climate of earth. However, the interrelation between climate change and alterations in soil quality are complicated and still under study.

Soil is a major factor in tackling climate change because of being the second largest carbon store after the oceans. Carbon sequestration in the soil cannot be the only solution due to the limited magnitude of its effect and potential reversibility. However, it may have an important role in mitigating the climate change in short term together with other measures, particularly due to its instant availability and the exceptionally low cost. Soil organic matter consists of over 50% carbon. Losing carbon rich organic matter from soils releases CO_2, a greenhouse gas that can enhance global warming but by regeneration of soils, more carbon can be sequestered underground and thus slowdown the global warming.

16.6 Role of Microbial Flora in the Degradation of PAHs

Microbial communities are the main contributors in elemental recycling and are essential in the bioremediation of polluted environments as well as turnover of waste substances. Despite the fact that volatilization, adsorption and chemical degradation are also involved in the removal process of PAHs from the contaminated soil, microbial degradation is the foremost degradation method of PAHs, which relies upon the microbial communities of contaminated site as well as environmental conditions. Hydrocarbon metabolizing microorganisms are extensively distributed in the contaminated environment and mainly act in controlling and determining their fate in all ecosystems (Ghazali et al. 2004; Brooijmans et al. 2009; Souza et al. 2014; Shen et al. 2015; Lamichhane et al. 2016).

Common petroleum hydrocarbon degrading microbes are found in all the domains of life such as bacteria, archaea and some eukaryotes like fungi and algae, they have the unique ability of degrading hydrocarbon components of petroleum by utilizing them as a sole carbon sources, thereby, liberating energy in subsequent cleavage of the hydrocarbon chains. So far, 79 bacterial, 9 cyanobacterial, 14 algal and 103 fungal genera have been recorded as hydrocarbon degrading and/or transforming microorganisms (Head et al. 2006; Hassanshahian et al. 2012).

The hydrocarbon degrading microbial population is dominant in petroleum contaminated ecosystems indicating their significance in the elimination of organic pollutants in such environments. It has been suggested that hydrocarbon degrading microbes interact with each other and non-hydrocarbon degraders within the communities in addition to the environmental factors, directly or indirectly (Head et al. 2006).

Biodegradation of petroleum hydrocarbons occurs only if microorganisms possess the appropriate enzyme systems (Atlas 1981; Fritsche and Hofrichter 2008; Singh et al. 2012). Native microbial population in aged contaminated soils are capable of successfully metabolising PAHs and can be exploited to enhance the PAHs removal from a newly contaminated soil. In situ conditions including temperature, water content, nutrient and aeration would possibly restrict the PAHs degradation (Li et al. 2009). Biosurfactant production is another important parameter in bioremediation processes which emulsify the petroleum hydrocarbons and thus reducing the surface tension and ultimately enhancing their bioavailability for easy accessibility to members of the microbial communities. The importance of biosurfactant producers in bioremediation processes has been well documented including certain strains of *Rhodococcus, Bacillus* and *Pseudomonas* species (Atlas 1995b; Colores et al. 2000; Iwabuchi et al. 2000; Ron and Rosenberg 2002; Pathak et al. 2008; Gudina et al. 2013).

16.6.1 Factors Affecting the Bacterial Degradation of PAHs

Various abiotic and biotic environmental variables can accelerate or relate the microbial activity are as follows (Fig. 16.3):

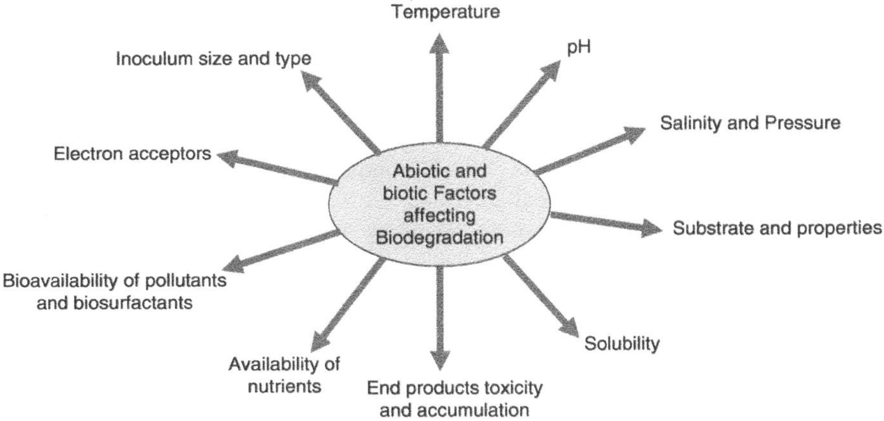

Fig. 16.3 Factors affecting bacterial degradation of PAHs

16.6.1.1 Abiotic Factors

Temperature

The temperature has a prominent role in the selection of microorganisms as well as in their growth and survival (Alexander 1999; Harmsen 2007) in addition to the biodegradation of PAHs in polluted sites since those sites have not the optimum temperature which is required for inhabitant microbial activity. Usually, microorganisms can survive within the much broader limit but for a particular microorganism optimum growth takes place within a narrow range of temperatures (Olson et al. 2003; Straube et al. 2003; Harmsen et al. 2007). With an increase in temperature, the solubility of PAHs increases and thus bioavailability of PAHs molecules also increases, while dissolved O_2 level decreases with increase in temperature that results in reduced metabolic activities of mesophilic aerobes. The optimum temperature range for the biodegradation is between 15 and 45 °C showing maximum rates of biodegradation in the range of 25–35 °C except for the psychrophiles that operate at low temperature (Brakstad and Bonaunet 2006) and thermophiles that function at high temperature (Feitkenhauer et al. 2003). Slower degradation is reported in cold climates as compared to temperate or tropical zones and the degradation rates usually decrease with depth through soil profile (Al-Turki and Dick 2003). Some pollutants may be transformed into new compound at higher temperature which usually tend to have more toxicity than the original one, which slows down or inhibits its rate of biodegradation (Müller et al. 1998).

pH

The pH of the contaminated sites has significant role to play in biodegradation of PAHs as it influences the solubility and bioavailability of nutrients and pollutants. Majority of bacteria are pH-sensitive, exhibit optimum growth at pH 6.5–7.5 and are favoured for normal microbial activities as several microbial processes such as enzyme activities and catalytic reaction balance as well as cell membrane transports. Laboratory-based studies on biodegradation have been done mostly in pH range of 5.0–9.0, (Yuan et al. 2001; Kim et al. 2003; Chang et al. 2008; Lu et al. 2011) as the acidic and alkaline conditions affect the normal microbial activities like cell membrane transports, enzymatic and catalytic reaction balance and ultimately creating the unfavourable conditions for microbial activities which leads to reduced biodegradation of PAHs (Bonomo et al. 2001). Soil pH also affects the mobility of nutrients and metals, e.g. maximum phosphorous solubilisation is at pH 6.5 and metal mobilization is minimized above pH 6.0 (Sims et al. 1990). Availability of some toxic metals can greatly reduce the rate of biodegradation of PAHs. Hence adjustment of pH at the contaminated sites is recommended by using chemicals such as $(NH_4)SO_4$ or NH_4NO_3 can be used for the treatment of basic soils while in acidic soils, $CaCO_3$ or $MgCO_3$ can be added to provide a favourable environment for effective biodegradation (Bowlen and Kosson 1995).

Salinity and Pressure

Salinity and pressure are distinct characteristics of a typical ecosystems, e.g. deep seas and saline lakes, representing particular environment that may be polluted by petroleum hydrocarbons (Atlas 1981). Effect of salinity and pressure on the biodegradation of petroleum hydrocarbons has been studied earlier (Atlas 1981; Minai-Tehrani et al. 2006; Das and Chandran 2011; Peixoto et al. 2011). The rate of hydrocarbon degradation decrease with the increasing level of salinity (Ward and Brock 1978). Chandra et al. (2013) also confirmed that the salinity and high temperatures have a negative impact on the microorganisms, hampering their growth and metabolic products. Minai-Tehrani et al. (2006) reported 41% crude oil degradation without NaCl and 12% degradation when 50 g/L NaCl was supplemented in soil samples in 4 months incubation period. Dastgheib et al. (2012) also found that at 5% salinity, 90% of phenanthrene was degraded but degradation as well as microbial growth was reduced at 15% and 17% salinity. Several researchers reported that salinity higher than 3% limits the microbial growth and degradation of PAHs (Bonfá et al. 2011; Cui et al. 2014; Guo et al. 2016). Schwarz et al. (1974, 1975) performed biodegradation of tetradecane, hexadecane and mixed hydrocarbons using deep sea sediment bacteria at high pressure (495 and 500 atm) and reported that in deep benthic zones they are degraded slowly and persist there for years or decades. Colwell et al. (1977) also reported that petroleum constituents may pollute deep benthic zones of the ocean on reaching deep-ocean environment and are biodegraded very slowly due to persistence of recalcitrant fractions of petroleum oil for longer time.

Substrate and Properties

Mineralization of petroleum hydrocarbons is also affected by its concentration (Leahy and Colwell 1990). As petroleum oil concentration increases, degradation rate and cumulative extents of mineralization increase while lag phase decreases (Towell et al. 2011). Extremely high concentrations of total petroleum hydrocarbon have been found to be lethal for microorganisms thus limiting their biodegradation potential, (Admon et al. 2001; Nzila 2018) while extremely low concentrations can reduce biodegradation due to the too low carbon supply that is not sufficient to support the growth of microorganisms (Leahy and Colwell 1990). Degradation ability of petroleum hydrocarbons also depends on its structure and composition (Atlas 1981; Varjani 2017): linear alkanes > branched alkanes > low molecular weight alkyl aromatics > mono-aromatics > cyclic alkanes > poly-aromatics > asphaltenes, indicating that hydrocarbon degrading microbes prefer less complex compounds to degrade (Atlas 1995a; Varjani and Upasani 2016b). Typically, the rate of biodegradation increases with decreasing molecular weight and chemical structure complexity of hydrocarbons (Varjani 2014). Biodegradation ability is innately affected by the composition of pollutants (Atlas 1981; Ruberto et al. 2003).

Solubility

The majority of the PAHs exhibit low aqueous solubility and do not dissolve effectively in water at normal temperature (Abbey et al. 2011; Lu et al. 2011). Solubility of PAHs in water is reliant upon temperature, pH, ionic quality and other organic compounds, such as dissolved organic carbon (Pierzynski et al. 2005). Rate of PAHs biodegradation vary depending upon the solubility of the PAHs (Alexander et al. 1999). Generally, low molecular weight PAHs are degraded more quickly than high molecular weight PAHs (Jarabak et al. 1997), e.g. benzo[e]pyrene, due to very low solubility is more recalcitrant to degradation (Cerniglia 1993; Wilson and Jones 1993; Juhasz and Naidu 2000). PAHs with lower solubility have been found to be more difficult to degrade in soil. They can be degraded only when they are released into the aqueous phase (i.e. desorption) which have PAHs degrading microbes (Rijnaarts et al. 1990; Volkering et al. 1992; Reid et al. 2000; Johnsen et al. 2005).

Availability of Nutrients

Several mineral nutrients such as nitrogen, phosphorus, potassium and iron along with easily metabolized carbon sources are essential for proper microbial growth and metabolism. Nutrients availability is considered another rate-limiting factor for successful bioremediation of PAHs contaminated environments. There is often lack of available nutrients that are essential for bacterial metabolism in contaminated sites due to high organic carbon contents (Breedveld and Sparrevik 2000). Thus, nutrient supplement is needed for the stimulation of the growth of indigenous microorganisms in the poor nutrient containing contaminated sites to enhance the bioremediation of pollutants (Atagana et al. 2003). Nutrient requirement varies site to site because of the differences in the type of soil, type and level of contaminants, type of microorganisms and environmental conditions which can be broadly categorized as micronutrients and macronutrients on the basis of quantity and need by the microorganisms. While studying the effects of addition of macro and micronutrients on bioremediation of PAHs and petroleum contaminated soil, Liebeg and Cutright (1999) reported the enhancement in the bioactivity of foreign consortium in high micronutrients and low macronutrient conditions. Therefore, to stimulate the autochthonous microbes to enhance bioremediation process, contaminated sites can be augmented with nutrients (Atagana et al. 2003). Nitrogen is the most commonly added nutrient in the form of ammonium chloride, ammonium nitrate, ammonia salt or urea for bioremediation as it is easily assimilated in bacterial metabolism and mainly utilized by microbes for their cellular growth. Sodium nitrate is preferable over ammonium nitrate because of its solubility and availability to microorganisms, whereas ammonium nitrate possesses adsorbent properties (Simarro et al. 2011). The next most commonly used nutrient is phosphorous which is also used by microorganisms as a source for their cellular growth and can be added as phosphate or polyphosphate and orthophosphoric salts. Floodgate (1984) reported poor biodegradation of petroleum hydrocarbons in marine environment because of low level of nitrogen and phosphorous. Bamforth and Singleton (2005) reported that growth of hydrocarbon degraders and rate of degradation can be strongly enhanced by the addition of N and P fertilizers. However, several authors have also reported that the high nutrient contents adversely influenced the biodegradation of organic contaminants especially PAHs, (Carmichael and Pfaender 1997; Chaineau et al. 2005) as excessive nutrient availability could inhibit the biodegradation process Chaillan et al. (2006). Bamforth and Singleton (2005) reported that microbial metabolism can be temporarily enhanced on addition of inorganic N and P but it may also inhibit the functionally important microbes leading to the failure of bioremediation.

End Products Toxicity and Accumulation

The ultimate goal of bioremediation is the detoxification or removal of pollutants from contaminated environments. However, sometimes there is a possibility of transformation of contaminants into more toxic end products. So confirmation of complete mineralization of pollutants becomes necessary at the end of the treatment (Mendonca and Picado 2002; Lundstedt et al. 2003). Lundstedt et al. (2003) studied in situ bioremediation of soil samples from gasworks contaminated with PAH using a bioreactor and removal of PAHs as well as accumulation of dead-end products, i.e. oxy-PAHs (PAH-ketones, quinines and coumarins which have more toxicity than the parent PAHs) was monitored; highlighting the importance of monitoring the metabolites during bioremediation processes, mainly the toxic dead-end products, and determining the toxicity of the metabolites both before and after treatment.

16.6.1.2 Biotic Factors

Inoculum Size and Type

Autochthonous microorganisms having the ability to utilize petroleum crude can be exploited for clean up of petroleum contaminated environment through bioremediation processes. Recalcitrant compounds are considered to be oxidized by the microbial consortium in multiple steps rather than to be used by a single microorganism as a growth substrate (Perry 1979). Hydrocarbon degrading microbial consortium has been proven better than the monoculture system, as combination of various enzymes converts more PAHs as compared to single isolate (Gupta et al. 2016). The optimum microbial population for petroleum hydrocarbons degradation was reported as 10^6 CFU/gram of pristine agricultural soil. The lag phase increases with the number of microbes increasing naturally or by the inoculation of hydrocarbon degrading microorganisms (Towell et al. 2011).

Electron Acceptors

Microbial degradation of petroleum constituents, e.g. PAHs can occur both via anaerobic and aerobic processes. Though, most of the researchers have focused on aerobic degradation of PAHs where O_2 is required to initially oxidize the aromatic rings by the action of monooxygenase as well as dioxygenase enzymes. O_2 acts as an electron acceptor as well as co-substrate and is considered as a rate-limiting factor, thereby increases the biodegradation activity (Gan et al. 2009; Abbasian et al. 2015; Waigi et al. 2015; Varjani 2017). Petroleum hydrocarbons limit the microbial growth by reducing soil air permeability (Meckenstock et al. 2016). The O_2 can be provided externally through mixing the O_2 liberating chemicals, drainage, tilling or by injecting the air into the contaminated site (Daniel et al. 1992; Atlas and Unterman 1999; Bewley and Webb 2001). However, sufficient O_2 supply for the complete mineralization of the hydrocarbons in soil is often troublesome and costly (Boopathy 2000).

Microbes also use sulphate (Heider et al. 1998), nitrate (Boll et al. 2014), iron (Weelink et al. 2009), manganese (Meckenstock et al. 2016) and CO_2 (Wilkes et al. 2016) as an alternative electron acceptors under anaerobic condition but they have not received much attention (Fuchs et al. 2011). These alternatives may be present at the polluted site and their supply is usually more economical than oxygen (Bertrand et al. 1990; Aeckersberg et al. 1991; Hutchins et al. 1991; Wilkes et al. 2016).

Bioavailability of Pollutants and Biosurfactants

Bioavailability is referred as the amount of substance that can be used or converted by organisms throughout the experiment and is considered as one of the critical factors in biodegradation of pollutants. It can differ under the impact of mass transfer parameters including physicochemical processes such as hydrological processes (e.g. mixing), dissolution, desorption, diffusion and biological processes including metabolism (Bosma et al. 1996; Semple et al. 2003). Low solubility of PAHs in water and their ability to adsorb onto soil organics, restrict their availability for microorganisms to degrade and thus making them persistent in the environment (Lai et al. 2009). The aqueous solubility of PAHs decreases with increasing molecular weight leading to their reduced bioavailability. PAHs bioavailability is also influenced by the soil properties such as texture, organic matter content and aggregation. Withal, heterogeneity of soil also have great influence on the bioavailability of PAHs (Eweis et al. 1998). The sorption may be more irreversible if the hydrophobic organic compounds (e.g. PAHs) are in contact with soil for longer duration and which results in lesser chemical and biological extractability of the contaminants.

This is referred to as "ageing of the contaminants" (Alexander 2000; Luo et al. 2012), which may significantly limit the rate of in situ bioremediation as extractability and bioavailability of PAHs extensively get reduced with time in ageing processes (Alexander 2000; Luo et al. 2012; Abdel-Shafy and Mansour 2016).

The growth rate of microorganisms on hydrocarbons becomes limiting because of interfacial surface area between water and oil, which ends up in less efficient degradation of hydrocarbons (Li and Chen 2009). The use of biosurfactants is a more favourable technique to enhance the bioavailability of PAHs (Si-Zhong et al. 2009; Santos et al. 2011; Zhao et al. 2011). Biosurfactants are amphiphilic molecules having hydrophilic and hydrophobic elements in their structure and are produced on living surfaces (Banat et al. 2010). Biodegradation of PAHs results in the production of a wide variety of biosurfactants which are produced particularly during the growth of microorganisms on the water-immiscible substrate which are either attached to surface of cell or released as extracellular molecules (Van Hamme et al. 2003; Sajna et al. 2015; Varjani and Upasani 2016a). Some biosurfactants producing bacteria such as *P. aeruginosa* (Nguyen et al. 2008), *Bacillus subtilis* (Vaz et al. 2012), *Bacillus subtilis* (Gudina et al. 2013), *Arthrobacter* spp. (Cameotra and Bollag 2003), *Lactobacillus paracasei* (Gudina et al. 2010), *Rhodococcus* spp. (Mutalik et al. 2008), *Flavobacterium* and *Staphylococcus* spp. (Ilori et al. 2005) have been reviewed (Varjani and Upasani 2017).

16.7 Genetics of PAHs Metabolism in Aerobic Bacteria

Fundamental knowledge of the PAH-degradative genes in various bacteria can provide valuable information related to the evaluation of enzyme structure-function relationships as well as evaluation and diversity of genes of catabolic pathway via horizontal gene transfer, DNA rearrangement, transposition events, point mutation, gene fusion, etc. From such information some molecular mechanisms may be advocated by means of which bacteria adapted to xenobiotics. In applied research, genetic information might be beneficial for monitoring of bacterial populations that degrade PAH in contaminated sites.

Microorganisms require the appropriate catabolic genes in order to be efficient degraders of a compound. Many of the PAHs degrading bacteria isolated from soil and sediments have comprehensive enzyme system encoded by the genes located on chromosomal DNA as well as plasmids (Johnsen et al. 2005). Catabolic or degradative plasmids possess structural genes, coding for degradation of several natural organic compounds as well as xenobiotics. By acquiring these genes through horizontal gene transfer (HGT), PAH-catabolic abilities can easily spread among bacteria in polluted soil (Wiedenbeck and Cohan 2011). A plasmid may encode a complete degradation pathway or partial degradative steps (Frantz and Chakrabarty 1986; Obayori and Salam 2010). There are a few plasmids that are specific for many substrates, e.g. NAH plasmids of many pseudomonads having genes that code for the upper as well as lower pathways of naphthalene, allowing the bacteria to grow on various two and three-ring PAHs, as sole carbon and energy sources (Foght and Westlake 1996). Some plasmid containing PAH degrading bacteria are listed in Table 16.1.

Table 16.1 Some plasmids bearing polycyclic aromatic hydrocarbons degrading bacterial strains

Substrate	Plasmid involved	Bacteria	References
Naphthalene	NAH7	*Pseudomonas putida* strains	Dunn and Gunsalus (1973)
	pDTG1	*Pseudomonas putida* NCIB9816	Kurkela et al. (1988)
	pLP6a	*Pseudomonas fluorescens* strain LP6a	Foght and Westlake (1996)
	pUC18	*Pseudomonas sp.* strain 112	Fuenmayor et al. (1998)
		Burkholderia strain RP007	Laurie and Lloyd-Jones (1999)
	pOV17 NPL-41	*P. aureofaciens* BS1393 *P. aureofaciens* BS1393	Zyakun et al. (2015)
Phenanthrene	pUC18	*Pseudomonas* strain C18	Denome et al. (1993)
	pNL	*Sphingomonas aromaticivorans* F199	Romine et al. (1999)
	pKS14	*Sphingomonas* strain KS14	Cho and Kim (2001)
	pPNY	*Staphylococcus* sp. PN/Y	Mallick et al. (2007)
	p4	*Sphingomonas* sp. HS362	Hwa et al. (2005)
		Pseudomonas sp. ARP26	Coral and Karagol (2005)
		Pseudomonas sp. ARP28	Coral and Karagol (2005)
	pKG2	*Beijerinckia* sp	Kiyohara et al. (1983)
	pLP6a	*Pseudomonas fluorescens* strain LP6a	Foght and Westlake (1996)
Anthracene	p4	*Pseudomonas* sp.	Kumar et al. (2010)
	pLP6a	*Pseudomonas fluorescens* strain LP6a	Foght and Westlake (1996)
Fluorene	pDBF1	*Terrabacter* sp. DBF63	Habe et al. (2005)
Pyrene	p4	*Mycobacterium* sp. KMS	Miller et al. (2007)
		Mycobacterium sp. PYR-GCK	Miller et al. (2007)
		Mycobacterium sp. MC	Miller et al. (2007)
	Megaplasmid	*Diaphorobacter*	Klankeo et al. (2009)
		Pseudoxanthomonas	
Mixed PAHs	pLA1	*Novosphingobium pentaromativorans* US6–1	Luo et al. (2012)

Source: modified from Obayori and Salam (2010)

16.8 Kinetics of Bacterial Degradation of PAHs

Kinetics of degradation by bacteria in natural and devised systems is crucial for estimating the risks from exposure of PAH and to implement the effective strategies for their removal. Several models for kinetics of PAH degradation have been tested such as a zero order, a first order, a kinetic Freundlich, and a two coupled first-order kinetic model (Thiele-Bruhn and Brümmer 2005). Zero order reaction has been more often found in case of negligible degradation and varying concentrations of PAHs, revealing linear degradation rates that are associated with the initial concentration of PAH. When low molecular weight PAHs are used as single substrate, degradation kinetics has been found to be best explained by a single first-order reaction in which initial lag phase could not be observed (Crampon et al. 2014). Many researchers demonstrated that PAHs degradation kinetics can be better represented by a biphasic process, where rapid initial degradation phase in which PAHs are easily desorbed and readily bioavailable, is followed by next phase of slower degradation rate which shows slow desorption from geosorbents (Carmichael and Pfaender 1997; Ghosh et al. 2001; Thiele-Bruhn and Brümmer 2005; Crampon et al. 2014).

Degradation of a PAH in a mixture can be significantly influenced by interacting with other compounds of the mixture (Guha et al. 1999). However, majority of the research done on biodegradation of PAH have emphasized either on modelling degradation of individual PAHs or describing their effect of interactions qualitatively (Dimitriou-Christidis and Autenrieth 2007; Desai et al. 2008; Hennessee and Li 2016). Kinetics of PAHs mixture has been reported to be either similar or slower than those observed for the individual PAHs. Kinetics have been reported to influence by the composition of mixture as well as by the kinetic properties of the individual compounds (Dimitriou-Christidis and Autenrieth 2007).

16.9 Enzymology of PAHs Metabolism in Aerobic Bacteria

Because of the complex nature of PAHs, their degradation becomes a complex process which employs a chain of enzyme systems. The mechanisms and enzymatic systems may differ according to the organism and substrate but overall degradation pathway for polycyclic aromatic hydrocarbon degradation includes initial oxidation that triggers the series of conversions of several hydrocarbons to Acetyl Co-A which ultimately enters into the Kreb's cycle to generate energy.

16.9.1 Hydroxylation-Activation of PAHs by Dioxygenases to Produce Cis-Dihydrodiols

The first step in the aerobic degradation of PAHs normally occurs via the incorporation of molecular O_2 into the aromatic nucleus by a multicomponent dioxygenase enzyme system and cis-dihydrodiol is formed. Despite the fact that, the insertion of oxygen is thermodynamically favoured, there is a need of NADH as a reductant. Dioxygenases, which converts arenes of PAH substrate into cis-dihydrodiol, appear to be the most prevalent in bacteria, having multicomponent enzyme systems that involve various proteins, non-heme Fe-atoms and also need NADH (Gibson and Subramanian 1984; Gibson et al. 1990; Resnick et al. 1994). The multicomponent dioxygenase enzyme usually consists of ferredoxin, reductase and terminal oxygenase subunits (Mallick et al. 2011). During the catalytic reactions, usually two electrons are transferred through reductase from the reduced pyridine nucleotide to the ferredoxin and the Rieske-centre and finally to the Fe (II) ion at the active site. The reducing equivalents then allow the activation of molecular O_2, a required step for the dihydroxylation of the substrate (Carredano et al. 2000; Ferraro et al. 2005, 2006). The dihydroxylated intermediates further can be broken down by dioxygenase either through ortho or meta-cleavage pathway or ultimately forms the (tricarboxylic acid) TCA cycle intermediates (Kanaly and Harayama 2000). Enzymatic cleavage of aromatic rings by highly selective dioxygenases converts protocatechuates/catechols to TCA cycle intermediates. It is also evident that some other enzymes like alkane hydroxylase, methane monooxygenase and lignin peroxidase can be important in the PAHs metabolism (Van Beilen and Funhoff 2007).

16.9.2 Rearomatization-Conversion of Cis-Dihydrodiols by Dehydrogenase to Diol Intermediates

The next step is the re-aromatization of cis-dihydrodiol via a cis-diol dehydrogenase to yield dihydroxylated intermediates. For example, in naphthalene biodegradation, 1,2-dihydroxynaphthalene is formed by dehydrogenation of cis-naphthalene dihydrodiol by cis-naphthalene dihydrodiol dehydrogenase. This 1,2-dihydroxynaphthalene is further cleaved by 1,2-dihydroxynaphthalene dioxygenase via meta-cleavage pathway. The resulting ring-cleavage product spontaneously recyclizes and 2-hydroxy-2H-chromene-2-carboxylate is formed.

16.9.3 Cleavage of Diol Intermediate to Catechols by Ring-Cleaving Dioxygenases

The dihydroxylated intermediates can be broken down by intra-diol or extra-diol ring-cleavage dioxygenases using either ortho or meta-cleavage pathway, forming central intermediates like catechols and protocatechuates which ultimately convert into TCA cycle intermediates (Cerniglia 1993; Eaton and Chapman 1992; Gibson and Parales 2000).

16.10 Conclusion and Future Perspective

As global environmental disintegration is unavoidable, developed nations should work with developing countries to ensure that industrialization should not further add to the world's environmental problems and development be sustainable. It is crucial to develop a cost effective, manageable and feasible technology that can be exploited to clean up the contaminated sites as well as oil spills. The pollution of soil with PAHs is mostly greater than the capabilities of microorganisms to disperse them. Microbes not only regulate the rate of climate change but can also contribute to a greater extent in its effective mitigation and adaptation tools. Microorganisms, if motivated and manipulated properly can prove to be an effective tool in mitigating greenhouse gas emissions. By understanding microbial interactions, the measures for mitigating and controlling the climate change and its effects can be designed and developed.

References

Abbasian F, Lockington R, Mallavarapu M, Naidu R (2015) A comprehensive review of aliphatic hydrocarbon biodegradation by bacteria. Appl Biochem Biotechnol 176(3):670–699

Abbey FW, Nicholas WH, Beshr FS, Kathrine CH, Lee D (2011) Remediation of PAH-contaminated soils and sediments: a literature review. Virginia Tech, Blacksburg, VA

Abdel-Shafy HI, Mansour MS (2016) A review on polycyclic aromatic hydrocarbons: source, environmental impact, effect on human health and remediation. Egypt J Pet 25(1):107–123

Admon S, Green M, Avnimelech Y (2001) Biodegradation kinetics of hydrocarbons in soil during land treatment of oily sludge. Biorem J 5(3):193–209

Aeckersberg F, Bak F, Widdel F (1991) Anaerobic oxidation of saturated hydrocarbons to CO_2 by a new type of sulfate-reducing bacterium. Arch Microbiol 156(1):5–14

Alexander M (1999) Biodegradation and bioremediation. Gulf Professional Publishing, Houston, TX

Alexander M (2000) Aging, bioavailability, and overestimation of risk from environmental pollutants. Environ Sci Technol 34(20):4259–4265

Alexander DL, Zhang L, Foroozesh M, Alworth WL, Jefcoate CR (1999) Metabolism-based polycyclic aromatic acetylene inhibition of CYP1B1 in 10T1/2 cells potentiates aryl hydrocarbon receptor activity. Toxicol Appl Pharmacol 161(2):123–139

Al-Turki AI, Dick WA (2003) Myrosinase activity in soil. Soil Sci Soc Am J 67(1):139–145

Atagana HI, Haynes RJ, Wallis FM (2003) Optimization of soil physical and chemical conditions for the bioremediation of creosote-contaminated soil. Biodegradation 14(4):297–307

Atlas RM (1981) Microbial degradation of petroleum hydrocarbons: an environmental perspective. Microbiol Rev 45(1):180–209

Atlas RM (1995a) Bioremediation of petroleum pollutants. Int Biodeterior Biodegradation 35 (1–3):317–327

Atlas RM (1995b) Petroleum biodegradation and oil spill bioremediation. Mar Pollut Bull 31 (4–12):178–182

Atlas RM, Unterman R (1999) Bioremediation in manual of industrial microbiology and biotechnology, 2nd edn. American Society for Microbiology Press, Washington DC

ATSDR (2017) Agency for Toxic Substances and Disease Registry. Atlanta, GA, Department of Health and Human Services

Bamforth SM, Singleton I (2005) Bioremediation of polycyclic aromatic hydrocarbons: current knowledge and future directions. J Chem Technol Biotechnol 80(7):723–736

Banat IM, Franzetti A, Gandolfi I, Bestetti G, Martinotti MG, Fracchia L, Smyth TJ, Marchant R (2010) Microbial biosurfactants production, applications and future potential. Appl Microbiol Biotechnol 87(2):427–444

Bardgett RD, Usher MB, Hopkins DW (2005) Biological diversity and function in soil. Cambridge Univ. Press, Cambridge

Barra R, Popp P, Quiroz R, Bauer C, Cid H, von Tümpling W (2005) Persistent toxic substances in soils and waters along an altitudinal gradient in the Laja River Basin, Central Southern Chile. Chemosphere 58(7):905–915

Bertrand JC, Almallah M, Acquaviva M, Mille G (1990) Biodegradation of hydrocarbons by an extremely halophilic archaebacterium. Lett Appl Microbiol 11(5):260–263

Bewley RJ, Webb G (2001) In situ bioremediation of groundwater contaminated with phenols, BTEX and PAHs using nitrate as electron acceptor. Land Contam Reclam 9(4):335–347

Blasco C, Picó Y (2009) Prospects for combining chemical and biological methods for integrated environmental assessment. TrAC Trends Anal Chem 28(6):745–757

Blum WEH (2005) Functions of soil for society and the environment. Rev Environ Sci Biotechnol 4 (3):75–79

Boll M, Löffler C, Morris BE, Kung JW (2014) Anaerobic degradation of homocyclic aromatic compounds via arylcarboxyl-coenzyme A esters: organisms, strategies and key enzymes. Environ Microbiol 16(3):612–627

Bonfá MR, Grossman MJ, Mellado E, Durrant LR (2011) Biodegradation of aromatic hydrocarbons by Haloarchaea and their use for the reduction of the chemical oxygen demand of hypersaline petroleum produced water. Chemosphere 84(11):1671–1676

Bonomo RP, Cennamo G, Purrello R, Santoro AM, Zappala R (2001) Comparison of three fungal laccases from *Rigidoporuslignosus* and *Pleurotusostreatus*: correlation between conformation changes and catalytic activity. J Inorg Biochem 83(1):67–75

Boopathy R (2000) Factors limiting bioremediation technologies. Bioresour Technol 74(1):63–67

Bosma TN, Middeldorp PJ, Schraa G, Zehnder AJ (1996) Mass transfer limitation of biotransformation: quantifying bioavailability. Environ Sci Technol 31(1):248–252

Bowlen GF, Kosson DS (1995) In situ processes for bioremediation of BTEX and petroleum fuel products. In: Young LY, Cerniglia CE (eds) Microbial transformations and degradation of toxic organic chemicals. Wiley-Liss, New York

Brakstad OG, Bonaunet K (2006) Biodegradation of petroleum hydrocarbons in seawater at low temperatures (0-5 degrees C) and bacterial communities associated with degradation. Biodegradation 17:71–82

Breedveld GD, Sparrevik M (2000) Nutrient-limited biodegradation of PAH in various soil strata at a creosote contaminated site. Biodegradation 11(6):391–399

Brooijmans RJ, Pastink MI, Siezen RJ (2009) Hydrocarbon-degrading bacteria: the oil-spill clean-up crew. Microb Biotechnol 2(6):587–594

Cameotra SS, Bollag JM (2003) Biosurfactant-enhanced bioremediation of polycyclic aromatic hydrocarbons. Crit Rev Environ Sci Technol 33(2):111–126

Carmichael LM, Pfaender FK (1997) The effect of inorganic and organic supplements on the microbial degradation of phenanthrene and pyrene in soils. Biodegradation 8(1):1–13

Carredano E, Karlsson A, Kauppi B, Choudhury D, Parales RE, Parales JV, Lee K, Gibson DT, Eklund H, Ramaswamy S (2000) Substrate binding site of naphthalene 1, 2-dioxygenase: functional implications of indole binding. J Mol Biol 296(2):701–712

Cerniglia CE (1993) Biodegradation of polycyclic aromatic hydrocarbons. Curr Opin Biotechnol 4 (3):331–338

Chaillan F, Chaineau CH, Point V, Saliot A, Oudot J (2006) Factors inhibiting bioremediation of soil contaminated with weathered oils and drill cuttings. Environ Pollut 144(1):255–265

Chaineau CH, Rougeux G, Yepremian C, Oudot J (2005) Effects of nutrient concentration on the biodegradation of crude oil and associated microbial populations in the soil. Soil Biol Biochem 37(8):1490–1497

Chandra S, Sharma R, Singh K, Sharma A (2013) Application of bioremediation technology in the environment contaminated with petroleum hydrocarbon. Ann Microbiol 63(2):417–431

Chang BV, Chang IT, Yuan SY (2008) Anaerobic degradation of phenanthrene and pyrene in mangrove sediment. Bull Environ Contam Toxicol 80(2):145–149

Cho JC, Kim SJ (2001) Detection of mega plasmid from polycyclic aromatic hydrocarbon-degrading *Sphingomonas* sp. strain KS14. J Mol Microbiol Biotechnol 3(4):503–506

Colborn T, VomSaal FS, Soto AM (1993) Developmental effects of endocrine-disrupting chemicals in wildlife and humans. Environ Health Perspect 101(5):378–384

Colores GM, Macur RE, Ward DM, Inskeep WP (2000) Molecular analysis of surfactant-driven microbial population shifts in hydrocarbon-contaminated soil. Appl Environ Microbiol 66 (7):2959–2964

Colwell RR, Walker JD, Cooney JJ (1977) Ecological aspects of microbial degradation of petroleum in the marine environment. CRC Crit Rev Microbiol 5(4):423–445

Coral G, Karagol S (2005) Isolation and characterisation of phenanthrene degrading bacteria from a petroleum refinery soil. Ann Microbiol 55(4):255–259

Crampon M, Bureau F, Akpa-Vinceslas M, Bodilis J, Machour N, Le Derf F, Portet-Koltalo F (2014) Correlations between PAH bioavailability, degrading bacteria, and soil characteristics during PAH biodegradation in five diffusely contaminated dissimilar soils. Environ Sci Pollut Res 21(13):8133–8145

Cui T, Martin J, Marzouk YM, Solonen A, Spantini A (2014) Likelihood-informed dimension reduction for nonlinear inverse problems. Inverse Probl 30(11):114015

Daniel LP, Edward JB, Alan TS (1992) Hydrogen peroxide use to increase oxidant capacity for in situ bioremediation of contaminated soils and aquifers. J Contam Hydrol 9:221–242

Das N, Chandran P (2011) Microbial degradation of petroleum hydrocarbon contaminants: an overview. Biotechnol Res Int 2011:941810

Dastgheib SMM, Amoozegar MA, Khajeh K, Shavandi M, Ventosa A (2012) Biodegradation of polycyclic aromatic hydrocarbons by a halophilic microbial consortium. Appl Microbiol Biotechnol 95(3):789–798

Denome SA, Stanley DC, Olson ES, Young KD (1993) Metabolism of dibenzothiophene and naphthalene in *Pseudomonas* strains: complete DNA sequence of an upper naphthalene catabolic pathway. J Bacteriol 175(21):6890–6901

Desai AM, Autenrieth RL, Dimitriou-Christidis P, McDonald TJ (2008) Biodegradation kinetics of select polycyclic aromatic hydrocarbon (PAH) mixtures by *Sphingomonas paucimobilis* EPA505. Biodegradation 19(2):223–233

Dimitriou-Christidis P, Autenrieth RL (2007) Kinetics of biodegradation of binary and ternary mixtures of PAHs. Biotechnol Bioeng 97(4):788–800

Dunn NW, Gunsalus IC (1973) Transmissible plasmid coding early enzymes of naphthalene oxidation in Pseudomonas putida. J Bacteriol 114(3):974–979

Eaton RW, Chapman PJ (1992) Bacterial metabolism of naphthalene: construction and use of recombinant bacteria to study ring cleavage of 1, 2-dihydroxynaphthalene and subsequent reactions. J Bacteriol 174(23):7542–7554

Eweis JB, Ergas SJ, Chang DP, Schroeder ED (1998) Bioremediation principles. McGraw-Hill Book Company Europe, Maidenhead

FAO & ITPS (2015) Status of the World's Soil Resources (SWSR) - main report. Rome, Italy, Wood and Agriculture Organization of the United Nations and Intergovernmental Technical Panel on Soils. http://www.fao.org/3/a-i5199e.pdf

Feitkenhauer H, Müller R, MAuml H (2003) Degradation of polycyclic aromatic hydrocarbons and long chain alkanes at 60-70 °C by *Thermus* and *Bacillus* spp. Biodegradation 14(6):367–372

Ferraro DJ, Gakhar L, Ramaswamy S (2005) Rieske business: structure–function of Rieske non-hemeoxygenases. Biochem Biophys Res Commun 338(1):175–190

Ferraro DJ, Okerlund AL, Mowers JC, Ramaswamy S (2006) Structural basis for regioselectivity and stereoselectivity of product formation by naphthalene 1, 2-dioxygenase. J Bacteriol 188 (19):6986–6994

Floodgate G (1984) The fate of petroleum in marine ecosystem. In: Atlas RM (ed) Petroleum microbiology. Macmillan, New York, pp 355–398

Foght JM, Westlake DW (1996) Transposon and spontaneous deletion mutants of plasmid-borne genes encoding polycyclic aromatic hydrocarbon degradation by a strain of *Pseudomonas fluorescens*. Biodegradation 7(4):353–366

Frantz B, Chakrabarty AM (1986) Degradative plasmids in pseudomonas. In: The biology of pseudomonas. Academic Press, New York, pp 295–323

Fritsche W, Hofrichter M (2008) Aerobic degradation by microorganisms. Chapter 6. In: Biotechnology: environmental processes II, vol 11, pp 147–155

Fuchs G, Boll M, Heider J (2011) Microbial degradation of aromatic compounds—from one strategy to four. Nat Rev Microbiol 9(11):803

Fuenmayor SL, Wild M, Boyes AL, Williams PA (1998) A gene cluster encoding steps in conversion of naphthalene to gentisate in *Pseudomonas* sp. strain U2. J Bacteriol 180:2522–2530

Gan S, Lau EV, Ng HK (2009) Remediation of soils contaminated with polycyclic aromatic hydrocarbons (PAHs). J Hazard Mater 172(2–3):532–549

Gevao B, Semple KT, Jones KC (2000) Bound pesticide residues in soils: a review. Environ Pollut 108(1):3–14

Ghazali FM, Rahman RNZA, Salleh AB, Basri M (2004) Biodegradation of hydrocarbons in soil by microbial consortium. Int Biodeterior Biodegradation 54(1):61–67

Ghosh U, Talley JW, Luthy RG (2001) Particle-scale investigation of PAH desorption kinetics and thermodynamics from sediment. Environ Sci Technol 35(17):3468–3475

Gibson DT, Parales RE (2000) Aromatic hydrocarbon dioxygenases in environmental biotechnology. Curr Opin Biotechnol 11(3):236–243

Gibson DT, Subramanian V (1984) Microbial degradation of aromatic hydrocarbons. In: Gibson DT (ed) Microbial degradation of organic compounds. Marcel Dekker, New York, pp 181–252

Gibson DT, Zylstra GJ, Chauhan S (1990) Biotransformations catalyzed by toluene dioxygenase from *Pseudomonas putida* F1. In: Silver S, Chakrabarty AM, Iglewski B, Kaplan S (eds) *Pseudomonas:* biotransformations, pathogenesis, and evolving biotechnology. American Society for Microbiology, Washington, DC, pp 121–132

Gudina EJ, Teixeira JA, Rodrigues LR (2010) Isolation and functional characterization of a biosurfactant produced by *Lactobacillus paracasei*. Colloids Surf B: Biointerfaces 76 (1):298–304

Gudina EJ, Pereira JF, Costa R, Coutinho JA, Teixeira JA, Rodrigues LR (2013) Biosurfactant-producing and oil-degrading *Bacillus subtilis* strains enhance oil recovery in laboratory sand-pack columns. J Hazard Mater 261:106–113

Guha S, Peters CA, Jaffe PR (1999) Multisubstrate biodegradation kinetics of naphthalene, phenanthrene, and pyrene mixtures. Biotechnol Bioeng 65(5):491–499

Guo G, He F, Tian F, Huang Y, Wang H (2016) Effect of salt contents on enzymatic activities and halophilic microbial community structure during phenanthrene degradation. Int Biodeterior Biodegradation 110:8–15

Gupta G, Kumar V, Pal AK (2016) Biodegradation of polycyclic aromatic hydrocarbons by microbial consortium: a distinctive approach for decontamination of soil. Soil Sediment Contam Int J 25(6):597–623

Habe H, Chung JS, Ishida A, Kasuga K, Ide K, Takemura T, Nojiri H, Yamane H, Omori T (2005) The fluorene catabolic linear plasmid in *Terrabacter* sp. strain DBF63 carries the β-ketoadipate pathway genes, pcaRHGBDCFIJ, also found in proteobacteria. Microbiology 151 (11):3713–3722

Harmsen J (2007) Measuring bioavailability: from a scientific approach to standard methods. J Environ Q 36:1420–1428

Harmsen J, Rulkens WH, Sims RC, Rijtema PE, Zweers AJ (2007) Theory and application of landfarming to remediate polycyclic aromatic hydrocarbons and mineral oil-contaminated sediments; beneficial reuse. J Environ Qual 36(4):1112–1122

Harner T, Bidleman TF, Jantunen LM, Mackay D (2001) Soil—air exchange model of persistent pesticides in the United States cotton belt. Environ Toxicol Chem 20(7):1612–1621

Hassanshahian M, Tebyanian H, Cappello S (2012) Isolation and characterization of two crude oil-degrading yeast strains, *Yarrowia lipolytica* PG-20 and PG-32, from the Persian Gulf. Mar Pollut Bull 64(7):1386–1391

Head IM, Jones DM, Röling WF (2006) Marine microorganisms make a meal of oil. Nat Rev Microbiol 4(3):173–182

Heider J, Spormann AM, Beller HR, Widdel F (1998) Anaerobic bacterial metabolism of hydrocarbons. FEMS Microbiol Rev 22(5):459–473

Hennessee CT, Li QX (2016) Effects of polycyclic aromatic hydrocarbon mixtures on degradation, gene expression, and metabolite production in four *Mycobacterium* species. Appl Environ Microbiol 82(11):3357–3369

Hutchins SR, Sewell GW, Kovacs DA, Smith GA (1991) Biodegradation of aromatic hydrocarbons by aquifer microorganisms under denitrifying conditions. Environ Sci Technol 25(1):68–76

Hwa KS, Jeong KH, Jin-cchul A, Hoon JJ, Seung-Yeol SL (2005) Characterisation of phenanthrene degradation by *Sphingomonas* sp. HS 362. Korean J Microbiol 41:201–207

Ilori MO, Amobi CJ, Odocha AC (2005) Factors affecting biosurfactant production by oil degrading *Aeromonas* spp. isolated from a tropical environment. Chemosphere 61(7):985–992

Iwabuchi N, Sunairi M, Anzai H, Nakajima M, Harayama S (2000) Relationships between colony morphotypes and oil tolerance in *Rhodococcus rhodochrous*. Appl Environ Microbiol 66 (11):5073–5077

Jarabak R, Harvey RG, Jarabak J (1997) Redox cycling of polycyclic aromatic hydrocarbono-quinones: reversal of superoxide dismutase inhibition by ascorbate. Arch Biochem Biophys 339 (1):92–98

Johnsen AR, Wick LY, Harms H (2005) Principles of microbial PAH-degradation in soil. Environ Pollut 133(1):71–84

Juhasz AL, Naidu R (2000) Bioremediation of high molecular weight polycyclic aromatic hydrocarbons: a review of the microbial degradation of benzo [a] pyrene. Int Biodeterior Biodegradation 45(1–2):57–88

Kanaly RA, Harayama S (2000) Biodegradation of high-molecular-weight polycyclic aromatic hydrocarbons by bacteria. J Bacteriol 182(8):2059–2067

Khan MAI, Biswas B, Smith E, Mahmud SA, Hasan NA, Khan MAW, Naidu R, Megharaj M (2018) Microbial diversity changes with rhizosphere and hydrocarbons in contrasting soils. Ecotoxicol Environ Saf 156:434–442

Kim TJ, Lee EY, Kim YJ, Cho KS, Ryu HW (2003) Degradation of polyaromatic hydrocarbons by *Burkholderia cepacia* 2A-12. World J Microbiol Biotechnol 19(4):411–417

Kiyohara H, Sugiyama M, Mondello FJ, Gibson DT, Yano K (1983) Plasmid involvement in the degradation of polycyclic aromatic hydrocarbons by *Beijerinckia* species. Biochem Biophys Res Commun 111:939–945

Klankeo P, Nopcharoenkul W, Pinyakong O (2009) Two novel pyrene-degrading *Diaphorobacter* sp. and *Pseudoxanthomonas* sp. isolated from soil. J Biosci Bioeng 108(6):488–495

Kumar G, Singlar R, Kumar R (2010) Plasmid associated anthracene degradation by *Pseudomonas sp.* isolated from filling station site. Nat Sci 8(4):89–94

Kurkela S, Lehväslaiho H, Palva ET, Teeri TH (1988) Cloning, nucleotide sequence and characterization of genes encoding naphthalene dioxygenase of *Pseudomonas putida* strain NCIB9816. Gene 73(2):355–362

Lai CC, Huang YC, Wei YH, Chang JS (2009) Biosurfactant-enhanced removal of total petroleum hydrocarbons from contaminated soil. J Hazard Mater 167(1–3):609–614

Lamichhane S, Krishna KB, Sarukkalige R (2016) Polycyclic aromatic hydrocarbons (PAHs) removal by sorption: a review. Chemosphere 148:336–353

Laurie AD, Lloyd-Jones G (1999) The phn genes of *Burkholderia* sp. strain RP007 constitute a divergent gene cluster for polycyclic aromatic hydrocarbon catabolism. J Bacteriol 181 (2):531–540

Leahy JG, Colwell RR (1990) Microbial degradation of hydrocarbons in the environment. Microbiol Mol Biol Rev 54(3):305–315
Li JL, Chen BH (2009) Effect of non-ionic surfactants on biodegradation of phenanthrene by a marine bacteria of *Neptunomonas naphthovorans*. J Hazard Mater 162(1):66–73
Li X, Lin X, Li P, Liu W, Wang L, Ma F, Chukwuka KS (2009) Biodegradation of the low concentration of polycyclic aromatic hydrocarbons in soil by microbial consortium during incubation. J Hazard Mater 172(2–3):601–605
Liebeg EW, Cutright TJ (1999) The investigation of enhanced bioremediation through the addition of macro and micro nutrients in a PAH contaminated soil. Int biodeterior biodegrad 44(1):55–64
Lombi E, Hamon RE (2005) Remediation of polluted soils. In: Hillel D (ed) Encyclopedia of soils in the environment. Elsevier, Oxford, pp 379–385
Lu XY, Zhang T, Fang HHP (2011) Bacteria-mediated PAH degradation in soil and sediment. Appl Microbiol Biotechnol 89(5):1357–1371
Lundstedt S, Haglund P, Öberg L (2003) Degradation and formation of polycyclic aromatic compounds during bioslurry treatment of an aged gasworks soil. Environ Toxicol Chem 22 (7):1413–1420
Luo YR, Kang SG, Kim SJ, Kim MR, Li N, Lee JH, Kwon KK (2012) Genome sequence of benzo (a) pyrene-degrading bacterium *Novosphingobium pentaromativorans* US6-1. J Bacteriol 194 (4):907
Mallick S, Chatterjee S, Dutta TK (2007) A novel degradation pathway in the assimilation of phenanthrene by *Staphylococcus* sp. strain PN/Y via meta-cleavage of 2-hydroxy-1-naphthoic acid: formation of trans-2,3-dioxo-5-(20-hydroxyphenyl)- pent-4-enoic acid. Microbiology 153 (7):2104–2115
Mallick S, Chakraborty J, Dutta TK (2011) Role of oxygenases in guiding diverse metabolic pathways in the bacterial degradation of low-molecular-weight polycyclic aromatic hydrocarbons: a review. Crit Rev Microbiol 37(1):64–90
McMahon S, Parnell J (2014) Weighing the deep continental biosphere. FEMS Micribiol Ecol 87:113–120
Meckenstock RU, Boll M, Mouttaki H, Koelschbach JS, Tarouco PC, Weyrauch P, Dong X, Himmelberg AM (2016) Anaerobic degradation of benzene and polycyclic aromatic hydrocarbons. J Mol Microbiol Biotechnol 26(1–3):92–118
Mendonca E, Picado A (2002) Ecotoxicological monitoring of remediation in a coke oven soil. Environ Toxicol 17(1):74–79
Miller CD, Child R, Hughes JE, Benscai M, Der JP, Sims RC, Anderson AJ (2007) Diversity of soil mycobacterium isolates from three sites that degrade polycyclic aromatic hydrocarbons. J Appl Microbiol 102(6):1612–1624
Minai-Tehrani D, Herfatmanesh A, Azari-Dehkordi F, Minuoi S (2006) Effect of salinity on biodegradation of aliphatic fractions of crude oil in soil. Pak J Biol Sci 9(8):1531–1535
Moschella PS, Abbiati M, Åberg P, Airoldi L, Anderson JM, Bacchiocchi F, Bulleri F, Dinesen GE, Frost M, Gacia E, Granhag L (2005) Low-crested coastal defence structures as artificial habitats for marine life: using ecological criteria in design. Coast Eng 52(10–11):1053–1071
Müller R, Antranikian G, Maloney S, Sharp R (1998) Thermophilic degradation of environmental pollutants. In: Antranikian G (ed) Biotechnology of extremophiles. Advances in biochemical engineering/bio-technology, vol 61. Springer, Berlin, Heidelberg, pp 155–169
Mutalik SR, Vaidya BK, Joshi RM, Desai KM, Nene SN (2008) Use of response surface optimization for the production of biosurfactant from *Rhodococcus* spp. MTCC 2574. Bioresour Technol 99(16):7875–7880
Naidu R, Bolan NS, Megharaj M, Juhasz AL, Gupta SK, Clothier BE, Schulin R (2008) Chemical bioavailability in terrestrial environments. In: Hartemink AE, McBratney AB, Naidu R (eds) Developments in soil science, vol 32. Elsevier, Amsterdam, pp 1–6
Nguyen TT, Youssef NH, McInerney MJ, Sabatini DA (2008) Rhamnolipid biosurfactant mixtures for environmental remediation. Water Res 42(6–7):1735–1743

Nielsen UN, Wall DH, Six J (2015) Soil biodiversity and the environment. Annu Rev Environ Resour 40:63–90

Nwazue NR (2011) The effect of crude oil spill on the ascorbic acid content of some selected vegetable species: *Spinaceaoleraceae*, *Solanummelongena* and *Talinumtriangulare* in an oil polluted soil. Pak J Nutr 10(3):274–281

Nzila A (2018) Current status of the degradation of aliphatic and aromatic petroleum hydrocarbons by thermophilic microbes and future perspectives. Int J Environ Res Public Health 15(12):2782

Obayori OS, Salam LB (2010) Degradation of polycyclic aromatic hydrocarbons: role of plasmids. Sci Res Essays 5(25):4093–4106

Okoh AI (2006) Biodegradation alternative in the cleanup of petroleum hydrocarbon pollutants. Biotechnol Mol Biol Rev 1(2):38–50

Olson PE, Reardon KF, Pilon-Smits EAH (2003) Ecology of rhizosphere bioremediation. In: McCutcheon SC, Schnoor JL (eds) Phytoremediation: transformation and control of contaminants. Wiley, Hoboken, NJ, pp 317–353

Pathak H, Jain PK, Jaroli DP, Lowry M (2008) Degradation of phenanthrene and anthracene by *Pseudomonas* strain, isolated from coastal area. Bioremed J 12(2):111–116

Peixoto RS, Vermelho AB, Rosado AS (2011) Petroleum-degrading enzymes: bioremediation and new prospects. Enzyme Res 2011:475193

Perry JJ (1979) Microbial cooxidations involving hydrocarbons. Microbiol Rev 43(1):59–72

Pierzynski GM, Vance GF, Sims JT (2005) Soils and environmental quality. CRC press, Boca Raton, FL

Poggio L, Vrščaj B, Schulin R, Hepperle E, Marsan FA (2009) Metals pollution and human bioaccessibility of topsoils in Grugliasco (Italy). Environ Pollut 157(2):680–689

Reid BJ, Jones KC, Semple KT (2000) Bioavailability of persistent organic pollutants in soils and sediments—a perspective on mechanisms, consequences and assessment. Environ Pollut 108 (1):103–112

Resnick SM, Torok DS, Lee K, Brand JM, Gibson DT (1994) Regiospecific and stereoselective hydroxylation of 1-indanone and 2-indanone by naphthalene dioxygenase and toluene dioxygenase. Appl Environ Microbiol 60(9):3323–3328

Rhind SM (2009) Anthropogenic pollutants: a threat to ecosystem sustainability? Philos Trans R Soc B Biol Sci 364(1534):3391–3401

Rijnaarts HH, Bachmann A, Jumelet JC, Zehnder AJ (1990) Effect of desorption and intra particle mass transfer on the aerobic biomineralization of alpha-hexachlorocyclohexane in a contaminated calcareous soil. Environ Sci Technol 24(9):1349–1354

Romine MF, Stillwell LC, Wong KK, Thurston SJ, Sisk EC, Sensen C, Gaasterland T, Fredrickson JK, Saffer JD (1999) Complete sequence of a 184-kilobase catabolic plasmid from *Sphingomonas aromaticivorans* F199. J Bacteriol 181(5):1585–1602

Ron EZ, Rosenberg E (2002) Biosurfactants and oil bioremediation. Curr Opin Biotechnol 13 (3):249–252

Ruberto L, Vazquez SC, Mac Cormack WP (2003) Effectiveness of the natural bacterial flora, biostimulation and bioaugmentation on the bioremediation of a hydrocarbon contaminated Antarctic soil. Int Biodeterior Biodegradation 52(2):115–125

Saha JK, Selladurai R, Coumar MV, Dotaniya ML, Kundu S, Patra AK (2017) Soil pollution - an emerging threat to agriculture. Environmental chemistry for a sustainable world. Springer Singapore, Singapore. http://link.springer.com/10.1007/978-981-10-4274-4

Sajna KV, Sukumaran RK, Gottumukkala LD, Pandey A (2015) Crude oil biodegradation aided by biosurfactants from *Pseudozyma* sp. NII 08165 or its culture broth. Bioresour Technol 191:133–139

Santos HF, Carmo FL, Paes JE, Rosado AS, Peixoto RS (2011) Bioremediation of mangroves impacted by petroleum. Water Air Soil Pollut 216(1–4):329–350

Schwarz JR, Walker JD, Colwell RR (1974) Deep-sea bacteria: growth and utilization of hydrocarbons at ambient and in situ pressure. Appl Environ Microbiol 28(6):982–986

Schwarz JR, Walker JD, Colwell RR (1975) Deep-sea bacteria: growth and utilization of n-hexadecane at in situ temperature and pressure. Can J Microbiol 21(5):682–687

Semple KT, Morriss AWJ, Paton GI (2003) Bioavailability of hydrophobic organic contaminants in soils: fundamental concepts and techniques for analysis. Eur J Soil Sci 54(4):809–818

Shen T, Pi Y, Bao M, Xu N, Li Y, Lu J (2015) Biodegradation of different petroleum hydrocarbons by free and immobilized microbial consortia. Environ Sci: Processes Impacts 17(12):2022–2033

Simarro R, González N, Bautista LF, Sanz R, Molina MC (2011) Optimisation of key abiotic factors of PAH (naphthalene, phenanthrene and anthracene) biodegradation process by a bacterial consortium. Water Air Soil Pollut 217(1–4):365–374

Sims JL, Sims RC, Matthews JE (1990) Approach to bioremediation of contaminated soil. Hazard Waste Hazard Mater 7(2):117–149

Singh SN, Kumari B, Mishra S (2012) Microbial degradation of alkanes. In: Microbial degradation of xenobiotics. Springer, Berlin Heidelberg, pp 439–469

Si-Zhong Y, Hui-Jun JIN, Zhi W, Rui-Xia HE, Yan-Jun JI, Xiu-Mei LI, Shao-Peng YU (2009) Bioremediation of oil spills in cold environments: a review. Pedosphere 19(3):371–381

Souza EC, Vessoni-Penna TC, de Souza Oliveira RP (2014) Biosurfactant-enhanced hydrocarbon bioremediation: an overview. Int Biodeterior Biodegradation 89:88–94

Straube WL, Nestler CC, Hansen LD, Ringleberg D, Pritchard PH, Jones-Meehan J (2003) Remediation of polyaromatic hydrocarbons (PAHs) through landfarming with biostimulation and bioaugmentation. Acta Biotechnol 23(2–3):179–196

Swartjes FA (2011) Dealing with contaminated sites. Springer, Dordrecht. http://link.springer.com/10.1007/978-90-481-9757-6

Swift MJ, Heal OW, Anderson JM, Anderson JM (1979) Decomposition in terrestrial ecosystems, vol 5. Univ of California Press, Berkeley, CA

Thiele-Bruhn S, Brümmer GW (2005) Kinetics of polycyclic aromatic hydrocarbon (PAH) degradation in long-term polluted soils during bioremediation. Plant Soil 275(1–2):31–42

Towell MG, Paton GI, Semple KT (2011) The biodegradation of cable oil components: impact of oil concentration, nutrient addition and bioaugmentation. Environ Pollut 159(12):3777–3783

Trofimov SY, Rozanova MS (2003) Transformation of soil properties under the impact of oil pollution. Eurasian Soil Sci 36:S82–S87

Van Beilen JB, Funhoff EG (2007) Alkane hydroxylases involved in microbial alkane degradation. Appl Microbiol Biotechnol 74(1):13–21

Van Hamme JD, Singh A, Ward OP (2003) Recent advances in petroleum microbiology. Microbiol Mol Biol Rev 67(4):503–549

Varjani SJ (2014) Hydrocarbon degrading and biosurfactants (bio-emulsifiers) producing bacteria from petroleum oil wells. Doctoral dissertation, PhD thesis, KadiSarvaVishwavidyalaya, Gandhinagar, India

Varjani SJ (2017) Microbial degradation of petroleum hydrocarbons. Bioresour Technol 223:277–286

Varjani SJ, Upasani VN (2016a) Core flood study for enhanced oil recovery through ex-situ bioaugmentation with thermo-and halo-tolerant rhamnolipid produced by *Pseudomonas aeruginosa* NCIM 5514. Bioresour Technol 220:175–182

Varjani SJ, Upasani VN (2016b) Biodegradation of petroleum hydrocarbons by oleophilic strain of *Pseudomonas aeruginosa* NCIM 5514. Bioresour Technol 222:195–201

Varjani SJ, Upasani VN (2017) A new look on factors affecting microbial degradation of petroleum hydrocarbon pollutants. Int Biodeterior Biodegradation 120:71–83

Vaz DA, Gudina EJ, Alameda EJ, Teixeira JA, Rodrigues LR (2012) Performance of a biosurfactant produced by a *Bacillus subtilis* strain isolated from crude oil samples as compared to commercial chemical surfactants. Colloids Surf B: Biointerfaces 89:167–174

Volkering F, Breure AM, Sterkenburg A, van Andel JV (1992) Microbial degradation of polycyclic aromatic hydrocarbons: effect of substrate availability on bacterial growth kinetics. Appl Microbiol Biotechnol 36(4):548–552

Waigi MG, Kang F, Goikavi C, Ling W, Gao Y (2015) Phenanthrene biodegradation by sphingomonads and its application in the contaminated soils and sediments: a review. Int Biodeterior Biodegradation 104:333–349

Ward DM, Brock TD (1978) Hydrocarbon biodegradation in hypersaline environments. Appl Environ Microbiol 35(2):353–359

Wasi S, Tabrez S, Ahmad M (2013) Toxicological effects of major environmental pollutants: an overview. Environ Monit Assess 185(3):2585–2593

Weelink SA, Van Doesburg W, Saia FT, Rijpstra WIC, Röling WF, Smidt H, Stams AJ (2009) A strictly anaerobic beta proteobacterium Georgfuchsiatoluolica gen. nov., sp. nov. degrades aromatic compounds with Fe (III), Mn (IV) or nitrate as an electron acceptor. FEMS Microbiol Ecol 70(3):575–585

Wiedenbeck J, Cohan FM (2011) Origins of bacterial diversity through horizontal genetic transfer and adaptation to new ecological niches. FEMS Microbiol Rev 35:957–976

Wilkes H, Buckel W, Golding BT, Rabus R (2016) Metabolism of hydrocarbons in n-alkane-utilizing anaerobic bacteria. J Mol Microbiol Biotechnol 26(1–3):138–151

Wilson SC, Jones KC (1993) Bioremediation of soil contaminated with polynuclear aromatic hydrocarbons (PAHs): a review. Environ Pollut 81(3):229–249

Yaron B, Dror I, Berkowitz B (2012) Soil-subsurface change: chemical pollutant impacts. Springer, Berlin

Yuan SY, Chang JS, Yen JH, Chang BV (2001) Biodegradation of phenanthrene in river sediment. Chemosphere 43(3):273–278

Zenker A, Cicero MR, Prestinaci F, Bottoni P, Carere M (2014) Bioaccumulation and biomagnification potential of pharmaceuticals with a focus to the aquatic environment. J Environ Manag 133:378–387

Zhao Z, Selvam A, Wong JWC (2011) Synergistic effect of thermophilic temperature and biosurfactant produced by *Acinetobacter calcoaceticus* BU03 on the biodegradation of phenanthrene in bioslurry system. J Hazard Mater 190(1–3):345–350

Zyakun AM, Kochetkov VV, Zakharchenko VN, Baskunov BP, Laurinavichius KS, Peshenko VP, Siunova TV, Anokhina TO, Boronin AM (2015) Use of mass spectrometry for characterizing the growth of *Pseudomonas* bacteria bearing naphthalene degradation plasmids. J Anal Chem 70(13):1569–1575

Biosurfactant-Producing Bacteria as Potent Scavengers of Petroleum Hydrocarbons

17

Zarreena Siddiqui, Mohammad Anas, Khalida Khatoon, and Abdul Malik

Abstract

Pollution with petroleum hydrocarbons is far-reaching and thus, a problem for the environment as well as human health. The pristine environment has continuously been influenced by anthropogenic activities. Due to the globalization of various industries; their waste materials are being discharged untreated or partially treated into the ecosystem and having adverse impact on different life forms. Petroleum despite being a priceless resource and central to human life on Earth today, extraction and transportation of petroleum products has a number of ecological repercussions. The most consequential effect of petroleum use leads to the environmental pollution, adversely affecting air, soil and water quality. Petroleum product's spill and leakage are also a major threat to the environment because petroleum products can rigorously destroy the surrounding ecosystem. So, the removal of petroleum products is imperative using eco-friendly methods, and microorganisms are the cheaply available option for doing so. Biosurfactants are extracellular amphiphilic, surface-active compounds produced by microorganisms. These microbially produced multifunctional biomolecules are versatile products having vast applications in various aspects related to clean up of environmental contaminants inclusive of enhanced oil recovery (EOR), controlling oil spills, detoxification and biodegradation of oil contaminated wastewater, soil or sediments. Biosurfactant works by reducing interfacial and surface tension by collecting at the interface of immiscible liquids and thus improve the bioavailability, solubility and subsequent biodegradation of the insoluble or hydrophobic organic compounds. This chapter summarizes the role of biosurfactant-producing bacteria in the bioremediation of petroleum hydrocarbons polluted environment.

Z. Siddiqui (✉) · M. Anas · K. Khatoon · A. Malik
Department of Agricultural Microbiology, Faculty of Agricultural Sciences, Aligarh Muslim University, Aligarh, Uttar Pradesh, India

S. A. Lone, A. Malik (eds.), *Microbiomes and the Global Climate Change*,
https://doi.org/10.1007/978-981-33-4508-9_17

Keywords

Petroleum hydrocarbons · Biosurfactants · Pollution · Biodegradation · Environment · Scavengers

17.1 Introduction

Growing population, rapid industrialization and high global demands for energy is on the rise and is satisfied by the rapid exploitation of different natural resources, such as petroleum (Al-Maamary et al. 2017). Globally, crude oil consumption grew by an above-average 1.4 million barrels per day (b/d) (Dudley 2019) and around 1.7 and 8.8 million metric tons of oil are emitted into the aquatic environment and soil per year, respectively (Sihag et al. 2014). The exploitation or in fact over-exploitation of petroleum and petroleum based products is responsible for release of enormous amounts of hydrocarbon-based waste into the environment, endangering the ecosystem's safety and biodiversity (Darvishi et al. 2011; Al-Hawash et al. 2018).

Petroleum is one of the major sources of energy to sustain a country's economic and social development. Petroleum based hydrocarbon pollution is one of the major concerns and is induced through shipping and recycling by the key leaks in underground storage facilities and unintentional spills. Such leaks mostly accidental ones not only contaminate the water bodies but also the air (volatile hydrocarbons) (Shi et al. 2018). Inputs of hydrocarbon into river sediments, seas and oceans are induced by both natural and anthropogenic behaviours (Head et al. 2006). These hydrocarbon-polluted environments are known to harbour an array of microorganisms capable of thriving in such hostile conditions, many of such microorganisms have developed the potential to utilize hydrocarbons and convert them into lesser toxic forms (Krishnani et al. 2019). Some microorganisms inhabiting these hydrocarbon-rich ecosystems are also explored for their ability to produce biosurfactants (an extracellular amphiphilic, surface-active compounds), which have a number of applications in the sectors of bioremediation, oil exploration and increased recycling, food processing and pharmaceutical industry (Sen 2008; Banat et al. 2010). Industrially, biosurfactants also known as surface-active amphiphilic substances are known to have great importance because of their biodegradable and non-toxic nature in contrast to manufactured chemical ones. This chapter emphasizes the types of microbially produced biosurfactant and its role in bioremediation and degradation of petroleum hydrocarbons in the polluted environment.

17.2 Petroleum Hydrocarbons

Petroleum is also known as rock oil and can be designated as primarily a complex mixture of hydrocarbons, consisting of complicated assemblage of paraffinic, alicyclic and aromatic hydrocarbons and a minute fractions of non-hydrocarbon compounds present as metalloporphyrins, heterocyclic nitrogen, phenols, thiol, sulphur compounds, naphthenic acids and asphalt compounds (Chandra et al. 2013; Varjani et al. 2015). Partial decomposition of both plant and animal waste over a long period of time creates crude oil as a fluid mixture. Crude oil once drawn out from below the earth surface undergoes distillation in refineries that leads to formation of various petroleum based products, such as Diesel, Gasoline, Heavy oil and kerosene (Speight 2007; Varjani 2014). Total Petroleum Hydrocarbons (TPH) is entitled to characterize a large group of chemical compounds derived from the crude oil. These are called hydrocarbons because they all are almost made entirely of hydrogen and carbon. Some are readily evaporating translucent and light-coloured fluids, while others that do not have evaporating tendency are dense, dark liquids and semi-solids. Majority of these compounds are distinguished by odours like petrol and kerosene. According to Balba et al. (1998), Widdel and Rabus (2001), Speight (2007) and Chandra et al. (2013) petroleum hydrocarbons can be classified as:

17.2.1 Saturates (Aliphatics)

These constitutes major portion of hydrocarbons and are without double bonds. These are saturated or unsaturated and linear or branched open-chain structures. On the basis of their chemical structures, aliphatics can be grouped into:

1. *Alkanes*, e.g. n-alkanes, iso-alkanes.
   n-alkanes can be categorized into four on the basis of molecular mass,
   (a) gaseous alkanes,
   (b) aliphatic hydrocarbons with lower molecular weight (C8–C16),
   (c) aliphatic hydrocarbons with medium molecular weight (C17–C28) and,
   (d) aliphatic hydrocarbons with high molecular weight (>C28) (Abbasian et al. 2015).
2. *Cycloalkanes*. e.g. (naphthenes), terpenes and steranes (Rahman et al. 2003; Abbasian et al. 2015).

17.2.2 Aromatics (Ringed Hydrocarbons)

Aromatics contain one or several aromatic rings that are substitutes with various alkyl groups (Meckenstock et al. 2016). They are mainly categorized as (a) Monocyclic aromatic hydrocarbons, examples are xylenes, ethylbenzene, toluene and benzene (Farhadian et al. 2008; Costa et al. 2012). (b) Polycyclic aromatic hydrocarbons (Chandra et al. 2013). PAHs contain two or more benzene rings, for example naphthalene (two-ringed), anthracene and phenanthrene (three-ringed) known as low molecular weight PAHs (Wilkes et al. 2016), and compounds like pyrene and chrysenes (four-ringed), fluoranthene and benzo[a]pyrene (five-ringed) are referred as high molecular weight PAHs (Farhadian et al. 2008; Costa et al. 2012; Macaulay and Rees 2014).

17.2.3 Resins

Resins comprises of multiple functional polar groups formed by Nitrogen, Oxygen, Sulphur and trace elements like Ni, Fe and V, etc. It constitutes very complex and mostly unknown carbon structure (Harayama et al. 2004; Chandra et al. 2013). These are amorphous solids and are completely soluble in oil (Balba et al. 1998; Speight 2007). In addition, resins possess long alkyl chain aromatic compounds which are soluble in both C_7H_{16} (n-heptane) and C_5H_{12} (n-pentane) and are structurally identical to surface-active molecules (Jada and Salou 2002; Parra-Barraza et al. 2003; Chandra et al. 2013).

17.2.4 Asphaltenes

Asphaltenes also contain number of polar functional groups like resins. They are large complex molecules, dark brown in colour, viscous and consist of polycyclic clusters with the substitution of various alkyl groups which impart resistance to biodegradation (Chandra et al. 2013).

17.3 Impact of Petroleum Hydrocarbons on Environment

Contamination due to petroleum hydrocarbons is one of the most important threats to the environment, having severe deteriorative effects on plants and animal ecosystems including the human health. Oil spills pose a significant threat to the aquatic and terrestrial habitats because they adversely deteriorate the environment (Head et al. 2006). Oil spills affect plants by creating conditions that decrease availability of essential nutrients such as nitrogen and oxygen which are necessary for plant growth. According to the Environmental Protection Agency (USEPA 2011), leaks of petroleum hydrocarbons into the environment can endanger people's health and safety due to contaminated drinking water, triggering fire and explosion risks, reducing the quality of air and water, damaging forestry, degrading recreational areas, killing ecosystems and livestock and wasting non-renewable resources.

Pollution due to crude oil at different levels resulted in a significant reduction in plant growth as assessed by reduction in plant height, fresh weight and leaf area and the impact was proportional to pollution levels (Fig. 17.1). It also has adverse effects on soil fertility and production of plants. As a result of forming a physical barrier and coating the roots, it could reduce or stop plant growth leading to death.

Diesel fuels are pure petroleum mid distillates isolated by fractional distillation which includes about 64% of aliphatic hydrocarbons, l-2% of olefinic hydrocarbons and 35% of aromatic hydrocarbons. Pollution due to diesel oil can cause plants to experience various detrimental effects like reduction in photosynthesis rate, growth inhibition, reduced nutrient uptake, etc. (Fig. 17.2). Interference in crop hydric ties is induced by emissions from diesel oil. Diesel oil is known to prevent nitrification much more than gasoline (Kucharski et al. 2010).

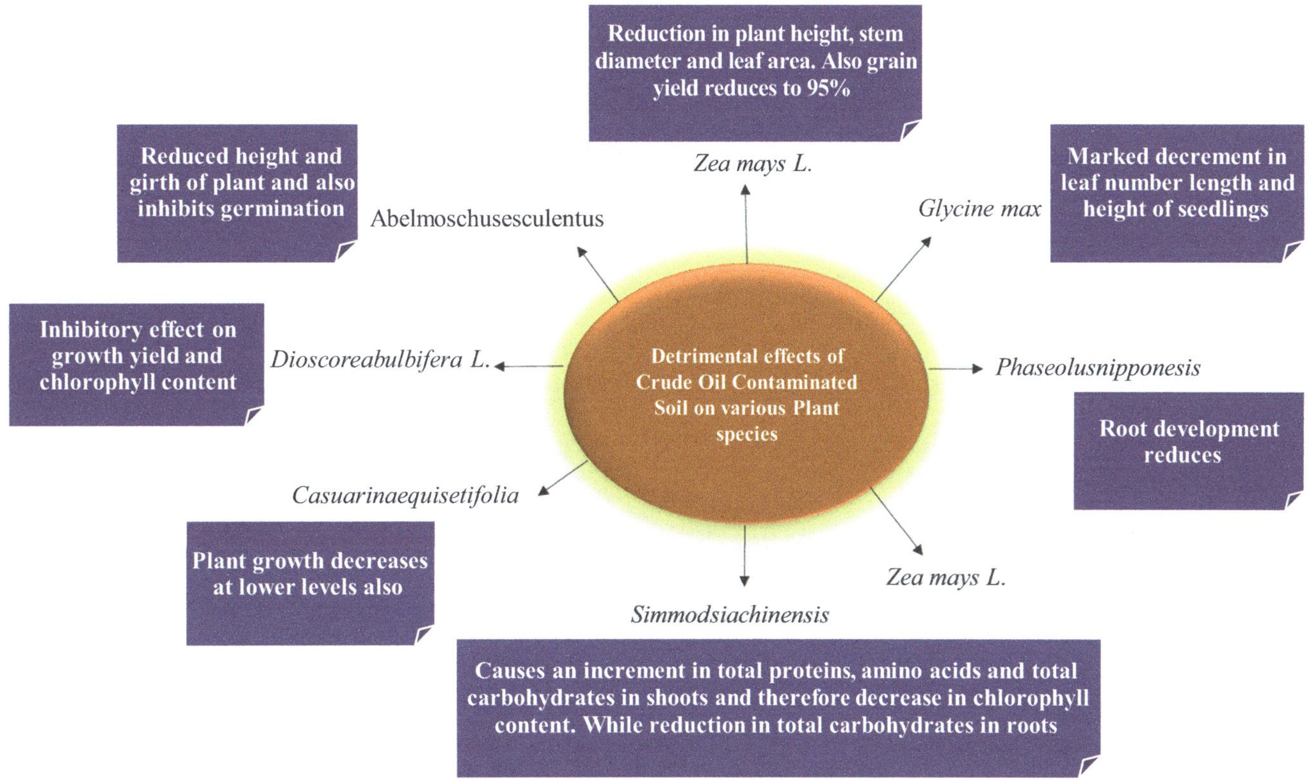

Fig. 17.1 Effects of crude oil contaminated soil on various plant species

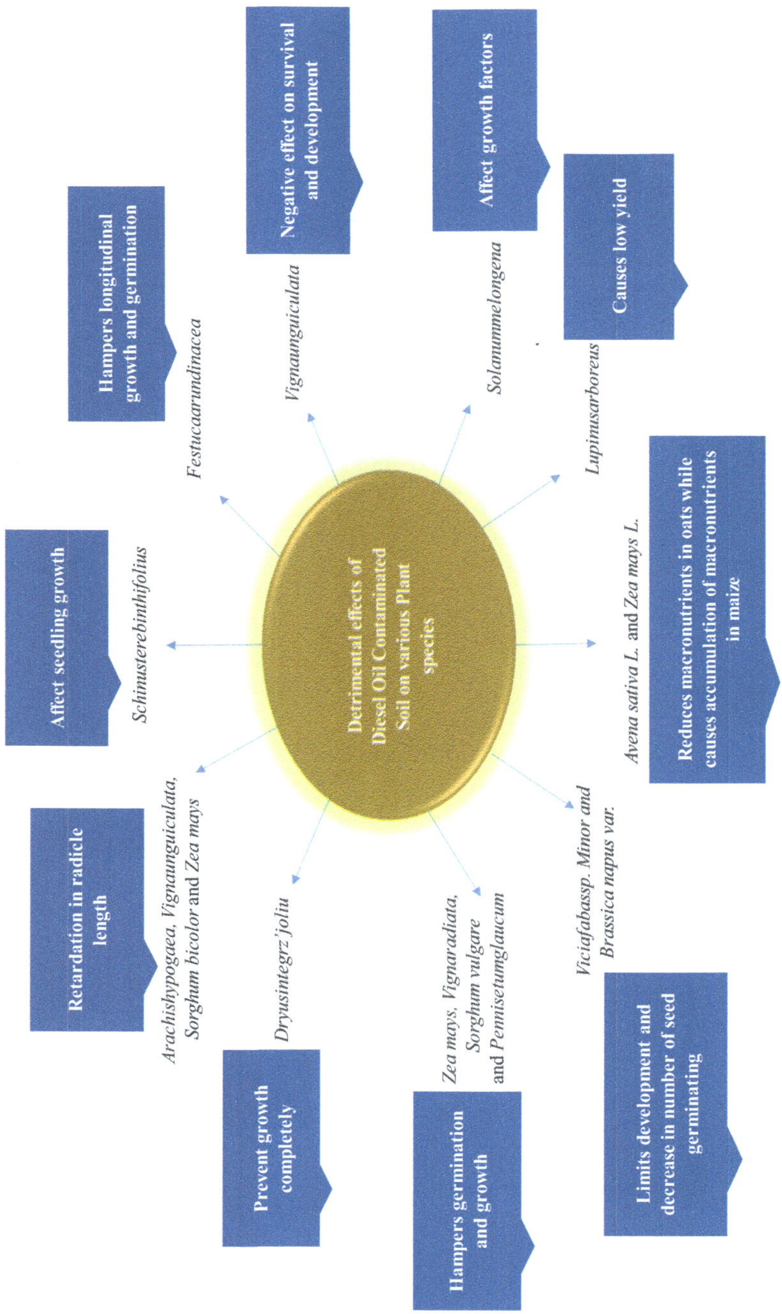

Fig. 17.2 Effects of diesel infected soil on multiple plant species

17.3.1 Impacts on Humans

Hydrocarbon pollutants can affect human health by suppressing the role of nerve synapse, protein synthesis, plasma membrane destruction with implications for membrane transport systems (Onwurah et al. 2007; Afuwale and Modi 2012). Light oils have a high saturated hydrocarbon ratio; thus, they may be more dangerous than heavy oils (Kauppi 2011). Effects of petroleum hydrocarbons on human health are mentioned in Fig. 17.3.

17.3.2 Impact on Microorganisms

The damage to the ecological community caused by petroleum hydrocarbons pollution is well known (Sikkema et al. 1995). For example, the Deep Water Horizon oil spill which occurred in the Gulf of Mexico left a significant impact on the economy and the protection of the environment, which is still the centre of attention for citizens (Xue et al. 2015). While people are increasingly worried about the toxic effects of oil emissions in affected areas on humans and animals (Díez Salvador et al. 2007; Mason et al. 2012), the significant toxic effects of petroleum hydrocarbons on microbial communities are often ignored. Petroleum hydrocarbons limit microbial biomass as it has been reported in many diesel exposure studies, that the main consequences of diesel contamination are decline in species number, evenness and phylogenetic diversity, with the new populations becoming heavily dominated by a

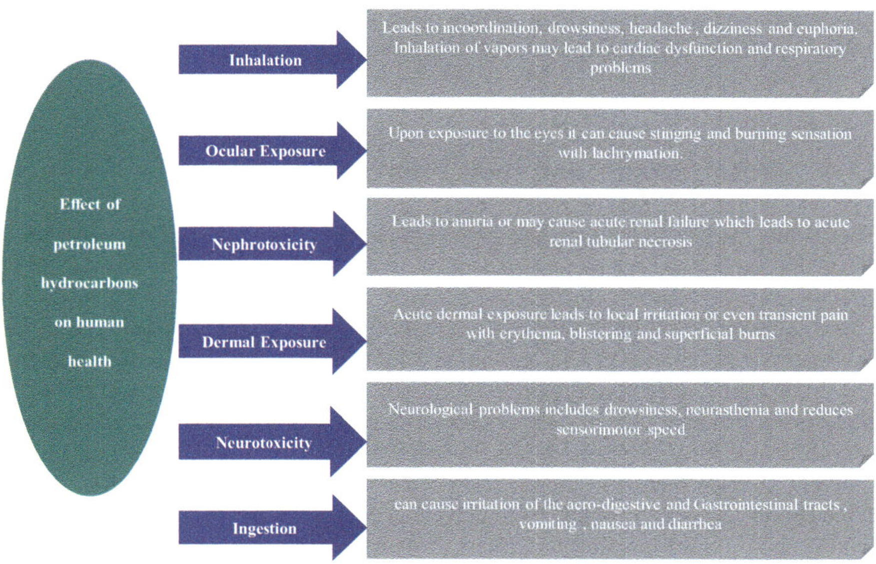

Fig. 17.3 Effects of petroleum hydrocarbons on human health

few organisms, particularly *Pseudomonas* (Labud et al. 2007; Rivers et al. 2013; Overholt et al. 2015). It is observed that the reduction in resources and phylogenetic diversity was partly related to the nitrogen cycle disturbance, with nitrification affecting organisms and active genes significantly reduced.

Cerniglia et al. (1983) and Van Dorst et al. (2014) investigated the toxicity and oxygenated versions of Naphthalene, 1-methylnaphthalene and 2-methylnaphthalene in *Agmenellum quadruplicatum* bacterial cells and noticed that these chemicals did not have inhibitory effects on microbial growth. However, bacterial development was hindered by the phenolic and quinonic naphthalene derivatives. This could be clarified by greater solubility of phenols and quinones, facilitating molecular mass transfer in bacterial cells, resulting in higher toxic effects than the other substances. It has also been documented in several studies that certain metabolic intermediates with relatively high solubility formed by bacterial degradation of petroleum hydrocarbons may have higher cytotoxicity than the parent molecules and therefore damage the bacteria (Hou et al. 2018). While some petroleum hydrocarbon susceptible bacteria are significantly impaired when subjected to petroleum hydrocarbons, others that can effectively degrade petroleum hydrocarbons, as well as bacteria that can take advantage of cytotoxic intermediate metabolites, can prosper.

Oil contamination is difficult to treat; petroleum hydrocarbon-degrading bacteria have evolved as a consequence of the presence of petroleum hydrocarbons in the atmosphere in close proximity (Margesin et al. 2003; Ron and Rosenberg 2014; Lea-Smith et al. 2015). However, it will take a long time for petroleum oil emissions to be cleaned up by relying solely on the capacity of these native microorganisms; thus, policy needs to be developed to speed up the process which includes use of additional microorganisms like bioaugmentation or engineered bioremediation involves biostimulation process or surfactant based biodegradation.

17.4 Biosurfactants and their Properties

Microbially produced surfactants are a collection of surface-active compounds which are also known as green surfactants or biosurfactants. These are formed either directly on microbial cell surfaces or secreted extracellularly by microorganisms. These compounds are biodegradable and non-toxic and thus do not build up in the environment (Saharan et al. 2011; Sharma 2016). They have amphipathic structure. The hydrophilic moiety comprises of mono, oligo and polysaccharides, while peptides or proteins that usually contains saturated, unsaturated or hydroxylated fatty acids or fatty alcohols in the hydrophobic moiety. Such heterogeneity of structure enables them to naturally mix fluid interfaces to form micelles and to demonstrate specific properties such as detergence, foaming, lubrication, solubilisation, emulsification and phase dispersion (Banat et al. 2000). The most significant application of biosurfactants was studied in bioremediation, for example, in removing heavy metals (Franzetti et al. 2009; Aşçı et al. 2010) and hydrocarbons from the contaminated sites (Kosaric et al. 1987). In addition, biosurfactants' physiological functions involve:

1. Increase the surface area of hydrophobic growth substrates; decrease the surface tension and interfacial pressure between the cell wall and the atmosphere.
2. Desorbing surface hydrophobic substances by increasing their solubility.
3. Induce membrane modifications such as increased hydrophobicity by reducing the content of lipopolysaccharides in the cell wall; enhance hydrocarbon dispersion by encapsulation into micelles (Aparna et al. 2011).

17.4.1 Classification and Properties of Biosurfactants

Biosurfactants are amphiphilic metabolites comprising of hydrophobic and hydrophilic molecules that allows them to partition with different degrees of polarity at the interface between fluid phases. These compounds consist of a wide variety of structures and can be divided into two main groups: low molecular weight and high molecular weight Surface-active compounds (SAC). Biosurfactants of lower molecular weight are usually classified based on their chemical composition and microbial origin. Hydrophilic part comprises of amino acids and anions or cations in peptides; mono-, di- and polysaccharides; and hydrophobic part consists of unsaturated, condensed and fatty acids. According to Hausmann and Syldatk (2014) and Gudiña et al. (2015a) the key groups of biosurfactants include glycolipids, lipopeptides and lipoproteins, phospholipids, polymeric surfactants and surfactants with particulate matter.

17.4.1.1 Glycolipids

These are the most common types of all the biosurfactant groups. Carbohydrate portion comprises of mono, di, tri and tetrasaccharides such as fructose, mannose, galactose, glucuronic acid, rhamnose and galactose-sulphate that are linked to the fatty acid component composed of long chain aliphatic acids and hydroxyl-aliphatic acids through an ether or ester unit (Jennings and Tanner 2000). *Rhodococcus erythropolis, Nocardia erythropolis, Arthrobacter sp., Mycobacterium sp.* are known as glycolipid producers. They are categorized as follows:

17.4.1.2 Rhamnolipids

Some members of genus *Pseudomonas* are known to produce rhamnose-containing glycolipid surfactant and decanoic acid β-hydroxy. The β-hydroxydecanoic acid group –OH is involved in glycosidic contact with the reduction end of rhamnose sugar and the second b-hydroxydecanoic acid group –OH is involved in ester production. *Pseudomonas aeruginosa*-produced rhamnolipids are widely studied and identified as a mixture of homologous species RL1 ($RhC_{10}C_{10}$), RL2 (RhC_{10}), RL3 ($Rh_2C_{10}C_{10}$) and RL4 (Rh_2C_{10}) (Rahman et al. 2002; Li et al. 2011; Gudiña et al. 2015b). These have a low average minimum surface tension (30–32 mN/m) (mili Newton per metre); high average emulsifying capacity (10.4–15.5 U/mL filtrate), weak CMC (5–65 mg/L) and good resistance for hydrophobic organic molecules (Dyke et al. 1993).

17.4.1.3 Sophorolipids

Sophorolipids are the diverse mixture of 6-9 different hydrophobic entities composed of a dimeric sugar (sophorose) and a β-glycosidic bond-connected hydroxyl fatty acid, generally produced by *Torulopsis* sp. There are two types: an acid with free carboxylic acid as a tail and a lactone form consisting of a lactone ring (macrolactones) produced through esterification reaction in between the carboxylic end of the fatty acid and 4″-hydroxy band of sophorous sugar (Van Bogaert et al. 2014). Commonly, lactonic sophorolipids have good surface tension control properties and which can possibly decrease the interfacial pressure of n-hexadecane and water from 40 to 5 mN/m, while acidic sophorolipids have much ability to shape foam and have marvellous solubility properties (Mulligan and Gibbs 2004). *Torulopsis bombicola, Torulopsis apicola, Torulopsis petrophilum and other Torulopsis sp.* are the examples of Sophorolipids producers.

17.4.1.4 Trehalose Lipids

Trehalose lipids typically formed by Gram-positive bacteria, consist of a disaccharide, trehalose, connected by an ester bond with a-branched β-hydroxy fatty acids in the carbohydrate form C6 and C6 in the case of trehalose dimycolates and C6 in the case of monomycolates (Lang and Philp 1998). Biosynthesis of trehalose was mainly studied in *Corynebacterium, Escherichia coli, Propionibacterium freudenreichii* and *Mycobacterium* (Ruhal et al. 2013).

17.4.1.5 Lipoproteins and Lipopeptides

Microorganisms produce a number of cyclic lipopeptides, entailing antibiotics for decapeptides (gramicidins) and antibiotics for lipopeptides (polymyxins). Generally speaking, bacterial lipopeptides are basically cyclic peptides acylated with fatty acids. They comprise of a lipid anchored to a chain of polypeptides formed by *Bacillus* and *Pseudomonas* sp. Cyclic lipopeptides of *Bacillus* comprise of three major groups classified as surfactin, iturin and fengycin. Surfactin synthesized by *Bacillus subtilis* is one of the most important possible biosurfactants. This consists of a cyclic heptapeptide (Glu-Leu-D-Leu-Val-Asp-D-Leu-Leu) bound to a fatty acid motion of 13–15 carbon lactone and is capable of reducing water surface tension from 72 to 27 mN/m (Peypoux et al. 1999; Mnif and Ghribi 2015).

17.4.1.6 Fatty Acids and Phospholipids

Acinetobacter, Candida, Corynebacterium, Aspergillus, Micrococcus and *Thiobacillus* were identified as potent producers of phospholipids and fatty acids (Liu et al. 2015). The hydrophilic or lipophilic equilibrium (HLB) of their configurations is directly associated with the whole of the hydrocarbon backbone. *Acinetobacte*r sp. strain HO1-N, produces vesicles rich in phosphatidylethanolamine which shape optically clear micro-emulsions of alkanes in water. Phosphatidylethanolamine formed by *Rhodococcus erythropolis* grown on alkanes results in a reduction of the interfacial pressure between water and hexadecane at a CMC (Critical Micelle Concentration) of 30 mg/L to <1 mN/m (Sanket and Yagnik 2013).

17.4.1.7 Polymeric Biosurfactants

These are heavy weight molecular biopolymers composed of polysaccharides, proteins, lipopolysaccharides, lipoproteins and mixtures of these. Biodispersan, mannoprotein, liposan and emulsan formed by *Candida lipolytica, Ustilago maydis, Acinetobacter* sp., *Schizonella malanogramma* and *Pseudomonas* sp. are the best known polymeric biosurfactants. Various physical properties of polymeric surfactants include shearing resistance, low viscosity and tensile power. Sophorolipids are produced by *Torulopsis bombicola* which helps in lowering surface area and surface tension. Polymeric biosurfactants can form highly stable emulsions additionally they are made up of oil coat droplets and tend to form oil/water emulsions for food and cosmetics that are stable.

17.4.1.8 Particulate Biosurfactants

These are extracellular membrane vesicles which cause hydrocarbon partitioning from microemulsion, which plays a very significant role in bacterial cell alkane absorption. For example, *Acinetobacter* sp. vesicles. H01-N is 20–50 nm in diameter and 1158 g/cm^3 in buoyant mass, composed of proteins, phospholipids and lipopolysaccharides (Gharaei-Fathabad 2011).

17.5 Biodegradation of Petroleum Hydrocarbons by Biosurfactant-Producing Bacteria in the Contaminated Environment

Biodegradation is broadly involving breakdown of organic compounds by microorganisms (e.g. bacteria, fungi and actinomycetes). Petroleum hydrocarbon-degrading (bacteria) have evolved as a consequence of the presence of natural petroleum hydrocarbons in their habitat. These species are targets for petroleum pollutant control (Margesin et al. 2003; Ron and Rosenberg 2014; Lea-Smith et al. 2015). The majority of petroleum hydrocarbons present in the atmosphere are eventually destroyed and metabolized by native bacteria owing to their energy and growth requirements, as well as the need to alleviate physiological stress caused by the presence of petroleum hydrocarbons in the microbial bulk system (Hazen et al. 2010; Kleindienst et al. 2015). Several experiments have shown that there are huge number of hydrocarbon reducing bacteria in oil-rich areas, such as oil spills and oil reserves (Hazen et al. 2010; Yang et al. 2015), and that their frequency and amount is closely linked to the forms of petroleum hydrocarbons and the environmental factors affecting them (Fuentes et al. 2015; Varjani and Gnansounou 2017).

Recent studies have established that more than 79 bacterial genera can scavenge the petroleum hydrocarbons through oxidation from contaminated regions (Tremblay et al. 2017); some of them are as, *Marinobacter, Alteromonas, Kocuria, Pandoraea, Pseudomonas, Streptococcus, Rhodococcus, Enterobacter, Alcaligenes,*

Arthrobacteria, Staphylococcus, Achromobacter, Streptobacillus and Acinetobacter (Margesin et al. 2003; Chaerun et al. 2004; Jin et al. 2012; Nie et al. 2014; Varjani and Upasani 2016; Sarkar et al. 2017; Varjani 2017; Xu et al. 2017). Some biosurfactant-producing bacteria are listed in Table. 17.1 which play an important role in biodegradation of petroleum hydrocarbons.

In addition, many bacteria can only degrade or use some components of petroleum hydrocarbons successfully, while many are completely unavailable (Chaerun et al. 2004; Varjani 2017). This can be imputed to the fact that numerous native bacteria have unique catalytic enzymes; so, their functions in polluted oil sites are diverse. This also suggests that the petroleum hydrocarbons remediation requires the combined efforts of several active bacteria in order to accomplish the greatest effect on the environment (Dombrowski et al. 2016). Based on this scenario, Varjani et al. (2015) used a halo-tolerant Hydrocarbon Utilizing Bacterial Consortium (HUBC) including *Pseudomonas aeruginosa, Ochrobactrum sp. and Stenotrophomonas maltophilia,* which was found to be very effective for crude oil degradation (3% v/v), with a degradation rate as high as 83.49%.

The identified co-culture of the indigenous bacterial consortium and exogenous *Bacillus subtilis* was used to effectively speed up crude oil degradation (Tao et al. 2017). Wang et al. (2018) noticed that the indigenous bacterial consortium focused on the Penglai oil spill incident (China) had higher oil degradation potency compared to single bacteria dispersed in the marine ecosystem. A field study found that, *Xylosoxidans, Pseudomonas putida, Gordonia sp., Stenotrophomonas maltophilia, Pseudomonas fluorescens, Xanthomonas., Alcaligenes, Rhodococcus equi* and *Aeromonas hydrophila* had good biodegradation performance (89%) in 365 days of treatment (Szulc et al. 2014). Considered together, these studies indicate that a rational and feasible approach for improving the extraction capacity of petroleum hydrocarbons from polluted ecosystems is to boost the biodegradation ability through the application of bacterial consortia with several catabolic genes.

For the degradation of petroleum hydrocarbons bacteria are selected according to their ability to degrade and this involves several mechanisms for break down (Das and Chandran 2011). According to Fritsche and Hofrichter (2005) the mechanism involved are based on two processes: (1) growth and (2) co-metabolism. With regard to growth, bacteria use organic carbon as a source of energy and carbon (mineralization of organic pollutants) (Angelidaki and Sanders 2004). Co-metabolism is the metabolism of an organic compound in the presence of a growth substrate that is used as the primary carbon and energy source (Fritsche and Hofrichter 2005). The susceptibility and subsequent use of various hydrocarbons to microbial attacks depends on the nature and configuration of the hydrocarbons present and varies from 0.003 to 100% for marine, 0.13–50% for soil bacteria and bacteria 6–82% for soil fungi (Das and Chandran 2011). Most of these hydrocarbon contaminants can be distinguished by low water solubility and high solid-liquid distribution ratios, reducing their contacts with bacterial cells that utilize soluble molecules in the liquid phase (Johnsen et al. 2005; Cameotra and Makkar 2010).

Some polycyclic aromatic hydrocarbons (PAHs) require emulsifiers for solubilisation because of their recalcitrant nature in order to enhance their

Table 17.1 List of some biosurfactant-producing bacteria

Bacteria	Source of isolation	Type of biosurfactant	References
Acinetobacter sp.	Petroleum contaminated soil	Lipopeptide	Bao et al. (2014)
Acinetobacter junii	Petroleum reservoir, Oil contaminated soil	Rhamnolipids	Dong et al. (2016), Ohadi et al. (2017)
Acinetobacter baylyi	Crude oil contaminated soil	Lipopeptide	Zou et al. (2014)
Agrobacterium sp.	Contaminated soil	Ornithine lipids	Desai and Banat (1997)
Bacillus sp.	Hydrocarbon contaminated medium	Rhamnolipids	Banat (1993)
Bacillus subtilis	Crude oil samples, Tunisian soil Oil contaminated desert	Surfactin Lipopeptide	Pereira et al. (2013), Arguelles-Arias et al. (2009), Ghribi and Ellouze-Chaabouni (2011), Jha et al. (2016)
Bacillus licheniformis	Petroleum reservoir	Peptide lipids	Begley et al. (2009)
		Lipopeptides	Yakimov et al. (1995)
Bacillus pumilus A1,2A	Soil samples	Surfactin	Morikawa et al. (1992), Marchut-Mikolajczyk et al. (2018)
Lactobacillus fermentum	Sediment	Diglycosyl diglycerides	Mulligan et al. (2001)
Leuconostoc mesenteroides		Viscosin	Banat et al. (2010)
Pseudomonas aeruginosa	Hydrocarbon contaminated soil	Rhamnolipids	Patowary et al. (2017)
	Crude oil contaminated soil	Glycolipids	Zhang et al. (2012)
Pseudomonas fluorescens		Viscosin	Banat et al. (2010)
	Oil contaminated soil	Lipopeptide	Neu et al. (1990)
		Carbohydrate–lipid complex	Nerurkar et al. (2009)
Rhodococcus sp.	Contaminated soil	Glycolipid Polysaccharide Trehalolipids	Drouin and Cooper (1992), Neu et al. (1992), Kuyukina et al. (2015)
Serratia rubidea	Textile dye contaminated soil	Rhamnolipids	Jadhav et al. (2011)
Serratia marcescens	Petroleum contaminated soil	Serrawettin	Lai et al. (2009)
Thiobacillus thiooxidans		Ornithine lipids	Desai and Banat (1997)

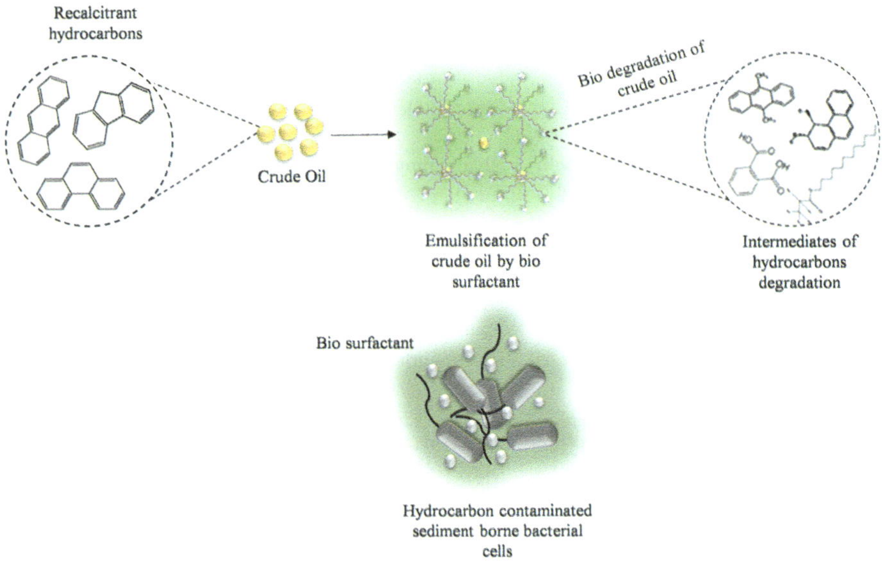

Fig. 17.4 Degradation of crude oil by Biosurfactant-producing bacteria

degradation. Biosurfactants improve the bioavailability of organic hydrophobic compounds, thereby making them a good agent for environmental cleaning. They can work in one of the following ways: emulsify the non-aqueous liquid phase pollutants or increase their solubility. Such features promote export of pollutants from the solid phase and enable access to the contaminant molecule for the microorganisms adsorbed on the soil particles (Bustamante et al. 2012; Hausmann and Syldatk 2015). The enhanced mobility of contaminants allows it to more easily associate with bacterial cells and is therefore rendered more prone to microbial degradation (Fig. 17.4) (Assadi et al. 2004; Adams et al. 2011; Saharan et al. 2011; Sharma 2016). By reducing surface and interfacial tensions, biosurfactants increase the surface areas of insoluble compounds, leading to increased mobility and bioavailability of hydrocarbons. The capability of biosurfactants and biosurfactant-producing bacterial strains to enhance availability and biodegradation rates of organic contaminants was reported by many authors (Rahman et al. 2003; Inakollu et al. 2004). Obayori et al. (2009) investigated the biodegradative properties of biosurfactants produced by *Pseudomonas* sp. Kang et al. (2010) used sophorolipids in studies of biodegradation of aliphatic and aromatic hydrocarbons and Iranian light, crude oil under laboratory conditions. Dong et al. (2016) and Ohadi et al. (2017) reported a rhamnolipids type biosurfactant-producing *Acinetobacter junii* play a vital role in biodegradation of petroleum hydrocarbons. Pereira et al. (2013) and Jha et al. (2016) isolated *Bacillus subtilis* from crude oil samples and oil contaminated desert producing surfactin and lipopeptide which enhance the biodegradation of petroleum hydrocarbons.

Addition of these biosurfactants to soil also increases biodegradation of tested hydrocarbons with the rate of degradation ranging from 85% to 97% of the total amount of hydrocarbons. Their results indicated that sophorolipids may have potential for facilitating the bioremediation of sites contaminated with hydrocarbons having limited water solubility and increasing the bioavailability of microbial consortia for biodegradation. The usefulness of biosurfactant-producing strains in bioremediation of highly contaminated sites with crude petroleum oil hydrocarbons was confirmed by Das and Mukherjee (2007) and El-sheshtawy et al. (2013). Due to environmental considerations, the use of surfactants or biosurfactants developed by microbiology has been increased. It has numerous advantages, such as wide pH, temperature and salinity tolerance, low toxicity, high biodegradability; marvellous low concentration surface and interfacial activity (Soberón-Chávez and Maier 2011).

17.6 Mechanism of Hydrocarbon Biodegradation

Microorganisms utilize hydrocarbons to gain energy and assimilate into cell biomass. There are various ways how microorganisms carry out degradation of petroleum hydrocarbon pollutants (a) Phototrophic, anoxygenic; (b) Chemotrophic, aerobic and (c) Chemotrophic, anaerobic (Leahy and Colwell 1990; Hendrickx et al. 2006; Abbasian et al. 2015; Meckenstock et al. 2016; Wilkes et al. 2016). Petroleum hydrocarbon degradation process has been reported to be strictly aerobic process. But few microorganisms are known to carry out anaerobic biodegradation (Widdel and Rabus 2001; Abbasian et al. 2015). Aerobic biodegradation primarily includes terminal oxidation and subterminal oxidation pathways for conversion of petroleum hydrocarbons. It also includes omega oxidation and beta oxidation pathways (Salleh et al. 2003; Abbasian et al. 2015). Initial step includes oxidation of methyl group leading to the formation of alcohol, then alcohol undergoes oxidation to produce aldehydes, aldehydes undergo dehydrogenation to produce carboxylic acids which enters into beta oxidation pathway (Das and Chandran 2011). Terminal oxidation is the most common pathway involved in utilization of n-alkane hydrocarbons which leads to the formation of primary alcohol via oxidation of terminal methyl group in initial step (Li and Liu 2002; Abbasian et al. 2015). Primary alcohol then undergoes oxidation to produce aldehyde, dehydrogenation of aldehyde leads to the formation of monocarboxylic acid which enters beta oxidation pathway resulting in the formation and removal of Acetyl Co enzyme A. The process begins by the formation of secondary alcohol by di-terminal oxidation of methyl group, secondary alcohol undergoes oxidation to produce methyl ketone which leads to the formation of acetyl ester which is then hydrolysed to form alcohol and fatty acid further it undergoes beta oxidation (Rojo 2009) and finally enters into tricarboxylic acid cycle to form other intermediate compounds. On the other hand, aromatic hydrocarbons are recalcitrant to degradation so they are the priority pollutants for bioremediation process (Salleh et al. 2003). Initial step involve in biodegradation of aromatic hydrocarbon includes oxidative attack which leads to breakage of benzene ring (Hendrickx et al. 2006). Further microorganisms involve series of enzymatic reactions to break these aromatic hydrocarbons into diols and or catechols. Benzene ring can be cleaved by two ways one is ortho-cleavage and the other is meta-cleavage. These cleavages involve two different enzymes intradiol dioxygenases and extradiol dioxygenases, respectively. Ortho and meta-cleavage leads to the formation of protocatechuates and catechols which further converted to tricarboxylic acid cycle intermediates (Li and Liu 2002; Abbasian et al. 2015).

Anaerobic biodegradation is more recent in terms of understanding as compared to the aerobic biodegradation pathway (Jaekel et al. 2013; Meckenstock et al. 2016). Very scanty literature is available about enzymes and factors involved in anaerobic biodegradation (Leahy and Colwell 1990; Wilkes et al. 2016). There are number of electron acceptors have been reported to be involved in this pathway, e.g. manganese or sulphate ions, ferrous iron, nitrate ions (Widdel and Rabus 2001; Foght 2008; Abbasian et al. 2015). Anaerobic biodegradation of alkane hydrocarbon involves carboxylation and addition of fumarate (Wilkes et al. 2016). Aromatic hydrocarbons are first transformed into phenols by hydroxylation and organic acids via carboxylation which are further converted into long chain fatty acids by β-oxidation pathway and finally to methane and carbon dioxide (Heider et al. 1998; Wilkes et al. 2016).

17.7 Factors Affecting the Biodegradation of Hydrocarbon

The biodegradation process of hydrocarbons is influenced by many parameters. Many environmental conditions such as pH, temperature, oxygen, nutrients and moisture from the soil can affect the results of biodegradation (Gavrilescu 2010; Msmg and Msmd 2012). Furthermore, microbial population and pollutant availability to the microbial population play a key role in the process of degradation. The main monitoring parameters are water quality, pH, nutrient levels and moisture content (Boopathy 2000; Chaillan et al. 2006). As in general, the degradation rate is highest at pH between 6.5 and 7.5 and temperature between 20 and 30 °C (depending on the microbial species) because microbial growth and activity can be easily influenced by temperature, pH and humidity (Harekrushna and Kumar 2012). Performance of biodegradation process depends on optimal parameters to enhance the process.

17.7.1 Oxygen (O_2)

Whether the system is anaerobic or aerobic determines the requirement of oxygen. Hydrogen peroxide (H_2O_2) and magnesium peroxide may be used in some situations to raise the amount of reactive O_2 in the soil. The oxidation of the surface by oxygenase is a first step in the breaking down of hydrocarbons by bacteria. The amount of oxygen in the soil depends on the rate of absorption of microbial O_2 in the region of the surface used (Bartha and Bossert 1984). Aerobic bacteria were known to quickly utilize hydrocarbons, such as *Mycobacterium*, *Rhodococcus*, *Pseudomonas* and *Sphingomonas* (Juwarkar et al. 2010).

These microbes were often reported to degrade pesticides and hydrocarbons, both alkanes and polyaromatic compounds (Msmg and Msmd 2012). Pseudomonas sp. BS2201, BS2203 and Brevibacillus sp. BS2202 (Nitrate-reducing) were able to degrade petroleum hydrocarbons in both aerobic and anaerobic environment. As a result 20–25% of the organic material and 90–95% of the alkanes were degraded by bacteria effectively in 10 day aerobic experiment. Also in another 50-day anaerobic experiment, nearly 15–18% of the organic material and only about 20–25% of the alkenes were reported to be degraded (Van 2011). It may be difficult to deliver oxygen to soils which are polluted because of many reasons: soil porosity may not be favourable and therefore bulk transfer from the gas phase to the water phase will be restricted.

17.7.2 Soil Properties and Nutrient Availability

It is also an important factor in accelerating the decay cycle, in order to improve the supply of heat, water and nutrition, other parameters likewise organic carbon is also useful in soil structure reclamation. It has been observed that volatile organic compound (VOCs) can be evaporated from all types of soil environments, particularly at freshly contaminated areas. Oil concentration seems to be reduced due to microbial breakdown rather than volatilization. Attachment to soil particles greatly influences the level of biodegradation, but this factor is hard to measure (Kauppi 2011).

Nutrients supply has direct effects on microbial activity and biodegradation (Jain et al. 2011). Nutrients are the only source of donors for electrons, carbon and heat. Bacteria basically require inorganic nutrients, amino acids and vitamins. Elements like phosphorus and nitrogen, and in some cases iron, are necessary for cellular metabolism that could become a limiting factor and therefore hinder the processes of biodegradation (Das and Chandran 2011). Atlas and Bragg (2009) suggested that the nutrient accessibility, especially N and P, was significant in oil spill degradation, regardless of whether it occurred in freshwater or marine ecosystems. Furthermore, for the reclamation and enhancement of biodegradation, nutrients were introduced (Zhu et al. 2004). Several researchers have studied the effects on the degradation process of organic fertilizers, composts, manure from poultry, banana peels, melon shell, wood chips, rice husk mixtures, soy cake, sewage sludge, mushroom and animal droppings (Park et al. 2001; Adesodun and Mbagwu 2008; Hickman and Reid 2008; Abioye et al. 2009). Bento et al. (2005) reported that adding N and P increased the rate of degradation in contaminated soil from 16% to over 90%.

17.7.3 Water Content and Temperature

Water availability is a life prerequisite and impacts the development and growth of microbes. To achieve optimal amounts of moisture, irrigation is often needed (Harekrushna and Kumar 2012). Increase in the soil moisture has had a positive impact on soil degradation of polycyclic aromatic hydrocarbon. The considerable effect of moisture was demonstrated in an experiment on the extraction of PAHs from sunflowers grown in soil nearby manufactured gas plant (Gong et al. 2005). Temperature influences biodegradation process due to changes in microbes' metabolic activity at varying temperatures (Eriksson et al. 2001). This variable plays an important role in the microbial flora's diversity, morphology and metabolism. Temperature not only influences the biodegradation process but oil's physical and chemical properties as well (Margesin and Schinner 2001a, 2001b). Microbial activity typically reduces at lower temperature (Delille 2000; Gibb et al. 2001; Baraniecki et al. 2002; Eckford et al. 2002). Diffusion rates of organic compound rises with increasing temperature (Northcott and Jones 2000). The highest degradation rate in soil and water occurs in the range of 30–40 °C and 20–30 °C, respectively (Das and Chandran 2011). In cold climates many microorganisms generally survive and grow at or below 5 °C, optimum temperature is required for successful progress to be useful in bioremediation at the polluted sites. Moreover, as temperature increases, the contaminant's solubility and bioavailability will also increase, but the solubility of oxygen will be reduced (Margesin and Schinner 1999). However, optimum temperatures are also species-specific. For instance, a variety of *Rhodococcus sp.* extracted from Antarctic soil succeeded in degrading a range of n-alkanes at −2 °C, but several were inhibited at elevated temperatures (Bej et al. 2001).

17.7.4 Hydrocarbon Bioavailability

Bioavailability is characterized as an accessibility of pollutants for assimilation and toxicity (Puglisi et al. 2007). It was also elaborated as the level at which a substance has liberty to move into an organism, and bioavailability is known to differ between species. Organic pollutants' bioavailability is the key factor that determines their existence, toxicity, ecological danger and environmental losses (Oleszczuk 2009). This comes up with information about the real risk of contamination being present.

The frequency of transition of the substance to the living cell from the soil and the level of absorption and metabolism are two key parameters that determine the volume of a chemical compound that is bioavailable (Semple et al. 2003). Bioavailability regulates biodegradation as bacterial cells need to absorb fuel to promote catabolic genes involved in biodegradation (Madsen 2002). Hydrocarbon bioavailability often relies on their hydrophobicity; hydrocarbon solubility, soil particle sorption, physical state and volatilization have a major impact on the degree of biodegradation. A large amount of hydrocarbon appears undegraded, under optimal conditions of hydrocarbon biodegradation. Mainly, upon reaching the soil, an environmental contaminant may be lost by biodegradation, leaching and volatilization, or may persist in the soil biota or be sequestered and complexes within the fractions of mineral and organic matter in the soil (Sihag et al. 2014). Number of physicochemical processes, such as dissolution, diffusion and sorption/desorption that regulate the pace of transport to the cell and the level of absorption and metabolism by the microorganisms. Since the fate of hydrocarbons in the subsurface depends on transportation processes, multi-phase flow, volatilization, dissolution, geochemical reactions, biodegradation and sorption, an interdisciplinary investigation of these processes is critical to the successful assessment of the potential for hydrocarbon migration to the subsurface. Bacteria convert or oxidize harmful contaminants into less harmful compounds by utilizing oxygen from their surrounding environment, producing CO_2 and H_2O.

In addition, photosynthesis produces oxygen that can be used by microbes to further oxidize toxins. One of the major factors affecting hydrocarbon microbial degradation is oil hydrophobicity, limiting its transfer to microorganism cell surfaces (Lee et al. 2006; Schein et al. 2009). This constraint can either be solved by increasing microorganisms producing biosurfactant or by introducing surface-active agents/compounds. Several constraints can affect bioavailability of organic compounds in the environment. These include low water solubility, sorption, elimination from micro pores and organic carbon content (Huesemann et al. 2004; Froehner et al. 2012). Puglisi et al. (2007) documented that, due to higher adsorption and lower water diffusion, phenanthrene degradation is usually lower in high organic soils. Biodegradation differences in soils with low levels of organic matter may be related to the amount of clay in the soil. The frequency of transmission is calculated in the bulk phase and aqueous phase by the equilibrium and real concentration. As it applies to biodegradation, this is essential to the theory of bioavailability.

17.7.5 Concentration of Petroleum Hydrocarbon

Petroleum hydrocarbon concentration determines to a greater extent the rate of environmental breakdown of hydrocarbons (Abioye et al. 2012). High hydrocarbon concentration can influence microorganism activities. In particular, levels of 1–100 g/ml of water and 1–100 g/kg of soil or sand not found to be harmful to bacteria. Ijah and Antai (2003) suggested fast oxidation level of hydrocarbons in low concentration soil (10–20%) relative to those polluted by high concentration (30–40%) of crude oil over 1 year. By analogy, Rahman et al. (2002) stated that a fall in oil concentration from 1 to 10% causes reduction of biodegradation process from 78 to 52%. Indeed, high hydrocarbon concentrations can be associated with increased amounts of crude oil in polluted sites resulting in biodegradation inhibition due to nutrient limitation.

17.8 Conclusion and Future Outlook

Because of their extreme risk to human health and the environment, petroleum hydrocarbons are one of the most troubling contaminants. Bioremediation of environment with hydrocarbon-degrading bacteria from petroleum has been considered to be an effective and environmentally friendly process. A large number of bacterial species with the ability to degrade petroleum hydrocarbons have been exploited and used for bioremediation.

During the practical application process, however, numerous problems were found that slow down the rate of biodegradation. This chapter addressed such limiting factors, including the toxic effects of petroleum hydrocarbons; pollutant bioavailability, environmental restrictions, metabolic constraints and time use, and summarized the latest countermeasures against these issues. Many approaches have been researched and introduced, such as controlling environmental factors and designing microbial inoculants. A series of studies still need to be carried out before the successful application of bioremediation for the recovery of polluted habitats of petroleum oil, based on the current state of knowledge examined here. The inferences are as follows:

1. According to the concepts the interface mechanism between bacteria and petroleum hydrocarbons needs to resolve barriers in the microbial absorption of petroleum derived hydrocarbons.
2. Create new biocomputational surfactants to improve interaction between bacteria and petroleum derived hydrocarbons.
3. Using modern biotechnology to investigate undiscovered tools of petroleum hydrocarbon reducing bacteria.
4. Advancing the approach of artificial microbial consortia, such as enriching metagenomics and creating chosen consortia.
5. Explore the novel functional genes that control hydrocarbon degradation pathway to provide new insights into the molecular mechanism and microbial remediation process.
6. Create genetically engineered bacteria utilizing synthetic biology technologies to improve their capacity to degrade petroleum hydrocarbons.

References

Abbasian F, Lockington R, Mallavarapu M, Naidu R (2015) A comprehensive review of aliphatic hydrocarbon biodegradation by bacteria. Appl Biochem Biotechnol 176(3):670–699. https://doi.org/10.1007/s12010-015-1603-5

Abioye OP, Alonge OA, Ijah UJJ (2009) Biodegradation of crude oil in soil amended with melon shell. AU J Technol 13(1):35–38

Abioye OP, Agamuthu P, Abdul Aziz AR (2012) Biodegradation of used motor oil in soil using organic waste amendments. Biotechnol Res Int 2012:587041

Adams RH, Castillo-Acosta O, Escalante-Espinosa E, Zavala-Cruz J (2011) Natural attenuation and phytoremediation of petroleum hydrocarbon impacted soil in tropical wetland environments. In: Torres LG, Bandala ER (eds) Remediation of soils and aquifers. Nova Publishers, New York, pp 1–24

Adesodun JK, Mbagwu JSC (2008) Biodegradation of waste-lubricating petroleum oil in a tropical alfisol as mediated by animal droppings. Bioresour Technol 99(13):5659–5665

Afuwale C, Modi HA (2012) Study of bacterial diversity of crude oil degrading bacteria isolated from crude oil contaminated sites. Life Sci Leafl 6:13–23

Al-Hawash B, Dragh A, Li S et al (2018) Principles of microbial degradation of petroleum hydrocarbons in the environment. Egypt J Aquat Res 44:71–76

Al-Maamary HM, Kazem HA, Chaichan MT (2017) The impact of oil price fluctuations on common renewable energies in GCC countries. Renew Sust Energ Rev 75:989–1007

Angelidaki I, Sanders W (2004) Assessment of the anaerobic biodegradability of macropollutants. Re/Views Environ Sci Bio/Technol 3(2):117–129

Aparna A, Srinikethan G, Hedge S (2011) Effect of addition of biosurfactant produced by Pseudomonas ssp. on biodegradation of crude oil. Int Proc Chem Biol Environ Eng 6:71–75

Arguelles-Arias A, Ongena M, Halimi B, Lara Y, Brans A, Joris B, Fickers P (2009) Bacillus amyloliquefaciens GA1 as a source of potent antibiotics and other secondary metabolites for biocontrol of plant pathogens. Microb Cell Factories 8(1):1–12

Aşçı Y, Nurbaş M, Açıkel YS (2010) Investigation of sorption/desorption equilibria of heavy metal ions on/from quartz using rhamnolipid biosurfactant. J Environ Manag 91(3):724–731

Assadi MM, Rostamza M, Noohi AS, Levin M, Shahamati M (2004) Rhamnolipid production by Pseudomonas aerugiosa MM1011 from sugar beet molasses. Asian J Microbiol Biotechnol Environ Sci 6(2):203–207

Atlas R, Bragg J (2009) Bioremediation of marine oil spills: when and when not–the Exxon Valdez experience. Microb Biotechnol 2(2):213–221

Balba MT, Al-Awadhi N, Al-Daher R (1998) Bioremediation of oil-contaminated soil: microbiological methods for feasibility assessment and field evaluation. J Microbiol Methods 32(2):155–164

Banat IM (1993) The isolation of a thermophilic biosurfactant producing Bacillus sp. Biotechnol Lett 15(6):591–594

Banat IM, Makkar SR, Cameotra SS (2000) Potential commercial application of microbial surfactants. Appl Microbiol Biotechnol 53(5):495–508

Banat IM, Franzetti A, Gandolfi I, Bestetti G, Martinotti MG, Fracchia L, Smyth TJ, Marchant R (2010) Microbial biosurfactants production, applications and future potential. Appl Microbiol Biotechnol 87:427–444

Bao M, Pi Y, Wang L, Sun P, Li Y, Cao L (2014) Lipopeptide biosurfactant production bacteria Acinetobacter sp. D3-2 and its biodegradation of crude oil. Environ Sci: Processes Impacts 16 (4):897–903

Baraniecki CA, Aislabie J, Foght JM (2002) Characterization of Sphingomonas sp. Ant 17, an aromatic hydrocarbon-degrading bacterium isolated from Antarctic soil. Microb Ecol 43:44–54

Bartha R, Bossert I (1984) Petroleum microbiology. In: Atlas RM (ed) The treatment and disposal of petroleum refinery wastes. Macmillan, New York, pp 553–557

Begley M, Cotter PD, Hill C, Ross RP (2009) Identification of a novel two-peptide lantibiotic, lichenicidin, following rational genome mining for Lan M proteins. Appl Environ Microbiol 75:5451–5460

Bej AK, Saul D, Aislabie J (2001) Cold-tolerant alkane-degrading Rhodococcus species from Antarctica. Polar Biol 23:100–105

Bento FM, Beech IB, Gaylarde CC, Englert GE, Muller IL (2005) Degradation and corrosive activities of fungi in a diesel–mild steel–aqueous system. World J Microbiol Biotechnol 21 (2):135–142

Boopathy R (2000) Factors limiting bioremediation technologies. Bioresour Technol 74(1):63–67

Bustamante M, Durán N, Diez MC (2012) Biosurfactants are useful tools for the bioremediation of contaminated soil: a review. J Soil Sci Nutr 12:667–687. https://doi.org/10.4067/S0718-95162012005000024

Cameotra SS, Makkar RS (2010) Biosurfactant-enhanced bioremediation of hydrophobic pollutants. Pure Appl Chem 82(1):97–116

Cerniglia CE, Freeman JP, Althaus JR, van Baalen C (1983) Metabolism and toxicity of 1-and 2-methylnaphthalene and their derivatives in cyanobacteria. Arch Microbiol 136(3):177–183

Chaerun SK, Tazaki K, Asada R, Kogure K (2004) Bioremediation of coastal areas 5 years after the Nakhodka oil spill in the sea of Japan: isolation and characterization of hydrocarbon-degrading bacteria. Environ Int 30(7):911–922

Chaillan F, Chaineau CH, Point V, Saliot A, Oudot J (2006) Factors inhibiting bioremediation of soil contaminated with weathered oils and drill cuttings. Environ Pollut 144(1):255–265

Chandra S, Sharma R, Singh K, Sharma A (2013) Application of bioremediation technology in the environment contaminated with petroleum hydrocarbon. Ann Microbiol 63:417–431

Costa AS, Romao LP, Araujo BR, Lucas SC, Maciel ST, Wisniewski A Jr, Alexandre MR (2012) Environmental strategies to remove volatile aromatic fractions (BTEX) from petroleum industry wastewater using biomass. Bioresour Technol 105:31–39

Darvishi P, Ayatollahi S, Mowla D, Niazi A (2011) Biosurfactant production under extreme environmental conditions by an efficient microbial consortium ERCPPI-2. Colloids Surf B Biointerfaces 84:292–300

Das N, Chandran P (2011) Microbial degradation of petroleum hydrocarbon contaminants: an overview. Biotechnol Res Int 2011:941810

Das K, Mukherjee AK (2007) Crude petroleum oil biodegradation efficiency of Bacillus subtilis and Pseudomonas aeruginosa strains isolated from a petroleum-oil contaminated soil from North East India. Bioresour Technol 98(7):1339–1345

Delille D (2000) Response of Antarctic soil bacterial assemblages to contamination by diesel fuel and crude oil. Microb Ecol 40(2):159–168

Desai JD, Banat IM (1997) Microbial production of surfactants and their commercial potential. Microbiol Mol Biol Rev 61(1):47–64

Díez Salvador S, Jover Comas E, Bayona Termens JM, Albaigés Riera J (2007) Prestige oil spill III, the fate of a heavy oil in the marine environment. Environ Sci Technol 41(9):3075–3082

Dombrowski N, Donaho JA, Gutierrez T, Seitz KW, Teske AP, Baker BJ (2016) Reconstructing metabolic pathways of hydrocarbon-degrading bacteria from the Deepwater Horizon oil spill. Nat Microbiol 1(7):1–7

Dong H, Xia W, Dong H, She Y, Zhu P, Liang K, Zhang G (2016) Rhamnolipids produced by indigenous Acinetobacter junii from petroleum reservoir and its potential in enhanced oil recovery. Front Microbiol 7:1710

Drouin CM, Cooper DG (1992) Biosurfactant and aqueous two phase fermentation. Biotechnol Bioeng 40:86–90

Dudley B (2019) BP statistical review of world energy statistical review of world. bp.Com

Dyke MV, Couture P, Brauer M, Lee H, Trevors JT (1993) Pseudomonas aeruginosa UG2 rhamnolipid biosurfactants: structural characterization and their use in removing hydrophobic compounds from soil. Can J Microbiol 39(11):1071–1078

Eckford R, Cook FD, Saul D, Aislabie J, Foght J (2002) Free-living heterotrophic nitrogen-fixing bacteria isolated from fuel-contaminated Antarctic soils. Appl Environ Microbiol 68 (10):5181–5185

El-Sheshtawy HS, El-Tabei AS, Kobisy AS, Doheim MM (2013) Application of biosurfactant produced by Bacillus lichneformis and chemical surfactant in biodegradation of crude oil: part I. Biosci Biotechnol Res Asia 10(2):515–526

Eriksson M, Ka JO, Mohn WW (2001) Effects of low temperature and freeze-thaw cycles on hydrocarbon biodegradation in Arctic tundra soil. Appl Environ Microbiol 67(11):5107–5112

Farhadian M, Vachelard C, Duchez D, Larroche C (2008) In situ bioremediation of monoaromatic pollutants in groundwater: a review. Bioresour Technol 9:5296–5308

Foght JM (2008) Anaerobic biodegradation of aromatic hydrocarbons: pathways and prospects. J Mol Micobiol Biotechnol 15:93–120

Franzetti A, Caredda P, La Colla P, Pintus M, Tamburini E, Papacchini M, Bestetti G (2009) Cultural factors affecting biosurfactant production by Gordonia sp. BS29. Int Biodeterior Biodegradation 63(7):943–947

Fritsche W, Hofrichter M (2005) Aerobic degradation of recalcitrant organic compounds by microorganisms. In: Environmental biotechnology concepts and applications. Wiley, Weinheim

Froehner S, Dombroski LF, Machado KS, Fernandes CS, Bessa M (2012) Estimation of bioavailability of polycyclic aromatic hydrocarbons in river sediments. Int J Environ Sci Technol 9 (3):409–416

Fuentes S, Barra B, Caporaso JG, Seeger M (2015) From rare to dominant: a fine-tuned soil bacterial bloom during petroleum hydrocarbon bioremediation. Appl Environ Microbiol 82:888–896. https://doi.org/10.1128/AEM.02625-15

Gavrilescu M (2010) Environmental biotechnology: achievements, opportunities and challenges. Dyn Biochem Process Biotechnol Mol Biol 4(1):1–36

Gharaei-Fathabad E (2011) Biosurfactants in pharmaceutical industry: a mini-review. Am J Drug Discov Dev 1(1):58–69

Ghribi D, Ellouze-Chaabouni S (2011) Enhancement of Bacillus subtilis lipopeptide biosurfactants production through optimization of medium composition and adequate control of aeration. Biotechnol Res Int 2011:653654

Gibb A, Chu A, Wong RCK, Goodman RH (2001) Bioremediation kinetics of crude oil at 5 C. J Environ Eng 127(9):818–824

Gong Z, Wilke BM, Alef K, Li P (2005) Influence of soil moisture on sunflower oil extraction of polycyclic aromatic hydrocarbons from a manufactured gas plant soil. Sci Total Environ 343 (1–3):51–59

Gudiña EJ, Fernandes EC, Rodrigues AI, Teixeira JA, Rodrigues LR (2015a) Biosurfactant production by Bacillus subtilis using corn steep liquor as culture medium. Front Microbiol 6:1–7

Gudiña EJ, Rodrigues AI, Alves E, Domingues MR, Teixeira JA, Rodrigues LR (2015b) Bioconversion of agro-industrial by-products in rhamnolipids toward applications in enhanced oil recovery and bioremediation. Bioresour Technol 177:87–93

Harayama S, Kasai Y, Hara A (2004) Microbial communities in oil-contaminated seawater. Curr Opin Biotechnol 15(3):205–214

Harekrushna S, Kumar DC (2012) A review on: bioremediation. Int J Res Chem Environ 2 (1):13–21

Hausmann R, Syldatk C (2014) Types and classification of microbial surfactants. In: Biosurfactants: production and utilization—processes, technologies, and economics, 1st edn. CRC Press, Boca Raton, p 159

Hausmann R, Syldatk C (2015) Types and classifcation of microbial surfactants. In: Biosurfactants. Production and utilization-processes, technologies, and economics. CRC Press, Boca Raton. https://doi.org/10.1201/b17599-3

Hazen TC, Dubinsky EA, DeSantis TZ, Andersen GL, Piceno YM, Singh N et al (2010) Deep-sea oil plume enriches indigenous oil-degrading bacteria. Science 330:204–208

Head IM, Jones DM, Röling WF (2006) Marine microorganisms make a meal of oil. Nat Rev Microbiol 4(3):173–182

Heider J, Spormann AM, Beller HR, Widdel F (1998) Anaerobic bacterial metabolism of hydrocarbons. FEMS Microbiol Rev 22(5):459–473

Hendrickx B, Junca H, Vosahlova J, Lindner A, Ruegg I, Bucheli-Witschel M, Faber F, Egli T, Mau M, Pieper DH, Top EM, Dejonghe W, Bastiaens L, Springael D (2006) Alternative primer sets for PCR detection of genotypes involved in bacterial aerobic BTEX degradation: distribution of the genes in BTEX degrading isolates and in subsurface soils of a BTEX contaminated industrial site. J Microbiol Methods 64(2):250–265

Hickman ZA, Reid BJ (2008) The co-application of earthworms (Dendrobaena Veneta) and compost to increase hydrocarbon losses from diesel contaminated soils. Environ Int 34 (7):1016–1022

Hou N, Zhang N, Jia T, Sun Y, Dai Y, Wang Q, Li C (2018) Biodegradation of phenanthrene by biodemulsifier-producing strain Achromobacter sp. LH-1 and the study on its metabolisms and fermentation kinetics. Ecotoxicol Environ Saf 163:205–214

Huesemann MH, Hausmann TS, Fortman TJ (2004) Does bioavailability limit biodegradation? A comparison of hydrocarbon biodegradation and desorption rates in aged soils. Biodegradation 15(4):261–274

Ijah UJJ, Antai SP (2003) The potential use of chicken-drop micro-organisms for oil spill remediation. Environmentalist 23(1):89–95

Inakollu S, Hung HC, Shreve GS (2004) Biosurfactant enhancement of microbial degradation of various structural classes of hydrocarbon in mixed waste systems. Environ Eng Sci 21:463–469. https://doi.org/10.1089/1092875041358467

Jada A, Salou M (2002) Effects of the asphaltene and resin contents of the bitumens on the water-bitumen interface properties. J Pet Sci Eng 33:185–193

Jadhav M, Kalme S, Tamboli D, Govindwar S (2011) Rhamnolipid from Pseudomonas desmolyticum NCIM-2112 and its role in the degradation of Brown 3REL. J Basic Microbiol 51(4):385–396

Jaekel U, Musat N, Adam B, Kuypers M, Grundmann O, Musat F (2013) Anaerobic degradation of propane and butane by sulfate-reducing bacteria enriched from marine hydrocarbon cold seeps. ISME J 7:885–895

Jain PK, Gupta VK, Gaur RK, Lowry M, Jaroli DP, Chauhan UK (2011) Bioremediation of petroleum oil contaminated soil and water. Res J Environ Toxicol 5(1):1–26

Jennings EM, Tanner RS (2000) Biosurfactant-producing bacteria found in contaminated and uncontaminated soils. In: Proceedings of the 2000 conference on hazardous waste research, vol 306, pp 299–306

Jha SS, Joshi SJ, SJ G (2016) Lipopeptide production by Bacillus subtilis R1 and its possible applications. Braz J Microbiol 47(4):955–964

Jin HM, Kim JM, Lee HJ, Madsen EL, Jeon CO (2012) Alteromonas as a key agent of polycyclic aromatic hydrocarbon biodegradation in crude oil-contaminated coastal sediment. Environ Sci Technol 46(14):7731–7740

Johnsen AR, Wick LY, Harms H (2005) Principles of microbial PAH-degradation in soil. Environ Pollut 133(1):71–84

Juwarkar AA, Singh SK, Mudhoo A (2010) A comprehensive overview of elements in bioremediation. Rev Environ Sci Biotechnol 9(3):215–288

Kang SW, Kim YB, Shin JD, Kim EK (2010) Enhanced biodegradation of hydrocarbons in soil by microbial biosurfactant, sophorolipid. Appl Biochem Biotechnol 160(3):780–790

Kauppi S (2011) Bioremediation of diesel oil contaminated soil and water. Master'thesis, University of Helsinki, Lathi, Findland

Kleindienst S, Paul JH, Joye SB (2015) Using dispersants after oil spills: impacts on the composition and activity of microbial communities. Nat Rev Microbiol 13(6):388–396

Kosaric N, Cairns WL, Gray NC (1987) Biosurfactants and biotechnology. M. Dekker, New York

Krishnani KK, Kathiravan V, Meena KK, Sarkar B, Kumar S, Brahmane MP, Kailasam NKM (2019) Bioremediation of aquatic toxicants: application of multi-omic approaches. Narendra Publishing House, Delhi

Kucharski J, Tomkiel M, Boros E, Wyszkowska J (2010) The effect of soil contamination with diesel oil and petrol on the nitrification process. J Elem 15(1):111–118

Kuyukina MS, Ivshina IB, Baeva TA, Kochina OA, Gein SV, Chereshnev VA (2015) Trehalolipid biosurfactants from nonpathogenic Rhodococcus actinobacteria with diverse immunomodulatory activities. New Biotechnol 32(6):559–568

Labud V, Garcia C, Hernandez T (2007) Effect of hydrocarbon pollution on the microbial properties of a sandy and a clay soil. Chemosphere 66(10):1863–1871

Lai CC, Huang YC, Wei YH, Chang JS (2009) Biosurfactant-enhanced removal of total petroleum hydrocarbons from contaminated soil. J Hazard Mater 167:609–614

Lang S, Philp JC (1998) Surface-active lipids in rhodococci. Antonie Van Leeuwenhoek 74 (1–3):59–70

Leahy JH, Colwell R (1990) Microbial degradation of hydrocarbons in the environment. Microbiol Rev 54(3):305–315

Lea-Smith DJ, Biller SJ, Davey MP, Cotton CA, Sepulveda BMP, Turchyn AV, Howe CJ (2015) Contribution of cyanobacterial alkane production to the ocean hydrocarbon cycle. Proc Natl Acad Sci 112(44):13591–13596

Lee M, Kim MK, Singleton I, Goodfellow M, Lee ST (2006) Enhanced biodegradation of diesel oil by a newly identified Rhodococcus baikonurensis EN3 in the presence of mycolic acid. J Appl Microbiol 100(2):325–333

Li XW, Liu ZP (2002) Microbial biodegradation of petroleum hydrocarbon. Acta Microbiol Sin 42:764–767

Li AH, Xu MY, Sun W, Sun GP (2011) Rhamnolipid production by Pseudomonas aeruginosa GIM 32 using different substrates including molasses distillery wastewater. Appl Biochem Biotechnol 163(5):600–611

Liu JF, Mbadinga SM, Yang SZ, Gu JD, Mu BZ (2015) Chemical structure, property and potential applications of biosurfactants produced by Bacillus subtilis in petroleum recovery and spill mitigation. Int J Mol Sci 16(3):4814–4837

Macaulay BM, Rees D (2014) Bioremediation of oil spills: a review of challenges for research advancement. Ann Environ Sci 8:9–37

Madsen EL (2002) Report on bioavailability of chemical wastes with respect to the potential for soil bioremediation. Department of Microbiology, Cornell University, Ithaca

Marchut-Mikolajczyk O, Drożdżyński P, Pietrzyk D, Antczak T (2018) Biosurfactant production and hydrocarbon degradation activity of endophytic bacteria isolated from Chelidonium majus L. Microb Cell Factories 17(1):1–9

Margesin R, Schinner F (1999) Biodegradation of diesel oil by cold-adapted microorganisms in presence of sodium dodecyl sulfate. Chemosphere 38(15):3463–3472

Margesin R, Schinner F (2001a) Biodegradation and bioremediation of hydrocarbons in extreme environments. Appl Microbiol Biotechnol 56(5–6):650–663

Margesin R, Schinner F (2001b) Bioremediation (natural attenuation and biostimulation) of diesel-oil-contaminated soil in an alpine glacier skiing area. Appl Environ Microbiol 67(7):3127–3133

Margesin R, Labbe D, Schinner F, Greer CW, Whyte LG (2003) Characterization of hydrocarbon-degrading microbial populations in contaminated and pristine alpine soils. Appl Environ Microbiol 69(6):3085–3092

Mason OU, Hazen TC, Borglin S, Chain PS, Dubinsky EA, Fortney JL, Mackelprang R (2012) Metagenome, metatranscriptome and single-cell sequencing reveal microbial response to Deepwater Horizon oil spill. ISME J 6(9):1715–1727

Meckenstock RU, Boll M, Mouttaki H, Koelschbach JS, Tarouco PC, Weyrauch P, Dong X, Himmelberg AM (2016) Anaerobic degradation of benzene and polycyclic aromatic hydrocarbons. J Mol Microbiol Biotechnol 26:92–118

Mnif I, Ghribi D (2015) Review lipopeptides biosurfactants: mean classes and new insights for industrial, biomedical, and environmental applications. Pept Sci 104(3):129–147

Morikawa M, Ito M, Imanaka T (1992) Isolation of a new surfactin producer Bacillus pumilus A-1, and cloning and nucleotide sequence of the regulator gene, psf-1. J Ferment Bioeng 74 (5):255–261

Msmg N, Msmd D (2012) Review paper on—parameters affecting bioremedaition. Int J Life Sci Pharm Res 2(3):77–80

Mulligan CN, Gibbs BF (2004) Types, production and applications of biosurfactants. Proc Indian Natl Sci Acad Part B 70(1):31–56

Mulligan CN, Yong RN, Gibbs BF (2001) Remediation technologies for metal-contaminated soils and groundwater: an evaluation. Eng Geol 60(1–4):193–207

Nerurkar AS, Hingurao KS, Suthar HG (2009) Bioemulsifiers from marine microorganisms

Neu TR, Härtner T, Poralla K (1990) Surface active properties of viscosin: a peptidolipid antibiotic. Appl Microbiol Biotechnol 32(5):518–520

Neu TR, Dengler T, Jann B, Poralla K (1992) Structural studies of an emulsion-stabilizing exopolysaccharide produced by an adhesive, hydrophobic Rhodococcus strain. Microbiology 138(12):2531–2537

Nie Y, Liang JL, Fang H, Tang YQ, Wu XL (2014) Characterization of a CYP153 alkane hydroxylase gene in a gram-positive Dietzia sp. DQ12-45-1b and its "team role" with alkW1 in alkane degradation. Appl Microbiol Biotechnol 98(1):163–173

Northcott GL, Jones KC (2000) Experimental approaches and analytical techniques for determining organic compound bound residues in soil and sediment. Environ Pollut 108(1):19–43

Obayori OS, Ilori MO, Adebusoye SA, Oyetibo GO, Omotayo AE, Amund OO (2009) Degradation of hydrocarbons and biosurfactant production by Pseudomonas sp. strain LP1. World J Microbiol Biotechnol 25:1615–1623

Ohadi M, Dehghannoudeh G, Shakibaie M, Banat IM, Pournamdari M, Forootanfar H (2017) Isolation, characterization, and optimization of biosurfactant production by an oil-degrading Acinetobacter junii B6 isolated from an Iranian oil excavation site. Biocatal Agric Biotechnol 12:1–9

Oleszczuk P (2009) Application of three methods used for the evaluation of polycyclic aromatic hydrocarbons (PAHs) bioaccessibility for sewage sludge composting. Bioresour Technol 100 (1):413–420

Onwurah INE, Ogugua VN, Onyike NB, Ochonogor AE, Otitoju OF (2007) Crude oil spills in the environment, effects and some innovative clean-up biotechnologies. Int J Environ Res 1 (4):307–320

Overholt WA, Marks KP, Romero IC, Hollander DJ, Snell TW, Kostka JE (2015) Hydrocarbon degrading bacteria exhibit a species specific response to dispersed oil while moderating ecotoxicity. Appl Environ Microbiol 82:518–527

Park JA, Hur JM, Jang BK, Son BS (2001) Evaluation of compost addition and its effect on biodegradation of diesel-oil in contaminated soil composting. J Ind Eng Chem 7(3):127–136

Parra-Barraza H, Hernández-Montiel D, Lizardi J, Hernández J, Urbina RH, Valdez MA (2003) The zeta potential and surface properties of asphaltenes obtained with different crude oil/n-heptane proportions. Fuel 82(8):869–874

Patowary K, Patowary R, Kalita MC, Deka S (2017) Characterization of biosurfactant produced during degradation of hydrocarbons using crude oil as sole source of carbon. Front Microbiol 8:279

Pereira JF, Gudiña EJ, Costa R, Vitorino R, Teixeira JA, Coutinho JA, Rodrigues LR (2013) Optimization and characterization of biosurfactant production by Bacillus subtilis isolates towards microbial enhanced oil recovery applications. Fuel 111:259–268

Peypoux F, Bonmatin JM, Wallach J (1999) Recent trends in the biochemistry of surfactin. Appl Microbiol Biotechnol 51(5):553–563

Puglisi E, Cappa F, Fragoulis G, Trevisan M, Del Re AA (2007) Bioavailability and degradation of phenanthrene in compost amended soils. Chemosphere 67(3):548–556

Rahman KSM, Thahira-Rahman J, Lakshmanaperumalsamy P, Banat IM (2002) Towards efficient crude oil degradation by a mixed bacterial consortium. Bioresour Technol 85(3):257–261

Rahman KSM, Rahman TJ, Kourkoutas Y, Petsas I, Marchant R, Banat IM (2003) Enhanced bioremediation of n-alkane in petroleum sludge using bacterial consortium amended with rhamnolipid and micronutrients. Bioresour Technol 90:159–168

Rivers AR, Sharma S, Tringe SG, Martin J, Joye SB, Moran MA (2013) Transcriptional response of bathypelagic marine bacterioplankton to the Deepwater horizon oil spill. ISME J 7 (12):2315–2329

Rojo F (2009) Degradation of alkanes by bacteria. Environ Microbiol 11(10):2477–2490

Ron EZ, Rosenberg E (2014) Enhanced bioremediation of oil spills in the sea. Curr Opin Biotechnol 27:191–194

Ruhal R, Kataria R, Choudhury B (2013) Trends in bacterial trehalose metabolism and significant nodes of metabolic pathway in the direction of trehalose accumulation. Microb Biotechnol 6 (5):493–502

Saharan BS, Sahu RK, Sharma D (2011) A review on biosurfactants: fermenta- tion, current developments and perspectives. Genet Eng Biotechnol J 1:1–14

Salleh AB, Ghazali FM, Rahman RNZA, Basri M (2003) Bioremediation of petroleum hydrocarbon pollution. Indian J Biotechnol 2:411–425

Sanket KG, Yagnik BN (2013) Current trend and potential for microbial biosurfactants. Asian J Exp Biol Sci 4(1):1–8

Sarkar P, Roy A, Pal S, Mohapatra B, Kazy SK, Maiti MK, Sar P (2017) Enrichment and characterization of hydrocarbon-degrading bacteria from petroleum refinery waste as potent bioaugmentation agent for in situ bioremediation. Bioresour Technol 242:15–27

Schein A, Scott JA, Mos L, Hodson PV (2009) Oil dispersion increases the apparent bioavailability and toxicity of diesel to rainbow trout (Oncorhynchus mykiss). Environ Toxicol Chem 28 (3):595–602

Semple KT, Morriss AWJ, Paton GI (2003) Bioavailability of hydrophobic organic contaminants in soils: fundamental concepts and techniques for analysis. Eur J Soil Sci 54(4):809–818

Sen R (2008) Biotechnology in petroleum recovery: the microbial EOR. Prog Energy Combust 34:714–724

Sharma D (2016) Biosurfactants in food, part of the springer briefs in food, health, and nutrition book series (BRIEFSFOOD). Springer, Berlin, pp 1–19. https://doi.org/10.1007/978-3-319-39415-2

Shi K, Liu ZX, Xu HC, Xue JL, Liu YY, Wu YN, Xiao XF, Gao Y, Liu B (2018) Degradation characteristics and microbial community change of marine petroleum-degrading bacteria in different degradation environments. Pet Sci Technol 36:1361–1367

Sihag S, Pathak H, Jaroli DP (2014) Factors affecting the rate of biodegradation of polyaromatic hydrocarbons. Int J Pure Appl Biosci 2(3):185–202

Sikkema J, de Bont JA, Poolman B (1995) Mechanisms of membrane toxicity of hydrocarbons. Microbiol Mol Biol Rev 59(2):201–222

Soberón-Chávez G, Maier RM (2011) Biosurfactants: a general overview. Biosurfactants, pp 1–11

Speight JG (2007) The chemistry and technology of petroleum, vol 114. CRC, Taylor and Francis, Boca Raton, FL

Szulc A, Ambrożewicz D, Sydow M, Ławniczak Ł, Piotrowska-Cyplik A, Marecik R, Chrzanowski Ł (2014) The influence of bioaugmentation and biosurfactant addition on bioremediation efficiency of diesel-oil contaminated soil: feasibility during field studies. J Environ Manag 132:121–128

Tao K, Liu X, Chen X, Hu X, Cao L, Yuan X (2017) Biodegradation of crude oil by a defined co-culture of indigenous bacterial consortium and exogenous Bacillus subtilis. Bioresour Technol 224:327–332. https://doi.org/10.1016/j.biortech.2016.10.073

Tremblay J, Yergeau E, Fortin N, Cobanli S, Elias M, King TL, Greer CW (2017) Chemical dispersants enhance the activity of oil-and gas condensate-degrading marine bacteria. ISME J 11 (12):2793–2808

US Environmental Protection Agency (USEPA) (2011) Crude and petroleum products

Van RM (2011) Environmental degradation of petroleum hydrocarbons. Utrecht University, Binnenstad

Van Bogaert INA, Ciesielska K, Devreese B, Soetaert W (2014) Microbial synthesis and application. Biosurfactants: production and utilization-processes. Technol Econ 159:19

Van Dorst J, Siciliano SD, Winsley T, Snape I, Ferrari BC (2014) Bacterial targets as potential indicators of diesel fuel toxicity in subantarctic soils. Appl Environ Microbiol 80 (13):4021–4033

Varjani SJ (2014) Hydrocarbon degrading and biosurfactants (bioemulsifiers) producing bacteria from petroleum oil wells. Ph.D. thesis. Kadi Sarva Vishwavidyalaya

Varjani SJ (2017) Microbial degradation of petroleum hydrocarbons. Bioresour Technol 223:277–286. https://doi.org/10.1016/j.biortech.2016.10.037

Varjani SJ, Gnansounou E (2017) Microbial dynamics in petroleum oilfields and their relationship with physiological properties of petroleum oil reservoirs. Bioresour Technol 245:1258–1265. https://doi.org/10.1016/j.biortech.2017.08.028

Varjani SJ, Upasani VN (2016) Biodegradation of petroleum hydrocarbons by oleophilic strain of Pseudomonas aeruginosa NCIM 5514. Bioresour Technol 222:195–201

Varjani SJ, Rana DP, Jain AK, Bateja S, Upasani VN (2015) Synergistic ex-situ biodegradation of crude oil by halotolerant bacterial consortium of indigenous strains isolated from on shore sites of Gujarat, India. Int Biodeterior Biodegradation 103:116–124

Wang C, Liu X, Guo J, Lv Y, Li Y (2018) Biodegradation of marine oil spill residues using aboriginal bacterial consortium based on Penglai 19-3 oil spill accident. Chin Ecotoxicol Environ Saf 159:20–27. https://doi.org/10.1016/j.ecoenv.2018.04.059

Widdel F, Rabus R (2001) Anaerobic biodegradation of saturated and aromatic hydrocarbons. Curr Opin Biotechnol 12:259–276

Wilkes H, Buckel W, Golding BT, Rabus R (2016) Metabolism of hydrocarbons in n-Alkane utilizing anaerobic bacteria. J Mol Microbiol Biotechnol 26:138–151

Xu X, Zhai Z, Li H, Wang Q, Han X, Yu H (2017) Synergetic effect of bio-photocatalytic hybrid system: g-C3N4 and Acinetobacter sp. JLS1 for enhanced degradation of C16 alkane. Chem Eng J 323:520–529

Xue J, Yu Y, Bai Y, Wang L, Wu Y (2015) Marine oil-degrading microorganisms and biodegradation process of petroleum hydrocarbon in marine environments: a review. Curr Microbiol 71:220–228. https://doi.org/10.1007/s00284-015-0825-7

Yakimov MM, Timmis KN, Wray V, Fredrickson HL (1995) Characterization of a new lipopeptide surfactant produced by thermotolerant and halotolerant subsurface Bacillus licheniformis BAS50. Appl Environ Microbiol 61(5):1706–1713

Yang Y, Wang J, Liao J, Xie S, Huang Y (2015) Abundance and diversity of soil petroleum hydrocarbon-degrading microbial communities in oil exploring areas. Appl Microbiol Biotechnol 99(4):1935–1946

Zhang X, Xu D, Zhu C, Lundaa T, Scherr KE (2012) Isolation and identification of biosurfactant producing and crude oil degrading Pseudomonas aeruginosa strains. Chem Eng J 209:138–146

Zhu X, Venosa AD, Suidan MT (2004) Literature review on the use of commercial bioremediation agents for cleanup of oil-contaminated estuarine environments. EPA/600/R-04/075

Zou C, Wang M, Xing Y, Lan G, Ge T, Yan X, Gu T (2014) Characterization and optimization of biosurfactants produced by Acinetobacter baylyi ZJ2 isolated from crude oil-contaminated soil sample toward microbial enhanced oil recovery applications. Biochem Eng J 90:49–58

Potent Biotechnological Applications of Psychrozymes

18

Burhan Hamid

Abstract

Psychrophilic yeasts possess the ability to live in extreme environmental condition (i.e. low temperature). This property makes them more valuable and more significant as they are adapted to these harsh conditions. It is well-known that major area of the earth is covered with cold environments. Psychrophilic yeasts are molecularly adapted to these cold conditions; due to these reasons researchers are getting attracted towards exploring adaptability of psychrophilic yeasts. Psychrophilic yeasts are capable of producing extremozymes having utmost stability and activity at low temperatures. Psychrozymes produced by psychrophilic yeasts show maximal activity at lower temperatures, they are having potential application in different fields. Psychrophilic enzymes produced by yeasts have proven to be economically feasible at industrial level and also keep process contamination free. These enzymes possess utility in different industries like food, pharmaceutical, detergent, leather, textile, biomedical, brewing for multiple low temperature process. The few potential enzymes secreted by psychrophilic yeasts are amylases, proteases, pectinases, lipases, lactases, etc., they are known as a valuable tool for various biotechnological processes. Climate change has directly or indirectly influenced the diversity of psychrophilic microorganisms including yeasts. Global warming leads to the negative impact on the habitats of cold-loving microorganisms and that is a matter of concern for researchers. In this chapter, industrial and biotechnological aspect of psychrophilic yeasts and their cold-active enzymes are reviewed and discussed. The focus has been given to their application in different sectors like food, pharmaceutical, detergent, leather, textile, biomedical, molecular, brewing, waste management.

B. Hamid (✉)
Centre of Research for Development (CORD), University of Kashmir, Srinagar, Jammu and Kashmir, India

S. A. Lone, A. Malik (eds.), *Microbiomes and the Global Climate Change*,
https://doi.org/10.1007/978-981-33-4508-9_18

Keywords

Extremophiles · Psychrophilic yeast · Psychrozymes · Waste management · Pharmaceutical

18.1 Introduction

Extremophiles are strange or bizarre microorganisms as they thrive in extreme environmental conditions which are known to be hostile for survival of life. Among different groups of extremophiles, psychrophiles (cold-loving) are attracting much attention due to various industrial as well as biotechnological applications. "*Psychrophiles*" or "*cold-loving*" microorganisms are those microorganisms which are living in extreme cold environments. The term psychrophiles was first used by Schmidt (1902). As this group of extremophiles (psychrophiles) are thriving at extremely low temperatures, which include glaciers, polar areas, high altitude mountains, ocean depths, upper atmosphere, low temperature appliances and are found on exteriors of plants and animals existing in cold environs, where temperature is below 5 °C. Psychrophiles propagate fast at 15 °C or lower temperatures, but these microorganisms do not live above 20 °C. It has been found that, psychrotolerant microorganisms show significant growth at lower temperatures, but show maximal growth above 20 °C up to 40 °C (Madigan et al. 2003). It has been reported that cold-loving yeasts grow underneath 5 °C and display no growth beyond 20 °C (Shivaji and Bhadra 2008). Turchetti et al. (2008) reported the occurrence of psychrophilic yeasts in glacial residues, ice, and melted water collected from two glaciers of the Italian Alps. There are plenty of such findings which recommend that cold environs are measured as home of psychrophilic yeasts secreting scientifically significant psychrozymes. Both cold as well as high temperature loving microbial groups have fascinated great devotion due to their capability to yield extremozymes with significant biotechnological potential (Cavicchioli et al. 2002). The reason behind diverting attention of scientists toward psychrophilic yeasts or cold-adapted yeasts is their adaptation in cold climatic conditions, thus these yeast species ensure noteworthy use in different industries (Margesin and Miteva 2011; Buzzini and Margesin 2014; Alcaíno et al. 2015).

It has been found that both groups of microorganisms including psychrophilic and psychrotolerants as well as their cold-acclimation proteins, cold-shock proteins and enzymes (e.g. proteases, lipases, and cellulases) are known to have vast biotechnological applications (Gounot 1991). Hamid et al. (2014) reviewed the application of psychrophilic yeasts and their different enzymes (pectinases, phytases, lactases, amylases, proteases, lipases, etc.) in various industries. At lower and modest temperatures, enzymes secreted by psychrophilic yeasts display extraordinary activity and possess high financial profits (Allen et al. 2001; Margesin et al. 2002). Application of enzymes produced by these microorganisms is huge, and they are utilized in industries like textiles, pharmaceuticals, dairy, food processing, in cleansing agents, beautifying agents, and biodiesel industries, as well as in agrochemicals

(Saxena et al. 1999; Jaeger and Eggert 2002). The application of psychrophilic yeasts and their psychrozymes are numerous as they possess versatility and are economically feasible. Industrially important yeast enzymes are lipases, esterases, amylases, cellulases, and lactases (Laitila et al. 2006). Psychrophilic microorganisms have been least studied even though they reside in greater part of our planet in contrast to mesophilic or thermophilic groups.

Cold-loving or psychrophilic microorganisms have gripped courtesy as rich sources of industrial enzymes which are having applicability in lower temperature catalysis (Morita 1975). Marketplace importance of engineering biocatalysts has increased; it is predictable to reach approximately USD 6.3 billion by 2021 (BCC Research 2017). Psychrophilic yeasts have been considered as good source of cold-active enzymes; they possess wide range of industrial as well as biotechnological applications. Different studies have revealed that low temperature loving proteins display reduced ionic interfaces and hydrogen bonds; own fewer hydrophobic clusters besides additional charged clusters on their superficial side and lengthier superficial loops as compared to proteins from mesophiles. By such alterations, at lower temperatures cold-loving proteins drop their firmness and achieve amplified organizational elasticity for boosted catalytic utility (Feller and Gerday 2003). The ability of psychrophilic microbes to produce cold-shock and antifreeze proteins as the temperature dips, the more effective enzyme activity due to modifications in enzyme kinetics and the balance of microtubules empower the psychrophiles to continue their activities (Georlette et al. 2004). Other than cold-shock proteins, cold-acclimation proteins (CAPs) are additional significant and valuable character of psychrotolerant microorganisms (Hebraud et al. 2000). These unique molecular adaptabilities have unlocked number of unexplored areas of applications (Feller and Gerday 1997).

Psychrophilic yeasts thrive in harsh cold environments and are considered as potential source of efficient cold-active enzymes. In this chapter we have discussed psychrophilic yeasts and application of cold-enzymes in different sectors. This area of research has not got much attention as compared to bacteria and fungi. In this chapter we have reviewed potential application of cold-active enzymes secreted by psychrophilic yeasts in food sector, beverage sector, pharmaceutical sector, leather and textile sector, biotechnology and waste management sector, respectively. Psychrophilic yeasts have highest stability and adaptability at low temperatures; this feature of psychrophilic yeasts and their cold-active enzymes have made their utility spectrum wide. This chapter covers most of the areas in which cold-active yeast enzymes can be used for different processes. The capacity in enzymes to catalyze different processes at low temperature has unwrapped new areas of research. This chapter also covers the research work carried on psychrophilic yeasts outside Antarctica and Polar Regions.

18.2 Habitats of Psychrophilic Microorganisms

Psychrophilic yeast prevails growth at low temperature (cold) environments which include glacial and high mountainous soils and waters, glaciers, fresh and seawaters, plants, animals living in extreme environment conditions. Other than these natural environments some artificial low temperature atmospheres like refrigerators and freezers of industrial and domestic food storage are also known as habitats of psychrophilic microorganisms. From different environs like oceanic waters, frosty and mountainous glaciers, and polar environments psychrophilic yeasts were isolated their existence and richness in these surroundings have been reported (Vishniac 1999; Bergauer et al. 2005; Buzzini et al. 2005). There are two major categories of cold environments; one is "*psychrophilic*" that is permanently cold and other one is "*psychrotrophic*" which are seasonally cold or in later one temperature fluctuates into mesophilic range (Morita 2000). On mountains, snow or ice remains round the year and is also surrounded by glaciers that lead to decrease in temperature and becomes the natural habitat for psychrophiles (Figs. 18.1 and 18.2).

Fig. 18.1 Drass, J&K India (Worlds second coldest desert)

Fig. 18.2 Razdan Top Gurez, J&K India. (Cold-climate zone)

For a long time, Antarctica has been preferred by researchers especially microbiologists for reviewing and exploring the diversity of psychrophilic microorganisms, which also include yeasts (Vishniac 2006; Onofri et al. 2007; Shivaji and Prasad 2009). Environments having cold climatic conditions have also been studied and explored for microbial diversity. *Cystofellobasidium capitatum* SPY11 cold-loving yeast was isolated from northern region of India, Kashmir valley and exhibited growth up to 20 °C (Hamid et al. 2012). In cold ecosystems psychrophilic yeasts have displayed crucial character in cycling of nutrients and in biomass production (Margesin et al. 2002). Shivaji and Bhadra (2008) reported new psychrophilic yeast, namely *Rhodotorula Himalayensis* sp. Nov. which have been isolated from Roopkund lake of Himalayan peak ranges. Different yeasts have been isolated from soil and moss samples collected from Livingston Island, South Antarctica (Pavlova et al. 2001).

Psychrophilic microbes are huge contributors to the Earth's biomass, not only this but they also play essential roles in total biogeochemical cycles. With advent of time, both habitats and diversity of psychrophiles are getting greatly influenced by climate change. But these cold-loving microorganisms have established adaptable tactics to withstand life threating impact of low temperatures (Margesin and Miteva 2011).

18.3 Cold-Loving Enzymes

Psychrophilic yeast are attracting focus due to their potential applications in biotechnology. Several researchers have reported that psychrophilic microorganisms producing cold-active enzymes are capable of breaking down multiple substances like starch, cellulose, xylan, pectin, chitin, protein, and lipid; they also produce potential cold-active enzymes including amylase, cellulase, xylanase, pectinases, chitinase, protease, and lipases (Demirjian et al. 2001; Cavicchioli et al. 2002; Margesin et al. 2002; Van den Burg 2003; Georlette et al. 2004). About 91 psychrophilic yeasts were screened for extracellular amylolytic, proteolytic, lipolytic, esterolytic, pectinolytic, chitinolytic and cellulolytic activities, these isolates belong to the species *Cryptococcus, Rhodosporidium, Sporobolomyces, Sporidiobolus, Cystofilobasidium, Leucosporidiella, Dioszegia, Mrakia, Rhodotorula*, and *Udeniomyces* (Brizzio et al. (2007). With growing demand for microbial enzymes in numerous sectors and fields, there is massive concentration in production of cold-adaptive enzymes along better-quality characters suitable for industrial uses as well as economical enzyme production processes (Burhan et al. 2003).

A significant proportion of isolated yeasts displayed multiple extracellular biocatalytic activities (amylolytic, esterolytic, proteolytic, and pectinolytic activity) at 4 °C. The capabilities of psychrozymes were reviewed by Cavicchioli et al. (2002); Margesin et al. (2002); Feller and Gerday (2003); Georlette et al. (2004), and Hamid et al. (2014). It is expected that the utilization of cold-active enzymes produced by psychrophilic microorganisms especially yeasts will be of high demand in food, pharmaceutical, detergent, textile, biotechnology, brewing, and waste management sectors.

18.3.1 Food Sector

To avoid spoilage of foodstuffs and to maintain nutritional value of food along with maintenance of taste, food products are treated with cold-active enzymes at an ambient temperature. Thus psychrozymes are used in food processing (Russell and Hamamoto 1998; Gerday et al. 2000). Psychrozymes from yeasts are used by manufacturers in food processing (Table 18.1). These enzymes can upgrade process efficiency, decrease processing expenditures, reduce impact on environments and subsequently enrich the taste, nutritional value, demand, digestibility, excellence, and shelf-life of the food product. Cold-active enzymes are attractive and demanding in foodstuff and drink preparation because of their extraordinary catalytic activity at temperatures that lessen decay and they maintain taste as well as nutritional value of food products (Huston et al. 2000). Cold-active proteases can be used for tenderization, taste improvement, and improvement of nutritional and useful properties of frozen products in meat and fish processing industries (He et al. 2004; Venugopal 2016). Most predominant and important yeast species present in the milk are

Table 18.1 Utilization of psychrozymes in food processing industry

Cold-active enzymes	Application/utility	References
Lipases, proteases, phytases, glucanases, xylanases	Enhancement of digestibility and assimilation of animal feed psychrozymes are highly applicable.	Collins et al. (2005); Hatti-Kaul et al. (2005); Huang et al. (2009); Tutino et al. (2009); Ueda et al. (2010)
Protease	Are used for elimination of hemicellulosic matrix from feedstuff	Wang et al. (2010)
Chitinases	Applied for tenderization of meat.	Dahiya et al. (2006)
α-amylase, glucoamylase	They are used for production of single cell protein (SCP) from the waste of shellfish	Gerday et al. (2000)
Pectinases, xylanases	Having potential applications in starch hydrolysis, clarification of juices	Nakagawa et al. (2004); Collins et al. (2005)
Pectate lyase, pectinase	Applied for clearing up of fruit juices, vegetable juices, and wine (clarification)	Truong et al. (2001); Sahay et al. (2013); Singh et al. (2012)
α-Amylase, xylanase	Both of them are used in ripening of cheese and starch hydrolysis	Gerday et al. (2000); Collins et al. (2005); Hamid (2015)
β-galactosidase	Used in fermentation of dough, bakery products, dairy plants	Białkowska et al. (2009); Hamid et al. (2013)
Laccase	Having potential application in dairy industry in degradation and removal of lactose from milk	Kunamneni et al. (2008)
Feruloyl esterase	For wine and drink stabilization	Aurilia et al. (2008)

K. lactis, *K. marxianus* and its anamorph, *C. kefyr* (Gadaga et al. 2000). He et al. (2004) described improvement of bakery products using unstiffening of wool or washing of contact lenses by proteases.

Psychrozymes from yeasts are useful to the food processing sectors; they are also used in clarification of fruit juice. Yeasts are mostly screened for their efficacy of producing proteinase or pectinase as they are essentially used in clarifying juices from fruit (Braga et al. 1998). Pectinases are produced by bacteria, fungi, actinomycetes, and yeast. Psychrozymes, especially pectinases from psychrophilic yeast *Cystofilobasidium capitatum* SPY11 and psychrotolerant yeast *Rhodotorula mucilaginosa*, might be used in production of wine and clarification of juices at low temperature (Sahay et al. 2013). Nakagawa et al. 2002 isolated psychrophilic yeast *Cystofilobasidium capitatum* strain PPY-1 having potential of producing cold-active pectinase. These enzymes can be utilized in food processing (Margesin and Schinner 1994; Russell et al. 1998; Gerday et al. 2000). Yeast isolates (*Rhodotorula mucilaginosa* PT1 and *Cystofilobasidium capitatum* SPY11) producing cold-active α-amylase can be utilized in food, textile, and detergent productions (Hamid 2015). Other than these enzymes, two psychrophilic yeast strains secreting cold-active β-galactosidases; proficient to degrade lactose at lower temperature has been reported with potent application at cold-processing's in dairy sector (Hamid et al. 2013).

18.3.2 Brewage Sector

Psychrophilic yeast and psychrozymes might be used in cold fermentation and also help in maintenance of contamination free product (Kourkoutas et al. 2002; Pfeffer et al. 2006). Cold-active amylases produced by moulds are used in liquor making and possess application in fermentation industries (Van and Smith 1968). Industries like brewing and wine, utilize cold-active pectinases and hemicellulases for intensification of extraction, advance clarification, diminish thickness, and improve colour and taste (Semenova et al. 2006; Tu et al. 2013; Adapa et al. 2014). Cold-active lipases might be better substitute to mesophilic biocatalysts in brewing and wine industries (Collins et al. 2002) as they can maintain taste and quality of the end product, they can save energy, and can protect process from contamination. Furthermore, it has been reported that pectinases used in wine production must be actively working at pH 3–4 and below 10 °C (Van Rensburg and Pretorius 2000). Such properties of enzymes can be useful to calculate the pectinases for their correctness to wine making (Meriın et al. 2011). In brewing industries, cold fermentation (15–20 °C) is thought to intensify the production and withholding of volatile composites, thus boosting the scent character in wines (Molina et al. 2007) and improving quality of product.

18.3.3 Pharmaceutical Sector

Psychrophilic yeast producing cold-active enzymes can be used in heat sensitive or volatile processes which are carried in pharmaceutical, fine chemicals industries and are low temperature processes (Table 18.2). Cold-active microbial lipases possess potential application in various industries like health, medicines, and food processing (Joseph et al. 2007). For the preparation of bioactive peptides, cold-adapted proteases might be utilized as efficient antioxidants, antihypertensive agents, and immune-regulatory agents (Cazarin et al. 2015). For pharmaceutical sector psychrophiles are known to be rich source of polyunsaturated fatty acids.

Amylases are most essential industrial enzymes; they possess vast applications right from transformation of starch (carbohydrates) to sugar syrups, and furthermore used for production of cyclodextrins for pharmaceutical sector (Maarel et al. 2002; Satyanarayana et al. 2005). Amylases are collected from diverse bases including plants, animals, and microorganisms, but amylases from microorganisms are the maximum manufactured and are extensively applied in industry, because of their efficiency and thermo stability (Burhan et al. 2003). Cold-active enzymes have potentiality to save energy and guard the production from spoilage (Gerday et al. 2000) and thus are attractive for the industries (Deming 1998; Cavicchioli et al. 2002). Lipases produced by *Candida antarctica* are amongst the widely and distinctly used enzymes in organic synthesis.

Table 18.2 Utilization of psychrozymes in pharmaceutical and cosmetic industry

Cold-active enzymes	Application	References
Chitinase	They are used for breakdown of chitin into chitosan, chito-oligosaccharides, glucosamine	Dahiya et al. (2006)
Chitinase	Utilized for preparation of antifungal substances and improver for antifungal ointments	Dahiya et al. (2006)
Endo-chitinase and lipase	They are used in control of mosquito at larval stage	Dahiya et al. (2006)
Lipase	Applied in synthesis of citronellol laurate	Joseph et al. (2008)
Lipases and laccase	Used in cosmetics	Joseph et al. (2008); Kunamneni et al. (2008)
Lysozyme	It is used as antibacterial	Sotelo-Mundo et al. (2007)
Feruloyl esterase	Multiple applications like anti-microbial, antioxidant, photoprotectant	Aurilia et al. (2008)
β-lactamase	Used in degradation of antibiotic	Michaux et al. (2008)

18.3.4 Leather and Textile Sectors

Cold-active enzymes are of interest in multiple areas like cleansing processes, fabric and foodstuff industry, bioremediation and biocatalysts under low water circumstances (Choo et al. 1998). Application of cold-active enzymes retains versatile significance like flaking of leather or skin products by proteases in industries; this can be carried out under tap water by psychrozymes other than heating to 37 °C for the development to be accomplished by mesophilic biocatalysts. For dehairing of animal hides at low temperatures, cold-active proteases or keratinases are of much importance because they also decrease the effects of lethal chemicals applied while dehairing. Cold-active proteases (Hamamoto et al. 1994; Baghel et al. 2005) ensure multiple uses in industries like detergents, food processing, leather and textiles, cosmetics, drinks and bakery. Different cold-adapted including amylases, cellulases, and laccases are used in fabric industry. They have been advanced and improved in fast desizing or starch elimination of textiles, bio-finishing of cellulosic textiles, and a reduced amount of rough enzymatic stonewashing and bleaching of denims (Sarmiento et al. 2015).

18.3.5 Detergent Sector

Enzymes are used for the improved cleaning and their effectiveness enables in reduction of use of other added dangerous constituents, e.g. laundry detergent, polymers, alkaline creators (in washing), phosphates (in dishwashing), and organic solvents (for cleaning-in-place), as well as reduced water consumption (Damhus et al. 2013; Sarmiento et al. 2015). Psychrophilic enzymes offer benefits in washing of huge immobile substances where warming is not feasible, and enzymes like glucose oxidases, proteases, amylases, and lipases have proved their effectiveness in biofilm elimination from various structures (Webster and May 2006; Valentini et al. 2010).

18.3.6 Biotechnology

It has been reported that psychrozymes have capability to support biotechnological processes taking place at low temperatures (Goodchild et al. 2004). Yeast are widely used as model organisms for studying a variety of cell functions and are also used in gene cloning and for producing purified proteins (Jill 2007). Genetic modification offers different benefits by modifying properties of enzymes for enhancement of activities as it has been found in cold tolerant Lipase B secreted by *Candida antarctica*; a mutant exhibited an escalation in activity and stability (Zhang et al. 2003). Heat-labile alkaline phosphatase from an Antarctic bacterium has been purified and characterized (Kobori et al. 1984). In molecular biology and recombinant technology these potential enzymes are used for the removal of phosphate group of DNA vectors prior to cloning to avoid re-circularization, for removal of phosphate group of 5′-nucleic acid termini before 5′-end labelling by polynucleotide kinase or for exclusion of dNTPs and pyrophosphate from PCR reactions. It has also been found that heat sensitive uracil-DNA *N*-glycosylase from Atlantic cod (*Gadus morhua*), that offers cold adaptation properties (Leiros et al. 2003), is also utilized in removal of DNA contaminants in sequential PCR reactions.

18.3.7 Waste Management

Numerous yeast strains such as *Candida cylindracea*, *Candida rugosa*, *Yarrowia lipolytica*, *Saccharomyces cerevisiae* and *Candida boidinii* have been screened for their use in bio-treatment of manufacturing or industrial and in-house waste waters; yeasts have high potential to treat wastes through biological methods (Thanh and Simard 1973; Goncalves et al. 2009). For the management of dissolved organic substances at low temperatures, cold-loving *Candida sp.* were isolated from water samples of Lake Vanda in Antarctica (Ohno et al. 1991; Katayam et al. 1997). Cold-loving microorganisms and their cold-active enzymes (e.g. oxidases, peroxidases, and catalases) have been projected as substitutes to physico-chemical procedures for the bioremediation of solids and waste waters contaminated by hydrocarbons, oils, and lipids (Margesin et al. 2002). Psychrozymes are useful with their producing microorganisms in environmental bioremediations including processes such as digestion, composting, and oil degradation processes and molecular biology applications (Feller et al. 1996); these microorganisms secrete enzymes which degrade particular compounds or convert them into non-toxic compounds, they are also used for conversion of wastes into compost.

Numerous yeast species were screened and selected for phytase activity and yeasts are considered as vital source of phytases (Vohra and Stayanarayan 2001). Psychrophilic yeasts have been found effective in environmental biotechnology and biofuel industry (Table 18.3). Phytases are a group of potential biocatalysts that secretes palatable phosphorus, calcium, and nutrients from phytic acid and so

Table 18.3 Utilization of psychrozymes in environmental biotechnology and biofuel industry

Cold-active enzymes	Application	References
Lipase, protease, hydrocarbon degrading enzyme, xylanase, peroxidase.	Having vast applications in waste management like bioremediation, breakdown and removal of xenobiotics and poisonous complexes	Joseph et al. (2008); Wang et al. (2010); Margesin et al. (2003); Collins et al. (2005)
Lipase	Biofuel production	Tutino et al. (2009)
Chitinase and yeast	Used in transformation of chitin to bioethanol	Dahiya et al. (2006)
Cellulase-β-glucosidase complex	Conversion of cellulose to ethanol	Ueda et al. (2010)
β-galactosidase	Biofuel manufacturing from dairy leftover	Hildebrandt et al. (2009)

diminish ecological or environmental phosphorus contamination (Mllaney et al. 2000). Cold-loving yeasts producing cold-active phytases are potential source for sinking the phosphorus contamination in cold environments; previous reviews also indicate the constancy of yeast phytases (Quan et al. 2001) from *Candida krusei.* Phytases can, therefore, play an important role in controlling phosphate pollution. Lipases form psychrophiles are in high demand for production of biofuel (Tuffin et al. 2009) and utility of a cold-active lipases for executing trans-esterification catalysis in making of biofuel at low temperature (Luo et al. 2006).

18.4 Conclusions and Future Perspectives

Psychrophilic yeasts have attracted attention of scientists including microbiologists and biotechnologists due to their potential of producing cold-active enzymes. As discussed earlier these biocatalysts are used in several production units for low temperature processes to improve quality and quantity of the products. Psychrophilic yeast enzymes are valuable tool for industries including food, beverage, pharmaceutical, leather and textile, detergent and biotechnological sector. Cold-loving enzymes denote exceptionally resourceful collection of enzymes, being used in different processes carried out at low or moderate temperatures. In future, research should be carried on improvement of psychrophilic yeasts for secretion of potential cold-active enzymes by applying genetic engineering and recombined DNA technology. It is also important to study the mechanism of production of cold-active enzymes by yeasts and their adaptability to extreme cold climatic conditions. Due to these reasons psychrophilic yeasts and their enzymes are interesting fields for future research and it can be explored very well.

References

Adapa V, Ramya LN, Pulicherla KK, Sambasiva Rao KRS (2014) Cold active pectinases: advancing the food industry to the next generation. Appl Biochem Biotechnol 172(5):2324–2337

Alcaíno J, Cifuentes V, Baeza M (2015) Physiological adaptations of yeasts living in cold environments and their potential applications. World J Microbiol Biotechnol 31:1467–1473

Allen D, Huston AL, Wells LE, Deming JW (2001) Encyclopedia of environmental microbiology, vol 1. Wiley, New York, pp 1–17

Aurilia V, Parracino A, D'Auria S (2008) Microbial carbohydrate esterases in cold adapted environments. Gene 410:234–240

Baghel VS, Tripathi RD, Ramteke PW, Gopal K, Dwivedi S, Jain RK, Rai UN, Singh SN (2005) Psychrotrophic proteolytic bacteria from cold environment of Gangotri glacier, Western Himalaya, India. Enzym Microb Technol 36:654–659

BCC Research (2017) BCC research report. Global markets for enzymes in industrial applications. BIO030J. BCC Research LLC, Wellesley

Bergauer P, Fonteyne PA, Nolard N, Shinner F, Margesin R (2005) Biodegradation of phenol and phenol-related compounds by psychrophilic and cold-tolerant alpine yeasts. Chemosphere 59:909–918

Białkowska A, Cieśliński H, Nowakowska K, Kur J, Turkiewicz M (2009) A new β-galactosidase with a low temperature optimum isolated from the Antarctic Arthrobacter sp. 20B: gene cloning, purification and characterization. Arch Microbiol 191:825–835

Braga AA, De Morais PB, Linrdi VR (1998) Screening of yeasts from Brazilian Amazon rain forest for extracellular proteinases production. Syst Appl Microbiol 21:353–359

Brizzio S, Benedetta T, Virginia G (2007) Extra cellular enzymatic activity of Basidomycetes yeasts isolated from glacial and sub glacial waters of North West Patagonia (Argentina). Can J Microbiol 53:519–525

Burhan A, Nisa U, Gokhan C, Omer C, Ashabil A, Osman G (2003) Enzymatic properties of a novel thermostable, thermophilic, alkaline and chelator resistant amylase from an alkaliphilic Bacillus sp. isolate ANT-6, process. Biochemist 38:1397–1403

Buzzini P, Margesin R (2014) Cold-adapted yeasts: a lesson from the cold and a challenge for the XXI century. In: Cold-adapted yeasts. Springer, Berlin, pp 3–22

Buzzini P, Turchetti B, Diolaiuti G, D'Agata G, Martini A, Smiraglia C (2005) Culturable yeasts in melt waters draining from two glaciers in the Italian Alps. Ann Glaciol 4:119–122

Cavicchioli R, Siddiqui KS, Andrews D, Sowers KR (2002) Low temperature extremophiles and their applications. Curr Opin Biotechnol 13:253–261

Cazarin C, Lima G, da Silva J, Marostica M (2015) Enzymes in meat processing. In: Chandrasekaran M (ed) Enzymes in food and beverage processing. CRC, Boca Raton, pp 337–351

Choo DW, Kurihawa T, Suzuki T, Soda K, Esaki N (1998) A cold-adapted lipase of Alaskan psychrotroph, *Pseudomonas sp.* strain B11-1: gene cloning and enzyme purification and characterization. Appl Environ Microbiol 64:486–491

Collins T, Meuwis MA, Stals I, Claeyssens M, Feller G, Gerday C (2002) A novel family 8 xylanase. Functional and physicochemical characterization. J Biol Chem 277:35133–35139

Collins T, Gerday C, Feller G (2005) Xylanases, xylanase families and extremophilic xylanases. FEMS Microbiol Rev 29:3–23

Dahiya N, Tewari R, Hoondal GS (2006) Biotechnological aspects of chitinolytic enzymes: a review. Appl Microbiol Biotechnol 71:773–782

Damhus T, Kaasgaard S, Olsen HS (2013) Enzymes at work, 4th edn. Novozymes A/S, Bagsværd

Deming JW (1998) Deep ocean environmental biotechnology. Curr Opin Biotechnol 9:283–287

Demirjian DC, Moris VF, Cassidy CS (2001) Enzymes from extremophiles. Curr Opin Chem Biol 5:144–151

Feller G, Gerday C (1997) Psychrophilic enzymes: molecular basis of cold adaptation. Cell Mol Life Sci 53:830–841

Feller G, Gerday C (2003) Psychrophilic enzymes: hot topics in cold adaptation. Nat Rev Microbiol 1:200–208

Feller G, Narinx E, Arpigny JL, Aittaleb M, Baise E, Genicot S, Gerday C (1996) Enzymes from psychrophilic organisms. FEMS Microbiol Rev 18:189–202

Gadaga TH, Mytukumira AN, Narvhus JA (2000) Enumeration and identification of yeasts isolated from Zimbabwean traditional fermented milk. Int Dairy J 10(7):456–466

Georlette D, Blaise V, Collins T, Amico SD, Gratia E, Hoyoux A, Marx JC, Sonan G, Feller G, Gerday C (2004) Some like it cold: biocatalysis at low temperatures. FEMS Microbiol Rev 28:25–42

Gerday C, Aittaleb M, Bentahir M, Chessa JP, Claverie P, Collins T, Damico S, Dumont J (2000) Cold-adapted enzymes: from fundamentals to biotechnology. Trends Biotechnol 18:103–107

Goncalves C, Lopes M, Ferreira JP, Belo I (2009) Biological treatment of olive mill wastewater by non-conventional yeasts. Bioresour Technol 100(15):3759–3763

Goodchild A, Saunders NFW, Ertan H, Raftery M, Guilhaus M, Curmi PMG, Cavicchioli R (2004) A proteomic determination of cold adaptation in the Antarctic archaeon, *Methanococcoides burtonii*. Mol Microbiol 53:309–321

Gounot AM (1991) Bacterial life at low temperature: physiological aspects and biotechnological implications. J Appl Bacteriol 71:386–397

Hamamoto T, Kaneda M, Horikoshi K, Kudo T (1994) Characterization of protease from a psychrotroph, *Pseudomonas fluorescens*114. Appl Environ Microbiol 60(10):3878–3880

Hamid B (2015) Cold-active α-amylase from psychrophilic and psychrotolerant yeast. J Glob Biosci 4:2670–2677

Hamid B, Singh P, Rana RS, Sahay S (2012) Isolation and identification of psychrophilic yeast from the soil of northern region of India. Inventi Rapid Pharm Biotech Microbio 2012(1):1–3

Hamid B, Singh P, Mohiddin FA, Sahay S (2013) Partial characterization of cold-active β-galactosidase activity produced by *Cystophallobaidium capatitum* SPY11 and *Rhodotorula mucilaginosa* PT1. J Endocytobiosis Cell Res 24:23–26

Hamid B, Rana RS, Chauhan D, Singh P, Mohiddin FA, Sahay S, Abidi I (2014) Psychrophilic yeasts and their biotechnological applications - a review. Afr J Biotechnol 13(22):2188–2197

Hatti-Kaul R, Birgisson H, Mattiasson B (2005) Cold active enzymes in food processing. In: Shetty K, Paliyath G, Pometto A, Levin RE (eds) Food biotechnology. Taylor & Francis, Boca Raton, FL, pp 1631–1653

He H, Chen X, Li J, Zhang Y, Gao P (2004) Taste improvement of refrigerated meat treated with cold-adapted protease. Food Chem 84(2):307–311

Hebraud M, Dubois E, Potier P, Labadie J (2000) Effect of growth temperatures on the protein levels in a psychrotrophic bacterium, *Pseudomonas fragi*. J Bacteriol 176:4017–4024

Hildebrandt P, Wanarska M, Kur J (2009) A new cold-adapted beta-D-galactosidase from the Antarctic Arthrobacter sp. 32c – gene cloning, overexpression, purification and properties. BMC Microbiol 9:151

Huang J, Zhou W, Dong W, Watson AM, Hong Y (2009) Directed, efficient, and versatile modifications of the Drosophila genome by genomic engineering. Proc Natl Acad Sci U S A 106(20):8284–8289

Huston AL, Krieger-Brockett BB, Deming JW (2000) Remarkably low temperature optima for extracellular enzyme activity from Arctic bacteria and sea ice. Environ Microbiol 2:383–388

Jaeger KE, Eggert T (2002) Lipases for biotechnology. Curr Opin Biotechnol 13:390–397

Jill BK (2007) Microorganisms: applications in molecular biology. In: Encyclopedia of life sciences. Wiley, Hoboken, pp 1–8

Joseph B, Promad W, Ramtake, George T, Nitisha S (2007) Standard revive cold active microbial lipases: a versatile tool for industrial applications. Biotechnol Microbiol 2(2):39–48

Joseph B, Ramteke PW, Thomas G (2008) Cold active microbial lipases: some hot issues and recent developments. Biotechnol Adv 26:457–470

Katayam H, Nishikawa J, Hira Y (1997) Biological treatment of dissolved organic matter by Antarctic psychropylic yeast cells. Mizushori Gijutsu (Water Purif Liquid Wastes Treat) 38:1–4

Kobori H, Sullivan CW, Shizuya H (1984) Heat-labile alkaline phosphatase from Antarctic bacteria: rapid 5′ end labelling of nucleic acids. Proc Natl Acad Sci U S A 8:6691–6695

Kourkoutas Y, Koutinas A, Kanellaki M, Banat I, Marchant R (2002) Continuous wine fermentation using a psychrophilic yeast immobilized on apple cuts at different temperatures. Food Microbiol 19(2–3):127–134

Kunamneni A, Plou FJ, Ballesteros A, Alcalde M (2008) Laccases and their applications: a patent review. Recent Pat Biotechnol 2:10–24

Laitila A, Wilhelmson A, Kotavitta E, Olkku J, Home S, Juvonen R (2006) Yeasts in an industrial malting ecosystem. J Ind Microbiol Biotechnol 33:953–966

Leiros I, Moe E, Lanes O, Smalas AO, Willassen NP (2003) The structure of uracil-DNA glycosylase from Atlantic cod (*Gadus morhua*) reveals cold-adaptation features. Acta Crystallogr D Biol Crystallogr 59:1357–1365

Luo Y, Zheng Y, Jiang Z, Ma Y, Wei D (2006) A novel psychrophilic lipase from *Pseudomonas fluorescens* with unique property in chiral resolution and biodiesel production via transesterification. Appl Microbiol Biotechnol 73:349–355

Maarel MJEC, Veen B, Uitdehaag JCM, Leemhuis H, Dijkhuizen L (2002) Properties and applications of starch-converting enzymes of the α-amylase family. J Biotechnol 94:137–155

Madigan MT, Martinko JM, Parker J (2003) Thermophile: brock biology of microorganism. Prentice Hall. Pearson Education, London, pp 13–157
Margesin R, Miteva V (2011) Diversity and ecology of psychrophilic microorganisms. Res Microbiol 162:346–361
Margesin R, Schinner F (1994) Properties of cold-adapted microorganisms and their potential role in biotechnology. J Biotechnol 33:1–14
Margesin R, Feller G, Gerday C, Russell N (2002) Cold-adapted microorganisms: adaptation strategies and biotechnological potential. In: The encyclopedia of environmental microbiology. Wiley, New York, pp 871–885
Margesin R, Gander S, Zacke G, Gounot AM, Schinner F (2003) Hydrocarbon degradation and enzyme activities of cold-adapted bacteria and yeasts. Extremophiles 7:451–458
Meriın MG, Mendoza LM, Farias ME, de Ambrosini VIM (2011) Isolation and selection of yeasts from wine grape ecosystem secreting cold-active pectinolytic activity. Int J Food Microbiol 147:144–148
Michaux C, Massant J, Kerff F, Frère JM, Docquier JD, Vandenberghe I (2008) Crystal structure of a cold-adapted class C beta-lactamase. FEBS J 275:1687–1697
Mllaney EJ, Daly CB, Ullah AH (2000) Advances in phytase research. Adv Appl Microbiol 47:157–199
Molina A, Swiegers JH, Varela C, Pretorius IS, Agosin E (2007) Influence of wine fermentation temperature on the synthesis of yeast-derived volatile aroma compounds. Appl Microbiol Biotechnol 77:675–687
Morita RY (1975) Psychrophilic bacteria. Bacteriol Rev 39:144–167
Morita RY (2000) Low-temperature environments. Encycl Microbiol 3:93–98
Nakagawa T, Yamada K, Miyaji T, Tomizuka N (2002) Cold-active pectinolytic activity of psychrophilic basidiomycetous yeast *Cystofilobasidium capitatum* strain PPY-1. J Biosci Bioeng 94:175–177
Nakagawa T, Nagaoka T, Taniguchi S, Miyaji T, Tomizuka N (2004) Isolation and characterization of psychrophilic yeasts producing cold-adapted pectinolytic enzymes. Lett Appl Microbiol 38:383–387
Ohno S, Suzuki A, Shiumra K (1991) Treatment of highly concentrated organic wastewater by yeasts. Yosui to Haisui. J Water Wastes 33:664–668
Onofri S, Zucconi L, Tosi S (2007) Continental Antarctic fungi. IHW Verlag, Munchen
Pavlova K, Grigorova G, Hristozova T, Angelov A (2001) Yeast strains from the Livingston Island, Antarctica. Folia Microbiol 46(5):397–401
Pfeffer J, Richter S, Nieveler J, Hansen CE, Rhlid R, Schmid R, Rusnak M (2006) High yield expression of lipase A from *Candida antarctica* in the methylotrophic yeast *Pichia pastoris* and its purification and characterization. Appl Microbiol Biotechnol 72(5):931–938
Quan C, Zhang L, Wang Y, Ohta Y (2001) Production of phytase in low phosphate medium by a novel yeast *Candida krusei*. J Biosci Bioeng 92:154–160
Russell NJ, Hamamoto T (1998) In: Horikoshi K, Grant WD (eds) Psychrophiles in extremophiles: microbial life in extreme environments. Wiley-Liss, New York, pp 25–45
Russell RJM, Gerike U, Danson MJ, Hough DW, Taylor GL (1998) Structural adaptations of the cold-active citrate synthase from an Antarctic bacterium. Structure 6:351–361
Sahay S, Hamid B, Singh P, Ranjan K, Chauhan D, Rana RS, Chaurse VK (2013) Evaluation of pectinolytic activities for oenological uses from psychrotrophic yeasts. Lett Appl Microbiol 57:115–121
Sarmiento F, Peralta R, Blamey JM (2015) Cold and hot extremozymes: industrial relevance and current trends. Front Bioeng Biotechnol 3:148
Satyanarayana T, Rao JLUM, Ezhilvannan M (2005) α-amylases. In: Pandey A, Webb C, Soccol CR, Larroche C (eds) Enzyme technology. Asiatech Publishers, New Delhi, pp 189–220
Saxena RK, Ghosh PK, Gupta R, Davidson WS, Bradoo S, Gulati R (1999) Microbial lipases: potential biocatalysts for the future industry. Curr Sci 77:101–115

Schmidt NS (1902) Ueber einige psychrophile Mikroorganismen und ihr Vorkommen. Centr. Bakteriol. Parasitenk., Abt. II. 9. pp 145–147
Semenova M, Sinitsyna O, Morozova V (2006) Use of a preparation from fungal pectin lyase in the food industry. Appl Biochem Microbiol 42:598–602
Shivaji S, Bhadra B (2008) *Rhodotorula himalayensis sp. Nov.*, novel psychrophilic yeast isolated from Roop Kund Lake of Himalayan mountain ranges, India. Extremophiles 12:375–381
Shivaji S, Prasad GS (2009) Antarctic yeasts: biodiversity and potential applications. In: Satyanarayana T, Kunze G (eds) Yeast biotechnology: diversity and applications. Springer-Verlag, Berlin, pp 3–16
Singh P, Hamid B, Lone MA, Ranjan K, Khan A, Chaurse VK, Sahay S (2012) Evaluation of pectinase activity from the psychrophilic fungal strain Truncatella angustata-BPF5 for use in wine industry. J Endocytobiosis Cell Res 22:57–61
Sotelo-Mundo RR, López-Zavala AA, Garcia-Orozco KD, Arvizu-Flores AA, Velázquez-Contreras EF, Valenzuela-Soto EM (2007) The lysozyme from insect (Manduca sexta) is a cold-adapted enzyme. Protein Pept Lett 14:774–778
Thanh NC, Simard RE (1973) Biological treatment of wastewater by yeasts. J Water Pollut Control Fed 45:674–680
Truong LV, Tuyen H, Helmke E, Binh LT, Schweder T (2001) Cloning of two pectate lyase genes from the marine Antarctic bacterium Pseudoalteromonas haloplanktis strain ANT/505 and characterization of the enzymes. Extremophiles 5:35–44
Tu T, Meng K, Bai Y, Shi P, Luo H, Wang Y, Yang P, Zhang Y, Zhang W, Yao B (2013) Highyield production of a low-temperature-active polygalacturonase for papaya juice clarification. Food Chem 141(3):2974–2981
Tuffin M, Anderson D, Heath C, Cowan DA (2009) Metagenomic gene discovery: how far have we moved into novel sequence space? Biotechnol J 4:1671–1683
Turchetti B, Buzzini P, Goretti M, Branda E, Diolaiuti G, D'Agata C, Smiraglia C, Vaughan MA (2008) Psychrophilic yeasts in glacial environments of alpine glaciers. FEMS Microbiol Ecol 63:73–83
Tutino ML, Di Prisco G, Marino G, de Pascale D (2009) Cold-adapted esterases and lipases: from fundamentals to application. Protein Pept Lett 16:1172–1180
Ueda M, Goto T, Nakazawa M, Miyatake K, Sakaguchi M, Inouye K (2010) A novel cold-adapted cellulase complex from Eisenia foetida: characterization of a multienzyme complex with carboxymethylcellulase, beta-glucosidase, beta-1,3 glucanase, and beta-xylosidase. Comp Biochem Physiol B Biochem Mol Biol 157:26–32
Valentini F, Diamanti A, Palleschi G (2010) New bio-cleaning strategies on porous building materials affected by biodeterioration event. Appl Surf Sci 256(22):6550–6563
Van den Burg B (2003) Extremophiles as a source for novel enzyme. Curr Opin Microbiol 6:213–218
Van Rensburg P, Pretorius IS (2000) Enzymes in winemaking: harnessing natural catalysts for efficient biotransformations – a review. S Afr J Enol Vitic 21:52–73
Van LJM, Smith MB (1968) US patent 3, 418, and 211
Venugopal V (2016) Enzymes from seafood processing waste and their applications in seafood processing. In: Se-Kwon K, Fidel T (eds) Advances in food and nutrition research, vol 78. Academic, New York, pp 47–69
Vishniac HS (1999) Psychrophilic yeasts. In: Seckbach J (ed) Enigmatic microorganisms and life in extreme environments. Kluwer, The Netherlands, pp 317–324
Vishniac HS (2006) Yeast biodiversity in the Antarctic. In: Rosa CA, Gabor P (eds) The yeast handbook. biodiversity and ecophysiology of yeasts. Springer-Verlag, Berlin, pp 419–440
Vohra A, Stayanarayan T (2001) Phytase production by the yeast Pichia anomala. Biotechnol Lett 23:551–554
Wang F, Hao J, Yang C, Sun M (2010) Cloning, expression, and identification of a novel extracellular cold-adapted alkaline protease gene of the marine bacterium strain YS-80-122. Appl Biochem Biotechnol 162:1497–1505

Webster A, May E (2006) Bioremediation of weathered-building stone surfaces. Trends Biotechnol 24(6):255–260

Zhang N, Suen WC, Windsor W, Xiao L, Madison V, Zaks A (2003) Improving tolerance of *Candida antarctica* lipase B towards irreversible thermal inactivation through directed evolution. Protein Eng Des Sel 16:599–605

Role of Green Nanotechnology in Alleviating Climate Change

19

Ruqeya Nazir and Insha Amin

Abstract

Industrialization along with population explosion in the developed and developing countries has led to accelerated degradation of natural resources on a large scale, ultimately leading to climate change. Since environment is being loaded with a large quantity of pollutants and recalcitrant compounds, environmental remediation has become a major cause of concern today. The advancement of clean technologies for reclamation of environment and sustainable development of society is very vital nowadays. Nanotechnology can help in developing clean and green technology with noteworthy benefits for human health and environment. The field of nanotechnology is now being investigated for its possible solutions for the management and mitigation of polluted water, land and air, and also to enhance the working of traditional technologies which are helpful in the remediation of contaminated environment. Green nanotechnology is the specialized branch of nanotechnology that envisions sustainable development by means of several applications. Various nanoparticles are being utilized increasingly in many areas, but there is mounting interest in the biological and environmental safety linked to their production. Green nanotechnology offers tools and techniques for the transformation of biological systems to green methods to synthesis of nanomaterial, while averting any related toxicity. Due to the large number of harmful chemicals used in the physicochemical synthesis of these nanoparticles, green methods are now being employed which utilize biological sources, such as microorganisms. Through the integration of the principles of green chemistry, engineering and microbiology, green nanotechnology can yield safe and eco-friendly metal nanoparticles that do not use toxic substances in their synthesis. The present chapter highlights the role of

R. Nazir (✉) · I. Amin
Centre of Research for Development (CORD), University of Kashmir, Srinagar, India

Department of Environmental Science, University of Kashmir, Srinagar, India

S. A. Lone, A. Malik (eds.), *Microbiomes and the Global Climate Change*,
https://doi.org/10.1007/978-981-33-4508-9_19

microbiome in nanobioremediation for treating the polluted environments and thus alleviating the impact of climate change.

Keywords

Nanoparticles · Microorganisms · Climate change · Microbiome · Environment · Nanobioremediation

19.1 Introduction

Today's world experiences many ecological issues of impactful proportions. Threats to the planet's biodiversity are increasing every year. The developed and developing nations are striving to find the constructive answers for maintaining the ecological balance of the planet Earth. It may be noted that conventional measures to conserve the environment and prevent further pollution have not been found successful in obtaining the long-lasting solutions. For years, researchers have been warning about the possible detrimental effects of climate change, which is chiefly caused by the greenhouse gas emissions such as carbon dioxide into the atmosphere. Yet global emissions have continued their inexorable rise (Solomon et al. 2009). For the world, nanotechnology is a future innovation that is well equipped to handle the major environmental problems humanity is facing, but it has to be addressed immediately and become a standard. Pollution control strategies driven by nanotechnology and nanomaterials are emerging as most powerful solutions to nagging environmental issues and challenges (Hood 2004). The Earth is a microbial planet, with microbes controlling practically the whole ecosystem, ranging from soil to seas, from the subsurface to the atmosphere, which includes extreme environments like hydrothermal vents (Merino et al. 2019). Although microorganisms are plentiful and ubiquitous, we at present lack basic understanding of many of the important roles played by microorganisms in nature. It will require an extensive range of approaches to study the microbiome and its functions fully at various levels, which in turn will help in synthesizing the data into systematic models, and the latter would amplify our predictive knowledge. Particularly, we need to understand the appropriate scales to study the microbiological systems in order to gain knowledge about the operations from the intracellular to the planetary levels. This is daunting task due to the huge number and diversity of microorganisms. The biggest advantage of nanoscience and nanotechnology in studying microbiomes is that the nanoscale is the scale of function in biology (Biteen et al. 2016).

19.2 Role of Nanotechnology

Nanotechnology is considered as the most emerging and useful tool to address threats to the well-being of our planet. Nanomaterial is progressively establishing the basis of clean and green technologies that can be helpful in capturing toxic gases and chemicals from air and water, respectively, and break down solid waste into non-toxic components (Jeevanandam et al. 2018). Experts, scientists and innovators rely on this expertise to gradually mitigate the process of climate change. The extent of progress in this field of research is tremendous, because of which nanomaterials are now regarded as most reliable and efficient catalysts. These characteristics stimulated series of novel inventions in which the nanomaterials play a very fundamental and integral role (Iavicoli et al. 2014). Use of nanoparticles suggests a favourable method to deal with this since they are having higher surface area to volume proportion for reacting with carbon dioxide and characteristics that permit them to help in the transformation of carbon dioxide into other beneficial elements. Nanoparticles can be utilized to clean the contaminated water owing to noxious dyes which are used in various manufacturing units. Toxic dyes used in tanneries have a tendency of leaching into natural water resources such as groundwater or deep wells, and if this contaminated water is not treated, it builds up into a problematic situation that over a period of time becomes increasingly difficult to resolve. Nanomaterial can be broadly used to eradicate toxic contaminants from heavy metal polluted water (Yang et al. 2019). Using the identical perception, properly designed magnetic nanomaterial can be utilized to remove contaminants such as lead, arsenic, mercury and chromium in wastewater. Other than metals and dyes, nanotechnology can also help in cleaning the oil spills (Avila et al. 2014).

Nanotechnology can be utilized efficiently for the management of organic waste, which can contaminate water and soil if not tackled appropriately (Dermatas et al. 2018). The food and agricultural industries produce massive amounts of waste of biodegradable nature. Perhaps the most established technique for treating decomposable waste is by using specialized tanks known as digesters. The digesters are fully occupied by the anaerobic microorganisms that feed on the material, changing it into biogas and solids that can be utilized as manure. As digestion in absence of oxygen occurs very slowly, nanoparticles can hasten the digestion process of the sludge, making it more proficient as far as duration and improved production of the biogas is concerned (Faisal et al. 2019). The utilization of engineered nanomaterials (ENMs) for remediation of environment, called nanoremediation, represents a unique and advanced solution, guaranteeing an immediate and effective expulsion of contaminants from polluted sites. Though the mounting curiosity in nanotechnology-based solutions for pollution remedy, with huge monetary venture around the world, ecological and biological hazard evaluation related with the usage of ENMs is still debatable and nanoremediation is yet seen as developing expertise (Corsi et al. 2018). The only challenge is to make these nanomaterials environmentally and economically feasible, and in quest for that purpose, scientists have experimented with everything from synthetic to biological nanoparticles to reduce the cost on environment (Khandel et al. 2018).

19.3 Green Nanotechnology

Nanotechnology plays an efficient role in developing the efficiency of current technology. A novel idea of "green" nanotechnology comprises the progress of clean technology to reduce the possible health risks and environmental hazards associated with the use of nanoproducts and nano production, and the further development and design of new products or the replacement of current nanoproducts that are environment and mankind friendly during the course of their life cycle (Dhingra et al. 2010). The two fundamental purposes of green nanotechnology are: to manufacture engineered nanomaterials devoid of any ecological damage or any complications to human health, and engineering nanomaterials that offer a way out to the environmental complications. This perception utilizes the standards of green engineering and green chemistry to produce nanoparticles without toxicity, using limited resources and developmental thinking at all phases. Nano production methods are predicted to be eco-friendlier with the advent of Green nanotechnology (Ashoka et al. 2016). Biosynthesis of nanoparticles utilizing microbes has developed as a swiftly emerging study field in green nanotechnology throughout the world, with several organic components being utilized in production of nanoparticles persistently framing a substitute for physical and conservative chemical methods. Effective procedures can lead to nanoparticle production with preferred morphologies and precise sizes, rapidly and neatly (Iravani 2014). Nanotechnologies are a persistent way out vector in our economic setting. It is essential to progress with novel techniques to evaluate progress for improved knowledge of nanotechnology-based novelty. One such novelty is the process of nano bioremediation.

Nanobioremediation is a novel method for remedy of contaminants by means of biologically synthesized nanoparticles. It is still a new field of research yet developing rapidly in the area of nanotechnology. The synthesis of nanoparticles from microbes is rapidly evolving as nanofactories and prospective application in cleaning environmental contamination (Patil et al. 2016). Nanoparticles, mostly biogenic that have been tried so far have given better results. The biological pathway for nanoparticle production might be developed as an improved and safer substitute to conservative methods for bioremediation of pollutants while mitigating global warming and climate change (Yadav et al. 2017).

19.4 Role of Microbiome in Nanotechnology

Traditionally, nanoparticles were formed merely by chemical and physical means. The necessity for biological production of nanoparticles emerged, as chemical and physical processes were not cost effective. In the quest of economical pathways for production of nanoparticles, microbes and extracts from plants were also used for production. Biological production of nanoparticles is a basic method in which the principal reaction taking place is oxidation/reduction. Enzymes of microbes and plant phytochemicals having antioxidant and reducing nature are mostly responsible for the reduction of metallic complexes into their specific nanoparticles.

Nanoparticle synthesis is currently projected to be millions of tons globally and is likely to increase sooner rather than later (Pulit-Prociak and Banach 2016). For the biological synthesis of nanoparticles, microbes such as fungi, yeasts, bacteria, algae and actinomycetes can be used (Sastry et al. 2003).

Microbiological nanoparticle production will be a green chemistry technique and is measured as a stimulating field of exploration for future application (Narayan and Sakthivel 2010). With the help of simple microbial cultures, under cellular, biochemical and molecular mechanisms, the production rate and enhancement in the characteristics of nanoparticles can be accomplished. Metal and microbe interactions are well known in biotechnological processes, like biomineralization, bioremediation, bioleaching and biocorrosion. The characteristics of nano mineral crystals and metal nanoparticles formed by microorganisms are same as synthetic nanoparticles (Gericke and Pinches 2006). Extracellular or intracellular inorganic materials are produced by various microorganisms, like bacteria, fungi and viruses (Thakkar et al. 2010). By regulating the culture parameters, the formation of these inorganic compounds to a certain degree can be worked out for the size and shape. (Mohanpuria et al. 2008). The mechanism of extracellular and intracellular formation of nanoparticles is different across varied microorganisms (Hulkoti and Taranath 2014). The mechanism of intracellular formation includes the interaction between positive metal ions and negative metal ions through cell wall. Furthermore, the cell wall enzymes help in the reduction of metal ions into nanoparticles, and then these nanoparticles diffuse through the bacterial cell wall. In case of fungi, the formation of nanoparticles extracellularly is the amalgamation facilitated by nitrate reductase enzyme, which aids the bio-reduction of metal ions into nanoparticles (Hulkoti and Taranath 2014).

In case of mobilization and immobilization of metals, bacteria are very competent, bacteria that facilitate reduction of metal ions also show the capability to precipitate metals at the nanoscale. Bacteria are regarded as a prospective "bio factory" for the nanoparticle production such as gold, silver, palladium, platinum, titanium, etc. Using bacterial enzymes as a source is a lucid strategy for the biosynthesis of nanoparticles. The extracellular secretion of enzymes has the advantage that it produces a huge number of nanoparticles having size between 100–200-nm in a comparatively pure form, free of other proteins. Purification process of nanoparticles is further effectively accomplished by filtration. For technical purposes in nanotechnology and bioremediation, special metal-binding capabilities of bacterial cells and S-layers make them appropriate. The properties of nanoparticles are controlled by optimizing certain factors such as cellular activities, growth conditions and enzymatic processes. The extensive nanoparticle production using bacteria is interesting as it does not need any poisonous, harmful and expensive chemicals for synthesis and stabilization methods (Yadav et al. 2017).

Fungi are also an exceptional source of several extracellular enzymes that affect synthesis of nanoparticles. They have been widely used for the biological synthesis of nanoparticles and the mechanistic aspects governing the nanoparticle production have also been recognized for a few of them (Mohanpuria et al. 2008). In addition to monodispersity, nanoparticles with well-defined dimensions can be acquired using fungi. Compared with bacteria, fungi could be utilized as a source for the production of larger amounts of nanoparticles. This is due to the fact that fungi secrete a larger volume of proteins which directly mean higher productivity of nanoparticle formation (Yadav et al. 2017). Instead of fungal strains, proteins are also utilized effectively for nanoparticle synthesis. The utilization of explicit enzymes produced by fungi in the nanoparticle production is encouraging. It is likewise significant to understand the nature of the biological nanoparticles. Microbial approaches yield nanoparticles at a far slow rate than when using plant extracts. Enzymatic proteins are produced that reduce a salt to its solid metal nanoparticles through the catalytic effect in biological synthesis of metal nanoparticles by fungi (Ovais et al. 2018). This is one of the main disadvantages of nanoparticle biosynthesis utilizing microbes and must be improved for competing with other techniques. For uses in the industrial field, fungi must have several properties including a high growth rate, high production of certain enzymes, easy to manipulate during synthesis, and a low-cost prerequisite for the production processes as it provides advantage over other fungal methods (Vahabi et al. 2011). Fungi have an advantage due to wide-ranging diversity, easy growing methods, increased profitability, and reduced time. This offers an eco-friendlier approach to the production of nanoparticles. To improve the particle properties in future, genetic engineering procedures can be used (Saxena et al. 2014). Yeast strains have some benefits over bacteria in the production of various enzymes and prompt growth with the usage of simple nutrients, so the production of metallic nanoparticles using yeast is being considered.

19.5 Nanobioremediation

The eradication of environmental pollutants from polluted sites utilizing nanomaterials/nanoparticles produced by fungi, bacteria and plants is known as nano bioremediation. Nano bioremediation is the emergent technology for the exclusion of contaminants for cleaning up the environment. The current technologies for the reclamation of polluted resources include physical and chemical remediation, bioremediation and incineration. With new advancements, bioremediation suggests a viable ecological and economical alternative to eradicate toxins from the environmental settings (Singh and Walker 2006). Three main options of bioremediation include use of plants, microbes and enzymatic remediation. Likewise, nanoparticles are also being used for the remediation of heavy metal contaminated soil and water (Zhang 2003). In remediation mediated by enzymes, nanoparticles can also be utilized alongside phytoremediation (Singh 2010). For example, various multifaceted organic compounds like organochlorines and long-chain hydrocarbons are usually unaffected by plant and microbial degradation. A technique comprising both biotechnology and nanotechnology might remove this restraint: complex compounds (organic) would be reduced to simple compounds by nano-encapsulated enzymes, which then would be quickly reduced by the collective activities of microorganisms and plants. It was found that silver nanoparticles from *Aspergillus niger* (*A. niger* grows with $AgNO_3$) removed 85.8% of the dye efficiently in an incubation cycle of 24 h and after that the dye was entirely removed within incubation of 48 h. While the simple culture strain of *Aspergillus niger* degraded only 76% of the dye (Congo red) in the same condition of incubation and full discoloration was detected only after an entire 48 h incubation (Nithya and Ragunathan 2009).

Palladium is associated with Platinum Group Metals. It is an assemblage of catalytically very active metals and are presently utilized as reagents for hydrogenation and dehalogenation reactions. It was lately revealed that by means of bacteria which are found at Alpine sites, severely polluted with heavy metals, zero-valent palladium (Pd0) nanoparticles can be biosynthesized. Among all varieties of metal tolerant bacterial species that were found in the same environment, *Pseudomonas* cells formed nanoparticles that were efficiently utilized in dehalogenation reduction of dioxins (Schlüter et al. 2014). It has been proposed that with the help of hydrogenases found in *Escherichia coli*, Pd0 nanoparticles can also be produced (Lloyd et al. 1998). *Bacillus licheniformis* was found to produce intracellular silver nanoparticles (AgNPs). After adding silver ions, the culture changed its colour to brown signifying the existence of AgNPs (Kalimuthu et al. 2008). Although silver and gold NPs are vital because of their remarkable antimicrobial capabilities, other metals have also been studied like uranium (U). *Bacillus* species have been researched extensively because of the metal bioaccumulation abilities (Pollmann et al. 2006). Pollmann and team investigated the capability of *Bacillus sphaericus* to amass higher absorptions of toxic heavy metals like Pb, Al, Cu, U and Cd. They observed that the proteins (S-layer) of *B. sphaericus* help in the remediation of uranium from water resources. Parikh et al. (2008) recommended that the method of silver resistance also helps in the production of the AgNPs. The fungus *Fusarium oxysporum* has been used in many studies endeavouring to create metallic nanoparticles, specifically those which are made of Ag. Pure AgNPs were biosynthesized at a size range of 5–15 nm and it was proposed that they were capped in order to stabilize them by proteins secreted by the fungus (Ahmad et al. 2003). He et al. detected formation of gold nanoparticles extracellularly of size 10–20 nm by the *Rhodopseudomonas capsulata* and suggested that those nanoparticles were produced by means of an NADH-dependant reductase (He et al. 2007), an enzyme that has been revealed in the past to be significant in biological production of metal nanoparticles (Fig. 19.1) (Ahmad et al. 2003).

Even though cost effectiveness and simpler downstream processing make extracellular formation beneficial (Das et al. 2014), intracellular formation can also be of significant importance. By utilizing fungi that are capable of synthesizing nanoparticles intracellularly, it would be easy to eradicate the fungus as well as accumulated heavy metal toxins in the polluted sample. Zero-valent iron nanoparticles (NPs) and Carbon nano tubes (CNTs) are vital and favourable means of remediation of polluted water (Kitching et al. 2014; Saif et al. 2016). Biologically produced nanoparticles in combination with a metal-catalyst increase the rate of the reduction reaction. Biosynthesized iron nanoparticles and iron-Pd nanoparticles have shown widespread uses in removal of many hydrocarbons, dyes, dioxins, lindane, pesticides and polychlorinated biphenyls, etc., with help of metabolism of bacteria (Asmel et al. 2017).

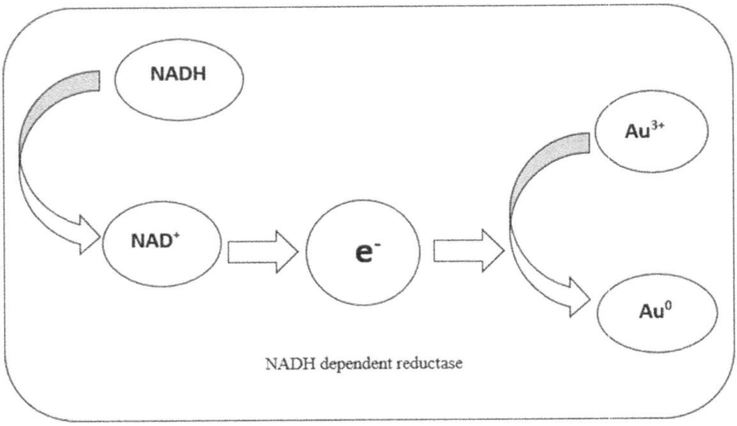

Fig. 19.1 Mechanism of reduction of gold ions into AuNPs (He et al. 2007)

19.6 Conclusion

Ecological imbalance due to growing population and industrialization is a grave cause of concern. The introduction of nanobioremediation using smarter engineered nanomaterials can provide time saving and inexpensive in-situ clean-up processes for large-scale polluted sites. Green nanotechnology has the potential of mitigating climate change significantly by reducing/eliminating fossil fuels, reducing energy consumption and growing the efficacy of current technologies. Changes in climate have been experienced by earth in the past as well but the prevailing changes in climate are more rapid and damaging. Green nanotechnology consistently involves the use of biologically synthesized nanomaterials. It is anticipated that combination of nanotechnology with biological arenas such as microbiology can produce new scientific approaches in health, agriculture, energy, remediation of polluted sites and other areas. Biogenic nanomaterials have helped in mitigating pollution since they are effective catalysts and can mostly be recyclable. Nanotechnology plays an economical role in commercialization and can completely replace present-day technologies. The use of microbiome in nanotechnology will influence the world-wide investments in science and technology and can transform the scientific world with innovative breakthroughs.

References

Ahmad A, Mukherjee P, Senapati S, Mandal D, Khan MI (2003) Extracellular biosynthesis of silver nanoparticles using the fungus *Fusarium oxysporum*. Colloids Surf B: Biointerfaces 28:313–318

Ashoka P, Meena RS, Gogoi N, Kumar S (2016) Green nanotechnology is a key for eco-friendly agriculture. J Clean Prod 142:4440–4441

Asmel NK, Yusoff ARM, Krishna LS, Majid ZA, Salmiati S (2017) High concentration arsenic removal from aqueous solution using nano-iron ion enrich material (NIIEM) super adsorbent. Chem Eng J 317:343–355

Avila AF, Munhoz V, Oliveira AM, Santos MCG (2014) Nano based systems for oil spills control and cleanup. J Hazard Mater 272:20–27

Biteen JS, Blainey PC, Cardon ZG, Chun M, Church GM, Dorrestein PC, Fraser SE, Gilbert JA (2016) Tools for the microbiome: Nano and beyond. ACS Nano 10:6–37

Corsi I, Winther-Nielsen M, Sethi R, Punta C, Della Torre C, Libralato G, Lofrano G, Sabatini L, Aeillo M, Fiordi L, Cinuzzi F, Caneschi A, Pellegrini D, Buttino I (2018) Ecofriendly nanotechnologies and nanomaterials for environmental applications: key issue and consensus recommendations for sustainable and ecosafe nanoremediation. Ecotoxicol Environ Saf 154:237–244

Das V, Thomas R, Varghese R, Soniya EV, Mathew J (2014) Extracellularsynthesis of silver nanoparticles by the Bacillus strain CS 11 isolated from industrialized area. 3 Biotech 4:121–126

Dermatas D, Mpouras T, Panagiotakis I (2018) Application of nanotechnology for waste management: challenges and limitations. Waste Manag Res 36(3):197–199

Dhingra R, Naidu S, Upreti G, Sawhney R (2010) Sustainable nanotechnology: through green methods and life-cycle thinking. Sustainability 2:3323–3338

Faisal S, Hafeez FY, Zafar Y, Majeed S, Leng X, Zhao S, Saif I, Malik K, Li X (2019) A review on nanoparticles as boon for biogas producers- Nano fuels and biosensing monitoring. Appl Sci 9:59

Gericke M, Pinches A (2006) Biological synthesis of metal nanoparticles. Hydrometallurgy 83 (1):132–140

He S, Guo Z, Zhang Y, Zhang S, Wang J (2007) Biosynthesis of goldnanoparticles using the bacteria *Rhodopseudomonas capsulata*. Mater Lett 61:3984–3987

Hood E (2004) Nanotechnology: looking as we leap. Environ Health Perspect 112(13):A740–A749

Hulkoti NI, Taranath T (2014) Biosynthesis of nanoparticles using microbes- a review. Colloids Surf B: Biointerfaces 121:474–483

Iavicoli I, Leso V, Ricciardi W, Hodson LL, Hoover MD (2014) Opportunities and challenges of nanotechnology in the green economy. Environ Health 13:78

Iravani S (2014) Bacteria in nanoparticle synthesis: current status and future prospects. Int Sch Res Not 2014:359316

Jeevanandam J, Barhoum A, Chan YS, Dufresne A, Danquah MK (2018) Review on nanoparticles and nanostructured materials: history, sources, toxicity and regulations. Beilstein J Nanotechnol 9:1050–1074

Kalimuthu K, Suresh Babu R, Venkataraman D, Bilal M, Gurunathan S (2008) Biosynthesis of silver nanocrystals by *Bacillus licheniformis*. Colloids Surf B: Biointerfaces 65:150–153

Khandel P, Yadaw RK, Soni DK, Kanwar L, Shahi SK (2018) Biogenesis of metal nanoparticles and their pharmacological applications: present status and application prospects. J Nanostruct Chem 8:217–254

Kitching H, Kenyon AJ, Parkin IP (2014) The interaction of gold and silver nanoparticles with a range of anionic and cationic dyes. *Phys Chem Chem Phys* 16:6050

Lloyd JR, Yong P, Macaskie LE (1998) Enzymatic recovery of elemental palladium byusing sulfate-reducing bacteria. Appl Environ Microbiol 64:4607–4609

Merino N, Aronson HS, Bojanova DP, Feyhl-Buska J, Wong ML, Zhang S, Giovannelli D (2019) Living at the extremes: extremophiles and the limits of life in a planetary context. Front Microbiol 10:780

Mohanpuria P, Rana NK, Yadav SK (2008) Biosynthesis of nanoparticles: technological concepts and future applications. J Nanopart Res 10(3):507–517

Narayan KB, Sakthivel N (2010) Biological synthesis of metal nanoparticles by microbes. Adv Colloid Interf Sci 156:1): 1–1):13

Nithya R, Ragunathan R (2009) Synthesis of silver nanoparticle using *pleurotus sajor caju* and its antimicrobial study. Dig J Nanomater Biostruct 4:623–629

Ovais M, Khalil AT, Ayaz M, Ahmad I, Nethi SK, Mukherjee S (2018) Biosynthesis of metal nanoparticles via microbial enzymes: a mechanistic approach. Int J Mol Sci 19(12):4100

Parikh RY, Singh S, Prasad BLV, Patole MS, Sastry M (2008) Extracellular synthesis of crystalline silver nanoparticles and molecular evidence of silver resistance from *Morganellasp.*: towards understanding biochemical synthesis mechanism. Chem Bio Chem 9:1415–1422

Patil SS, Shedbalkar UU, Truskewycz A, Chopade BA, Ball AS (2016) Nanoparticles for environmental clean-up: a review of potential risks and emerging solutions. Environ Technol Innov 5:10–21

Pollmann K, Raff J, Merroun M, Fahmy K, Selenska-Pobell S (2006) Metalbinding by bacteria from uranium mining waste piles and its technological applications. Biotechnol Adv 24:58–68

Pulit-Prociak J, Banach M (2016) Silver nanoparticles- a material of the future…? Open Chem 14:76–91

Saif S, Tahir R, Chen Y (2016) Green synthesis of iron nanoparticles and their environmental applications and implications. Nano 6(11):209

Sastry M, Ahmad A, Khan MI, Kumar R (2003) Biosynthesis of metal nanoparticles using fungi and actinomycete. Curr Sci 85:162–170

Saxena J, Sharma MM, Gupta S, Singh A (2014) Emerging role of fungi in nanoparticle synthesis and their applications. World J Pharm Pharm Sci 3:1586–1613

Schlüter M, Hentzel T, Suarez C, Koch M, Lorenz WG (2014) Synthesisof novel palladium (0) nanocatalysts by microorganisms from heavy-metal influenced high-alpine sites for dehalogenation of polychlorinated dioxins. Chemosphere 117C:462–470

Singh BK (2010) Exploring microbial diversity for biotechnology: the way forward. Trends Biotechnol 28:111–116

Singh BK, Walker A (2006) Microbial degradation of organophosphorus compounds. FEMS Microbiol Rev 30:428–471

Solomon S, Plattner G, Knutti R, Friedlingstein P (2009) Irreversible climate change due to carbon dioxide emissions. Proc Natl Acad Sci U S A 106(6):1704–1709

Thakkar KN, Mhatre SS, Parikh RY (2010) Biological synthesis of metallic nanoparticles. Nanomedicine 6(2):257–262

Vahabi K, Mansoori GL, Karimi S (2011) Biosynthesis of silver nanoparticles by fungus *Trichoderma Reesei* (a route for large-scale production of AgNPs). Insciences J 1:65–79

Yadav KK, Singh JK, Gupta N, Kumar V (2017) A review of nanobioremediation technologies for environmental cleanup: a novel biological approach. J Mater Environ Sci 8(2):740–757

Yang J, Hou B, Wang J, Tian B, Bi J, Wang N, Li X, Huang X (2019) Nanomaterials for the removal of heavy metals from wastewater. Nano 9(3):424

Zhang W (2003) Nanoscale Iron particles for environmental remediation: an overview. J Nanopart Res 5:323–332

The manufacturer's authorised representative in the EU is Springer
Nature Customer Service Centre GmbH, Europaplatz 3, 69115 Heidelberg,
Germany. If you have any concerns regarding our products, please
contact ProductSafety@springernature.com

Printed and bound by CPI Group (UK) Ltd, Croydon, CR0 4YY
30/09/2026
02225297-0018